TOPICS IN
LOCALLY CONVEX SPACES

NORTH-HOLLAND
MATHEMATICS STUDIES **67**

Notas de Matemática (85)
Editor: Leopoldo Nachbin
Universidade Federal do Rio de Janeiro
and University of Rochester

Topics in Locally Convex Spaces

MANUEL VALDIVIA
University of Valencia
Spain

1982

NORTH-HOLLAND PUBLISHING COMPANY – AMSTERDAM • NEW YORK • OXFORD

ISBN: 0 444 86418 0

Publishers
NORTH-HOLLAND PUBLISHING COMPANY
AMSTERDAM • NEW YORK • OXFORD

Sole distributors for the U.S.A. and Canada
ELSEVIER SCIENCE PUBLISHING COMPANY, INC.
52 VANDERBILT AVENUE
NEW YORK, N.Y. 10017

Library of Congress Cataloging in Publication Data

Valdivia, Manuel, 1928-
Topics in locally convex spaces.

(Notas de matemática ; 85) (North-Holland mathematics studies ; 67)
Bibliography: p.
Includes index.
1. Locally convex spaces. I. Title. II. Series: Notas de matemática (North-Holland Publishing Company) ; 85. III. Series: North-Holland mathematics studies ; 67.
QA1.N86 no. 85 [QA322] 510s [515.7'3] 82-6449
ISBN 0-444-86418-0 AACR2

PRINTED IN THE NETHERLANDS

To
Manuel, Nieves, Marta
and
Ma. Teresa

PREFACE

The choice of topics considered here are dictated by the author's own interest in the field and concentrated heavily on his own research work done in the last years. No claim for completeness is made for the bibliography at the end of the notes. Numbers in square brackets refer to it.

The notes are aimed to persons who already have an acquaintance with the general theory of locally convex spaces. Since the proofs are presented with detail and since some efforts have been made to give a number of simple arguments replacing some rather cumbersome construc - tions, most of the notes should be readable for graduate students but they can also serve as a reference for the more advanced mathematician.

These notes consist of three Chapters. Each chapter splits into several paragraphs and each paragraph in sections which are enumerated in consecuteve fashion. Gross references are usually u. v. w. z *meaning that reference is made to statement* z *of section* w *in paragraph* v *in chapter* u. *Gross references within the chapter are* v. w. z *and within the paragraph* w. z.

Nine paragraphs constitue the first chapter. Paragraphs 1, 2, 3, 5 *are dedicated to the study of classes of locally convex spaces which are used as domain class for the closed graph theorem. Paragraph* 4 *is devoted to the closed graph theorem when the range class is the quasi - Suslin,* K *- Suslin, Suslin or semi - Suslin spaces. Paragraph* 6 *studies the incidence of the duality theory on the closed graph theorem.* A *characterization of the locally convex spaces which are weakly realcom pact is included as well as a discussion on generalized countable inductive limits. Some properties on bounded sets in* (LN) *- spaces are given.*

The second chapter is concerned with sequence spaces which are studied along six paragraphs. A general study of the Kothe perfect spaces and echelon and co - echelon spaces is included. A characterization of echelon quasi - normable spaces is given as well as a discussion on echelon and co - echelon spaces of order p, $1 < p < \infty$, and of order zero. Paragraph 5 contains examples of sequences spaces which answer several questions on aspects of the general theory of locally convex spaces. An example of a Banach space which is an hyperplane of its strong bidual due to R. C. JAMES inspires the end of the chapter where a construction of some vector - valued sequence spaces can be found.

Chapter three has three paragraphs: the first includes easy representations of the more interesting spaces of infinitely differentiable functions and distributions. In the second paragraph representations of spaces of C^m - differentiable functions can be found. The last paragraph is a detailed exposition of Milutin' s representation theorem: if X and Y are non - countable compact metric spaces, then the Banach spaces $C(X)$ and $C(Y)$ are isomorphic.

I wish to acknowledge the help I have received from colleagues who have read parts of the manuscript: P. Pérez Carreras, J. Bonet, M. López Pellicer, M. Maestre and R. Crespo. I thank also my wife for her help in typing the manuscript.

Manuel Valdivia

TABLE OF CONTENTS

CHAPTER ONE
SOME CLASSES OF LOCALLY CONVEX SPACES

Certain classes of locally convex spaces are studied: Baire, convex-Baire, ordered convex-Baire, suprabarrelled, realcompact, Γ_r^- ,Γ- *and* (LB) *spaces. Two paragraph are dedicated to the closed graph and open mapping theorems.*

§ 1. BAIRE SPACES

1. TOPOLOGICAL SPACES OF SECOND CATEGORY. The topological spaces we shall use in this paragraph are supposed distinc from the void set. Let B be a subset of a topological space X. B is nowhere dense or rare if and only its closure has void interior. It is obvious that, if B is rare, every subset of B is also rare. B is of first category or meager if and only if it is the countable union of rare sets of X. Clearly, if B is of first category, every subset of B is also of first category. B is of second category if and only if it is not of first category. If B is of second category every subset of X containing B is of second category. If the subset X of X is of second category we say that X is a space of second category. If every non-void open subset of X is of second category, X is said to be a Baire space. It is immediate that if X is a Baire space, it is a space of second category.

In what follows R denotes the field of the real numbers. If we set

$$A = \{(x,0) : x \in R\} ,$$

$$B = \{(0,y) : y \text{ rational number}, y \neq 0\}$$

and if $Y = A \cup B$ is endowed with the topology induced by the euclidian space R^2, it is easy to show that Y is a space of second category which is not Baire, since B is an open subset of Y which is countable union of rare subsets which have only one element.

Given a subset M of a topological space X we set $\overline{M}$ to denote the closure of M and $\overset{\circ}{M}$ for its interior. A open set M is regular and if only if $M = \overline{\overset{\circ}{M}}$.

(1) *A topological space is Baire if and only if given any sequence* (A_m) *of dense open subsets of* X, *then* $\cap\{A_m : m = 1,2, \ldots\}$ *is dense in* X

Proof. Suppose that X is a Baire space. Let A be a non-void open set of X. For every positive integer m, $X \sim A_m$ is a closed subset of X without interior point and therefore $(X \sim A_m) \cap A$ is rare. Since A is of second category we have that

$$A \neq \cup\{(X \sim A_m) \cap A : m = 1,2, \ldots\}$$

and therefore there is a point X in A. which is not in $X \sim A_m$, m = 1,2,.., thus x belongs to $A_m \cap A$, m = 1,2, ..., and therefore $\cap\{A_m : m = 1,2,\ldots\}$ is dense in X.

Reciprocally, let A be a non-void open set of X. Let (M_m) be a sequence of rare subsets of X contained in A. For every positive integer m, we set A_m for $X \sim \overline{M}_m$. Then (A_m) is a sequence of dense open sets of X and therefore $\cap\{A_m : m = 1,2, \ldots\}$ is dense in X. Then

$$A \cap \{A_m : m = 1,2, \ldots\} \neq \emptyset$$

and thus A is not contained in

$$U \cup \{M_m : m = 1,2, \ldots\}.$$

Consequently, A is of second category. The proof is complete.

Result (1) can be stated as

(2) *The topological space* X *is Baire if and only if, given any sequence* (A_m) *of dense open subset of* X *and given a non-void open subset* A *of* X

$$A \cap (\cap\{A_m : m = 1,2,\ldots\})$$

is non-void.

(3) *Let* B *be a subset of a topological space* X. *Let* $A = \{A_i : i \in I\}$ *the family of all open sets of* X *such that* $A_i \cap B$ *is of first category,* $i \in I$ *Then* $A = U\{A_i : i \in I\}$ *is an open regular subset of* X *which intersects* B *in a set of first category.*

Proof. Let

(4) $\{\mathcal{P}_j : j \in J\}$

be the collection of all subfamilies of A such that if j belongs to J and P and Q are different elements of $\mathcal{P}_j$, then P and Q are disjoint.

We order the collection (4) by inclusion. We apply Zorn's lemma to obtain a maximal element $\mathcal{P} = \{M_h : h \in H\}$ in (4). Set $M = \bigcup\{M_h : h \in H\}$. For every h in H there is a sequence (M_h^n) of rare subsets of X such that

$$M_h \cap B = \cup \{M_h^n : m = 1,2,\ldots\}.$$

For every positive integer n, we set

$$M^n = \cup\{M_h^n : h \in H\},\ m = 1,2,\ldots$$

Suppose that the interior P of $\overline{M^n}$ is non-void. Then $P \cap M^n$ is non-void and therefore there is k in H such that $P \cap M_k^n$ is non-void. Since the elements of $\mathcal{P}$ are pairwise disjoint and since M_k is open we have that the closure Q of

$$\cup \{M_h^n : h \in H,\ h \neq k\}$$

is disjoint from M_k. Therefore

$$P \cap M_k \subset \overline{M^n} \cap M_k = (\overline{M_k^n} \cup Q) \cap M_k = \overline{M_k^n} \cap M_k \subset \overline{M_k^n}$$

and that is a contradiction. Thus M^n is a rare subset of X and since

$$M \cap B = \cup\{M_h \cap B : h \in H\} = \cup\{M_h^n : h \in H,\ n = 1,2,\ldots\}$$

$$= \cup \{M^n : n = 1,2,\ldots\}$$

it follows that $M \cap B$ is of first category. Since $\overline{M} \sim M$ is rare, it follows that $U \sim M$ is rare, U being the interior of $\overline{M}$ and from

$$U \cap B = ((U \sim M) \cap B) \cup (M \cap B)$$

we obtain that $U \cap B$ is of first category. We shall see now that U coincides with A. Let L be an element of A. If L is not contained in U, then $L \sim \overline{M}$ is a non-void open set which is disjoint from each of the elements of $\mathcal{P}$ and intersects B in a set of first category, contradicting the maximality of $\mathcal{P}$. Now the conclusion follows.

Using the same notations as in (3), we denote by $D(B)$ the set of all points x of X such that every neighbourhood of x meets B in a set of second category. Then $D(B)$ coincides with $X \sim A$. We set $O(B)$ to denote the interior of $D(B)$. We conclude from (3) the so called Banach's condesation theorem:

(5) *The set* $(X \sim D(B)) \cap B$ *is of first category and* $D(B)$ *coincides with* $\overline{O(B)}$.

(6) *For every subset* B *of a topological space* X, $B \sim O(B)$ *is of first category.*

Proof. Let A be the open set defined in (3). Then $A \cap B$ is of first category and $D(B) \sim O(B)$ is a rare set. Consequently,

$$B \cap (A \cup (D(B) \sim O(B)))$$

is of first category. Finally

$$B \sim O(B) = B \cap (X \sim O(B)) = B \cap (A \cup (D(B) \sim O(B)))$$

and the conclusion follows.

(7) *Let* (B_m) *be a sequence of subsets of a topological space* X *whose union is* B. *Then*

(8) $$D(B) \sim \cup \{O(B_m) : m = 1,2,\ldots\}$$

is rare.

Proof. Suppose that the closed set (8) has non-void interior S. Then $S \cap B$ is of second category and therefore there is a positive integer p such that $S \cap B_p$ is of second category. Consequently $S \cap D(B_p)$ is non-void and therefore $S \cap O(B_p)$ is non-void and that is a contradiction.

A subset B in a topological space X has the Baire property if there exists an open set U such that $U \sim B$ and $B \sim U$ are of first category.

(9) *A subset* B *in a topological space* X *has the Baire property if and only if* $O(B) \sim B$ *is of first category.*

Proof. According to (6), $B \sim O(B)$ is of first category. Therefore if $O(B) \sim B$ is of first category B has the Baire property.

Now suppose that B has the Baire property. Let U be an open subset of X such that $U \sim B$ and $B \sim U$ are of first category. Then $X \sim \bar{U}$ meets B in the set of first category $B \sim \bar{U}$. Therefore D(B) is contained in $\bar{U}$. Since $\bar{U} \sim U$ is rare we have that $\bar{U} \sim B$ is of first category. On the other hand,

$$O(B) \sim B \subset D(B) \sim B \subset \bar{U} \sim B$$

and therefore $O(B) \sim B$ is of first category. The proof is complete.

2. PRODUCTS OF BAIRE SPACES. In what follows N denotes the set of the positive integers. Let d be a metric on a topological space X. We say that d is compatible with the topology of X if this topology coincides with the topology of the metric space (X,d).

Let $\{X_i : i \in I\}$ be a family of topological spaces. For every i of I let d_i a metric on X_i compatible with the topology of X_i such that (X_i, d_i) is a complete metric space. Then we have the following result due to BOURBAKI:

(1) *The topological product* $X = \Pi\{X_i : i \in I\}$ *is a Baire space.*

Proof. Let A be a non-void open set of X. Let (A_m) be a sequence of dense open sets of X. Since $A \cap A_1$ is non-void there is a finite subset I_1 of I and a closed ball A_i^1 in (X_i, d_i), of radius less than $\frac{1}{2}$, $i \in I_1$, such that

$$\Pi\{A_i^1 : i \in I_1\} \times \Pi\{X_i : i \in I \sim I_1\} \subset A \cap A_1$$

Proceeding by recurrence suppose that, for a positive integer n, we have selected a finite subset I_n of I and a closed ball A_i^n in (X_i, d_i) of positive radius less than $\frac{1}{2^n}$, $i \in I_n$. Since

$$M_n = \Pi\{A_i^n : i \in I_n\} \times \Pi\{X_i : i \in I \sim I_n\}$$

has non-void interior we can find a finite subset I_{n+1} in I, $I_n \subset I_{n+1}$, and a closed ball A_i^{n+1} in (X_i, d_i) of positive radius less than $\frac{1}{2^{n+1}}$, $i \in I_{n+1}$, such that

$$\Pi\{A_i^{n+1} : i \in I_{n+1}\} \times \Pi\{X_i : i \in I \sim I_{n+1}\}$$

is contained in the interior of $M_n \cap A_{n+1}$. We set $J = \cup\{I_n : n = 1,2,\ldots\}$. For every i in $I \sim J$ take a point x_i in X_i and set $x_i^n = x_i$, $n = 1,2,\ldots$

If i belongs to J and n to N we take x_i^n in A_i^n. The sequence (x_i^n) is obviously a Cauchy sequence in (X_i, d_i) and therefore converges in this space to a point x_i belonging to

$$\cap \{A_i^n : n = 1,2,\ldots\}.$$

Consequently the sequence $((x_i^n : i \in I))$ of X converges in this space to $(x_i : i \in I)$ and

$$(x_i : i \in I) \in \cap \{M_n : n = 1,2,\ldots\} \subset A \cap (\cap \{A_n : n = 1,2,\ldots\})$$

and the conclusion follows.

If we suppose that the former index set I has only one element we obtain from (1) the classical theorem of Baire:

(2) *If there is a metric* d *in a topological space* X *compatible with its topology and such that* (X,d) *is complete, then* X *is a Baire space.*

Now suppose that Y_i is a topological space, $i \in I$. Denote by Y the topological product $\Pi\{Y_i : i \in I\}$. A cylinder in Y is a subset of Y of the form

$$\Pi \{A_i : i \in I\}$$

where $A_i = Y_i$ save a finite number of indices i.

(3) *If the cardinal of* I *is less or equal than the cardinal of* R *and if* Y_i *is separable,* i C I, *then* *is separable.*

Proof. For every i in I let (x_{im}) be a sequence in Y_i whose elements form a dense subspace Z_i of Y_i. The topological space

$$Z = \Pi\{Z_i : i \in I\}$$

is dense in Y and therefore it is enough to show that Z is separable. Now suppose that N has the discrete topology. The mapping T from the topological space N^I onto Z such that

$$T(n_i : i \in I) = (x_{in_i} : i \in I)$$

is obviously continuous and therefore it is enough to show that N^I is separable. Let J be a non-void set, $J \cap I = \emptyset$, such that $I \cup J$ has the cardinality of R. Then $N^I \times N^J$ is homeomorphic to N^R. On the other hand, the projection of $N^I \times N^J$ onto N^I is continuous and therefore it is enough to show that N^R is separable.

Let P be the set of all the functions defined on R which are characteristic functions on intervals of rational ends. If k is the element of N^R which takes the value one in every point of R we set

$$H = \{k + \sum_{j=1}^{p} (n_j-1) f_j : n_j, p \in N, f_j \in P, j = 1,2,\dots,p\}.$$

H is a countable subset of N^R and we shall show that it is dense in N^R. Let U be a neighborhood of an element f of N^R. We find pairwise distinct real numbers $x_1, x_2,\dots,x_q$ such that

$$\{g \in N^R : g(x_j) = f(x_j), j = 1,2,\dots,q\} \subset U.$$

Take pairwise disjoint intervals $A_1, A_2,\dots,A_q$ of rational ends such that x_j is in A_j and set h_j to denote the characteristic function of A_j, j=1,2, ...q. Then

$$k + \sum_{j=1}^{q} (f(x_j)-1)h_j \in H \cap U$$

and the conclusion follows.

(4) *Let*

(5) $\{A_j : j \in J\}$

be a family of pairwise disjoint non-void open sets of Y. *If for every* i *in* I, Y_i *is separable then* J *is a countable set.*

Proof. Suppose the property is not true. Take a subfamily of (5), which we denote by (5) again, such that the cardinality of J is less or equal than the cardinality of R, and J is not countable.

For every j in J we find a finite subset I_j in I and a non-void open subset B_i of Y_i, $i \in I_j$, such that

$$\Pi\{B_i : i \in I_j\} \times \Pi\{Y_i : i \in I \sim I_j\} \subset A_j$$

If we set $L = \cup\{I_j : j \in J\}$ we have that the cardinality of L is less or equal than the cardinality of R. We write

$$Z = \Pi\{Y_i : i \in L\}$$

According to (3), Z is separable. On the other hand,

$$\Pi\{B_i : i \in I_j\} \times \Pi\{Y_i : i \in L \sim I_j\},\ j \in J$$

is a non-countable family of pairwise disjoint non-void open subsets of Z and that contradicts the separability of Z.

(6) *Let A be a non-void open set of Y. If for every i in I, Y_i is separable, there is a countable family $\{D_j : j \in J\}$ of pairwise disjoint open cylinders of Y contained in A whose union D is dense in A.*

Proof. Let

(7) $\{P_h : h \in H\}$

be the collection of all families of non-void open cylinders contained in A such that h belongs to H and if P and Q are distinct elements of P_h, then P and Q are disjoint. It is obvious that the collection (7) can be ordered by inclusion and therefore Zorn's lemma can be applied. Let $P = \{D_j : j \in J\}$ be a maximal element of (7). By (4), J is a countable set. Let D be the closure of $\cup\{D_j : j \in J\}$. If D does not contain A, it follows that $A \sim D$ is a non-void open cylinder B, which is in contradiction with the maximality of P. The proof is complete.

If V and W are sets and B is a subset of $V \times W$, we set

$$B(y) = \{z \in W : (y,z) \in B\}$$

for every y of V.

(8) *Let V and W topological spaces. Let (G_m) be a sequence of dense open subsets of $V \times W$. If W is metrizable and separable, there is a subset A of V which is of first category such that $G_m(y)$ is dense in W, $m = 1,2,\ldots$, for every y in $V \sim A$.*

Proof. There is a sequence

(9) $U_m,\ m = 1,2,\ldots,$

of non-void open sets of W such that every non-void open set of this space is union of elements of the sequence (9). The projection H_m from

$$(V \times U_m) \cap G_1 \cap \ldots \cap G_m,\ m = 1,2,\ldots,$$

onto V is obviously dense in V and therefore $V \sim H_m$ is a closed subset of V with void interior. Consequently

$$A = V \sim \bigcap_{m=1}^{\infty} H_m = \bigcup_{m=1}^{\infty} (V \sim H_m)$$

is a subset of V of first category. If y belongs to $V \sim A = \bigcap_{m=1}^{\infty} H_m$ and if p is a positive integer, let us see that $G_p(y)$ is dense in W. If B is a non-void open subset of W we find a positive integer q larger than p such that U_q is contained in B. Then y belongs to H_q and therefore $B \cap G_p(y)$ is non-void and the conclusion follows.

(10) *Let* V *and* W *be Baire spaces. If* W *is metrizable and separable, then* V x W *is a Baire space* .

Proof. Let (G_m) be a sequence of dense open subset of V × W. Let S be a non-void open subset of V × W. Let A be the set of first category of V we constructed in (8). We set P to denote the projection of S in V. Since V is a Baire space it follows that $P \cap (V \sim A)$ is non-void. Take a point y in this space. By (8), $(G_m(y))$ is a sequence of dense open subsets of W and, since W is Baire the intersection G of the sets $G_m(y)$, m = 1,2,... is dense in W, according to 1.(1). S(y) is a non-void open subset of W and therefore there is a point z in $G \cap S(y)$. Then (y,z) belongs to $S \cap (\bigcap\{G_m : m = 1,2,\ldots\})$ and the conclusion follows.

Let (Z_m) be a sequence of topological spaces. Given the integer m and n, with $0 \leq m < n$, we write

$$Z^{(n)} = \prod_{j=1}^{n} Z_j,\quad Z^{(m,n)} = \prod_{j=m+1}^{n} Z_j,\quad V^{(n)} = \prod_{i=n+1}^{\infty} Z_i$$

(11) *If* (Z_m) *is a sequence of metrizable, separable and Baire topological spaces, then* $Z = \prod\{Z_m : m = 1,2,\ldots\}$ *is a Baire space.*

Proof. Let (G_m) be a sequence of dense open subsets of Z. Let G be a non-void open subset of Z. We find a positive integer n_1 such that $G \cap G_1$ contains a non-void open cylinder $U_1 \times V^{(n_1)}$, U_1 being a non-void open subset of $Z^{(n_1)}$. By (10), U_1 is a subset of second category of $Z^{(n_1)}$ and, by (8) there is a point z_1 of U_1 with $G_m(z_1)$ dense and open in $V^{(n_1)}$ for every positive integer n. Therefore

$$\{z_1\} \times V^{(n_1)} \subset G \cap G_1$$

Proceeding by recurrence suppose that we have obtained the integers $0 = n_0 < n_1 < \ldots < n_k$ and the points z_j in $Z^{(n_{j-1}, n_j)}$, $j = 1,2,\ldots, k$, such that $G_m(z_1, z_2, \ldots, z_k)$ is a dense open subset of $V^{(n_k)}$, $m = 1, 2,\ldots,$ being $G(z_1, z_2,\ldots, z_j) = G(z_1, z_2, \ldots, z_{j-1})(z_j)$, $j = 2,3, \ldots, k$, and

$$(12) \qquad \{(z_1, z_2, \ldots, z_k)\} \times V^{n_k} \subset G \cap G_1 \cap G_2 \cap \ldots \cap G_k.$$

We find an integer $n_{k+1} > n_k$ such that $G_j(z_1, z_2, \ldots, z_k)$ $j = 1,2,\ldots,k+1$ contains a non-void open cylinder $U_{k+1} \times V^{(n_{k+1})}$, U_{k+1} being a non--void open subset of $Z^{(n_k\ n_{k+1})}$. According to (10) applied to $Z^{(n_k, n_{k+1})} \times V^{(n_{k+1})}$, there is a point z_{k+1} in U_{k+1} such that $G_m(z_1, z_2, \ldots, z_{k+1}) = G_m(z_1, z_2, \ldots, z_k)(z_{k+1})$ is a dense open subset of $V^{(n_{k+1})}$, $m = 1, 2, \ldots,$ and

$$\{z_{k+1}\} \times V^{(n_{k+1})} \subset G_j(z_1, z_2, \ldots, z_k), \quad j = 1,2, \ldots, k+1$$

and therefore

$$\{z_1, z_2, \ldots, z_{k+1})\} \times V^{(n_{k+1})} \subset G_1 \cap G_2 \cap \ldots \cap G_{k+1}$$

from where it follows, according to (12), that

$$\{z_1, z_2, \ldots, z_{k+1}) \times V^{(n_{k+1})} \subset G \cap G_1 \cap G_2 \cap \ldots \cap G_{k+1}$$

Therefore (12) is satisfied taking $k + 1$ instead of K. Then $(z_1, z_2, \ldots, z_k, \ldots)$ is a point of

$$Z^{(n_1)} \times Z^{(n_1, n_2)} \times Z^{(n_1, n_2)} \times \ldots = Z$$

which belongs to G $(\cap\{G_n : n = 1,2,\ldots\})$ and the conclusion follows.

(13) *If the topological spaces* Y_i, $i \in I$, *are metrizable, separable and Baire, the topological product* $Y = \Pi\{Y_i : i \in I\}$ *is a Baire space.*

Proof. Let (G_m) be a sequence of dense open subsets of Y. For every positive integer n we apply (6) to obtain a countable set J_n and a family

$\{H_j^n : j \in J_n\}$ of open cylinders of Y whose union is contained and dense in G_n. We set $H^n = \cup \{H_j^n : j \in J_n\}$. For every j of J_n there is a finite subset I_{jn} in I such that

$$H_j^n = \Pi\{A_i^n : i \in I_{jn}\} \times \Pi\{Y_i : i \in I \sim I_{jn}\}$$

where A_i^n is an open subset of Y_i, $i \in I_{jn}$. We set

$$L = \cup \{I_{jn} : j \in J_n, n = 1,2,\ldots\}$$

$$U = \Pi \{Y_i : i \in L\}, \quad V = \Pi\{Y_i : i \in I \sim L\}$$

Then H^n can be written as $M^n \times V$, M_n being a dense open subset of U. By (11), $M = \cap \{M^n : n = 1,2,\ldots\}$ is dense in U and therefore $\cap\{G_n : n=1,2,\ldots\}$ which contains $M \times V$, is dense in Y and the conclusion follows.

The results on products of Baire spaces included here can be found in a more general context in OXTOBY [1]. This author constructs a Baire space Z such that $Z \times Z$ is not a Baire space using the continuum hypothesis. An example of this situation, where the continuum hypothesis is not used, can be found in FLEISSNER and KUNEN [1].

3. LOCALLY CONVEX BAIRE SPACES. The linear spaces we shall use are defined over the field K of the real or complex numbers. The locally convex spaces are supposed to be Hausdorff. A locally convex space is said to be normable if its topology can be derived from a norm. A locally convex space is a Banach space if it is normable and complete. A Fréchet space is a metrizable complete locally convex space. Results (1) and (2) are particular cases of 2.(1) and 2.(13) respectively.

(1) *The topological product of Fréchet spaces is a Baire space.*

(2) *The topological product of metrizable, separable Baire locally convex spaces is a Baire space.*

(3) *Let* A *be a subset of a locally convex space* E. *Let* h *be a non-zero element of* K *and let* z *be a point of* E. *Then*

a) *If* A *is rare, then* hA *and* z+A *are rare;*

b) *if* A *is of first category, then* hA *and* z+A *are of first category;*

c) *if* A *is of second category, then* hA *and* z+A *are of second category;*

d) D(x+A) = x + D(A) *and* O(x+A) = x + O(A)

Proof. For every x of E, we set Tx = hx, Sx = z+x. Then T and S are homeomorphism from E onto E. The conclusion follows.

(4) *Let* E *be a locally convex space. If* E *is a space of second category, then* E *is a Baire space.*

Proof. Let A be a non-void open set of E. Take a point z in A and a neighbourhood U of the origin in E such that z+U is contained in A. Since E is a space of second category and

$$E = \cup \{m\, U : m = 1,2,\ldots\}$$

there is a positive integer q such that q U is of second category. Apply (3) to obtain that z+U is of second category. Consequently A is a subset of E of second category and the conclusion follows.

(5) *Let* E *be a locally convex space. Let* F *be a dense subspace of* E. *If* F *is a Baire space, then* E *is a Baire space.*

Proof. Let (G_m) be a sequence of dense open sets of E. Then $(G_m \cap F)$ is a sequence of dense open sets of F and therefore

$$\cap \{G_m \cap F : m = 1,2,\ldots\}$$

is dense in F. Thus $\cap \{G_m : m = 1,2,\ldots\}$ is dense in E and the conclusion follows.

(6) *Let* E *be an infinite dimensional locally convex space. If* E *is a Baire space, there is a one-codimensional dense subspace of* E *which is a Baire space.*

Proof. Let (x_n) be a sequence of linearly independent vectors. Find a family A of elements of E such that $A \cup \{x_n : n = 1,2,\ldots\}$ is an algebraic basis of E. Set E_n to denote the linear hull of $A \cup \{x_j : j = 1,2,\ldots,n\}$. Since the sequence (E_n) covers E, there is a positive integer q such that E_q is a subset of E of second category. Then E_q is a Baire dense proper subspace of E. Find an hyperplane F of E containing E_q. We apply (5) to obtain that F is a Baire space and the conclusion follows.

(7) *Let* E *be a locally convex space. Let* F *be a closed subspace of* E. *If* E *is a Baire space, then* E/F *is a Baire space.*

Proof. Let f : ⟶ E/F be the canonical mapping. f is continuous and maps every open set of E in an open set of E/F. Let (G_m) be a sequence of dense open subsets of E/F. Then $(f^{-1}(G_n))$ is a sequence of open dense subsets of E and therefore $H = \cap \{f^{-1}(G_n) : n = 1,2,\ldots\}$, is dense in E. Consequently $f(H) = \cap\{G_n : n = 1,2,\ldots\}$ is dense in E/F and the conclusion follows.

(8) *Let* E *be a locally convex space. Let* U *be an absolutely convex subset of* E. *If* U *is of second category and has the Baire property, then* U *is a neighbourhood of the origin.*

Proof. According to 1.(6), O(U) is non void and therefore the origin belongs to the open set O(U)-O(U). $W = O(U) \cap U$ is of second cathegory and, since U has the Baire propery, we apply 1.(9) to obtain that $A = O(U) \sim W$ is of first category.

Suppose that there is a point x in O(U) - O(U) which is not in 2 U. If there is a point z in $(x + U) \cap U$ we can find a vector y in U such that $x + y = z$ and therefore

$$x = z - y \in U + U = 2U$$

which is a contradiction. Thus $(x + U) \cap U$ is void and consequently $(x + W)$ W is also void. We have that

$$(x + O(U)) \cap O(U) = (x + W \cup A) \cap (W \cup A)$$

$$= ((x + W) \cup (x + A) \cap (W \cup A) = D \cup ((x + W) \cap W)$$

where D is of first category. Since U is of second category we have that E is also of second category and therefore a Baire space. If the open set $(x + O(U)) \cap O(U)$ is non void, then it is of second category and consequently $(x + W) \cap W$ is of second category and that is a contradiction. Thus $(x + O(U)) \cap O(U)$ is void. On the other hand, there are vectors u and v in O(U) such that x = u-v and therefore x-v = u and consequently $(x + O(U)) \cap O(U) \neq \emptyset$ which is a contradiction. Thus O(U) - O(U) is contained in 2 U and therefore U is a neighbourhood of the origin in E.

(9) *Let* E *be a locally convex space. Let* F *be a dense subspace of* E. *If* F *is a Baire space, then* F *is a subset of* E *of second category.*

Proof. Suppose that F is a subset of E of first category. There is a sequence (A_n) of rare closed subsets of E covering F. Then $F \cap (E \sim A_n)$ is open and dense in F, n = 1,2, ..., and, since F is a Baire space.

$$F \cap (\cap \{ E \sim A_n : n = 1,2,.. \})$$

is non void and therefore the sequence (A_n) does not cover F and that is a contradiction.

(10) *Let* E *be a locally convex space. Let* F *be a dense subspace of* E. *If* F *has the Baire property and if* F *is a Baire space, then* E *coincides with* F.

Proof. By (9), F is a subset of E of second category and, according to (8), a neighbourhood of the origin in E. Clearly F coincides with E.

In (6) we have seen that if E is an indinite dimensional locally convex space which is a Baire spaces there is an one-codimensional dense subspace of E which is a Baire space. By (9), F is an hyperplane of E of second category. If E is an infinite dimensional separable Banach space and if the continuum hypotesis is verified, ARIAS de REINA [1] has proved the existence of a dense hyperplane of E of first category.

§ 2. CONVEX - BAIRE SPACES

1. PROPERTIES OF COUNTABLE FAMILIES OF CONVEX SETS. The results contained in this section will we used in the three forthcoming sections.

Let E be a locally convex space A ray in E coming from x ∈ E is a set $\{ x + hz = h \geq 0 \}$, being a non-zero element of E. Let us suppose in the stament of result (1) that

$$\{ A_n : n = 1,2, ...\}$$

is a family of convex subsets of E covering E.

(1) *If* A_n *is rare,* n = 1,2, ..., *there is a family* $\{ B_n : n = 1,2,..\}$ *of convex subsets of* E *covering* E *such that* B_n *is rare and contains the origin of* E, n = 1,2, ...

Proof. Let P be the subset of N defined by p ∈ P if and only if the re is a ray in E coming from the origin meeting A_p in more than one point. For every $p \in P$ there is a vector $x_p \neq 0$ and a number $h_p > 1$ such that x_p and $h_p x_p$ are in A_p. Let $n(p)$ be an integer with $n(p)(h_p-1) > 2h_p$. We set

$$C_p = n(p)\,(A_p - \frac{1+h_p}{2} x_p) + \frac{1+h_p}{2} x_p$$

Since A_p is rare it is obvious that C_p is rare. If z belongs to A_p we have

$$\frac{1}{n(p)} (z - \frac{1+h_p}{2} x_p) \in A_p - \frac{1+h_p}{2} x_p,$$

since $A_p - \frac{1+h_p}{2} x_p$ is a convex set containing the origin, and therefore

$$z \in n(p)\,(A_p - \frac{1+h_p}{2} x_p) + \frac{1+h_p}{2} x_p,$$

and thus A_p is contained in C_p. Since x_p belongs to A_p it follows that

$$\frac{2h_p}{n(p)(h_p-1)} (x_p - \frac{1+h_p}{2} x_p) \in A_p - \frac{1+h_p}{2} x_p$$

and therefore

$$(2) \qquad -h_p x_p \in n(p)\,(A_p - \frac{1+h_p}{2} x_p)$$

and thus

$$(3) \qquad -x_p \in n(p)\,(A_p - \frac{1+h_p}{2} x_p)$$

By (2) and (3),

$$\frac{1-h_p}{2} x_p \in C_p, \quad \frac{h_p-1}{2} x_p \in C_p$$

and, since C_p is convex, we have that

$$0 = \frac{1}{2} \frac{1-h_p}{2} x_p + \frac{1}{2} \frac{h_p-1}{2} x_p \in C_p$$

We consider the family $\{m\, C_p : m = 1,2,\ldots,\ p \in P\}$ we denote it by

$$(4) \qquad \{B_n : n = 1,2,\ldots\}$$

It is obvious that B_n is rare, convex and contains the origin, $n = 1,2,\ldots$ Every ray coming from the origin has a non-countable infinity of points and therefore P is non-void and thus (4) covers the origin. We shall see that (4) covers E. We suppose the existence of a point x in E which is not in B_n $n = 1,2,\ldots$ Then $x \neq 0$. For every p in P the set A_p is contained in some member of (4) and therefore there is a number r in $N \sim P$ such that x belongs to A_r. We set

$$M = \{hx : h \geqslant 0\}$$

and, since M is a ray in E coming from the origin, the set

$$M \cap (\cup \{A_n : n \in N \sim P\})$$

is countable and therefore there is s in P and $k > 0$ such that kx belongs to A_s. Then

$$M \subset \bigcup_{m=1}^{\infty} m\, C_s$$

and thus x belongs to some member of (4). That is a contradiction and the conclusion follows.

In results (5) and (9) we suppose that E is a locally convex space with the following property: if $\{E_n : n = 1,2,\ldots\}$ is any countable family of subspaces of E covering E, there is a positive integer p such that E_p is dense in E. Let $\{A_n : n = 1,2,\ldots\}$ be a family of convex subset of E covering E.

(5) *If* A_n *is rare,* $n = 1,2,\ldots$, *there is a family* $\{B_n : n = 1,2,\ldots\}$ *of convex subset of* E *covering* E *such that the origin of* E *belongs to* B_n, B_n *is rare and the linear hull* F_n *of* B_n *is dense in* E, $n = 1,2,\ldots$

Proof. According to result (1), we can suppose that the origin of E belongs to A_n, $n = 1.2\ldots.$ Let E_n be the linear hull of A_n, $n = 1,2\ldots.$ Let P be the subset of N defined by $p \in P$ if and only if E_p is dense in E. By hypothesis

$$F = \cup \{E_n : n \in N \sim P\}$$

is distinct from E. Take x in $E \sim F$. Let Q be the subset of P defined by $q \in Q$ if and only if there is a ray in E coming from x meeting A_q in at least two distinct points. It is obvious that Q is non-void. For every q in Q we find a vector $x_q \neq 0$ and a number $h_q > 1$ such that $x+x_q$ and $x+h_q x_q$ are in A_q. Let $n(q)$ be a positive integer such that $n(q)(h_q-1) > 2 h_q$. We set

$$C_q = n(q)\ (A_q - (x + \frac{h_q+1}{2}\ x_q)) + x + \frac{h_q+1}{2}\ x_q.$$

Obviously C_q is rare and A_q is contained in C_q and thus the origin of E lies in C_q. Since $x+x_q$ is in A_q, it follows that

$$\frac{2h_q}{n(q)(h_q-1)}\ (x+x_q - (x + \frac{1+h_q}{2} x_q) \in A_q - (x + \frac{1+h_q}{2}\ x_q)$$

and therefore

(6) $$-h_q x_q \in n(q)\ (A_q-(x+ \frac{1+h_q}{2}\ x_q))$$

and thus

(7) $$-x_q \in n(q)\ (A_q - (x + \frac{1+h_q}{2}\ x_q)).$$

From (6) and (7) we obtain

$$x + \frac{1-h_q}{2}\ x_q \in C_q,\ x + \frac{h_q-1}{2}\ x_q \in C_q,$$

and, since C_q is convex,

$$x = \frac{1}{2}\ (x + \frac{1-h_q}{2}\ x_q) + \frac{1}{2}\ \ (x + \frac{h_q-1}{2}\ x_q \in C_q,$$

We consider the family $\{m(C_q-x)+x\colon q \in Q,\ m = 1,2,\ldots\}$ and we denote it by

(8) $$\{B_n : n = 1,2,\ldots\}$$

B_n is convex, contains the origin, is rare and its linear hull F_n is dense in E, $n = 1,2,\ldots$ We shall see that (8) covers E. We suppose the existence of a point z in E which is not in B_n, $n = 1,2,\ldots$ Then $z \neq x$. We set

$$M = \{x + h\ (z-x) : h \geq\ 0\}.$$

Let m be an element of $N \sim Q$. If m belongs to $N \sim P$, we suppose that

$$x+h_1(z-x) \in E_m,\ x + h_2(z-x) \in E_m,\ h_1,h_2, \geq 0,\ h_1 \neq h_2.$$

Since E_m is a linear space we have that

$$\frac{h_2}{h_2-h_1}(x + h_1 (z - x)) + \frac{h_1}{h_1-h_2} (x + h_2 (z - x)) = x \in E_m$$

and that is a contradiction. Consequently M meets A_m in at most one point. If m belongs to P, M meets A_m in at most one point according to the definition of Q. Thus

$$M \cap (\cup \{A_n : n \in N \sim Q\})$$

is countable and consequently there is $s \in Q$ and $k > 0$ such that

$$x + k (z - x) \in A_s$$

Then if r is an integer such that $r k > 1$ we have that

$$z - x \in k^{-1} (A_s - x) \subset r (C_s - x)$$

and therefore z belongs to $r (C_s - x) + x$ and thus z belongs to a number of (8) which is a contradiction. The proof is complete.

(9) *If*

(10) $$\{E_n : n = 1,2,\ldots\}$$

is a family of subspaces of E *covering* E *there is a subfamily* $\{F_n : n = 1, 2,\ldots\}$ *of* (10) *covering* E *such that* F_n *is dense in* E, $n = 1,2,\ldots$

Proof. We can repeat the construction in the proof of (5) considering (10) instead of $\{A_n : n = 1,2,\ldots\}$. Then C_q coincides with E_q and the set (8) is a subfamily $\{F_n : n = 1,2,\ldots\}$ of (10) covering E and such that F_n is dense in E, $n = 1,2,\ldots$

(11) *Let* E *be a locally convex space. Let* A *be a closed convex subset of* E *with void interior. If the convex hull* B *of* $\{0\} \cup A$ *has an interior point there is a closed real hyperplane of* E *containing* A.

Proof. It is obvious that $0 \notin A$ and therefore there is a closed real hyperplane T in E containing 0 such that A lies in a closed halfspace L with boundary T. Then B is contained in L and consequently the origin of E is not an interior point of B.

We suppose the existence of $x \in A$ and $h > 1$ such that $h x \in A$. A positive integer m can be found with $m (h - 1) > 2h$. The method of proof of (1) shows that

$$M = m\,(A - \frac{h+1}{2}\,x) + \frac{h+1}{2}\,x$$

contains A and the origin of E and therefore contains B. On the other hand, it is obvious that M has void interior. That is a contradiction. Then the rays coming from the origin of E meet A in at most one point. Consequently, since B coincides with

$$\{\lambda x : 0 \leq \lambda \leq 1,\ x \in A\}$$

it follows that no point of A is interior to B.

We find k, $0 < k < 1$, and $z \in A$ such that hz is interior to B. Since z is not interior to B there is a continuous real form u on B such that $u(z) = 1$, $u(t) \leq 1$, for each $t \in B$. We set

$$H = \{x \in E : u(x) = 1\}$$

We shall see that H contains A. If $A = \{z\}$, then $A \subset H$. If $A \neq \{z\}$, let y be a point of A distinct from z. Since

$$\lim_{s \to 0}\ k(z + s(z - y)) = k\,z$$

there is $p > 0$ such that $k\,(z + p(\,z - y))$ is interior to B and therefore we can find a real number r, $0 < r < 1$, and a point $x \in A$ such that

$$(12)\qquad k\,(z + p\,(z - y)) = rx.$$

We suppose $y \notin H$. Then $u(y) < 1$ and thus

$$u\,(z + p\,(z - y)) = 1 + p\,(1 - u(y)) > 1$$

and therefore $ru(x) > k$. Since $u(x) \leq 1$, it follows that $r > k$. Setting

$$q = \frac{k\,(1+p)}{kp + r}$$

we have that $0 < q < 1$ and, according to (12),

$$qz = \frac{qp}{1 + p}\,y + \frac{qr}{k\,(1 + p)}\,x.$$

Since

$$\frac{qp}{1+p} > 0, \quad \frac{qr}{k(1+p)} > 0, \quad \frac{qp}{1+p} + \frac{qr}{k(1+p)} = 1,$$

and since $x, y \in A$ it follows that qz belongs to A and therefore the ray coming from the origin containing z meets A in $qz \neq z$. That is a contradiction and thus y belongs to H. Therefore A is contained in H.

(13) *Let* E *be a locally convex space. Let* $\{H_n : n = 1,2,\ldots\}$ *be a family of closed hyperplanes of* E *covering* E. *Then there is a family* $\{K_n : n = 1,2,\ldots\}$ *of closed two-codimensional subspaces of* E *covering* E.

Proof. We select from $\{H_n \cap H_m : n, m = 1,2,\ldots\}$ the subfamily

(14) $\{K_n : n = 1,2,\ldots\}$

of all those elements having codimension two in E.

If z is in E there is a positive integer p such that z belongs to H_p. Let x be a point of E which is not in H_p. Let F be the linear hull of $\{x,z\}$ with the topology induced by E. Since

$$\{H_n \cap F : n = 1,2,\ldots\}$$

covers F there is a positive integer q such that $H_q \cap F$ has interior point in F and since H_q is a linear space, it follows that H_q contains F and thus z belongs to $H_p \cap H_q$. Since x does not belong to H_p it follows that $H_p \cap H_q$ is a closed hyperplane of H_p and consequently $H_p \cap H_q$ is an element of (14). The (14) covers E and the proof is complete.

Let $\{E_i : i \in I\}$ a family of locally convex spaces. If $E = \Pi\{E_i : i \in I\}$ and if H is a subset of I we set $E(H)$ to denote the subspace of E of all those elements which have zero in the coordinate positions indexed by $I \sim H$. Let

$$\mathcal{B} = \{B_n : n = 1,2,\ldots\}$$

be a family of closed convex subsets of E covering E and such that the origin lies in each B_n.

(15) *There is a finite subset* J *of* I *and a positive integer* s *such that* B_s *contains* $E(I \sim J)$.

Proof. Given a subset M of I and an element A of $\mathcal{B}$ containing $E(\{i\})$,

$i \in M$, it follows that A contains also $E(M)$: indeed, A contains the subspace G of $E(M)$ of all those vectors having zero coordinates save in a finite number of them. Since G is dense in $E(M)$ and A is closed, A contains $E(M)$.

If (15) is not true we can find a sequence (i_{1p}) of different elements of I such that

$$E(\{i_{1p}\}) \not\subset B_1, \; p = 1,2,\ldots$$

By recurrence, let $(i_{rp})_{p=1}^{\infty}$ be a sequence of different elements of I such that

$$E(\{i_{rp}\}) \not\subset B_r, \; r = 1,2,\ldots, n; \; p = 1,2,\ldots$$

We find a sequence $(i_{(n+1)p})$ of different elements of I such that

$$i_{(n+1)p} \neq i_{rq}, \; r, q = 1,2,.,n, \text{ and } E\{i_{(m+1)p}\} \not\subset B_{n+1} \; p = 1,2\ldots,$$

For each pair of positive integers n and p we find an element $x_{np} \in E(\{i_{np}\})$ with

$$(16) \qquad x_{np} \not\in B_n$$

Let L_{np} be the linear hull of $\{x_{np}\}$. Let L be the closed linear hull of

$$\{x_{np} : n,p = 1,2,\ldots\}$$

with the topology induced by E. We order all different elements of $\{L_{np} : n, p = 1,2,\ldots\}$ in a sequence (F_n) and we suppose F_n endowed with the topology induced by E. If j belongs to J and $E(\{j\})$ contains some F_n there are positive integer r and s with $j = i_{rs}$. In every sequence $(i_{mp})_{p=1}^{\infty}$ there is at most one element which equals i_{rs}, $m = 1,2,\ldots$ On the other hand, if $n > r + s$ we have that $i_{np} \neq i_{rs}$, $p = 1,2,\ldots$ Therefore the number of subspaces F_n contained in $E(\{j\})$ is finite and thus L is isomorphic to the Fréchet space $\Pi \{F_n : n = 1,2,\ldots\}$. Since

$$\bigcup_{m=1}^{\infty} (B_m \cap L) = L$$

there is a positive integer q such that $B_q \cap L$ has an interior point x in L. We can find a finite subset P of I and a neighbourhood of the origin U_i in E_i, $i \in P$, such that

$$x + (\Pi\{U_i : i \in P\} \times \Pi\{E_i : i \in I \sim P\}) \cap L \subset B_q.$$

We select in the sequence $(i_{qp})_{p=1}^{\infty}$ an element i_{qm} which is not in P. Then $x + n\,x_{qm}$ belongs to B_q, $n = 1,2,\ldots$, and, since the origin of E lies in B_q, we have that

$$\frac{1}{n} x + x_{qm} \in B_q$$

and, remembering that B_q is closed,

$$\lim_{n\to\infty} \left(\frac{1}{n} x + x_{qm}\right) = x_{qm} \in B_q$$

which is in contradiction with (16). The conclusion follows.

2. CONVEX-BAIRE SPACES. Let E be a locally convex space. We say that E is a convex-Baire space if and only if given any sequence (A_n) of closed convex subsets of E having void interior then $\bigcup_{n=1}^{\infty} A_n$ has void interior.

It is obvious that every Baire locally convex space is a convex-Baire space. The convex-Baire spaces enjoy better stability properties than the Baire spaces and, on the other hand, the concept of convexity is widely used in applications of Baire's theorem in Functional Analysis. These considerations justify the introduction and the forthcoming study of our convex-Baire spaces.

(1) *A locally convex space* E *is convex-Baire if and only if given any sequence* (A_n) *of closed convex subsets of* E *covering* E *there is a positive integer* p *such that* A_p *has non-void interior.*

Proof. We suppose E convex-Baire. If the sequence (A_n) of closed convex subset of E covers E, then $\bigcup_{n=1}^{\infty} A_n$ has non-void interior and therefore there is a positive integer p such that A_p has an interior point.

Reciprocally, let (B_n) be a sequence of closed convex subsets of E whose union has an interior point z. We consider the countable family of closed convex subsets of E

(2) $\{m(B_n - z)+z : m = n, 1,2,\ldots\}$

If x is any point in E, a positive integer p can be select such that

$$\frac{1}{p}(x-z) \in \bigcup_{n=1}^{\infty} B_n - z = \bigcup_{n=1}^{\infty} (B_n - z)$$

since $\bigcup_{n=1}^{\infty} B_n - z$ is a neighbourhood of the origin. Therefore there is a positive integer q such that

$$\frac{1}{p}(x-z) \in B_q - z$$

from where it follows that

$$x \in p\ (B_q - z) + z$$

and thus the family (2) covers E. Two positive integer r and s can be chosen such that $r\ (B_s - z) + z$ has non-void interior. Thus B_s has non-void interior and E is a convex-Baire space.

(3) *Every separated quotient of a convex-Baire space is convex-Baire.*

Proof. Let E be a convex-Baire space, F a closed subspace of E and $f : E \longrightarrow E/F$ the canonical surjection. We consider a sequence (A_n) of closed convex subsets of E/F covering E/F. Clearly the sequence $(f^{-1}(A_n))$ covers E and is constitued by closed convex subsets of E. Since E is convex-Baire there is a positive integer p such that $f^{-1}(A_p)$ has non-void interior. Since A_p coincides with $f(f^{-1}(A_p))$ we have that A_p is a subset of E/F with non-void interior. The conclusion follows.

(4) *Let* E *be a locally convex space. Let* F *be a dense subspace of* E. *If is a convex-Baire space, then* E *is convex-Baire.*

Proof. It is a straightforward conclusion from the definition.

(5) *Let*

(6) $\{E_n : n = 1,2,\ldots\}$

be a family of subspaces of a locally convex space E *covering* E. *If* E *is a convex-Baire space there is a positive integer* p *such that* E_p *is dense in* E *and convex-Baire.*

Proof. Let G_n be the closure of E_n in E, n = 1,2... The sequence of closed convex subsets (G_n) of E covers E and therefore there is a positive integer r such that G_r has non-void interior, i.e., G_r coincides with E. We apply 1.(9) to obtain a subfamily $\{F_n : n = 1,2,\ldots\}$ of (6) covering E such that F_n is dense in E, n = 1,2,... We suppose now that F_n is not convex -Baire space, n = 1,2, ... For every positive integer n we find in F_n a family of subsets.

$$\{ B_{np} : p = 1, 2, \dots , \}$$

covering F_n such that B_{np} is closed an convex and has void interior. If A_{np} denotes the closure in E of B_{np} we have that A_{np} has void interior in E, p = 1, 2, ..., and

$$\cup \{A_{np} : n, p = 1, 2, \dots\} = E$$

which is a contradiction. Then there is a positive integer q such that F_q is a convex-Baire space. The conclusion follows.

3. COUNTABLE CODIMENSIONAL SUBSPACES IN A CONVEX - BAIRE SPACE. For the proof of result (1) we suppose that E is a locally convex space such that if $\{E_n : n = 1, 2, \dots\}$ is any countable family of subspaces of E covering E, then there is a positive integer p such that E_p is dense in E.

(1) *Let* F *be an hyperplane of* E *and let* $\{F_n : n = 1, 2, \dots\}$ *be a family of subspace of* F *covering* F. *Then there is a positive integer* q *such that* F_q *is dense in* F.

Proof. We suppose that the result is not true. For every positive integer n we can find a closed hyperplane H_n in F containing F_n. Since

$$F = \cup \{H_n : n = 1, 1, \dots \}$$

We apply 1.(13) to obtain a family

$$\{ K_n : n = 1, 2, \dots \}$$

of closed subspaces of F covering F and of codimension two.

Let L be an algebraic complement of F in E and let G_n be the closure in E of K_n, n = 1, 2, ... The codimension of G_n in E is two or three. If $E_n = G_n + L$, then E_n is closed in E and has codimension one or two in E. Therefore E_n is not dense in E. On the other hand, it is obvious that

$$E = \cup \{E_n : n = 1, 2, \dots\}$$

which is a contradiction. The conclusion follows.

(2) *Let* E *be a convex-Baire space and let* F *be a countable codimensional subspace of* E. *Then* F *is a convex-Baire space.*

Proof. We suppose first that F has infinite codimension in E. Let $\{x_n : n = 1,2,\ldots\}$ be a cobasis of F in E. If F_n denotes the linear hull of $F \cup \{x_1, x_2, \ldots x_n\}$ we have that

$$E = \cup \{F_n : n = 1,2,\ldots\}.$$

We apply 2.(5) to obtain a positive integer p such that F_p is a convex-Baire space. Therefore it is enough to prove the theorem being E real and F an hyperplane of E. Let z be a vector of E which is not in F. If F is closed in E and if L is the linear hull of $\{z\}$ it is clear that F is isomorphic to E/L and thus convex-Baire by virtue of 2.(3). We suppose now that F is dense in E and F is not convex-Baire. Let (A_n) be a family of closed convex subsets of F covering F such that A_n has void interior, n = 1,2,... According to (1) and 1.(5), we obtain in F a family $\{B_n : n = 1,2,\ldots\}$ of closed convex subsets containing $\{0\}$, covering F, with void interior and such that their linear hull are dense in F.

If M_n denotes the closure of B_n in E it is clear that M_n has void interior in E. Let P_{nm} be the convex hull of

$$M_n \cup \{mz\}, \; m = 1, -1, 2, -2, \ldots$$

It is obvious that P_{nm} is closed in E. On the other hand, if x belongs to E, there is a real number h and a positive integer p such that

$$2x = hz + y, \; y \in B_p$$

If h is larger or equal than zero there is a positive integer q with $h < q$. Then x belongs to P_{pq}. If h is less than zero there is a negative interior q with $q < h$. Then x belongs to P_{pq}. Therefore the family

$$\{P_{nm} : m = 1, -1, 2, -2, \ldots ; n = 1,2,\ldots\}$$

covers E and therefore there are integers r and s such that P_{rs} has interior point in E. If u is a point of $P_{rs}-sz$ we can find a real number k, $0 < K < 1$, and a element v in M_r such that

$$u = ksz + (1-k)\, v - sz = (1 - k)\, (v - sz)$$

and therefore u belongs to the convex hull M of $\{0\} \cup \{M_r - sz\}$. Then M has interior point in E and since $M_r - sz$ has void interior in E, we apply 1(11) to obtain a closed hyperplane H in E containing $M_r - sz$. Then $S = H + sz$ is a closed hyperplane in E containing M_r. Since F is dense in E, $F \cap S$ is a closed hyperplane in F containing B_r. We arrive to a contradiction observing that the linear hull of B_r is dense in F.

(3) *Let* H *be a dense hyperplane in a Fréchet space* E. *Let* (A_n) *be a sequence of closed convex sets in* E *covering* H. *Then there is a positive integer* p *such that* A_p *has non-void interior in* E.

Proof. Since E is convex-Baire, we apply (2) to obtain that H is convex-Baire. Then there is a positive integer p such that $A_p \cap H$ has non-void interior in H and therefore A_p has non-void interior in E.

4. PRODUCTS OF CONVEX - BAIRE SPACES. The main results of this section is that the product of convex-Baire spaces is a convex-Baire space. We shall need some previous results.

(1) *Let* F *be a subspace of a locally convex space* E. *Let* A *be a convex subsets of* E. *Let* x *and* z *be a points in* E *such that* $(x + F) \cap A$ *has an interior point* v *in* $x + F$ *and* $(z + F) \cap A$ *is non-void. Then there is a positive integer* m *such that* z *belongs to* $m(A - z) + v$.

Proof. Let w be a point of $(z + F) \cap A$. Since $(x + F) \cap A - v$ is a neighbourhood of the origin in F and since $z - w$ is in F, there is a real number r, $0 < r < 1$, with

$$r(z - w) \in (x + F) \cap A - v$$

and thus

$$r(z - w) + v \in (x + F) \cap A \subset A$$

and since A is convex

$$\frac{r}{r+1} w + \frac{1}{r+1} (r(z - w) + v) = \frac{rz + v}{r + 1} \in A.$$

Finally, if m is an integer larger than $\frac{r + 1}{r}$ we have that

$$z = \frac{r + 1}{r} \left(\frac{rz + v}{r + 1} - v\right) + v \in \frac{r + 1}{r} (A-v) + v \subset m(A-v) + v.$$

Let P be a set of positive integers. Let T be a countable subset of a locally convex space E. Given a subspace $F \neq \{0\}$ of E we consider a family

$$\beta = \{B_n : n \in P\}$$

of closed convex subsets of E such that for every x in T and for every n in P the set $(x + F) \cap B_n$ has void interior in x + F.

(2) *If* F *is a convex-Baire space there is an one-dimensional real subspace* L *of* F *such that the set*

(3) $$\cup \{(x + L) \cap B_n : x \in T, n \in P\}$$

is countable.

Proof. Given any point x of T, let P(x) be the subset of P such that p belongs to P(x) if and only if every ray coming from x, which is contained in x + F, meets B_p at most one point. For every n of $P \sim P(x)$ we find a point z (n, x) in F and a real number h (n, x) > 1 such that

$$x + z(n, x) \text{ and } x + h(n,x)\, z(n,x)$$

belong to B_n. Let m(n, x) be an integer with

$$m(n, x)\,(h(n, x)-1) > h(n, x) + 1$$

We set

$$M(n, x) = m(n, x)\left((x + F) \cap B_n - \left(x + \frac{1 + h(n, x)}{2} z(n, x)\right)\right)$$

Then M(n, x) is in F, has no interior point in F and $0 \in M(n, x)$. Since

$$x + z(n, x) \in (x + F) \cap B_n$$

we have that

$$\frac{1-h(n, x)}{2} z(n, x) = x + z(n, x) - \left(x + \frac{1 + h(n, x)}{2} z(n, x)\right)$$

$$\in \frac{1}{m(n, x)} M(n, x)$$

and, since $0 \in M(n, x)$, it follows that

$$-\frac{1 + h(n, x)}{2m(n, x)} z(n, x) = \frac{1 + h(n, x)}{m(n, x)(h(n, x)-1)} \frac{1 - h(n, x)}{2} z(n, x)$$

$$\in \frac{1}{m(n, x)} M(n, x)$$

and thus

(4) $$-\frac{1 + h(n, x)}{2} z(n, x) \in M(n, x)$$

We set

$$P(n, x) = M(n, x) + x + \frac{1 + h(n, x)}{2} z(n, x).$$

It is clear that $P(n, x)$ is contained in $x + F$, has no interior point in $x + F$, contains $(x + F) \cap B_n$ and, according to (4), x belongs to $P(n, x)$. Since F is convex-Baire.

(5) $$\cup \{m(P(n, x)-x) : n \in P \sim P(x), x \in T, m \text{ integer}\}$$

does not cover F and, accordingly, there is a vector u in F which is not in (5). It follows that the real linear hull L of u meets (5) only in the origin.

We shall see now that (3) is countable. If n belongs to $P \sim P(x)$, then L meets (5) in the origin and consequently L meets $P(n, x)-x$ in the origin and therefore

$$(x + L) \cap P(n, x) = \{x\}$$

and thus $(x + L) \cap B_n$ contains, at most, the set $\{x\}$. If n belongs to $P(x)$ every ray coming from x and contained in $x + F$ meets B_n, $n \in P(x)$ at most one point and, since B_n is convex, $(x + L) \cap B_n$ has at most one point. The proof is complete.

Let B be a subset of a locally convex space E and let F be a subspace of E. We take a point z in E and a positive integer m. We set $D = m(B - z) + z$.

(6) *If for every* x *of* E, *the set* $(x + F) \cap B$ *has no interior point in* $x+F$, *then* $(x + F) \cap D$ *has no interior point in* $x + F$.

Proof. Suppose a point y in E such that $(y + F) \cap D$ has non-void interior in $y + F$. Then

$$\frac{1}{m}((y + F) \cap D) = (\frac{1}{m} y + F) \cap \frac{1}{m} D$$

has non-void interior in $\frac{1}{m} y + F$ and consequently the set

$$(\frac{1}{m} y + F) \cap \frac{1}{m} D + \frac{m-1}{m} z = (\frac{1}{m} y + \frac{m-1}{m} z + F) \cap (\frac{1}{m} D + \frac{m-1}{m} z)$$

$$= (\frac{1}{m} y + \frac{m-1}{m} z + F) \cap B$$

has non-void interior in $\frac{1}{m} y + \frac{m-1}{m} z + F$, which is a contradiction.

In order to state results (8), (11), (12), (13) and (14) we shall consider a family $\{E_i : i \in I\}$ of convex-Baire spaces. If $E = \Pi\{E_i : i \in I\}$ and if H is a subset of I, E(H) has the same meaning as in section 1.

Let $\mathcal{P}$ be a countable family of closed convex subsets of E. Let $\mathcal{B}$ be a subfamily of $\mathcal{P}$ such that B belongs to $\mathcal{B}$ if and only if there is an index i in I, depending on B, such that $(x + E(\{i\})) \cap B$ has void interior in $x + E(\{i\})$ for every x of E. We set

$$\mathcal{B} = \{B_n : n \in M\}$$

being $M = \{1,2,\ldots,s\}$ if $\mathcal{B}$ is finite and non-void and $M = \{1,2,\ldots\}$ if $\mathcal{B}$ is infinite.

If $\mathcal{B}$ is non-void we select i_1 in I such that for every x of E the set

$$(x + E(\{i_1\}) \cap B_1$$

has void interior in $x + E(\{i_1\})$. Let $\mathcal{B}_1$ be the subfamily of $\mathcal{B}$ of all those B of $\mathcal{B}$ such that

$$(x + E(\{i_1\}))\cap B$$

has void interior in $x + E(\{i_1\})$ for every x of E. By recurrence let us suppose that we have obtained $i_r \in I$ and $\mathcal{B}_r \subset \mathcal{B}$, $1 \leq r \leq n$. If

$$\mathcal{B} \sim \cup \{\mathcal{B}_r : r = 1,2,\ldots,n\} \tag{7}$$

is non-void, let p be the first integer such that B_p is in (7). We take i_{n+1} in I such that

$$(x + E(\{i_{n+1}\}))\cap B_p$$

has void interior in $x + E(\{i_{n+1}\})$ for every x of E. We denote by $\mathcal{B}_{n+1}$ the subfamily of (7) such that B belongs to $\mathcal{B}_{n+1}$ if and only if

$$(x:E(\{i_{n+1}\}))\cap B$$

has void interior in $x + E(\{i_{n+1}\})$ for every x of E.

In such way, we determine a set P of positive integers, which coincides with N when P is infinite, verifying

$$\mathcal{B}= \cup \{\mathcal{B}_n : n \in P\}.$$

(8) *Given a point* x *in* E *there is, for every* p *of* P, *an one-dimensional real subspace* L_p *of* $E(\{i_p\})$ *such that, if* L *denotes the real closed linear hull of* $\cup \{L_p : p \in P\}$,

$$(x + L)\cap B_n,\ n \in M,$$

has void interior in x + L.

Proof. For every n in P, let M_n be the family of all those D in $\mathcal{B}_n$ such that

$$(z+E(\{i_p\})\cap D$$

has void interior in $z + E(\{i_p\})$ for every p in P and every z in E. If B belongs to $\mathcal{B}_n \sim M_n$ let H(n,B) be the subset of P of all those positive integers s such that there is x(s,B) in B interior to

$$(x(s,B) + E(\{i_s\}))\cap B$$

in $x(s,B) + E(\{i_s\})$.Let N_n be the set

$$M_n \cup \{m(B-x(s,B)) + x(s, B): B \in \mathcal{B}_n \sim M_n,\ s \in H(n, B),\ m = 1,2,\ldots\}$$

Obviously $\mathcal{B}_n$ is contained in N_n, $n \in P$, and, according to (6), if A belongs to N_n, then

$$(z+E(\{i_n\}))\cap A$$

has no interior point in $z + E(\{i_n\})$ for every z in E.

Applying result (2) for T = {x} and $F = E(\{i_1\})$ we obtain a real one-dimensional subspace L_1 in $E(\{i_1\})$ such that the set

(9) $\cup \{(x+L_1) \cap B : B \in \mathcal{N}_1\}$

is countable. We take in $x + L_1$ a dense countable subset S_1 disjoint from (9). Proceeding by recurrence, suppose we have obtained the one-dimensional real subspace L_m of $E(\{i_m\})$, $1 \leq m \leq n$, and a dense subset S_n of $x + L_1 + L_2 + \ldots + L_n$ distinct from

$$\cup\{(x+L_1+L_2+\ldots+L_n) \cap B : B \in \mathcal{N}_m, m = 1,2,\ldots,n\}$$

We take u in S_n and A in $\mathcal{N}_m$ for a positive integer m, $1 \leq m \leq n$. If

$$(u + E(\{i_{n+1}\})) \cap A$$

has interior point in $u + E(\{i_{n+1}\})$, the following two cases can occur: a) A belongs to $\mathcal{B}_m$; then n+1 belongs to H(m,A) and x(n+1, A) is interior to

$$(x(n+1,A) + E(\{i_{n+1}\})) \cap A$$

in $x(n+1,A) + E(\{i_{n+1}\})$ and, applying result (1), there is a positive integer p such that

$$u \in p(A-x(n+1,A)) + x(n+1,A) \in \mathcal{N}_m$$

which is a contradiction. b) A does not belong to $\mathcal{B}_m$; then there is B in $\mathcal{B}_m$ and positive integers r and s, s in P, with

$$A = r(B-x(s,B)) + x(s,B)$$

Since

$$(u + E(\{i_{n+1}\})) \cap A$$

has interior point in $u + E(\{i_{n+1}\})$, we apply (6) to obtain w in E such that

$$(w + E(\{i_{n+1}\}) \cap B$$

has interior point in $w + E(\{i_{n+1}\})$. We proceed as in case a) to reach a contradiction.

From what has been said it follows that, for every u in S_n and A in $\mathcal{N}_m$, $1 \leq m \leq n+1$,

$$(u + E(\{i_{n+1}\})) \cap A$$

has no interior point in $u + E(\{i_{n+1}\})$. We apply result (2) for $T = S_n$ and $F = E(\{i_{n+1}\})$ to obtain an one-dimensional real subspace L_{n+1} in $E(\{i_{n+1}\})$ such that

$$\cup \{(u + L_{n+1}) \cap A : u \in S_n,\ A \in \mathcal{N}_m,\ m = 1,2,\dots, n+1\}$$

is countable. Then there is a dense countable subset R_n in L_{n+1} such that

(10) $\quad S_{n+1} \cap A = \emptyset,\ A \in \mathcal{N}_m,\ m = 1,2,\dots, n+1,$

being

$$S_{n+1} = \{u + v : u \in S_n,\ v \in R_n\}.$$

Obviously S_{n+1} is countable and dense in $x + L_1+L_2+\dots+L_{n+1}$. Let L be the closed real linear hull of $\cup \{L_n : n \in P\}$. We suppose the existence of B in $\mathcal{B}$ such that $(x + L) \cap B$ has an interior point z in $x + L$. We find a finite subset J in I and an open neighbourhood of the origin U_i in E_i, $i \in J$, such that if $U_i = E_i$, $i \in I \sim J$, and

$$U = \Pi\{U_i : i \in I\},$$

then $(z + U) \cap (x + L)$ is contained in B. If P is finite, let q be the largest of its elements. If P is infinite, let q be a positive integer such that

$$B \in \cup \{\mathcal{N}_n : 1 \leq n \leq q\}$$

and $i_n \notin J$, $n \geq q$. Then

$$z + L_n \subset (z + U) \cap (x + L) \subset B,\ n \geq q.$$

Therefore, being P finite or not,

$$B \cap (x + L_1 + L_2 + \dots + L_q)$$

has an interior point in $x + L_1 + L_2 + \dots + L_q$ and, since S_q is dense in $x+L_1+L_2+ \dots + L_q$, $S_q \cap B$ is non-void which is in contradiction with (10). The proof. is complete.

(11) *If* $\mathcal{B}$ *belongs to* $P \sim B$, *given any finite subset* J *of* I, *there is* $x \in E$ *interior to*

$$(x + \Sigma\{E(\{i\}) : i \in J\}) \cap B$$

in $x + \Sigma\{E(\{i\}) : i \in J\}$.

Proof. The property is obvious if J has a single element. We suppose now the property true if J has n elements. We take $j_1, j_2, \ldots, j_{n+1}$ of I, every index being different from the others. We find an element u in E interior to

$$(u + E(\{j_{n+1}\})) \cap B$$

in $u + E(\{j_{n+1}\})$. By hypothesis there is $v \in E$ interior to

$$(v + \Sigma\{E(\{j_p\}) : p = 1,2,\ldots,n\}) \cap B$$

in $v + \Sigma\{E(\{j_p\}) : p = 1,2,\ldots,n\}$. Then

$$(u + E(\{j_{n+1}\})) \cap B - u$$

is a neighbourhood of the origin in $E(\{j_{n+1}\})$. Consequently $\frac{1}{2}(u+v)$ is an interior point to

$$(12) \qquad \frac{1}{2}((u + E(\{j_{n+1}\})) \cap B-u) + \frac{1}{2}(u + v)$$

in $\frac{1}{2}(u + v) + E(\{j_{n+1}\})$. If z is an element of (12) there is $w \in E(\{j_{n+1}\})$ such that u + w belongs to B and

$$z = \frac{1}{2}(u + w - u) + \frac{1}{2}(u + w) = \frac{1}{2}(u + v) + \frac{1}{2}w$$

and thus z is in $\frac{1}{2}(u + v) + E(\{j_{n+1}\})$. On the other hand, since u + w and v belong to B, we have that

$$\frac{1}{2}(u + w) + \frac{1}{2}v = \frac{1}{2}(u + v) + \frac{1}{2}w \in B$$

and then

$$z \in (\frac{1}{2}(u + v) + E(\{j_{n+1}\})) \cap B$$

and thus $\frac{1}{2}(u + v)$ is an interior point to

$$(\frac{1}{2}(u + v) + E(\{j_{n+1}\})) \cap B$$

in $\frac{1}{2}(u + v) + E(\{j_{n+1}\})$. Analogously we change u by v and $(\{j_{n+1}\})$ by $\Sigma\{E(\{j_p\}) : p = 1,2,\ldots, n\}$ to obtain that $\frac{1}{2}(u + v)$ is an interior point to

$$(\tfrac{1}{2}(u + v) + \Sigma\{E(\{j_p\}) : p = 1,2,\ldots,n\}) \cap B$$

in $\frac{1}{2}(u + v) + \Sigma\{E(\{j_p\}) : p = 1,2,\ldots,n\}$. Finally if U_1 and U_2 are absolutely convex neighbourhoods of the origin in $E(\{j_{n+1}\})$ and in $\Sigma\{E(\{j_p\}) : p = 1,2,\ldots,n\}$ respectively, such that

$$\tfrac{1}{2}(u + v) + U_1 \subset B \text{ and } \tfrac{1}{2}(u + v) + U_2 \subset B,$$

then

$$\tfrac{1}{2}(\tfrac{1}{2}(u + v) + U_1) + \tfrac{1}{2}(\tfrac{1}{2}(u + v) + U_2)$$

$$= \tfrac{1}{2}(u + v) + \tfrac{1}{2}U_1 + \tfrac{1}{2}U_2 \subset B$$

and, since $E(\{j_{n+1}\})$ is a topological complement of $\Sigma\{E(\{j_p\}) : p = 1,2,\ldots,n\}$ in $\Sigma\{E(\{j_p\}) : p = 1,2,\ldots, n + 1\}$, then $\frac{1}{2}U_1 + \frac{1}{2}U_2$ is a neighbourhood of the origin in the former space. Thus $\frac{1}{2}(u + v)$ is interior to

$$(\tfrac{1}{2}(u + v) + \Sigma\{E(\{j_p\}) : p = 1,2,\ldots,n+1\}) \cap B$$

in $\frac{1}{2}(u + v) + \Sigma\{E(\{j_p\}) : p = 1,2,\ldots,n+1\}$, which completes the proof.

For every B in $P\sim\mathcal{B}$ and every n in P, we can find a vector u(n, B) in E interior to

$$(u(n, B) + \Sigma\{E(\{i_p\}) : p = 1,2,\ldots,n\}) \cap B$$

in $u(n, B) + \Sigma\{E(\{i_p\}) : p = 1,2,\ldots,n\}$ according to result (11). We set

$$R = \cup\{m(B - u(n, B)) + u(n, B) : n \in P,\ B \in P\sim\mathcal{B},\ m = 1,2,\ldots\}.$$

(13) *If P covers* E *then R covers* E.

Proof. We suppose the existence of $x \in E$ which is not in any element of R. Applying (8) we obtain, for every p in P, an one-dimensional real subspace L_p in $E(\{i_p\})$ such that

$$(x + L) \cap B_n,\ n \in M,$$

has no interior point in x + L, L being the closed real linear hull of

$\cup \{L_p : p \in P\}$.

It is obvious that L is isomorphic to the Fréchet space $\Pi\{L_p : p \in P\}$ and, since P covers E, we apply Baire's theorem to obtain an element B in P such that

$$(x + L) \cap B$$

has an interior point z in x + L. Then B belongs to $P \sim \mathcal{B}$. As we did in the proof of (8) we can find a positive integer q such that

$$(x + L_1 + \dots + L_q) \cap B$$

has an interior point in $x + L_1 + \dots + L_q$. Therefore

$$(x + \Sigma\{E(\{i_p\}) : p = 1,2,\dots,q\}) \cap B \supset (x + L_1 + \dots + L_q) \cap B \neq \emptyset$$

On the other hand, u (q, B) is interior to

$$(u\,(q, B) + \Sigma\{E\ (\{i_p\}) : p = 1,2,\dots,q\}) \cap B$$

in $u\,(q, B) + \Sigma\{E\,(\{i_p\}) : p = 1,2,\dots,q\}$. We apply (1) for $F = \Sigma\{E\,(\{i_p\}) : p = 1,2,\dots,q\}$ and we obtain a positive integer m such that

$$x \in m\,(B - u\,(q, B)) + u(q, B) \in \mathcal{R},$$

which is a contradiction.

(14) *The product* $E = \Pi\{E_i : i \in I\}$ *is a convex-Baire space.*

Proof. We suppose that E is not convex-Baire. We can choose the family P covering E and such that every element A of P has void interior and the origin belongs to A. The elements of $\mathcal{R}$ have void interior, contain {0} and cover E by virtue of (13).

Applying 1.(15) we obtain a finite subset J of I and an element D of $\mathcal{R}$ such that D contains $E(I \sim J)$. According to (11) there is a vector x in E interior to

$$(x + \Sigma\{E(\{i\}) : i \in J\}) \cap D$$

in $x + \Sigma\{E\,(\{i\}) : i \in J\}$. Consequently,

$$(15) \qquad (x + \Sigma\{E\,(\{i\}) : i \in J\}) \cap D - x$$

$$= (\Sigma\,\{E\,(\{i\,) : i \in J\}) \quad (D - x)$$

is a neighbourhood of the origin in $\Sigma\{E(\{i\}) : i \in J\}$. On the other hand, if i belongs to $I \sim J$ and if z belongs to $E(\{i\})$, we have that

$$n z \in E(\{i\}) \subset E(I \sim J) \subset D$$

for every positive integer n and, since x belongs to D, it follows that

$$\frac{1}{n} n z + \frac{n-1}{n} x = z + \frac{n-1}{n} x \in D$$

and, since D is closed,

$$z + x = \lim_n (z + \frac{n-1}{n} x) \in D$$

and thus $x + E(\{i\})$ is contained in D. Then $E\{i\}$ is contained in $D - x$, $i \in J$, and, since $D - x$ is closed and convex, $E(I \sim J)$ is contained in D-x. Since (15) is a neighbourhood of the origin in $\Sigma\{E(\{i\}) : i \in J\}$ and,since E is the topological direct sum of that space and $E(I \sim J)$, it is obvious that $D - x$ is a neighbourhood of the origin in E and thus D has non-void interior in E and that is a contradiction. The proof is complete.

5. NOTE. Following TODD and SAXON [1], a locally convex space E is said to be unordered Baire-like if given a sequence (A_n) of closed absolutely convex subsets of E covering E, then there is a positive integer p such that A_p is a neighbourhood of the origin in E. It is obvious that a convex-Baire space is unordered Baire-like. TODD and SAXON [1] show that the unordered Baire-like spaces are stable by products and countable codimensional subspaces.

6. CONVEX-BAIRE SETS. Let E be a locally convex space. A subset B of E is convex-Baire if and only if it is convex and can not be covered by a countable union of rare convex sets. By taking B as E we have the concept of convex-Baire space.

Results (1) and (2) are immediate.

(1) *Let* A *be a convex-Baire subset of a locally convex space* E *and let* x *be a point in* E. *Then* x + A *is convex-Baire.*

(2) *Let* A *be a convex-Baire subset of a locally convex space* E *and let* m *be a non-zero escalar. Then* mA *is convex-Baire.*

(3) *Let* A *be a convex-Baire subset of a locally convex space* E. *Let* (A_m)

be a sequence of convex subsets of E. *If*

$$A \subset \cup \{A_m : m = 1,2, \dots\}$$

there is a positive integer p *such that* A_p *is convex-Baire.*

Proof. Suppose the property not true. For every positive integer m, we find a sequence $(A_{m\,p})$ of rare convex subsets of E covering A_m. Then A is the union of the countable family

$$\{A_{m\,p} \cap A : m, p = 1,2, \dots\}$$

of rare convex subsets of E. That is a contradiction.

Let A be a conve-Baire subset of a locally convex space E. We consider a family of subsets of A

$$\{A_{m_1,m_2, \dots, m_k} : m_1, m_2, \dots, m_k, \quad k \in N\}$$

such that

(4) $$A = \cup\{A_m : m = 1,2,\dots\},$$

$$A_{m_1, m_2, \dots, m_k} = \cup \{A_{m_1,m_2,m_k,m} : m = 1,2,\dots\}$$

We set $M_{m_1, m_2, \dots, m_k}$ to denote the convex hull of $A_{m_1,m_2, \dots, m_k}$.

(5) *There is a sequence of positive integers* (n_k) *such that* $M_{n_1,n_2,\dots, n_k}$ *is a convex-Baire subset of* E.

Proof. Let *P* be the family of all finite sequences $\{m_1, \dots, m_2, \dots, m_k\}$ of N such that $M_{m_1,m_2, \dots, m_k}$ is not convex-Baire. Since *P* is countable, we apply (3) to obtain a point z in

$$A \sim \cup \{M_{m_1, m_2, \dots, m_k} : \{m_1, m_2, \dots, m_k\} \in P\}$$

According to (4), there is a sequence of positive integers (n_k) such that

$$z \in A_{n_1, n_2, \dots, n_k}, \quad k = 1,2,\dots$$

Then $M_{n_1, n_2, \dots, n_k}$ is a convex-Baire subset of E, k = 1,2, ...

§ 3. QUASIBARRELLED, BARRELLED, BORNOLOGICAL AND ULTRABORNOLOGICAL SPACES

1. INCREASING SEQUENCES OF CONVEX SUBSETS IN LOCALLY CONVEX SPACES. Let F be a locally convex space. If B is a bounded absolutely convex subset of F, we write F_B to denote the linear hull of B endowed with the topological derived from the gauge or Minkowski functional of B (cf KÖTHE [1], Chapter Three, §15, section 10). If B is complete, it is obvious that F_B is a Banach space. A sequence (y_n) in F is said to be a Cauchy (convergent) sequence in the sense of Mackey if there is a bounded absolutely convex subset B of F such that (y_n) is a Cauchy (convergent) sequence in F_B. F is locally complete if and only if for every closed bounded absolutely convex subset A of F, F_A is a Banach space. F is dual locally complete if $F'[\sigma(F',F)]$ is locally complete.

If the topology of the locally convex space F is the Mackey topology, we say that F is a Mackey space. A barrel in F is an absorbent closed absolutely convex subset of F. F is barrelled if every barrel in F is a neighbourhood of the origin. Equivalently, F is barrelled if every bounded subset of $F'[\sigma(F', F)]$ is an equicontinuous set. A linear form u on F is locally bounded, shortly bounded, if it maps every bounded subset of F in a bounded subset of the field K. If A is a subset of the topological dual F' of F we set A* to denote the closure of A in the algebraic dual F* of F endowed with the weak topology $\sigma(F^*,F)$. We identify F with a subspace of the algebraic dual F'* of F' by means of the canonical injection. If B is a subset of F we set B* to denote the closure of B in $F'^*[\sigma(F'^*, F')]$ and $\hat{B}$ stand for the closure in the completion $\hat{F}$ of F.

(1) *Let E be a locally convex space. Let* (A_n) *be an increasing sequence of convex subsets of* E' *covering* E'. *If* u *is a linear form on* E *which is bounded, then* u *belongs to* mA^*_m *for some positive integer* m.

Proof. For every positive integer n suppose that u is not in $n A^*_n$ and $0 \in A^*_n$. Let B_n be the absolutely convex hull of $A_n \cap (-A_n) \cap i A_n \cap (-i A_n)$. Then $\bigcup_{n=1}^{\infty} B_n = E'$ and $B^*_n \subset 2 A^*_n$. We obtain x_n in E with

$$\langle x_n, u\rangle = 1, \quad |\langle x_n, v\rangle| < 1, \ v \in nB^*_n$$

Given any element w of E' there is a positive integer p such that w belongs to B_n, $n > p$ Then

$$|\langle x_n, nw\rangle| = |\langle nx_n, w\rangle| < 1, n > p,$$

from where it follows that the sequence (nx_n) is bounded in E. On the other hand,

$$\langle nx_n, u\rangle = n, n = 1,2,...,$$

and this is a contradiction.

(2) *Let* E *be a locally convex space. If every sequence in* $E'[\sigma(E', E)]$ *which converges to the origin in the sense of Mackey is equicontinuous, then every element of the completion* $\hat{E}$ *of* E *is a linear bounded form on* $E'[\sigma(E,E)]$.

Proof. Suppose the property not true. Then there is an element x in $\hat{E}$ and a bounded closed absolutely convex subset A of $E'[\sigma(E', E)]$ such that

$$\sup\{|\langle x, u\rangle| : u \in A\} = \infty.$$

For every positive integer n take u_n in A with $\langle x, u_n\rangle = n^2$. Then the sequence $(\frac{1}{n} u_n)$ converges to the origin in the sense of Mackey in $E'[\sigma(E',E)]$ and therefore is equicontinuous. Consequently it is an equicontinuous in E and thus $\sigma(E', \hat{E})$- bounded. Since

$$\langle x, \frac{1}{n} u_n\rangle = n, n = 1,2,...,$$

we obtain a contradiction. Thus x is locally bounded.

(3) *Let* (A_n) *be an increasing sequence of convex subsets of a locally convex space* E *covering* E. *If every sequence in* $E'[\sigma(E', E)]$ *which converges to the origin in the sense of Mackey is equicontinuous, then* $\hat{E}$ *coincides with* $\bigcup_{n=1}^{\infty} n \hat{A}_n$.

Proof. Let x be an element of $\hat{E}$. We apply (2) and (1) to obtain that x belongs to mA^*_m for a certain positive integer m. Since $m\hat{A}_m$ coincides $mA^*_m \cap \hat{E}$, it follows that x belongs to $m \hat{A}_m$ and the conclusion follows

(4) *Let* E *be a barrelled space. Let* (A_n) *be an increasing sequence of closed convex subset of* E *whose union is* E. *If* $\hat{E}$ *is a Baire space there is a positive integer* p *such that* A_p *has non-void interior.*

Proof. Since E is barrelled, we apply (3) to obtain that $\hat{E}$ coincides with $\bigcup_{n=1}^{\infty} n\hat{A}_n$. Since $\hat{E}$ is Baire space there is a positive integer p such that p $\hat{A}_p$ has non-void interior in $\hat{E}$. Since A_p coindices with $\hat{A}_p \cap E$ it follows that A_p has non-void interior in E.

(5) *Let E be a metrizable barrelled space. Let* (A_n) *be an increasing sequence of closed absolutely convex subsets of E convering E. Then there is a positive integer p such that* A_p *has non-void interior.*

Proof. Since E is metrizable its completion $\hat{E}$ is a Fréchet space and therefore Baire. Apply (4) to reach the conclusion.

(6) *Let E be a Mackey space which is dual locally complete. Let* (A_n) *be an increasing sequence of complete convex subsets of E. If E coincides with* $\bigcup_{n=1}^{\infty} A_n$, *then E is complete.*

Proof. Let (u_n) be a sequence in $E'[\sigma(E', E)]$ which converges to the origin in the sense of Mackey. Let B be a closed bounded absolutely convex subset of $E'[\sigma(E', E)]$ such that (u_n) is a sequence in E'_B converging to the origin. Since E is dual locally complete. E'_B is a Banach space and, if we denote by M the closed absolutely convex hull of $\{u_1, u_2, \ldots, u_n, \ldots\}$ in E'_B, M is compact in $E'[\sigma(E', E)]$. Since E is a Mackey space, M is equicontinuous. We apply (3) to reach the conclusion.

(7) *Let E be a locally convex space such that every sequence in* $E'[\sigma(E', E)]$ *which converges to the origin in the sense of Mackey is equicontinuous. Then E is barrelled if* $\hat{E}$ *is barrelled.*

Proof. Let U be a barrel in E. The sequence (n U) is increasing and covers E. According to (3), $\hat{E}$ coincides with $\bigcup_{n=1}^{\infty} n\hat{U}$. Thus $\hat{U}$ is a barrel in $\hat{E}$ and therefore a neighbourhood of the origin. Since U coincides witn $\hat{U} \cap E$, U is a neighbourhood of the origin in E and the conclusion follows.

(8) *Let E be a dual locally complete Mackey space. Let* (E_n) *be an increasing sequence of subspaces of E covering E. Then E is the inductive limit of the sequence* (E_n).

Proof. Let T be the topology on E such that E $[T]$ is the inductive limit of the sequence (E_n). It is obvious that T is finer than the original topology of E. Let u be a continuous linear form on E $[T]$. We set u_n to de note the restriction of u on E_n. Applying Hahn-Banach's theorem we obtain an element v_n in E' which coincides with u_n in E_n. We set

$$w_1 = v_1, \; w_{n+1} = (n + 1)^2 (v_{n+1} - v_n), \; n = 1,2,\ldots$$

Given any point x of E we find a positive integer p such that x belongs to E_p. Then

$$<x, w_{n+1}> = (n + 1)^2 <x, v_{n+1}> - (n + 1)^2 <x, v_n> = 0$$

for n > p and therefore the sequence (w_n) is bounded in E'[σ (E' E)] . We find a subset B in E'[σ (E', E)] which is closed bounded and absolutely convex such that w_n is in B, n = 1,2,... Since E'_B is a Banach space the series $\Sigma \frac{1}{n^2} w_n$ converges to an element w in E'_B. Then we have

$$<z, w> = \Sigma \frac{1}{n^2} <z, w_n> = \lim <z, v_n> = \lim <z, u_m> = <z, u>$$

if z denotes any point of E and therefore w coincides with u. Consequently the topological dual of E [T] is E'. Since the topology of E is the Mackey topology, it follows that E [T] coincides with E. The proof is complete.

Next result is corollary of (8).

(9) *Let* E *be a barrelled space. Let* (E_n) *be an increasing sequence of subspaces of* E *covering* E. *Then* E *is the inductive limit of the sequence* (E_n).

(10) *Let* E *be a dual locally complete Mackey space. Let* E *be a closed subspace of* E *of countable codimension. If* G *is an algebraic complement in* E *of* F, *then* G *is a topological complement of* F.

Proof. If F is of finite codimension in E the result is obvious. Suppose that F is of infinite countable codimension. Let $\{x_1, x_2,\ldots, x_n,\ldots\}$ be an algebraic basis of G. We set F_n to denote the linear hull of $\{x_1, x_2,\ldots,x_n\}$. Let E_n be the space F + F_n, n = 1, 2,... If T : E ⟶ F is the projection onto F along G and if T_n is the restriction of T to E_n, then T_n is the projection of E_n along F_n and therefore T_n is continuous, n =1,2,... Since E is the inductive limit of the sequence (E_n), it follows that T is continuous and the conclusion follows.

The exposition given above follows VALDIVIA [1]. In VALDIVIA [2] result (4) can be found considereing A_n as an absolutely convex set, n = 1, 2,... Result (5) is due to AMEMIYA and KŌMURA [1]. Related results can be found in VALDIVIA [2], DE WILDE and HOUET [1], ROELCKE [1], RUESS [1], TSIRULNIKOV [1].

In JARCHOW [1], HORVÁTH [2] and PEREZ CARRERAS and BONET [1] results on increasing sequences of absolutely convex subset of locally convex spaces and barrelledness properties derived from them are given.

2. COUNTABLE CODIMENSIONAL SUBSPACES OF CERTAIN LOCALLY CONVEX SPACES. Let E be a locally convex space. Let $\mathcal{A}$ be the family of all closed bounded absolutely convex subsets of E. The family $\mathcal{A}$ is directed by inclusion, i.e., if A_1 and A_2 are in $\mathcal{A}$ there is an element A_3 in $\mathcal{A}$ containing A_1 and A_2. If $\mathcal{T}$ is the topology of E, $\mathcal{T}^x$ denotes the locally convex topology on E such that $E[\mathcal{T}^x]$ is the inductive limit of the family of normable spaces $\{E_A : A \in \mathcal{A}\}$. We write $\mathcal{T}_0$ to denote the topology on E' of the uniform convergence on every sequence of E which converges to the origin in the sense of Mackey. E is a bornological space if every absolutely convex subset of E which absorbs the bounded subsets of this space is a neighbourhood of the origin. Equivalently E is bornological if and only if it is a Mackey space and $E'[\mathcal{T}_0]$ is complete. The space $E[\mathcal{T}^x]$ is always bornological and we refer to it as the associated bornological space to $E[\mathcal{T}]$. The space $E[\mathcal{T}]$ is bornological if and only if coincides with $E[\mathcal{T}^x]$ (cf. KÖTHE [1], Chapter Six, §28).

A locally convex space E is quasibarrelled if every barrel in E which absorbs the bounded subsets of E is a neighbourhood of the origin. Obviously every bornological space is quasibarrelled.

(1) *Let* E *be a locally convex space. Let* F *be a subspace of* E *of codimension one. If* V *is a barrel in* F *which absorbs the bounded subsets of* F *there is a barrel* U *in* E *absorbing the bounded subsets of* E *suchthat* $U \cap F$ *coincides with* V.

Proof. Let z be a vector in E which is not in F. We set

$$B = \{\lambda z : \lambda \in K, |\lambda| \leq 1\}$$

G denotes the linear hull of z in E. Let W be the closure of V in E. We set $\mathcal{T}$ for the topology of E. $\mathcal{U}$ denotes the topology on F induced by $\mathcal{T}^x$.

First suppose F closed in $E[\mathcal{T}^x]$. Since every separated quotient of a bornological space is bornological and since $E[T^x]$ is the topological direct sum of $F[\mathcal{U}]$ and G, it follows that $F[\mathcal{U}]$ is bornological and therefore V is a neighbourhood of the origin in $F[\mathcal{U}]$. If W is distinct from V we set U = W. Then U is absorbing in E and therefore $U \cap G$ is a neighbourhood of the origin in G. Then $\frac{1}{2} U \cap G + \frac{1}{2} V$ is a neighbourhood of the origin in $E[T^x]$ and, since U contains $\frac{1}{2} U \cap G + \frac{1}{2} V$, it follows that U is a neighbourhood of the origin in $E[T^x]$. Consequently U absorbs the bounded sets of $E[T^x]$. Since $U \cap F = V$ and since every bounded subset of E is bounded in $E[T^x]$, the conclusion follows. If W coincides with V we set U=V+B. Then U is a neighbourhood of the origin in $E[T^x]$ and therefore absorbs the bounded subsets of E. U is closed in E and $U \cap F$ coincides with V.

Now suppose F dense in $E[T^x]$. From the definition of T^x it follows the existence of a bounded absolutely convex subset A of E so that $F \cap E_A \neq E_A$ is dense in E_A. We set U = W. Then $U \cap F = W$. Let M be a bounded subset of E. We set B to denote the closed absolutely convex hull of $A \cup M$ in E. Since $F \cap E_A$ is dense in E_A it follows that $F \cap E_B$ is dense in E_B and therefore B coincides with the closure of $B \cap F$ in E_B. We find $k > 0$ such that $k(F \cap B)$ is contained in V. Then kB is contained in U and therefore

$$k M \subset K B \subset U.$$

Thus U absorbs the bounded subsets of E. The proof is complete.

(2) *Let* E *be a quasibarrelled space. Let* F *be a finite codimensional subspace of* E. *Then* F *is quasibarrelled.*

Proof. It is enough to consider the case of F being an hyperplane of E. Let V be a barrel of F which absorbs the bounded subsets of F. We apply (1) to obtain a barred U in E which absorbs the bounded subsets of E and such that $U \cap F = V$. Since E is quasibarrelled, U is a neighbourhood of the origin in E and consequently V is a neighbourhood of the origin in F. The conclusion follows.

(3) *Let* E *be a bornological space. Let* F *be a finite codimensional subspace of* E. *Then* F *is bornological.*

Proof. It is enough to consider the case of F being an hyperplane of E. If F is closed in E then F has a topological complement of dimension one in E and therefore F is isomorphic to a separated quotient of E. Consequently F is bornological. Now suppose F dense in E. Let T be the topology of E and let $\mathcal{U}$ be the topology on F induced by $\mathcal{T}$. Since F is dense in E we identify E' with the topological dual of F. According to (2), F is quasibarrelled and therefore a Mackey space. Clearly $\mathcal{U}_o$ is coarser than $\mathcal{T}_o$. On the other hand, let (x_n) be a sequence in E which converges to the origin in the sence of Mackey. We can find in E a closed bounded absolutely convex subset A such that (x_n) is a sequence in E_A converging to the origin and such that $F \cap E_A$ is dense in E_A. We can find a sequence (y_n) in the subspace $F \cap E_A$ of E_A converging to the origin such that (x_n) is contained in the closed absolutely convex hull of $\{y_1, y_2, \ldots, y_n, \ldots\}$ in E_A (cf. A. P. ROBERTSON and W. ROBERTSON [1], Chapter VII, §2, p. 133). Therefore $\mathcal{U}_o$ coincides with $\mathcal{T}_o$ in E' and therefore $E'[\mathcal{U}_o]$ is complete. Consequently F is bornological.

(4) *Let* F *be a locally convex space. Let* F *be a subspace of* E *of codimension one. Let* V *be an absolutely convex subset of* F *which absorbs the bounded subsets of* F*. Then there exists an absolutely convex subset* U *of* E *which absorbs the bounded subsets of* E *such that* $U \cap F$ *coincides with* V.

Proof. Let $\mathcal{T}$ be the topology of E and let $\mathcal{U}$ be the topology on F induced by $\mathcal{T}^\times$. By (3) $F[\mathcal{U}]$ is bornological and thus V is a neighbourhood of the origin in $F[\mathcal{U}]$. Consequently the exists an absolutely convex neighbourhood of the origin W in $E[\mathcal{T}^\times]$ such that $W \cap F$ is contained in V. If U denotes the absolutely convex bull of $V \cup W$, then U absorbs the bounded subsets of E and $U \cap F$ coincides with V.

Given the locally convex space E, let $\mathcal{B}$ be the family of all bounded absolutely convex subsets of E such that B is an element of $\mathcal{B}$ if and only if E_B is a Banach space. If B_1 and B_2 belongs to $\mathcal{B}$, let T_1 and T_2 be the canonical injections from E_{B_1} and E_{B_2} into E respectively. Let T be the mapping from $E_{B_1} \times E_{B_2}$ into E such that

$$T(x, y) = x + y, \qquad x \in E_{B_1},\ y \in E_{B_2}.$$

Then $E_{B_1 + B_2}$ can be identified with the quotient $E_{B_1} \times E_{B_2}/T^{-1}(o)$ and therefore $E_{B_1 + B_2}$ is a Banach space. Consequently $\mathcal{B}$ is directed by inclusion. E is ultrabornological if and only if coincides with the inductive limit of the family of Banach spaces $\{E_B : B \in \mathcal{B}\}$. It is immediate that if E is ultrabornological, then E is barrelled and bornological.

(5) *Let E be an ultrabornological space. Let F be a countable codimensional subspace of E. Then F is bornological.*

Proof. Let (x_n) be a sequence of vectors of E such that E coincides with the linear hull of $F \cup \{x_1, x_2,\ldots,x_n,\ldots\}$. Let F_n be the subspace of E linear hull of $F \cup \{x_1, x_2,\ldots, x_n\}$, $n = 1,2,\ldots$ Let V be an absolutely convex subset of F absorbing the bounded subsets of F. According to (4), we can find an absolutely convex subset V_1 in F_1 absorbing the bounded subsets of F_1 such that $V_1 \cap F$ coincides with V. Proceeding by recurrence suppose that, for a positive integer n, we have found an absolutely convex subset V_n of F_n absorbing the bounded subsets of F_n. Then we find in F_{n+1} an absolutey convex subset V_{n+1} absorbing the bounded subsets of F_{n+1} such that $V_{n+1} \cap F_n$ coincides with V_n. We set $U = \bigcup_{n=1}^{\infty} V_n$. Then

$$U \cap F = V,\ U \cap F_n = V_n,\ n = 1,2,\ldots$$

Let B be a bounded absolutely convex subset of E such that E_B is a Banach space. We set E_n to denote the subspace $F_n \cap E_B$ of E_B. According to 1.(9), E_B is the inductive limit of the family of normable spaces $\{E_n: n = 1,2,\ldots$ Since E_n is bornological and $U \cap E_n$ absorbs the bounded subsets of E_n, then $U \cap E_n$ is a neighbourhood of the origin in E_n, $n = 1,2,\ldots$, and therefore $U \cap E_B$ is a neighbourhood of the origin in E_B. Consequently U is a neighbourhood of the origin in E and therefore V is a neighbourhood of the origin in F. The proof is complete.

(6) *Let E be a locally convex space and let (A_n) be an increasing sequence of closed convex subset of $E'[\sigma(E',E)]$ covering E'. Let G be a countable codimensional subspace of E'. If for every positive integer n, $G \cap A_n$ is $\sigma(E', E)$-closed and if T is the topology of E, then G is closed in $E'[T_o]$*

Proof. Let v be any element of E' which is not in G. Let (u_n) be a sequence in E' such that the linear hull of $G \cup \{u_1, u_2,\ldots, u_n,\ldots\}$ is an hyperplane H of E' not containing v. Let B_n be the absolutely convex hull of $\{u_1, u_2,\ldots, u_n\}$. Take an element x_n in E such that

$$\langle x_n, v\rangle = 1, |\langle x_n, t\rangle| < 1$$

for every t of $n^2(C_n \cap G + n B_n)$, C_n being the closed absolutely convex hull of $A_n \cap (-A_n) \cap i\ A_n \cap (-i A_n)$. Given any element of E' we set $u = hv + w$, $h \in K$, $w \in H$, and we choose a positive integer q with

$$w \in C_n \cap G + n B_n, \quad n > q,$$

from where it follows

$$|< x_n, n^2 w >| < 1 \quad \text{for every} \quad n > q.$$

Let z be the element of E'* which takes the value zero on H and $< z,v > = 1$. For $n > q$ we have that

$$|< n(x_n - z), u >| = |< n(x_n - z), hv + w >|$$

$$= |< n(x_n - z), hv> + <n(x_n - z), w >|$$

$$= |< n x_n, w >| < \frac{1}{n}$$

and therefore the sequence $(n(x_n - z))$ converges to the origin in $E'*[\sigma(E'*,E)]$; thus (x_n) converges to z in $E'*[\sigma(E'*,E')]$ in the sense of Mackey. We can find a closed bounded absolutely convex subset B of E such that (x_n) is a Cauchy sequence in E_B. Then there is a sequence (y_n) in E_B converging to the origin whose closed absolutely convex hull contains x_n, $n = 1,2,\ldots$, (cf. A.P. ROBERTSON and W. ROBERTSON [1], Chapter VII, § 3, p. 133). Therefore z is continuous in $E'[T_0]$ and thus H is closed there. Then G is closed in $E'[T_0]$.

(7) *Let* E *be a dual locally complete space. Let* F *be a closed subspace of* E. *Then* E/F *is dual locally complete.*

Proof. Let G be the subspace of $E'[\sigma(E',E)]$ orthogonal to F. Then G can be identified with $(E/F)'[\sigma((E/F)', E/F)]$ (cf. HORVÁTH [1], Chap. 3, § 13, p. 263). Let A be a closed bounded absolutely convex subset of G. Then A is a closed bounded absolutely convex subset of $E'[\sigma(E',E)]$ and therefore E'_A is a Banach space. Since E'_A coincides with G_A the conclusion follows.

(8) *Let* E *be a locally convex space. Let* F *be a countable codimensional*

subspace of E. *If* E *is dual locally complete, then* F *is dual locally complete.*

Proof. Let G be the closure of F in E. Let H be an algebraic complement of G in E. Suppose H endowed with the topology induced by $\mu(E, E')$. Let $\mathcal{U}$ be the topology induced by $\mu(E, E')$ on G. Since the dimension of H is countable we apply 1.(10) to obtain that $E[\mu(E, E')]$ is the topological direct sum of $G[\mathcal{U}]$ and H and therefore $G[\mathcal{U}]$ is isomorphic to a separated quotient of $E[\mu(E, E')]$. We apply (7) to obtain that $G[\mathcal{U}]$ is dual locally complete and therefore G is dual locally complete. Consequently it is enough to carry the proof for F being dense in E. We identify E' with the topological dual of F by restricting every element of E' to F. Let A be a closed bounded absolutely convex subset of $E'[\sigma(E', F)]$. Let V be the polar set of A in F. W denotes the closure of V in E. Let L be the linear hull of W. Suppose $L \neq E$. If L is of finite codimension in E and if $\{x_1, x_2, \ldots, x_p\}$ is a cobasis of L, we set B_p to denote the absolutely convex hull of $\{x_1, x_2, \ldots, x_p\}$ and we write W_n instead of $n(W + B_p)$, $n = 1,2,\ldots$ If L is of infinite codimension in E and if $\{x_1, x_2, \ldots, x_n, \ldots\}$ is a cobasis of L we set B_n to denote the absolutely convex hull of $\{x_1, x_2, \ldots, x_n\}$ and we write. W_n instead of $n(W + B_n)$, $n = 1,2,\ldots$ In any case (W_n) is an increasing sequence of closed convex subsets of E covering E such that.

$$W_n \cap L = n\,W, \quad n = 1,2,\ldots$$

from where it follows that L is closed, according to (6). This is a contradiction. Thus L coincides with E and therefore A is $\sigma(E', E)$-bounded. Clearly E'_A is a Banach space and the conclusion follows.

(9) *Let* E *be a barrelled space. If* F *is a countable codimensional subspace of* E, *then* F *is barrelled.*

Proof. Let H be the closure of F in E. Since E is a dual locally complete Mackey space we apply the first part of the proof (8) to obtain that H is isomorphic to a separated quotient of E. Therefore H is barrelled. Consequently it is enough to carry the proof for F being dense in E. Let A be a closed bounded absolutely convex subset of $E'[\sigma(E', F)]$. In the proof of (8) we saw taht A is $\sigma(E', E)$-bounded. Therefore A is equicontinuous in E and thus A is equicontinuous in F. Now the conclusion follows.

Results (1) and (2) are taken from VALDIVIA [3]. Result (3) is due to DIEUDONNÉ [1]. Theorem (5) is found in VALDIVIA [4] . (9), for finite codimensional subspaces, is due to DIEUDONNÉ [1]. (5) is due to VALDIVIA [2] and LEVIN and SAXON [1] independently. For more results concerning countable codimensional subspaces of certain locally convex spaces we refer to VALDIVIA [5].

There are bornological spaces with countable codimensional subspaces which are not even quasibarrelled (cf. VALDIVIA [6]). In VALDIVIA [7] the following result can be found: *Let* E *be an ultrabornological space which does not have the strongest locally convex topology. Then there is an hyperplane in* E *which is not ultrabornological.*

S. DIEROLF and P. LURJE [1] construct a barrelled bornological space with a non bornological subspace of countable codimension. In the same article a normed ultrabornological space which has an infinity of dense ultrabornological hyperplanes is constructed.

§ 4. CLOSED GRAPH THEOREM

1. MAPPING WITH CLOSED GRAPH BETWEEN TOPOLOGICAL SPACES. We remind the reader that, in the whole book, "locally convex space" stands for "Hausdorff locally convex topological linear space ".

In this section E and F are Hausdorff topological spaces and f is a mapping from E into F. The graph G(f) of f is the set $\{(x, f(x)) : x \in E\}$. If G(f) is a closed subset of the topological space E x F we say that f has closed graph. G(f) is sequentially closed if given any sequence (x_n) in G(f) converging to x in E x F, then x belongs to G(f).

(1) *If* f *is injective, onto and has closed graph, then* f^{-1} *has closed graph.*

Proof. Let h be the mapping from E x F onto F x E such that h(x, y) = (y, x), $x \in E$, $y \in F$. It is obvious that h is an homeomorphism. On the other hand, $h(G(f)) = \{(f(x), x) : x \in E\} = \{(y, f^{-1}(y)) : y \in F\} = G(f^{-1})$ from where the conclusion follows.

(2) *If* f *is injective, onto and has sequentially closed graph, then* f^{-1} *has sequentially closed graph.*

Proof : Proceed as in (1)

Let T and U be topologies on E and F respectively finer than the original ones. The two following results are immediate:

(3) *If* f *has closed graph, then* $f : E[T] \longrightarrow F[U]$ *has closed graph.*

(4) *If* f *has sequentially closed graph, then* $f : E[T] \longrightarrow F[U]$ *has sequentially closed graph.*

(5) *Let* $\{x_i : i \in I, \geq\}$ *be a net in* E *converging to a point* x. *If* f *has closed graph and if the net* $\{f(x_i) : i \in I, \geq\}$ *converges to* y *in* F, *then* $y = f(x)$.

Proof. The net $\{(x_i, f(x_i)) : i \in I \geq\}$ converges in $E \times F$ to (x, y). Since f has closed graph there is a point z in E such that $(z, f(z))$ coincides with (x,y). Then $z = x$ and $f(z) = y$.

(6) *If given any net* $\{x_i : i \in I, \geq\}$ *converging to* x *in* E *such that* $\{f(x_i) : i \in I, \geq\}$ *converges to* y *in* F *it follows that* $y = f(x)$, *then* f *has closed graph.*

Proof. Let (x, y) be a point of the closure of $G(f)$ in $E \times F$. Take a net $\{(x_i, f(x_i)) : i \in I, \geq\}$ in $G(f)$ converging to (x, y) in $E \times F$. Then $\{x_i : i \in I \geq\}$ converges to x in E and $\{f(x_i) : i \in I, \geq\}$ converges to y in F and therefore $y = f(x)$. Thus (x, y) belongs to $G(f)$.

(7) *Let* (x_n) *be a sequence in* E *converging to a point* x. *If* $G(f)$ *is sequentially closed and if* $(f(x_n))$ *converges to* y *in* F, *then* $y = f(x)$.

Proof. Proceed as in (5).

(8) *If given any sequence* (x_n) *converging to* x *in* E *such that* $(f(x_n))$ *converges to* y *in* F *it follows that* $y= f(x)$, *then* f *has sequentially closed graph.*

Proof. Proceed as in (6), taking (x,y) such that there is a sequence $((x_n, f(x_n))$ in $G(f)$ converging to (x,y) in $E \times F$.

(9) *Let* H *be a Hausdorff topological space and let* h *be a continuos mapping from* H *into* E. *If* f *has closed graph, then* $f \circ h$ *has closed graph.*

Proof. Let $\{x_i : i \in I, \geq\}$ a net in H converging to x such that

(10) $$\{f \circ h(x_i) : i \in I, \geq\}$$

converges to z in F. Since h is continuous the net $\{h(x_i) : i \in I, \geq\}$ converges to h(x) in E. According to (5) the net (10) converges to f o h(x) in F. Apply (6) to reach the conclusion.

The mapping f is said to be sequentially continuous if given any sequence (x_n) in E converging to x, then the sequence $(f(x_n))$ converges to f (x) in F.

(11) *Let* H *be a Hausdorff topological space and let* h *be a sequentially continuous mapping from* H *into* E. *If* f *has sequentially closed graph, then* f o h *has sequentially closed graph.*

Proof. Proceed as in (9) using sequences instead of nets and apply (7) and (8) instead (5) and (6) respectively.

(12) *If* f *is continuous, then* f *has closed graph.*

Proof. See (6).

(13) *If* f *is sequentially continuous, then* f *has sequentially closed graph.*

Proof. See (8).

The mapping $f : E \longrightarrow F$ is open if every open set of E is mapped in an open set of the topological subspace f(E) of F. If E and F are locally convex spaces and if f is linear continuous and open, f is said to be an homomorphism from E into F. Moreover if f is injective f is an isomorphism from E into F.

(14) *Let* H *be a Hausdorff topological space and let* h *be an open continuous mapping from* H *onto* E. *If* f o h *has closed graph, then* f *has closed graph.*

Proof. Let

(15) $\{x_i : i \in I, \geq\}$ be a net in E converging to x such that

(16) $\{f(x_i) : i \in I, \geq\}$

converges to y in F. Take a point z in H with h(z) = x. Let $\{U_j : j \in J\}$ be a fundamental system of neighbourhoods of z in H with $H \in \{U_j : j \in J\}$. Since h is continuous and open $\{h(U_j)\ j \in J\}$ is a fundamental system of neighbourhoods of x in E. If i_1 and i_2 belong to I and if j_1 and j_2 are in J we set

$$(i_j, j_i) \geq (i_2, j_2) \quad \text{if} \quad i_1 \geq i_2 \quad \text{and} \quad U_{j1} \subset U_{j2}$$

If for indices $i \in I$ and $j \in J$, x_i belongs to $h(U_j)$ we write $x_{ij} = x_i$. Let $J_i = \{j \in J : x_i \in h(U_j)\}$. Then

(17) $\{x_{ij} : (i,j) \in I \times J_i,\ i \in I, \geq\}$

is a subnet of (15) *and therefore*

(18) $\{f(x_{ij}) : (i,j) \in I \times J_i,\ i \in I \geq\}$

is a subnet of (16). *Consequently* (18) *converges to* y *in* F.

For every element x_{ij} of the net (17) take a point z_{ij} in U_j such $h(z_{ij}) = x_{ij}$. Then

$$\{z_{ij} : (i,j) \in I \times J_i,\ i \in I, \geq\}$$

is a net in H converging to z. Since f o h has closed graph and

$$f \circ h(z_{ij}) = f(h(z_{ij})) = f(x_{ij}),\ (i,j) \in I \times J_i,\ i \in I$$

we apply (5) to obtain

$$y = f \circ h(z) = f(h(z)) = f(x)$$

Now we apply (6) to reach the conclusion.

(19) *If* E *and* F *are locally convex spaces and* f *is linear,* f *has closed graph if and only if given any net* $\{x_i : i \in I, \geq\}$ *in* E *converging to the origin such that* $\{f(x_i) : i \in I, \geq\}$ *converges to a point* y *in* F, *then* y=0.

Proof. If f has closed graph and if the net $\{x_i : i \in I, \geq\}$ in E converges to zero such that $\{f(x_i) : i \in I, \geq\}$ converges to a point y of F we apply (5) to obtain $y = f(0) = 0$.

Reciprocally, suppose that $\{z_i : i \in I, \geq\}$ is a net in E converging to a point z such that $\{f(z_i) : i \in I, \geq\}$ converges to a point u in F. Since the net $\{z_i - z : i \in I, \geq\}$ converges to the origin in E and $\{f(z_i - z) : i \in I, \geq\}$ converges to $u - f(z)$ in F it follows that $u = f(z)$. We apply (6) to obtain the conclusion.

(20) *If* E *and* F *are locally convex spaces and if* f *is linear,* f *has sequentially closed graph if and only if given any sequence* (x_n) *in* E *conver-*

ging to the origin such that $(f(x_n))$ *converges to a point* y *in* F, *then* $y = 0$.

Proof. Proceed as in (19) using (7) and (8) instead of (5) and (6) respectively.

(21) *If* E *and* F *are locally convex spaces and if* F *is linear and has closed graph there exists a Hausdorff topology* T *on* F, *coarser than the original one, such that* $F[T]$ *is a locally convex space and* $f : E \longrightarrow F[T]$ *is continuous.*

Proof. Let $\mathcal{U}$ and $\mathcal{V}$ be fundamental systems of neighbourhoods of the origin in E and F respectively which we suppose absolutely convex. Take the family of absolutely convex subsets of F

$$\{f(U) + V : U \in \mathcal{U}, V \in \mathcal{V}\}$$

which is a fundamental system of neighbourhoods of the origin for a linear topology T on F. Suppose T not Hausdorff. Then there is a point $y \neq 0$ such that

$$y \in \bigcap \{f(U) + V : U \in \mathcal{U}, V \in \mathcal{V}\}. \tag{22}$$

Since (o, y) does not belongs to $G(f)$ there are W and Z in $\mathcal{U}$ and $\mathcal{V}$ respectively such that

$$((0, y) + W \times Z) \cap G(f) = \emptyset \tag{23}$$

From (22) we deduce the existence of u in W and v in Z such that $y = f(u)+v$. Then $(u, y-v)$ belongs to $G(f)$. Since $-v$ belongs to Z we have that $(0, y) + (u,-v) = (u, y - v)$ belongs to $(0, y) + W \times Z$. Consequently

$$(u, y-v) \in ((0, y) + W \times Z) \cap G(f)$$

which is in contradiction with 23. The proof is complete.

2. QUASI-SUSLIN SPACES. Let X be a topological space. X is said to be a Polish space if it is separable and there is a metric d on X compatible with its topology such that the metric space (X, d) is complete.

Let E be a topological space. We denote by $\mathcal{P}(E)$ the family of all the parts of E. E is a quasi-Suslin space if it is Hausdorff and there exists a mapping T from X into $\mathcal{P}(E)$ satisfying

a) $\bigcup \{Tx : x \in X\} = E$;

b) If (x_n) is a sequence in X converging to x and if z_n belongs to

Tx_n for every positive integer n, then the sequence (z_n) has an adherent point in E belonging to Tx.

(1) *Let* E *be a quasi-Suslin space. If* F *is a closed subspace of* E, *then* F *is quasi-Suslin.*

Proof. Let X be a Polish space and T a mapping from X into $P(E)$ satisfying conditions a) and b). If F is the void set the conclusion is obvious. If F is distinct from the void set we write

$$Y = \{x \in X : F \cap Tx \neq \emptyset\}$$

where Y is endowed with the topology induced by X. For every x of Y we set $Sx = F \cap Tx$. Then S is a mapping from Y into $P(F)$ such that

$$\cup \{Sx : x \in Y\} = F.$$

If (y_m) is any sequence in Y converging to y in X we have that $S_{y_m} \neq \emptyset$, $m = 1,2,\ldots$, and therefore there is a point z_n in S_{y_m}. The sequence (z_m) has an adherent point z in E belonging to Ty. Since F is closed it follows that z belongs to $F \cap Ty = Sy$. Consequently Y is a closed subspace of X and therefore a Polish space. Now the conclusion follows.

(2) *Let* E *and* F *be Hausdorff topological spaces such that a continuous onto mapping* $f : E \longrightarrow F$ *exists. If* E *is a quasi-Suslin space the same is true for* F.

Proof. Let X be a Polish space and let T be a mapping from X into $P(E)$ verifying conditions a) and b). We set $S = f \circ T$. Then

$$\cup \{Sx : x \in X\} = \cup \{f(Tx) : x \in X\} = f(E) = F$$

If (x_n) is a sequence in X converging to x and z_n belongs to Sx_n for every positive integer n there is an element y_n in Tx_n such that $f(y_n) = z_n$. Since E is a quasi-Suslin space there is an adherent point y of (y_m) in E belonging to Tx. Consequently f(y) is and adherent point to (z_n) in F belonging to $f(Tx) = Sx$. The proof is complete.

(3) *Let* E *be a Hausdorff topological space. Let* (E_n) *be a sequence of subspaces of* E *covering* E. *If for every positive integer* n, E_n *is a quasi-Suslin space, then* E *is a quasi-Suslin space.*

Proof. For every positive integer n let X_n be a Polish space and T_n a mapping from X_n into $P(E_n)$ satisfying conditions a) and b) taking x_n, E_n,

T_n instead of x, E, T respectively. Let d_n be a metric on X_n compatible with its topological such that (X_n, d_n) is a complete metric space. If x and y are in X_n we set $\delta_n(x,y) = \inf(1, d_n(x,y))$.

It is immediate that δ_n is a metric on X_n compatible with its topology such that (X_n, δ_n) is a complete metric space. We set

$$Y_n = \{(x, n) : x \in X_n\}.$$

The sets Y_n, n = 1, 2,..., are pairwise disjoint. We set Y to denote the union of all of them. Let d be the mapping from Y x Y in the set of real numbers such that if (x, m) are in Y, then

$$d((x, m), (y, n)) = 1 \text{ if } m \neq n,$$

$$d((x, m), (y, n)) = \delta_n(x, y) \text{ if } m = n$$

It is not difficult to check that d is a metric Y. Suppose Y endowed with the topology derived from the metric d. Clearly Y_n, considered as subspace of (Y, d) is isometric to (X_n, δ_n) and therefore complete. Since X_n is separable it follows that Y is separable. On the other hand, if $((x_m, n(m)))$ is a Cauchy sequence in (Y, d) there are positive integers p and q with

$$n(m) = q \text{ for } m \geqslant p.$$

Consequently the sequence

$$(x_p, q), (x_{p+1}, q), \ldots, (x_{p+n}, q), \ldots$$

converges to a point x in Y_q. Then $((x_m, n(m)))$ converges to x in Y and therefore Y is a Polish space.

Let S be the mapping from Y into $\mathcal{P}(E)$ defined by

$$S(x, m) = T_m x, \quad (x, m) \in Y$$

If z is any point of E we find a positive integer p such that z belongs to E_p. Let x be a point of X_p such that z belongs to $T_p x$. Then

$$z \in T_p x = S(x, p)$$

and therefore

$$\cup \{S(x, m) : (x, m) \in Y\} = E.$$

If $((x_m, n(m)))$ is a sequence of Y converging to (x, p) there is a positive integer q such that $n(m) = p$ for m larger than q. Consequently if we set $y_r = y_{q+r}$, $r = 1,2,\ldots,$ the sequence (y_r) converges to x in X_p. Therefore, if z_m belongs to $S(x_m, n(m))$ we have that z_{q+r} belongs to $T_p y_r$, $r = 1,2,\ldots,$ and thus there is a point z in E_p adherent to (z_{q+r}) and belonging to $T_p x = S(x, p)$. The proof is complete.

(4) *Let* (E_n) *be a sequence of quasi-Suslin spaces. Then* $\prod_{n=1}^{\infty} E_n$ *is a quasi-Suslin space.*

Proof. For every positive integer n, X_n, T_n, d_n and δ_n have the same meaning as in the proof of (3). We set

$$E = \prod_{n=1}^{\infty} E_n, \quad X = \prod_{n=1}^{\infty} X_n,$$

$$d((x_1, x_2,\ldots,x_n,\ldots), (y_1, y_2,\ldots,y_n,\ldots)) = \Sigma \frac{1}{2^n} \delta_n (x_n, y_n),$$

$$x_n, y_n \in X_n, \quad n = 1,2,\ldots$$

It is immediate that d is a metric in the topological product X which is compatible with its topology and such that (X, d) is complete. Since X_n is separable, $n = 1,2,\ldots,$ it follows that X is separable. Thus X is a Polish space.

Let $S : X \longrightarrow \mathcal{P}(E)$ be the mapping defined by

$$Sx = \prod_{n=1}^{\infty} T_n x_n, \quad x = (x_1, x_2,\ldots,x_n,\ldots) \in X.$$

If $z = (z_1, z_2,\ldots,z_n,\ldots)$ is any point of E we find a point y_n in X_n such that z_n belongs to $T_n y_n$ for every positive integer n. Then

$$z \in S(y_1, y_2,\ldots, y_n,\ldots)$$

and consequently

$$\cup \{S_x : x \in X\} = E$$

Let $x^r = (x_1^r, x_2^r,\ldots, x_n^r,\ldots)$, be a sequence in X converging to $x = (x_1, x_2, \ldots, x_n,\ldots)$. For every positive integer r let $z^r = (z_1^r, z_2^r,\ldots,z_n^r,\ldots)$

be an element of S_x^r. Then

$$z_n^r \in T_n x_n^r, \quad r = 1,2,\ldots$$

and therefore there is an element z_n in E_n adherent to the sequence z_n^1, $z_n^2, \ldots, z_n^r, \ldots$ such that z_n^r belongs to $T_n x_n$. Consequently $z = (z_n)$ is a point of E adherent to (z^r) and belonging to Sx. Now the proof is complete.

(5) *Let* E *be a topological space. Let* (E_n) *be a sequence ofsubspace of* E. *If for every positive integer* n, E_n *is a quasi-Suslin space, then the subspace* $F = \bigcap\{E_n : n = 1,2,\ldots\}$ *of is quasi-Suslin.*

Proof. If F is the void set the result is obvious. If $F \neq \emptyset$ let g be the mapping from F into $\prod_{n=1}^{\infty} E_n$ such that

$$g(x) = (x, x, \ldots, x, \ldots), \; x \in F$$

It is immediate that the subspace g(F) of $\prod_{n=1}^{\infty} E_n$ is closed and that g is an homeomorphism from F onto g(F). It is enough to apply (1), (2) and (4) to reach the conclusion.

(6) *If* $E_1, E_2, \ldots, E_r$ *are quasi-Suslin spaces, then* $\prod_{n=1}^{r} E_n$ *is a quasi-Suslin space.*

Proof. Let E be a quasi-Suslin space, $E \neq \emptyset$. We set $E_n = E$, $n = r+1$, $r+2, \ldots$ Then $\prod_{n=1}^{r} E_n$ is homeomorphic to a closed subspace of $\prod_{n=1}^{\infty} E_n$ and therefore it is a quasi-Suslin space.

(7) *Let* E *be a quasi-Suslin locally convex space. If* F *is a closed subspace of* E, *then* E/F *is a quasi-Suslin space.*

Proof. It is obvious according to the continuity of the canonical mapping $f : E \longrightarrow E/F$ and (2).

(8) *Let* E *be a locally convex space. Let* (E_n) *be a sequence of subspaces of* E *covering* E. *For every positive integer* n *let* T_n *be a locally convex topology on* E_n *finer than the original one such that* $E_n[T_n]$ *is a quasi-Suslin space. If* E *is the locally convex hull of*

$$\{E_n[T_n] : n = 1,2,\ldots\}$$

then E *is a quasi-Suslin space.*

Proof. For every positive integer n the canonical injection from $E_n[T_n]$ in E_n is continuous. Apply (2) to obtain that E_n is a quasi-Suslin space. According to (3) the conclusion follows.

(9) *Let* G *be a metrizable topological space. Let* F *be a quasi-Suslin space. If* $f : G \longrightarrow F$ *is a mapping with closed graph and if* E *is a closed subset of* F, *then the subset* $f^{-1}(E)$ *of* G *has the Baire property.*

Proof. If we suppose that E is endowed with the topology induced by F, then E is a quasi-Suslin space. Let X be a Polish space and let $T : X \longrightarrow \mathcal{P}(E)$ be a mapping satisfying a) and b). Let d be a metric on X compatible with its topology and such that (X, d) is complete. Since X is separable there is a sequence (B_m) of closed balls in (X, d) of radii less than one whose union coincides with X. Suppose we have constructed the closed set $B_{m_1, m_2, \ldots, m_p}$ in X for the positive integers $m_1, m_2, \ldots, m_p$. We suppose this set endowed with the metric induced by d. We take in $B_{m_1, m_2, \ldots, m_p}$ a sequence of closed balls $(B_{m_1, m_2, \ldots m_p, m})$ of radii less than $\frac{1}{2^p}$ such that

$$B_{m_1, m_2, \ldots m_p} = \cup \{B_{m_1, m_2, \ldots m_p, m} : m = 1,2,\ldots.\}.$$

We set

$$A_{m_1, m_2, \ldots, m_p} = f^{-1}(T(B_{m_1, m_2, \ldots, m_p}))$$

Since

$$f^{-1}(E) = \cup \{A_m : m = 1,2,\ldots\}$$

We apply 1, §1.(8) to obtain that

$$O(f^{-1}(E)) \sim \cup \{O(A_m) : m = 1,2,\ldots\} = D$$

is a rare subset of G. Analogously

$$O(A_{m_1, m_2, \ldots, m_p}) \sim \cup \{O(A_{m_1, m_2, \ldots, m_p, m}) : m = 1,2,\ldots\}$$

$$= D_{m_1 m_2, \ldots m_p}$$

is a rare subset of G. We set

$$D(p) = \cup \{D_{m_1, m_2, \ldots, m_p} : m_1, m_2, \ldots, m_p = 1, 2, \ldots.\}$$

and

$$B = D \cup (\cup \{D(p) : p = 1,2,\ldots\})$$

for every positive integer p. It is obvious that is a subset of G of first category.

If z is a point of $O(f^{-1}(E)) \sim B$ there is a sequence (m_p) of positive integers such that

$$z \in O(A_{m_1, m_2, \ldots, m_p}), \; p = 1,2,\ldots$$

Let $\{U_p : p = 1,2,\ldots\}$ be a fundamental system of neighbourhoods of the point z in G. Then

$$U_p \cap A_{m_1, m_2, \ldots, m_p}$$

is a set of second category in G and therefore

$$U_p \cap A_{m_1, m_2, \ldots, m_p} \sim B \neq \emptyset.$$

For every positive integer p take

$$z_p \in U_p \cap A_{m_1, m_2, \ldots, m_p} \sim B,$$

$$x_p \in B_{m_1, m_2, \ldots, m_p}, \text{ with } f(z_p) \in Tx_p.$$

It is obvious that (x_p) is a Cauchy sequence in (X, d) and therefore converges to a point x in X. Consequently $(f(z_p))$ has an adherent point y in E. Take a subnet

(10) $\{z_j : j \in J, \geqslant\}$

of the sequence (z_p) such that $\{f(z_j) : j \in J, \geqslant\}$ converges to y. The net (10) converges to z in G and, since f has closed graph, it follows that $y = f(z)$. Therefore z belongs to $f^{-1}(E)$ from where it follows

$$O(f^{-1}(E)) \sim f^{-1}(E) \subset O(f^{-1}(E)) \sim (O(f^{-1}(E)) \sim B) \subset B$$

and therefore $O(f^{-1}(E)) \sim f^{-1}(E)$ is a subset if G of first category. Now apply §1, 1.(6) to get the conclusion.

Now we arrive to the following closed graph theorem:

(11) *Let* E *be a locally convex space which is the locally convex hull of Baire metrizable locally convex spaces. Let* F *be a quasi-Suslin locally convex spaces. If* $f : E \longrightarrow F$ *is a linear mapping with closed graph, then* f *is continuous.*

Proof. Let $\{E_i : i \in I\}$ a family of Baire metrizable locally convex spaces. For every i of I let A_i be a linear mapping from E_i into E such that the topology of E is the finest locally convex topology for which the mapping A_i, $i \in I$, are continuous.

Let U be an absolutely convex closed neighbourhood of the origin in E. For every i in I we apply 1.(9) to obtain that f o A_i has closed graph. According to (9), $(f \circ A_i)^{-1}(U)$ is a subset of E_i with the Baire property. We apply §1, 3.(5) to obtain that $A_i^{-1}(f^{-1}(U))$ is a neighbourhood of the origin in E_i. Consequently $f^{-1}(U)$ is a neighbourhood of the origin in E. The proof is complete.

Now we arrive to the following open-mapping theorem:

(12) *Let* E *be a quasi-Suslin locally convex space. Let* F *be a locally convex space which is the locally convex hull of metrizable Baire locally convex spaces. If* $f : E \longrightarrow F$ *is a linear onto mapping with closed graph, then* f *is open.*

Proof. According to 1.(21), $f^{-1}(0)$ is a closed subspace of E. Let h be the canonical mapping from E onto $E/f^{-1}(0)$. Let g be the linear injective mapping from $E/f^{-1}(0)$ onto F such that $f = g \circ h$. Since h is an homomorphism we apply 1.(14) to obtain that g has closed graph and thus g^{-1} has closed graph. The space $E/f^{-1}(0)$ is quasi-Suslin and, according to (11), g^{-1} is continuous. Consequently f is open.

The results included in this section can be found in VALDIVIA [8] and [9].

3. K-SUSLIN SPACES. A topological space is a K-Suslin space if it is Hausdorff and there exists a Polish space X and a mapping T from X into $K(E)$, $K(E)$ being the compact subsets of E, satisfying the following conditions.

a) $\cup \{Tx : x \in X\} = E$;

b) if x is any point of X and V is a neighbourhood of the set Tx in E there is a neighbourhood U of x in X such that T(U) is contained in V.

(1) *Let* E *be a Hausdorff topological space. Let* X *be a Polish space and*

T a mapping from X into K(E) satisfying the conditions

1) $\cup\{Tx : x \in X\} = E$;

2) *If* (x_n) *is a sequence in X converging to x and if* z_n *belongs to* Tx_n *for every positive integer n, then the sequence* (z_n) *has an adherent point in E belonging to Tx.*

Then E is a K-Suslin space.

Proof. Suppose that E is not a K-Suslin space. Then there is a point x in X and an open neighbourhood V of Tx in E such that T(U) is not contained in V for every neighbourhood U of x. Let $\{U_n : n = 1,2,\ldots\}$ be a fundamental system of neighbourhoods of x in X. For every positive integer n take a point x_n in U_n such that Tx_n is not contained in V. We select z_n in $Tx_n \sim V$. According to condition 2) the sequence (z_n) has an adherent point z in E belonging to Tx. Since V is open and z_n does not belongs to V, n= 1,2,...,it follows that z does not belong to Tx and that is a contradiction.

(2) *Let E be a K-Suslin space. Let X be a Polish space and let T be a mapping from X into K(E) verifying conditions a) and b). If* $\{x_i : i \in I, \geq\}$ *is a net in X converging to x and if* z_i *belongs to* Tx_i *for every i in I, then the net* $\{z_i : i \in I, \geq\}$ *has an adherent point in E belonging to Tx.*

Proof. Suppose the property not true. We find a net $\{x_i : i \in I, \geq\}$ in X converging to x and a point z_i in Tx_i, $i \in I$, such that the net $\{z_i : i \in I, \geq\}$ does have not any adherent point in E which belongs to Tx. If y is a point of Tx there is i(y) in I and an open neighbourhood V(y) of y such that

$$z_i \notin V(y),\ i \in I,\ i \geq i(y).$$

Since Tx is compact there is a finite subset $y_1, y_2, \ldots, y_q$ in Tx such that the open set

$$V = \cup\ \{V(y_p) : p = 1,2,\ldots,q\}$$

contains Tx. Since is is a K-Suslin space we find a neighbourhood U of x in X such that T(U) is contained in V. Now we find an element j in I, $j \geq i(y_p)$, $p = 1, 2, \ldots, q$, such that x_j belongs to U. Then z_j is in V and that is a contradiction.

(3) *If E is a K-Suslin space, then E is a quasi-Suslin space*

Proof. Let X be a Polish space and let T be a mapping from X into K(E) satisfying conditions a) and b). Let (x_n) be a sequence in X converging to x. If z_n belongs to Tx_n for every positive integer n we apply (2) to obtain that (z_n) has an adherent point in E belonging to Tx. Now the conclusion is immediate.

According to (1) and (2) the proof of (4), (5), (6), (7), (8), (9), (10) and (11) run analogously to 2.(1), 2.(2), 2.(3), 2.(4), 2.(5), 2.(6), 2.(7) and 2.(8), respectively.

(4) *Let* E *be a* K-*Suslin space. If* F *is a closed subspace of* E, *then* F *is a* K-*Suslin space.*

(5) *Let* E *and* F *be Hausdorff topological spaces. If* E *is a* K-*Suslin space and if there is a continuous mapping from* E *onto* F, *then* F *is a* K-*Suslin space.*

(6) *Let* E *be a Hausdorff topological space. Let* (E_n) *be a sequence of subspaces of* E *covering* E. *If for every positive integer* n, E_n *is a* K-*Suslin space, then* E *is a* K-*Suslin space.*

(7) *If* (E_n) *is a sequence of* K-*Suslin spaces, then* $\prod_{n=1}^{\infty} E_n$ *is a* K-*Suslin space.*

(8) *Let* E *be a topological space. Let* (E_n) *be a sequence of subspaces of* E. *If for every positive integer* n E_n *is a* K-*Suslin space, then the subspace* $\bigcap\{E_n : n = 1,2,\ldots\}$ *of* E *is a* K-*Suslin space.*

(9) *If* $E_1, E_2, \ldots, E_r$ *are* K-*Suslin spaces, then* $\prod_{n=1}^{r} E_n$ *is a* K-*Suslin space.*

(10) *Let* E *be a* K-*Suslin locally convex space. If* F *is a closed subspace of* E, *then* E/F *is a* K-*Suslin space.*

(11) *Let* E *be a locally convex space. Let* (E_n) *be a sequence of subspace of* E *covering* E. *For every positive integer* n *let* T_n *be a topology on* E_n *finer than original one such that* $E_n[T_n]$ *is the* K-*Suslin locally convex space. If* E *is the locally convex hull of*

$$\{E_n [T_n] : n = 1,2,\ldots\},$$

then E *is a* K-*Suslin space.*

A topological space E is a Lindelöf space if and only if every open cover of E has a countable subcover.

(12) *If* E *is a* K- *Suslin space, then* E *is a Lindelöf space.*

Proof. Let X be a Polish space and let T be a mapping from X into K (E) satisfying conditions a) and b). Let

(13) $\{ O_i : i \in I \}$

be an open cover of E. If x is any point of X there is a finite subfamily A(x) of (13) covering the compact subset Tx of E. We set A(x) to denote the union of the open sets of the family (13) which belongs to A(x). Since A(x) is a neighbourhood of Tx there is an open neighbourhood U(x) of x in X such that T(U(x)) is contained in A(x). Since X is a metrizable separable topological space its topology has a countable basis and therefore there exists a sequence (x_n) in X such that

$$\cup\{U(x_n) : n = 1,2, \dots \}$$

coincides with X. We have that

$$\cup\{A(x_n) : n = 1,2, \dots \}$$

is a countable subfalimy of (13). If z is any point of E there is a point x in X such that z belongs to Tx. We find a positive integer p such that x belongs to $U(x_p)$. Consequently z belongs to $A(x_p)$ and therefore $\cup\{A(x_n) : : n = 1,2, \dots\}$ covers E. The proof is complete.

(14) *Let* G *be a Hausdorff topological space. Let* F *be a* K-*Suslin space. If* f *is a mapping with closed graph from* G *into* F *and if* E *is a closed subset of* F, *then the subset* $f^{-1}(E)$ *of* G *has the Baire property.*

Proof. We proceeded has ve did in 2 (9) to obtain a Polish space X and a mapping T from X into K(E) satisfying conditions a) and b). Let d be a metric on X compatible with is topology and such that (X, d) is complete. Using the same notations as in the proof of 2. (9) let z be a point of $O(f^{-1}(E)) \sim B$. Then there is a sequence $m_1, m_2, \dots, m_p, \dots$ of positive integers such that

$$z \in O(A_{m_1, m_2, \dots, m_p}).$$

Let $\{ U_i : i \in I \}$ be a fundamental system of neighbourhoods of the point z in G. If i and j are in I and if p and q are positive integers we set $(i,p) \geq (j, q)$ when $i \geq j$ and $p \geq q$. Since

$$U_i \cap A_{m_1, m_2, \ldots, m_p}$$

is of second category in G we have that

$$U_i \cap A_{m_1, m_2, \ldots, m_p} \sim B \neq \emptyset$$

and therefore there exists a point z_{ip} in the former set. Let x_{ip} a point of $B_{m_1, m_2, \ldots, m_p}$ with $f(z_{ip}) \in Tx_{ip}$. It is obvious that the net

$$\{x_{ip} :(i, p) \in I \times N, \geq\}$$

converges to a point x of X. We apply (2) to obtain that the net

(15) $\{f(z_{ip}): (i, p) \in I \times N, \geq\}$

has an adherent point y in E. Take a subnet

(16) $\{z_j : j \in J, \geq\}$

from the net $\{z_{in} : (i, p) \in I \times N, \geq\}$ such that $\{f(x_j) : j \in J, \geq\}$ converges to y. The net (16) converges in G to z and since f has closed graph it results that y coincides with f(z). Therefore z belongs to $f^{-1}(E)$. To conclude the proof we proceed as in 2.(9).

Result (17) can be proved analogously to 2.(11).

(17) *Let* E *be a locally convex space which is the locally convex hull of locally convex Baire spaces. Let* F *be a* K*-Suslin locally convex space. If* f *is a linear mapping with closed graph from* E *into* F*, then* f *is continuous.*

(18) *Let* E *be a* K*-Suslin locally convex space. Let* F *be a locally convex space which is the locally convex hull of locally convex Baire spaces. If* f *is a linear mapping with closed graph from* E *onto* F*, then* f *is open.*

Proof. See 2.(12).

(19) *Let* G *and* F *be Hausdorff topological spaces. Let* E *be a subpace of* F *which is a* K*-Suslin space. If* f *is a continuous mapping from* G *into* F*, then the subset* $f^{-1}(E)$ *of* G *has the Baire property.*

Proof. See the proof of (14) where the property of "E being a closed subspace of F" is used only to conclude that is a K-Suslin space.

(20) *Let* F *be a Hausdorff topological space. If* E *is a subspace of* F *which is a* K-*Suslin space, the subset* E *of* F *has the Baire property.*

Proof. It is enough to consider G = F and f as the identity mapping from F into itself in (19).

(21) *Let* E *be a Baire locally convex space. If* E *is a* K-*Suslin space, then* E *is a separable Fréchet space.*

Proof. Let X be a Polish space and let T be a mapping from X into K(E) satisfying conditions a) and b). Let

(22) $\{A_m : m = 1,2,\ldots\}$

be a basis for the topology of X. We set

$$M_m = (O(T(A_m)) \sim T(A_m)) \cup (T(A_m) \sim O(T(A_m))),\ m = 1,2,\ldots$$

$$M = \cup \{M_m : m = 1,2,\ldots\}$$

For every positive integer m, A_m is an open subspace of X and therefore a Polish space (cf. BOURBAKI [2], Chap. 9, §6, Prop. 2) from where it follows easily that $T(A_m)$ is a K-Suslin space as subspace of E. We apply (20) to obtain that the subset $T(A_m)$ of E has the Baire property. Then M_m is a subset of E of first category and therefore M is a subset of E of first category. Since E is a Baire space $E \sim M$ is non-void and therefore a point x in $E \sim M$ can be extracted. Let y be a point of X such that x belongs to Ty. We select from (22) all the elements which contain {y} and we form a sequence (B_m). Since M_m is contained in M we have that

$$O(T(B_m)) \cap (E \sim M)$$

$$= ((T(B_m) \sim O(T(B_m))) \cup O(T(B_m))) \cap (E \sim M)$$

$$= T(B_m) \cup (O(T(B_m) \sim T(B_m))) \cap (E \sim M)$$

$$= T(B_m) \cap (E \sim M)$$

and therefore $O(T(B_m))$ is a neighbourhood of x.

For every positive integer p we find a neighbourhood ot the origin V in E such that

$$V_p \subset \frac{1}{p} \cap \{O(T(B_m)) - x : m = 1,2,\dots p\}.$$

Now take a balanced and closed neighbourhood of the origin W in E. Since Ty is a compact subset of E we find a positive integer q with $Ty - x \subset q\,\mathring{W}$. Since E is a K-Suslin space and $x + q\,\mathring{W}$ is a neighbourhood of Ty, there is a positive integer r such that $T(B_r)$ is contained in $x + q\,\mathring{W}$. Consequently $O(T(B_r))$ is contained in $x + q\,W$. Then

$$V_{q+r} \subset \frac{1}{q+r} \cap \{O(T(B_m)) - x : m = 1,2,\dots,q + r\}$$

$$\subset \frac{1}{q+r} (O(T(B_r)) - x) \subset \frac{1}{q+r} q\,W \subset W.$$

Therefore $\{V_m : m = 1,2,\dots\}$ is a fundamental system of neighbourhoods of the origin in E and thus E is metrizable.

Finally, if F denotes the completion of E we apply (20) to obtain that the subset E of F has the Baire property. We apply §1, 3.(10) to obtain that E coincides with F. The proof is complete.

(23) *If* E *is a Fréchet space, then its bidual* E" *endowed with the topology* $\sigma(E'', E')$ *is a quasi-Suslin space.*

Proof. We identify E as a subspace of E" by means of the canonical injection. Let $\{U_m : m = 1,2,\dots\}$ be a fundamental system of neighbourhoods of the origin in E which we suppose closed and absolutely convex. We set V_m to denote the closure of U_m in $E''[\sigma(E'', E')]$. Suppose N endowed with the discrete topology. We set X to denote the topological product $N \times N \times \dots$ If d is the mapping from $N \times N$ into R defined by

$$d(m, n) = 2 \text{ if } m \neq n, \quad d(m, n) = 0 \text{ if } m = n,$$

it is obvious that d is a metric on N compatible with its topology and such that (N, d) is a separable complete metric space and thus X is a Polish space.

Let $T : X \longrightarrow P(E'')$ be the mapping defined by

$$T(m_1, m_2,\dots,m_p,\dots) = \cap \{m_p V_p : p = 1,2,\dots\}$$

$$m_p \in N, \ p = 1,2,\ldots$$

$\{V_m : m = 1,2,\ldots\}$ is a fundamental system of neighbourhoods of the origin in the strong bidual $E''[\beta(E'', E')]$ of E and therefore each V_m is absorbing in E" and thus

$$\cup \{Tx : x \in X\} = E''$$

Let $x_r = (m_{1r}, m_{2r},\ldots,m_{pr},\ldots)$, $r = 1,2,\ldots$, be a sequence in X converging to the point $x = (m_1, m_2,\ldots,m_p,\ldots)$. Take a point z_r in Tx_r, $r = 1,2,\ldots$ For every positive integer p the sequence (m_{pr}) converges to m_p in N and therefore there is a positive integer q(p) such that $m_{pr} = m_p$, $r > q(p)$. Consequently there is a positive integer $n_p > m_{pr}$, $r = 1,2,\ldots$ We set $z = (n_1, n_2,\ldots,n_p,\ldots)$. Then z is a point of X such that Tx_r is contained in Tz and thus z_r belongs to Tz, $r = 1,2,\ldots$ It is obvious that T_z is a bounded subset of $E''[\beta(E'', E')]$. According to a result of GROTHENDIECK (cf. KÖTHE [1], Chapter Six, §29, Section 2), (z_r) is an equicontinuous sequence with respect to the topoloby $\beta(E',E)$ and therefore this sequence has an adherent point u in $E''[\sigma(E'',E')]$. We shall see that u belongs to Tx. Indeed, given a positive integer p we have that m_{pr} coincides with m_p for $r > q(p)$. Then z_r belongs to m_pV_p for those values of r. Since m_pV_p is $\sigma(E'', E')$-closed it follows that u belongs to m_pV_p and therefore

$$u \in \cap \{m_pV_p : p = 1,2,\ldots\} = Tx$$

which completes the proof.

(24) *Let* E *be a Fréchet space.* $E''[\sigma(E'', E')]$ *is a* K-*Suslin space if and only if* $E'[\mu(E', E'')]$ *is barrelled.*

Proof. We use the same notations of (23). If E' endowed with the topology $\mu(E', E'')$ is barrelled, then Tx is $\sigma(E'', E')$ compact for every x of X. Conditions 1) and 2) of (1) are satisfied and therefore $E''[\sigma(E'', E')]$ is a K-Suslin space.

Now suppose that $E''[\sigma(E'', E')]$ is a K-Suslin space. If A is a closed bounded subset of $E'[\sigma(E'', E')]$, then A endowed with the topology induced by $\sigma(E'', E')$ is a K-Suslin space and therefore Lindelöf. According to the mentioned result of GROTHENDIECK A is countably compact. Thus A is compact and the conclusion follows.

In Chapter Two, §5, Section 4 we give an example of a Fréchet space E such that $E'[\mu(E', E'')]$ is not bornological and therefore not barrelled. Then $E''[\sigma(E'', E')]$ is a quasi-Suslin space which is not K-Suslin.

Given a locally convex space E we denote by $\rho(E', E)$ the topology on E' of the uniform convergence on every compact subset of E and by $\rho(E'', E')$ the topology on E'' of the uniform convergence on every compact subset of $E'[\beta(E', E)]$.

(25) *If* E *is a Fréchet space, then* $E''[\rho(E'', E')]$ *is a quasi-Suslin space.*

Proof. We use the same notations of (23). Let (x_m) be a sequence in X converging to x. For every positive integer m, we take a point z_m in Tx_m. The sequence (z_m) has an adherent point z a in $E''[\sigma(E'', E')]$ which belongs to Tx. The net $\{z, z_1, z_2, \ldots\}$ is $\sigma(E'', E')$-bounded and therefore $\beta(E', E)$-equicontinuous (cf. KÖTHE [1], Chapter Six, §29, Section 2) and therefore $\sigma(E'', E')$ and $\rho(E'', E')$ coincide in this set (cf. KÖTHE [1] Chapter Four, §21, Section 7) and thus z is an adherente point of (z_m) in $E''[\sigma(E'', E')]$. Then, conditions a) and b) of the former section are satisfied by taking $E''[\rho(E'', E')]$ as E and therefore $E''[\rho(E'', E')]$ is a quasi-Suslin space.

(26) *Let* E *be a metrizable topological space. If* E *is a quasi-Suslin space then* E *is a* K-*Suslin space*

Proof. Let X be a Polish space and let T be mapping from X into $P(E)$ satisfying conditions a) and b) of the former section. If x is any point of X the sequence $x_m = x$, $m = 1, 2, \ldots$, converges to x and therefore, if z_m is a point of Tx, the sequence (z_m) has an adherent point in Tx. Consequently Tx is countably compact. Since E is metrizable, Tx is compact. Apply (1) to reach the conclusion.

(27) *Let* E *be a locally convex space. Let* A *be a compact subset of* E. *If* $E'[\rho(E', E)]$ *is a quasi-Suslin space, then* A *is metrizable.*

Proof. Let G be the linear hull of A endowed with the topology induced by E. If x belongs to E' we set Sx to denote the restriction of x to G. If z is an element of G' we apply Hahn-Banach's theorem to obtain an element y in E' such that Sy = z. Therefore $S : E' \longrightarrow G'$ is a linear onto mapping. If T denotes the topology on G' of the uniform convergence on A it is obvious that $S : E'[\rho(E', E)] \longrightarrow G'[T]$ is continuous and, applying 2.(2), $G'[T]$ is a quasi-Suslin space. Since $G'[T]$ is a normable space we apply (26) to obtain that $G'[T]$ is a K-Suslin space and consequently $G'[T]$ is a Lindelöf space. Then $G'[T]$ is separable (cf. JAMESON [1], Part I, Topology, 10).

Let H be a dense subspace of $G'[T]$ having countable algebraic basis. Then $\sigma(G, H)$ is a metrizable topology on G coarser than the original topology. Since A is compact both topologies coincide on A. Therefore A is metrizable.

(28) *Let* E *be a Fréchet space. If* A *is a compact subset of* $E'[\beta(E', E)]$ *, then* A *is metrizable.*

Proof. The conclusion follows from (25) and (27).

(29) *If* E *is a Fréchet-Montel space, then* E *is separable.*

Proof E is reflexive and $E'[\beta(E', E)]$ is a Montel space. Consequently E coincides with $E'[\rho(E'', E')]$. We apply (25) to obtain that E is a quasi-Suslin space. According to (26), E is K-Suslin and therefore Lindelöf. Thus E is separable.

The definition of K-Suslin space can be seen in MARTINEAU [1] and is an adaptation of a definition due to FROLÍK [1] and ROGERS [1]. The K-analitic spaces are K-Suslin (cf. CHOQUET [1]). In the quoted article of ROGERS every K-Suslin space which is completely regular is shown to be a K-analitic space. The properties (4), (5), (6), (7), (8) and (12) are contained in MARTINEAU [1] as well as the open mapping and closed graph theorems. Result (21) can be found in DE WILDE and SUNYACH [1];the proof given here follows VALDIVIA [8] and [9] . Results (23), (24), (25) and (27) appear here for the first time. Result (28) is contained in PFISTER [1]; there a different proof is provided and the result is considered in a more general context. Result (29) is due to DIEUDONNÉ [2].

4. SUSLIN SPACES. A topological space is Suslin if and only if it is Hausdorff and there exists a continuous mapping T defined on a Polish space X onto E. We suppose that the void set $\emptyset$ provided with the topology $\{\emptyset\}$ is a Suslin space. Obviously every Polish space is a Suslin space. In parti-

cular every separable Fréchet space is a Suslin space. If we denote by $M(E)$ the family of all the singleton of E, the definition of Suslin space coincides with the corresponding of K-Suslin space by taking $M(E)$ instead of $K(E)$. Since every element of $M(E)$ belongs to $K(E)$, result (1) is obvious.

(1) *Every Suslin space is a K-Suslin space.*

(2) *Every Suslin space is separable.*

Proof. Note that every Polish space is separable.

(3) *Let E be a Suslin space. Let F be a sequentially closed subspace of E. Then F is a Suslin space.*

Proof. If $E = \emptyset$ the conclusion is obvious. If $E \neq \emptyset$, let X be a Polish space and let T be a continuous mapping from X onto E. Then $T^{-1}(F)$ is a closed subspace of X and therefore a Polish space the restriction of T to $T^{-1}(F)$ is continuous from $T^{-1}(F)$ onto F and thus F is a Suslin space.

Results (4), (5), (6), (7), (8), (9) and (10) can be proved analogously to 2.(2), 2.(3), 2.(4), 2.(5), 2.(6), 2.(7) and 2.(8) respectively with the obvious modifications.

(4) *Let E and F be Hausdorff topological spaces. If E is Suslin and if there is a sequentially continuous mapping from E onto F, then F is Suslin.*

(5) *Let E be a Hausdorff topological space. Let (E_n) be a sequence of subspace of E covering E. If E_n is Suslin, n = 1,2,.., then E is Suslin.*

(6) *If (E_n) is a sequence of Suslin spaces, then $\prod_{n=1}^{\infty} E_n$ is a Suslin space.*

(7) *Let E be a topological space. Let (E_n) be a sequence of subspaces of E. If for every positive integer n, E_n is Suslin, then $\cap\{E_n : n = 1,2,\ldots\}$ is Suslin.*

(8) *If $E_1, E_2, \ldots, E_m$ are Suslin spaces, then $\prod_{j=1}^{m} E_j$ is a Suslin space.*

(9) *Let E be a Suslin locally convex space. If F is a closed subspace of E, then E/F is a Suslin space.*

(10) *Let E be a locally convex space. Let (E_n) be a sequence of subspaces of E covering E. For every positive integer n let T_n be a locally convex topology on E_n finer than the original one such that $E_n[T_n]$ is Suslin. If*

E is the locally convex hull of

$$\{E_n\ [T_n] : n = 1,2,\dots\}$$

then E is Suslin.

(11) *Let E be a Suslin space. If F is an open subspace of E, then F is Suslin.*

Proof. There is nothing to prove if $F = \emptyset$ or $F = E$. If $F \neq \emptyset$, $F \neq E$, let T be a continuous mapping from a Polish space X onto E. Let d be a metric on X compatible with its topology. We set d(x) to denote the distance from any point x of X to $X \sim T^{-1}(F)$. Fix a point z in $T^{-1}(F)$. Then $d(z) > 0$. For every positive integer n we set

$$X_n = \{x \in X ; d(x) \geq \frac{d(z)}{n}\}.$$

F_n denotes $T(X_n)$ with the topology induced by E and T_n is the restriction of T to X_n. Since X_n is closed in X, X_n is a Polish space. On the other hand, $T_n : X_n \longrightarrow F_n$ is continuous and thus F_n is a Suslin space. Clearly F coincides with $\cup \{F_n : n = 1,2,\dots\}$ and F is Suslin according (5).

Given the topological space E let $\{A_i : i \in I\}$ be the collection of all σ-algebras of subsets of E containing the family of the open sets. We set

$$\mathcal{B} = \cap \{A_i : i \in I\}.$$

$\mathcal{B}$ is the σ-algebra of Borel of E and every element of $\mathcal{B}$ is a Borel subset of E.

(12) *Let E be a topological space. Let F be a Suslin subspace of E. If A is a Borel subset of E, then $A \cap F$ is a Suslin subspace of E.*

Proof. Let A be the family of all subsets B of E such chat $B \cap F$ and $(E \sim B) \cap F$ are Suslin subspaces of E. According to (3) and (11), the open sets of E belong to A. Let (B_m) be a sequence of elements of A. We apply (5) and (7) to obtain that

$$(\cup \{B_m : m = 1,2,\dots\}) \cap F = \cup \{B_m \cap F : m = 1,2,\dots\}$$

and

$$(E \sim \cup \{B_m : m = 1,2,\dots\}) \cap F = \cap\{(E \sim B_m) \cap F : m = 1,2,\dots\}$$

are Suslin subspaces of E. Then it follows that A is a σ-algebra on E which contains the open sets of E and therefore every Borel subset of E belongs to A. Consequently, $A \cap F$ is Suslin.

(13) *Let* G *be a metrizable topological space. Let* F *be a hausdorff topological space. Let* $f : G \longrightarrow F$ *be a mapping with sequentially closed graph. If* E *is a Suslin subspace of* F, *then* $f^{-1}(E)$ *is a Subset of* G *with the Baire property.*

Proof. Let X be a Polish space and let T be a continuous mapping from X onto E. Let d be a metric on X compatible with its topology and such that (X, d) is a complete metric space. Using the same notations of 2.(9) we take a point z in $O(f^{-1}(E)) \sim B$ and, for every positive integer p,

$$z_p \in U_p \cap A_{m_1, m_2, \ldots, m_p} \sim B.$$

$$x_p \in B_{m_1, m_2, \ldots, m_p}, \text{ with } f(z_p) = Tx_p$$

The sequence (x_p) converges to a point x of X. Consequently $(f(z_p))$ converges to Tx and therefore z belongs to $f^{-1}(E)$. We reach the conclusion as in 2.(9).

(14) *Let* G *be a metrizable topological space. Let* F *be a Suslin space. If* $f : G \longrightarrow F$ *is a mapping with sequentially closed graph and if* E *is a closed subspace of* F, *then* $f^{-1}(E)$ *is a subset of* G *with the Baire property.*

Proof. According to (1), E is Suslin. It is enough to apply the forner result to obtain the conclusion.

(15) *Let* E *be a locally convex space which is the locally convex hull of Baire metrizable spaces. Let* F *be a Suslin locally convex space. If* $f:E \rightarrow F$ *is a linear mapping with sequentially closed graph, then* f *is continuous.*

Proof. Let $\{E_i : i \in I\}$ be a family of Baire metrizable spaces. For every i of I let A_i be a linear mapping from E_i into E such that the topology of E is the strongest locally convex topology for which all the mapping A_i, $i \in I$, are continuous.

Let U be an absolutely convex closed neighbourhood of the origin in F. For every i of I, we apply 1.(11) to obtain that $f \circ A_i$ has sequentially closed graph and therefore $A_i^{-1}(f^{-1}(U))$ is a subset of E_i with the Bai re property, according to (14). We apply §1,3.(8) to obtain that $A_i^{-1}(f^{-1}(U))$ is a neighbourhood of the origin in E_i. Consequently, $f^{-1}(U)$ is a neighbourhood of the origin in E and the conclusion follows.

(16) *Let* F *be a Suslin space. Let* G *be a metrizable topological space. Let* E *be a subspace of* F *and* $f : E \longrightarrow G$ *a mapping with sequentially closed graph in* F x G. *If* H *is a sequentially closed subspace of* E, *then* f(H) *is a subset of* G *with the Baire property.*

Proof. Let X be a Polish space and let T be a continuous mapping from X onto F. Let d be a metric on X compatible with its topology such that (X,d) is complete. Suppose $T^{-1}(H)$ endowed with the metric induced by d. We take a sequence of balls (B_m) of radii less than one whose union coincides with the metric space $T^{-1}(H)$. Suppose that, for the positive integers $m_1, m_2, \ldots, m_p$, we have constructed the subset $B_{m_1, m_2, \ldots, m_p}$ of X. We suppose this set endowed with the metric induced by d. We take a sequence of balls $(B_{m_1, m_2, \ldots, m_p, m})$ in $B_{m_1, m_2, \ldots, m_p}$ with radii less than $\frac{1}{2^p}$ such that

$$B_{m_1, m_2, \ldots, m_p} = \cup \{B_{m_1, m_2, \ldots, m_p, m} \quad m = 1, 2, \ldots \}$$

We set

$$A_{m_1, m_2, \ldots, m_p} = f\,(T(B_{m_1, m_2, \ldots, m_p}))$$

Since

$$f(H) = \cup \{A_m : m = 1, 2, \ldots\}$$

we apply §1,1.(7) to obtain that

$$O(f(H)) \sim \cup \{O(A_m) \quad : m = 1, 2, \ldots\} = D$$

is a rare subset of G. Analogously

$$O\ (A_{m_1, m_2, \ldots, m_p}) \sim \cup \{O(A_{m_1, m_2, \ldots, m_p, m})$$

$$: m = 1, 2, \ldots,\} = D_{m_1, m_2, \ldots, m_p}$$

is a rare subset of G. For every positive integer p, we set

$$D(p) = \cup \{D_{m_1, m_2,\ldots,m_p} : m_1, m_2, \ldots, m_p = 1,2,\ldots\}$$

and

$$B = D \cup (\cup \{D(p) : p = 1, 2,\ldots\}).$$

It is obvious that B is a subset of G of first category. If z is a point of $O(f(H)) \sim B$, then there is a sequence $m_1, m_2,\ldots,m_p,\ldots$ of positive integers such that

$$z \in O(A_{m_1, m_2,\ldots,m_p}),\ p = 1,2,\ldots$$

Let $\{U_p : p = 1,2,\ldots\}$ be a fundamental system of neighbourhoods of the point z in G. Since $U_p \cap A_{m_1, m_2,\ldots,m_p}$ is a subset of G of second category we have that

$$U_p \cap A_{m_1, m_2,\ldots,m_p} \sim B \neq \emptyset$$

and therefore, for every positive integer p, we take

$$z_p \in U_p \cap A_{m_1, m_2,\ldots,m_p} \sim B$$

$$x_p \in B_{m_1, m_2,\ldots,m_p},\ \text{with } f(Tx_p) = z_p.$$

Obviously (x_p) is a Cauchy sequence in (X, d) and therefore converges to a point x of X. Consequently (Tx_p) converges to Tx in F. The sequence (z_p) converges to z in G and, since f has sequentially closed graph in F x G, we have that f(Tx) = z and therefore Tx belongs to E. Since H is sequentially closed in E, Tx belongs to H and therefore z belongs to f(H). Thus

$$O(f(H)) \sim f(H) \subset O(f(H)) \sim (O(f(H)) \sim B) \subset B$$

and therefore $O(f(H)) \sim g(H)$ is a subset of G of first category. Now the conclusion follows from §1, 1.(9).

(17) *Let E be a Suslin locally convex space. Let F be a locally convex space which is the locally convex hull of Baire metrizable locally convex spaces. If f : E —> F is a linear onto mapping with sequentially closed graph, then f is open.*

Proof. Let $\{F_i : i \in I\}$ be a family of Baire metrizable locally convex spaces. For every i of I let A_i be a linear mapping from F_i into F such that F is endowed with the strongest locally convex topology for which the mappings A_i, $i \in I$, are continuous. We set G_i instead of $A_i(F_i)$ with the topology induced by F. Let T_i be the finest locally convex topology on $A_i(F_i)$ such that A_i is continuous from F_i into $A_i(F_i)[T_i]$. Then this space is isomorphic to the quotient $F_i/A_i^{-1}(0)$ and therefore is a metrizable Baire space.

Fix i in I and set $H_i = f^{-1}(G_i)$. Let f_i be the restriction of f to H_i. Let $((x_m, f(x_m)))$ a sequence in the graph of f_i converging to (x, y) in $E \times G_i$. Since the graph of f is sequentially closed we have that $f(x) = y$ and therefore x belongs to H_i. Therefore f_i has sequentially closed graph in $E \times G_i$ and consequently f_i has sequentially closed graph in $E \times A_i(F_i)[T_i]$.

Let U be a closed absolutely convex neighbourhood of the origin in E. According to (16), $f_i(U \cap H_i)$ is a subset of $A_i[F_i][T_i]$ with the Baire property and, by virtue of §1,3.(8), is a neighbourhood of the origin in $A_i[F_i][T_i]$. Consequently f(U) is a neighbourhood of the origin in F from where it follows that f is open.

Let E and F be topological spaces and let f be a mapping from E into F. We say that f has Borel graph if G(f) is a Borel subset of $E \times F$.

(18) *Let* E *be an ultrabornological space. Let* F *be a Suslin locally convex space. If* $f : E \longrightarrow F$ *is a linear mapping with Borel graph, then* f *is continuous.*

Proof. There is a family $\{E_i : i \in I\}$ of subspaces of E such that there is a topology T_i on E_i finer than the original topology such that $E_i[T_i]$ is a separable Banach space, $i \in I$, and E coincides with the inductive limit of $\{E_i[T_i] : i \in I\}$ (cf. JARCHOW [1], 13.2).

For every i in I, $E_i[T_i]$ is a Suslin and therefore E_i is Suslin. Let

f_i be the restriction of f to E_i. By (8), $E_i \times F$ is a Suslin space. We set G and G_i to denote the graphs of f and f_i respectively with the topologies induced by the topology of $E \times F$. We have that

$$G_i = G \cap (E_i \times F)$$

and, since G is a Borel subset of $E \times F$, we apply (12) to obtain that G_i is a Suslin subspace of $E_i \times F$. Let S be the mapping from G_i into E_i defined by

$$S\ (x, f(x)) = x,\ x \in E_i$$

Clearly S is linear continuous and bijective. Therefore S is a linear mapping with closed graph from the Suslin space G_i onto the Banach space $E_i\ [\mathcal{T}_i]$. We apply (17) to obtain that S is open. Consequently S^{-1} is a continuous mapping from $E_i\ [\mathcal{T}_i]$ into G_i. Then f_i is a continuous mapping from $E_i\ [\mathcal{T}_i]$ into F and the conclusion follows.

(19) *Let F be a Suslin locally convex space. Let E be an ultrabornological space. Let g : F ⟶ E be a linear onto mapping with Borel graph. Then g is open.*

Proof. Let $\{E_i : i \in I\}$ be the family of subspaces of E considered in the proof of (18). $\mathcal{T}_i$ has the same meaning as in (18). Given an index i of I let $g_i : g^{-1}(E_i) \longrightarrow E_i$ the mapping defined by $g_i(x) = g(x)$ for every x of $g^{-1}(E_i)$. Let H and H_i be the graphs of g and g_i respectively endowed with the topologies induced by the topology of $F \times E$. We have that

$$H_i = H \cap (F \times E_i)$$

Since $F \times E_i$ is a Suslin subspace of $F \times E$ and H is a Borel subset of $F \times E$ it follows that H_i is a Suslin subspace of $F \times E_i$. Let $W : H_i \longrightarrow E_i$ be the mapping defined by

$$W\ (x, g(x)) = g\ (x),\ x \in g^{-1}(E_i).$$

Obviously W is linear continuous and onto. Thus W is a linear mapping with closed graph from H_i onto the Banach space $E_i\ [\mathcal{T}_i]$. We apply (17) to obtain that $W : H_i \longrightarrow E_i\ [T_i]$ is open. Let U be an open subset of the subspace $g^{-1}(E_i)$ of F. We set

$$V = \{(x, g(x)) : x \in U\}.$$

V is an open subset of H_i. Therefore the set $g(U) = W(V)$ is open in $E_i[T_i]$. Now it is easy to show that g is open.

If E is a non-separable reflexive Banach space we apply (2) to obtain that E is not a Suslin space. On the other hand, $E[\sigma(E, E')]$ is K-Suslin according to 3.(26). Thus there are K-Suslin locally convex spaces which are not Suslin.

(20) *If* E *is a Fréchet-Montel space, then* $E'[\mu(E', E)]$ *is a Suslin space.*

Proof. According to 3.(29) there is in $E'[\mu(E', E)]$ a sequence (A_m) of metrizable compact subsets covering E'. A_m is a Suslin topological space and thus $E'[\mu(E', E)]$ is Suslin, according to (5).

Chapter Three includes the definition of all the function spaces we shall use in this section.

Let Ω be a non-void open subset of the n-dimensional euclidian space R^n. The space $\mathcal{E}(\Omega)$ is a Fréchet-Montel space and therefore separable (see 3.(28)) and clearly is a Suslin space. Its strong dual is also Suslin according to (20). If H is a compact subset of R^n with non-void interior, then $\mathcal{D}(H)$ is a Fréchet-Montel space. Therefore $\mathcal{D}(H)$ and its strong dual are Suslin.

(21) *Let* E *be a locally convex space. Let* (E_m) *be a sequence of subspaces of* E *covering* E. *If for every positive integer* m *there exists a topology* T_m *on* E_m *finer than the original topology such that* $E[T_m]$ *is a separable Fréchet space, then* E *is Suslin.*

Proof. It is an immediate consequence of (4) and (5).

If Ω is a non-void open subset of R^n, $\mathcal{D}(\Omega)$ is the inductive limit of a sequence of Fréchet-Montel spaces. We apply (21) to obtain that $\mathcal{D}(\Omega)$ is a Suslin space.

(22) *Let* E *be as in* (21). *If* A *is a compact absolutely convex subset of* $E[\sigma(E, E')]$, *there is a positive integer* p *such that* A *is a bounded subset of* $E_p\ [T_p]$.

Proof. Let G_m be the linear space $E_m \cap E_A$ with the topology induced by the topology of E_A. Since E_A is a Banach space and $\{G_m : m = 1,2,\dots\}$ covers E_A there is a positive integer p such that G_p is a subset of E_A of second category and therefore G_p is a Baire space dense in E_A. The canonical injection $T : G_p \longrightarrow E_p$ is obvious continuous and therefore T has closed graph in $G_p \times E_p\,[T_p]$. Since $E_p\,[T_p]$ is a Suslin space we apply (15) to obtain that $T : G_p \longrightarrow E_p\,[T_p]$ is continuous. Now take a point x in E_A. Let (x_n) be a sequence in G_p converging to x in E_A. Then $(Tx_n) = (x_n)$ is a Cauchy sequence in $E_p\,[T_p]$ and therefore converges in this space to a point z of E. Consequently (x_n) converges to the points x and z of E and therefore x = z. Then x belongs to G_p and therefore G_p coincides with E_A. Finally T maps the bounded set A of E_A in a bounded set of $E_p\,[T_p]$.

(23) *Let* E *be a locally convex space. Let* (E_m) *be a sequence of subspaces of* E *covering* E. *Suppose that, for every positive integer* m, *there is a topology* T_m *on* E_m *finer than the original topology such that* $E_m\,[T_m]$ *is a Fréchet-Montel space. If* E *is the locally convex hull of the family* $\{E_m[T_m]$ $m=1,2,\dots\}$, *then* $E'[\mu(E', E)]$ *is a Suslin space.*

Proof. We set G_m to denote the topological dual of $E_m\,[T_m]$ with the Mackey topology, m = 1,2,... We write

$$L = \bigoplus_{m=1}^{\infty} E_m\,[T_m] \;,\; G = \prod_{m=1}^{\infty} G_m$$

According to (6) and (20) G is a Suslin space and since G is isomorphic to $L'[\mu(L', L)]$ it follows that this last space is Suslin. If $x = (x_1, x_2, \dots, x_n, \dots)$ is any element of L we set $f(x) = \Sigma x_n$. Then f is an homomorphism from L onto E (cf. KÖTHE [1], Chapter Four, §19, Section 3). If $g : E'[\mu(E', E)] \longrightarrow L'[\mu(L', L)]$ is the transposed mapping of f, g is continuous and an isomorphism from $E'[\sigma(E', E)]$ into $L'[\mu(L', L)]$ such that $g(E')$ is $\sigma(L',L)$-closed (cf. KÖTHE [2], Chapter Seven, §32, Section 3).

Let U be a closed absolutely convex neighbourhood of the origin in

$E'[\mu(E', E)]$. If A is the polar set of U in E, A is weakly compact and absolutely convex and, applying (22), there is a positive integer p such that A is a bounded subset in $E_p [\mathcal{T}_p]$. Obviously A is closed in this space and thus A is $\mathcal{T}_p$-compact. If we denote by H_p the subspace of L of all those elements $(x_1, x_2, \ldots, x_m, \ldots)$ which $x_m = 0$, $m \neq p$, and if f_p is the restriction of f to H_p, then $f_p : H_p \longrightarrow E_p [\mathcal{T}_p]$ is an isomorphism and therefore the polar set V of $f_p^{-1}(A)$ in L' is a neighbourhood of the origin in $L'[\mu(L', L)]$. Let u be any point of $V \cap g(E')$. Let z be any point of A. We find a point x in $f_p^{-1}(A)$ such that $f_p(x) = z$. Then

$$|\langle z, g^{-1}(u)\rangle| = |\langle f_p(x), g^{-1}(u)\rangle|$$

$$= |\langle f(x), g^{-1}(u)\rangle| = |\langle x, u\rangle| \leq 1$$

and thus $g^{-1}(V)$ is contained in U. Therefore g is an homomorphism. Then $E'[\mu(E', E)]$ is isomorphic to the closed subspace $g(E')$ of $L'[\mu(L', L)]$. We apply (3) to obtain that $E'[\mu(E', E)]$ is a Suslin space.

Let Ω be a non-void open subset of R^n. We suppose the space of distributions $\mathcal{D}'(\Omega)$ endowed with the topology $\mu(\mathcal{D}'(\Omega), \mathcal{D}(\Omega))$ which is also its strong topology. We apply (23) to obtain that $\mathcal{D}'(\Omega)$ is a Suslin space.

Results (18) and (19) are due to SCHWARTZ [2]. Measure-theoretic ideas are used in their proof. MARTINEAU [1] and [2] extend the result of SCHWARTZ without using measure theory.

5. SEMI-SUSLIN SPACES. Let E be a locally convex space. A subset A of E is CS-compact if every series of the form $\Sigma a_m x_m, \ldots, x_m \in A$, $a_m \geq 0$, $m = 1, 2, \ldots$, $\Sigma a_m = 1$, converges to a point in A. The concept of CS-compact set can be seen in (JAMESON [1], p. 212). It is immediate that every CS-compact subset of E is bounded and convex. We denote by $\mathcal{S}(E)$ the family of all CS-compact subset of E.

A locally convex space E is semi-Suslin if there is a Polish space X and a mapping from X into $\mathcal{S}(E)$ verifying the following conditions:

a) $\cup \{Tx : x \in X\} = E$,

b) if (x_n) is a sequence in X converging to x there is an element A

of $\mathcal{S}(E)$ such that Tx_n is contained in A, n = 1,2,...

The semi-Suslin spaces are related to the spaces $\mathcal{D}_0$ of RAÍKOV [1] and to the webbed spaces introduced in DE WILDE [1] (see also HORVÁTH [2], DE WILDE [2], KÖTHE [2], Chapter Seven, §35, and JARCHOW [1], 5).

(1) *Let E be a semi-Suslin locally convex space. If F is a sequentially closed subspace of E, then F is a semi-Suslin space.*

Proof. Let X be a Polish space and let T : X —> E be a mappings satisfying conditions a) and b). Let x be a point of X. We set $Sx = F \cap Tx$ when $F \cap Tx$ is non-void. If $F \cap Tx$ is void, we write Sx = {0}. It is not difficult to check that S is a mapping from S into $\mathcal{S}(F)$ satisfying

$$\{Sx : x \in X\} = F$$

and such that, given a sequence (x_n) in X converging to x, there is an element A in $\mathcal{S}(F)$ with Sx_n in A, n = 1,2,...

(2) *Let E and F be locally convex spaces. Let f be a continuous linear mapping from E onto F. If E is semi-Suslin, then F is semi-Suslin.*

Proof. Let X be a Polish space and let T be a mapping from X into $\mathcal{S}(E)$ satisfying conditions a) and b). We set S = f o T. It is immediate that S is a mapping from X into $\mathcal{S}(F)$ such that

$$\cup \{Sx : x \in X\} = \cup \{f(Tx) : x \in X\} = f(E) = F.$$

On the other hand, if (x_n) is a sequence in X converging to x, there is an element A in $\mathcal{S}(E)$ such that Tx_n belongs to A, n = 1,2,... Then f(A) belongs to $\mathcal{S}(F)$ and Sx_n belongs to f(A) for every positive integer n. The proof is complete.

(3) *Let E be a locally convex space. Let (E_m) be a sequence of subspaces of E covering E. If E_m is a semi-Suslin space, m = 1,2,..., then E is semi-Suslin.*

Proof. For every positive integer m, let X_m be a Polish space and let a mapping from X_m into $\mathcal{S}(E_m)$ satisfying conditions a) and b) with X, E, T standing for X_m, E_m, T_m respectively. Proceeding as in the proof of 2.(3) we obtain a Polish space Y and a mapping S from Y into $\mathcal{P}(E)$ such that

$$\cup \{S(x, m) : (x, m) \in Y\} = E.$$

It is immediate that S applies Y in $\mathcal{S}(E)$. On the other hand, if $((x_m, n_m))$ is a sequence in Y converging to (x, p), there is a positive integer q such that $n_m = p$ for $m > q$ and therefore, if we set $y_r = y_{q+r}$, $r = 1,2, \ldots$, then (y_r) converges to x in X_p. Therefore there is an element A in $\mathcal{S}(E_p)$ with

$$S(x_m, n_m) = T_p x_m \subset A, \quad m = q+1, q+2$$

Then $B = A + \sum_{m=1}^{q} S_m$ is an element of S(E) such that

$$S(x_m, n_m) \subset B, \quad m = 1,2,$$

A' and S_m being the convex hull of A and $S(x_m, n_m)$ respectively.

(4) *If* (E_m) *is a sequence of semi-Suslin locally convex spaces, then* $\prod_{m=1}^{\infty} E_m$ *is semi-Suslin*

Proof. For every positive integer m, X_m and T_m have the same meaning as in (3). We proceed as we did in the proof of 2(4) to obtain the mapping S from X into $\mathcal{S}(E)$. The rest of the proof follows easily.

(5) *Let* E *be locally convex space. Let* (E_m) *be a sequence of subspaces of* E. *If for every positive integer* m, E_m *is semi-Suslin, then* $\bigcap\{E_m : m = 1, 2, \ldots\}$ *is semi-Suslin.*

Proof. Proceed as in 2.(5).

(6) *If* $E_1, E_2, \ldots, E_r$ *are semi-Suslin locally convex spaces, then* $\prod_{m=1}^{r} E_m$ *is semi-Suslin.*

Proof. We set

$$E_{r+1} = E_{r+2} = \ldots = E_{r+m} = \ldots = \{0\}.$$

Then $\prod_{m=1}^{r}$ is isomorphic to $\prod_{m=1}^{\infty} E_m$ which is a semi-Suslin space according to the former result. The conclusion follows from (2).

(7) *Let* E *be a quasi-Suslin locally convex space. If* E *is locally complete, then* E *is semi-Suslin.*

Proof. Let X be a Polish space and let T be a mapping from X into

$P(E)$ satisfying conditions a) and b) of Section 2. This last condition implies that, if (x_m) is a sequence of X converging to x, then

$$\cup \{Tx_m : m = 1,2,\ldots\}$$

is a relatively countably compact subset of E and therefore bounded. For every x of X, we set Sx to denote the closed absolutely convex hull of Tx. Since E is locally complete, Sx is CS-compact and therefore S is a mapping defined on X taking values in $S(E)$. We have that

$$\{Sx : x \in X\} \supset \{Tx : x \in X\} = E.$$

On the other hand, if (x_m) is a sequence in X converging to x, the closed absolutely convex hull A of $\cup \{Tx_m : m = 1,2,\ldots\}$ belongs to $S(E)$ and Sx_m belongs to A, m = 1,2,... The proof is complete.

(8) *If* E *is a Fréchet space, then* E *is semi-Suslin.*

Proof. Since every Fréchet space is a closed subspace of a countable product of Banach spaces (cf. KÖTHE [1], Chapter Four, §19, Section 9) it is enough to carry the proof for E being a Banach space (see (1) and (4)). Let B be the closed unit ball of E. Then mB belongs to $S(E)$, m = 1,2,... Let X be the set N endowed with the discrete topology. Clearly X is a Polish space. For every m in N we set Tm = mB. Then T is a mapping from X into $S(E)$ satisfying conditions a) and b). Thus E is semi-Suslin.

(9) *If* E *is a metrizable locally convex space, then its strong dual* $E'[\beta(E', E)]$ *is semi-Suslin.*

Proof. Let $\{U_m : m = 1,2,\ldots\}$ be a fundamental system of neighbourhoods of the origin in E. For every positive integer m, let V_m be the polar set in E' of U_m. Let G_m be the linear hull of V_m with the topology induced by $\beta(E', E)$. Let T_m be the topology on G_m defined by the fundamental system of neighbourhoods of the origin

$$\{\frac{1}{p} V_m : p = 1,2,\ldots\}.$$

Then $G_m[T_m]$ is a Banach space. Since T_m is finer than the topology of G_m, we apply (2) and (8) to obtain that G_m is semi-Suslin. Since

$$G = \cup \{G_m : m = 1,2,\ldots\}$$

we apply (3) to obtain that G is semi-Suslin.

(10) *Let* E *be a locally convex space. If* A *is a* CS-*compact subset of* $E'[\sigma(E', E)]$, *then* A *is a* CS-*compact subset of* $E'[\beta(E', E)]$.

Proof. Suppose that A is not bounded in $E'[\beta(E', E)]$. We find a bounded subset M in E such that, if U is the polar set of M in E', U does not absorb A. Take a point u_1 in A. We find a positive integer p such that u_1 belongs to p U and we set V = p U. For every positive integer n we find an element z_n in A which does not belong to $2^{4n+1}V$, which is $\sigma(E', E)$-closed. Then the set of real numbers b with

$$0 \leq b < 1 \text{ and } (1-b)\, u_1 + bz_n \in 2^{4n+1}\, V$$

has a maximum d. The point $(1-d)\, u_1 + dz_n$ is not interior to $2^{4n+1}V$ in $E'[\beta(E', E)]$ and, since A is convex, $(1-d)\, u_1 + dz_n$ belongs to A. Analogously let c be the maximum of the real numbers b with

$$0 \leq b < 1 \text{ and } (1-b)\, u_1 + bz_n \in 2^n\, V$$

Then $(1-c)\, u_1 + cz_n$ is an interior point of $2^{4n+1}V$ in $E'[\beta(E', E)]$ and belongs to A. Therefore $c < d$. If we take a real number h with $c < h < d$, it is immediate to check that the vector $x_n = (1-h)\, u_1 + hz_n$ verifies

$$x_n \in A,\ x_n \notin 2^{4n}V,\ x_n \in 2^{4n+1}V.$$

Consequently

$$\sum_{j=1}^{m} \frac{1}{2^j} x_j \in \sum_{j=1}^{m} 2^{3j+1}V \subset 2^{3m+2}V.$$

Now we suppose that

$$\sum_{j=1}^{m+1} \frac{1}{2^j} x_j \in 2^{3m+2}V.$$

Then

$$\frac{1}{2^{m+1}} x_{m+1} \in 2^{3m+2}V - \sum_{j=1}^{m} \frac{1}{2^j} x_j \subset 2^{3m+3}V$$

and therefore

$$x_{m+1} \in 2^{4(m+1)}V$$

and that is a contradiction. Thus

$$(11) \qquad \sum_{j=1}^{m+1} \frac{1}{2^j} x_j \notin 2^{3m+2}V, \ m = 1,2,.$$

We set $x = \Sigma \frac{1}{2^m} x_m$ in $E'[\sigma(E', E)]$ and

$$b_m = \Big(\sum_{j=1}^{m} \frac{1}{2^j} \Big)^{-1}, \ v_m = b_m \sum_{j=1}^{m} \frac{1}{2^j} x_j.$$

Then V_m belongs to A and we have in $E'[\sigma(E', E)]$

$$\lim v_m = x \lim b_m = x$$

and therefore the sequence $(v_m - x)$ converges to the origin in E'. The closed convex hull B of this sequence in $E^* [\sigma(E^*, E)]$ is a compact set and coincides with

$$\{\Sigma a_m(v_m - x) : a_m \geq 0, \ m = 1,2,\ldots, \ \Sigma a_m \leq 1\}.$$

Given $a_m \geq 0$, $m = 1,2,\ldots$ with $\Sigma a_m \leq 1$, $\Sigma a_m \neq 0$, we have that

$$\frac{1}{\Sigma a_m} \Sigma a_m v_m \in A$$

and therefore

$$\frac{1}{\Sigma a_m} \Sigma a_m(v_m - x) \in A - x$$

Since x belongs to A and A-x is a convex subset of E' we have that

$$\Sigma a_m (v_m - x) \in A - x.$$

Then B + x is a compact subset of $E'[\sigma(E', E)]$. Therefore every barrel in E absorbs B + x (cf. A. ROBERTSON and W. ROBERTSON [1], Chapter IV, p.66). In particular, V absorbs the sequence (v_m). Since $b_m > 1$, $m = 1,2,\ldots$, it follows from (11) that

$$v_{m+1} \notin 2^{3m+2}V, \ m = 1,2,\dots,$$

and therefore V does not absorb the sequence (v_m). That is a contradiction and therefore A is $\beta(E', E)$-bounded.

Let p be a continuous seminorm on $E'[\beta(E', E)]$. We consider a series $\Sigma a_m y_m$, with y_m in A, $a_m \geq 0$, $m = 1,2,\dots$, $\Sigma a_m = 1$. Since A is $\beta(E', E)$-bounded there is $k > 0$ such that $p(y_m) < k$, $m = 1,2,\dots$ Since $a_m p(x_m) < a_m k$, the series $\Sigma a_m p(y_m)$ is convergent. Then $(\sum_{j=1}^{m} a_j y_j)$ is a Cauchy sequence in $E'[\beta(E', E)]$ converging to a certain point y in $E'[\sigma(E', E)]$ from where it follows that $\Sigma a_m y_m$ converges to y in $E'[\beta(E', E)]$. The proof is complete.

(12) *Let E be a locally convex space. Let (E_m) be a sequence of subspaces of E covering E. For every positive integer m, let T_m be a topology on E_m finer than the original topology such that $E_m[T_m]$ is a Fréchet space. If E is the locally convex hull of the family $\{E_m[T_m] : m = 1,2,\dots\}$ then E and $E'[\beta(E', E)]$ are semi-Suslin spaces.*

Proof. According to (2), (3) and (8), E is a semi-Suslin space. We set G_m to denote the topological dual of $E_m[T_m]$ endowed with the topology $\sigma(G_m, E_m)$, $m = 1,2,\dots$ We write

$$G = \prod_{m=1}^{\infty} G_m, \ L = \bigoplus_{m=1}^{\infty} E_m[T_m]$$

Let $f : L \longrightarrow E$ the homomorphism onto defined by

$$f(x) = \Sigma x_n, \ x = (x_1, x_2, \dots, x_n, \dots) \in L$$

If $g : E'[\sigma(E', E)] \longrightarrow L'[\sigma(L', L)]$ is the transposed mapping of f, g is an isomorphism from $E'[\sigma(E', E)]$ in the closed subspace $g(E')$ of $L'[\sigma(L', L)]$. According (2), (4) and (9), G is semi-Suslin. Since G is isomorphic to $L'[\sigma(L', L)]$, we apply (1) to obtain that $E'[\sigma(E', E)]$ is semi-Suslin. We apply the former result to obtain that $E'[\sigma(E', E)]$ is a semi-Suslin space.

(13) *Let E and F be locally convex spaces. Let f be a linear mapping with*

sequentially closed graph from F *into* E. *If* F *is the locally convex hull of metrizable convex-Baire spaces and if* E *is semi-Suslin, then* f *is continuous.*

Proof. It is enough to carry the proof for F being a metrizable convex-Baire space. Let X be a Polish space and let T be a mapping from X into $\mathcal{S}(E)$ satisfying conditions a) and b). Let d be a metric on X compatible with its topology such that (X, d) is complete. We find a sequence (B_m) of balls in (X, d) of radii less than one covering X. Suppose that for the positive integers $m_1, m_2,\ldots,m_p$ the ball $B_{m_1, m_2,\ldots,m_p}$ has been constructed in X. We suppose this ball endowed with metric induced by d. Take in $B_{m_1, m_2,\ldots,m_p}$ a sequence of balls $(B_{m_1, m_2,\ldots,m_p,m})$ of radii less than $\frac{1}{2^p}$ such that

$$B_{m_1, m_2,\ldots,m_p} = \bigcup \{B_{m_1, m_2,\ldots,m_p, m} : m = 1,2,\ldots\}$$

Let U be an absolutely convex closed neighbourhood of the origin in E. We set $V = f^{-1}(U)$. Since F coincides with $\bigcup \{mV : m = 1,2,\ldots\}$, we apply §1, 6.(3) to obtain that V is a convex-Baire subset of F. For the positive integer $m_1, m_2,\ldots,m_p$ we write

$$A_{m_1, m_2,\ldots,m_p} = V \cap f^{-1} (T (B_{m_1, m_2,\ldots,m_p}))$$

and $M_{m_1, m_2,\ldots,m_p}$ for the convex hull of $A_{m_1, m_2,\ldots,m_p}$. We have that

$$V = \bigcup \{A_m : m = 1, 2,\ldots\},$$

$$A_{m_1, m_2,\ldots,m_p} = \bigcup \{A_{m_1, m_2,\ldots,m_p, m} : m = 1,2,\ldots\}$$

and therefore we apply §1, 6.(5) to obtain a sequence of positive integers (n_p) such that $M_{n_1, n_2,\ldots,n_p}$ is a convex-Baire subset of F, p = 1,2,... Let $\{V_p : p = 1, 2,\ldots\}$ be a fundamental system of neighbourhoods of the origin in F. For every positive integer p, we take a point z_p in $M_{n_1, n_2,}$

$\ldots, n_p$ and a neighbourhood of the origin U_p in F, contained in V_p, such that

$$z_p + U_p \subset \bar{M}_{n_1, n_2, \ldots, n_p}.$$

Let x be a point of $\bar{V}$. We find a point u in V such that

$$x - u \in \frac{1}{2} U_1.$$

Then

$$z_1 + 2 (x - u) \in z_1 + U_1 \subset \bar{M}_{n_1}$$

and therefore there is a point x_1 in M_{n_1} with

$$z_1 + 2 (x - u - \frac{1}{2} x_1) \in \frac{1}{2} U_2$$

Proceeding by recurrence, suppose that we have obtained x_j in $M_{n_1, n_2, \ldots, n_j}$ for j = 1,2,...,p, such that

$$y_p = z_p + 2 z_{p-1} + \ldots + 2^{p-1} z_1 + 2^p (x - u - \frac{1}{2} x_1 - \ldots - \frac{1}{2^p} x_p) \in \frac{1}{2} U_{p+1}$$

Then

(14) $$z_{p+1} + 2 y_p \in z_{p+1} + U_{p+1} \subset \bar{M}_{n_1, n_2, \ldots, n_{p+1}}$$

and therefore there is a point x_{p+1} in $M_{n_1, n_2, \ldots, n_{p+1}}$ such that

$$z_{p+1} + 2y_p - x_{p+1}$$

$$= z_{p+1} + 2z_p + \ldots + 2^p z_1 + 2^{p+1}(x - u - \frac{1}{2} x_1 - \ldots - \frac{1}{2^{p+1}} x_{p+1}) \in \frac{1}{2} U_{p+2}$$

Thus we obtain a sequence (x_p) of elements of F such that x_p belongs to $M_{n_1, n_2, \ldots, n_p}$, for p = 1, 2,..., and the sequence

(15) $$\left(\sum_{j=1}^{p} \frac{1}{2^j} z_j + x - u - \sum_{j=1}^{p} \frac{1}{2^j} x_j\right)$$

converges to the origin in F.

For every positive integers j there is a finite subset

$$A_j = \{x_{1, j}, x_{2, j}, \ldots, x_{n(j), j}\}.$$

in $A_{n_1, n_2, \ldots, n_j}$ such that x_j belongs to the convex hull of A_j. Take a point $z_{p, j}$ in $B_{n_1, n_2, \ldots, n_j}$ such that

$$f(x_{p, j}) \in Tz_{p, j}, \ p = 1,2,\ldots, n(j).$$

The sequence

$$z_{1, 1}, z_{2, 1}, \ldots, z_{n(1), 1}, \ldots, z_{1, j}, z_{2, j}, \ldots, z_{n(j), j}, \ldots$$

is a Cauchy sequence in (X, d) and therefore converges to a point v in X. Then there is an element A in $\mathcal{S}(E)$ such that

$$f(x_{p, j}) \in A, \ p = 1,2,\ldots, n(j) \text{ and } j = 1,2,\ldots,$$

and, since A is convex,

$$f(x_j) \in A, \ j = 1,2,\ldots,$$

from where it follows that the series $\Sigma \frac{1}{2^j} f(x_j)$ converges to an element y in E. Since U is closed and $f(x_j)$ belongs to U, j = 1,2,..., it follows that

$$\sum_{j=1}^{p} \frac{1}{2^j} f(x_j) \subset \sum_{j=1}^{p} \frac{1}{2^j} U$$

$$= \left(\sum_{j=1}^{p} \frac{1}{2^j}\right) U \subset U, \ p = 1,2,\ldots,$$

and therefore y belongs to U.

Analogously the series $\Sigma \frac{1}{2^j} f(x_j)$ converges to a point z in E which belongs to U.

The image by f of the sequence (15) is

$$\left(\sum_{j=1}^{p} \frac{1}{2^j} f(z_j) + f(x) - f(u) - \sum_{j=1}^{p} \frac{1}{2^j} f(x_j)\right)$$

which converges in E to z + f(x) - f(u) - y and, since the graph of f is se quentially closed, it follows that $z + f(x) - f(u) - y = 0$. Therefore

$$f(x) = f(u) + y - z \in U + U + U = 3U$$

and thus $\overline{V}$ is contained in $3V$. Therefore V is a neighbourhood of the origin in F. The proof is complete.

(16) *Let* E *and* F *be locally convex spaces. Let* G *be a subspace of* E. *Let* g *be a linear mapping with sequentially closed graph in* $E \times F$ *from* G *onto* F. *If* E *is semi-Suslin and* F *is a metrizable convex-Baire, then* g *is open.*

Proof. X, T and $B_{m_1, m_2,\ldots,m_p}$ have the same meaning as in the proof of (13). Let U be a closed absolutely convex neighbourhood of the origin of G. We set $V = g(U)$. Since F coincides with

$$\bigcup \{mV : m = 1,2,\ldots\}$$

we apply §1,6.(3) to obtain that V is a convex-Baire subset of F. For the positive integers $m_1, m_2,\ldots,m_p$ we write

$$A_{m_1, m_2,\ldots,m_p} = V \cap g(T(B_{m_1, m_2,\ldots,m_p}) \cap U)$$

and $M_{m_1, m_2,\ldots,m_p}$ for the convex hull of $A_{m_1, m_2,\ldots,m_p}$. We have that

$$V = \bigcup \{A_m : m = 1,2,\ldots\}$$

$$A_{m_1, m_2,\ldots,m_p} = \bigcup \{A_{m_1, m_2,\ldots,m_p, m} : m = 1,2,\ldots\}.$$

We apply §1, 6.(5) to obtain a sequence of positive integers (n_p) such that $M_{n_1, n_2,\ldots,n_p}$ is convex-Baire, $p = 1,2,\ldots$

Given a point x in $\overline{V}$ we follow the same path of (13) to obtain

$$u \in V,\ x_j,\ z_j \in M_{n_1, n_2,\ldots,n_p},\quad \text{for } j = 1,2,\ldots,$$

such that the sequence

$$(17)\qquad \left(\sum_{j=1}^{p} \frac{1}{2^j} z_j + x - u - \sum_{j=1}^{p} \frac{1}{2^j} x_j\right)$$

converges to the origin in F.

For every positive integer j, let A_j be the set defined in the proof of (13). For $p = 1,2,\ldots, n(j)$, take a point $v_{p,j}$ in $T(B_{n_1, n_2,\ldots, n_j} \cap U)$ such that $x_{p,j} = g(v_{p,j})$. Let $z_{p,j}$ be an element of $B_{n_1, n_2,\ldots,n_j}$ such that $v_{p,j}$ belongs to $Tz_{p,j}$. The sequence

$$z_{1,1}, z_{2,1},\ldots, z_{n(1),1},\ldots, z_{1,j}, z_{2,j},\ldots, z_{n(j),j},\ldots$$

is obviously a Cauchy sequence in (X, d) and therefore converges to a point y in X. There is an element A in $\mathcal{S}(E)$ such that

$$v_{p,j} \in A,\ p = 1,2,\ldots, p(j) \text{ and } j = 1,2,\ldots,$$

and, since A is convex, there is a point v_j in $A \cap U$ with $g(v_j)=x_j$, $j = 1, 2,\ldots$

The series $\Sigma \frac{1}{2^j} v_j$ converges to an element v of E. Since v_j belongs to U, $j = 1,2,\ldots$, it follows that

$$\sum_{j=1}^{p} \frac{1}{2^j} v_j \in \sum_{j=1}^{p} \frac{1}{2^j} U = \left(\sum_{j=1}^{p} \frac{1}{2^j}\right) U \subset U,\ p = 1,2,\ldots,$$

and therefore v belongs to the closure $\bar{U}$ of U in E. Analogously there is a point w_j in U such that $g(w_j) = z_j$ and the series $\Sigma \frac{1}{2^j} w_j$ converges in E to a point w which is in $\bar{U}$. Let t_1 and t_2 be points of U with $g(t_1) = x$ and $g(t_2) = u$. The sequence

$$\left(\sum_{j=1}^{p} \frac{1}{2^j} w_j + t_1 - t_2 - \sum_{j=1}^{p} \frac{1}{2^j} v_j\right)$$

converges in E to $w + t_1 - t_2 - v$ and its image by g coincides with (17) wich is convergent to the origin in F. Since the graph of g is sequentially closed in $E \times F$, it follows that $w + t_1 - t_2 - v$ belongs to G and $g(w + t_1 - t_2 - v) = 0$. Since t_1-t_2 belongs to G, it follows that $w-v$ belongs to G. Then

$$w - v \in 2\bar{U} \cap G = 2U.$$

Therefore

$$x = g(t_1) = g(t_1) - g(w + t_1 - t_2 - v)$$

$$= g(v - w) + g(t_2) = g(v - w) + u \in 2V + V = 3V$$

and thus $\tilde{V}$ is contained in 3V. Therefore V is a neighbourhood of the origin in F. The proof is complete.

(18) *Let* E *and* F *be locally convex spaces. Let* f *be a linear mapping with sequentially closed graph from* E *onto* F. *If* E *is semi-Suslin and* F *is the locally convex hull of metrizable convex-Baire spaces, then* f *is open.*

Proof. Let $\{F_i : i \in I\}$ a family of metrizable convex Baire spaces. For every i of I, let A_i be a linear mapping from F_i in F such that the topology of F is the finest locally convex topology for which all the mapping A_i, $i \in I$, are continuous. We set H_i to denote $A_i(F_i)$ endowed with the topology induced by F. Let T_i be the finest locally convex topology on $A_i(F_i)$ such that A_i is continuous from F_i onto $A_i(F_i)[T_i]$. Then $A_i(F_i)[T_i]$ is isomorphic to $F_i/A_i^{-1}(0)$ and therefore is metrizable and convex-Baire.

Fix i in I and set $G_i = f^{-1}(H_i)$. Let f_i be the restriction of f to G_i. Then f_i has sequentially closed graph in $E \times A_1(F_i)[T_i]$ (see the proof of 4.(17)). We apply (16) to obtain that f_i is open. The conclusion follows as in 4.(17).

(19) *Let* E *and* F *be locally convex spaces. Let* f *be a linear mapping with closed graph from* F *into* E. *If* F *is the locally convex hull of convex-Baire spaces and if* E *is a semi-Suslin space, then* f *is continuous.*

Proof. It is enough to carry the proof for F being a convex-Baire space. We use the same notations of the proof of (13), but here $\{V_p : p = 1,2,\ldots\}$ is a system of neighbourhoods of the origin in F which is not necessarily fundamental.

Let $\{W_i : i \in I\}$ be a fundamental system of neighbourhoods of the origin in F. By (14) we know the existence of an element $w_{p,i}$ in $M_{n_1, n_2,\ldots,n_p}$ with

$$2^p u_{p,i} = z_{p,i} + 2y_p - w_{p,i} \in W_i,\ i \in I,\ p = 1,2,\ldots$$

If m and n are positive integers and if i and h belong to I we set $(m, i) \geq (n, h)$ when $m \geq n$ and $W_i \subset W_h$. Then

(20) $\{u_{p,\,i} : (p, i) \in N \times I, \geq\}$

is a net in F which converges to the origin. Suppose that the net

(21) $\{\frac{1}{2^p} f(w_{p,\,i}) : (p, i) \in N \times I, \geq\}$

does not converge to the origin in E. We find a neighbourhood of the origin W in E, a sequence of positive integers $m_1 < m_2 < \dots < m_p < \dots$ and a sequence (i_p) of elements of I such that, for every positive integer p,

(22) $\frac{1}{2^p} f(w_{mp,ip}) \notin W.$

Since w_{mp,i_p} belongs to $M_{n_1, n_2, \dots, n_{m_p}}$, we proceed as we did in the proof of (13) with the sequence (x_j) to obtain that the series $\Sigma \frac{1}{2^p} f(w_{m_p, i_p})$ converges to a point in E and therefore the sequence $(\frac{1}{2^p} f(w_{m_p, i_p}))$ converges to the origin in E. That is in contradiction with (22) and therefore the net (21) converges to the origin in E.

The image of the net (20) by f is the net

$$\{\sum_{j=1}^{p} \frac{1}{2^j} f(z_j) + f(x) - f(u) - \sum_{j=1}^{p} \frac{1}{2^j} f(x_j) - \frac{1}{2^p} f(w_{p,\,i})$$

$$: (p, i) \in N \times I, \geq\}$$

which converges in E to $z + f(x) - f(u) - y$. Since the graph of f is closed, it follows that $z + f(x) - f(u) - y = 0$. The conclusion follows that in (13).

(23) *Let* E *and* F *be locally convex spaces. Let* f *be a linear mapping with closed graph from* E *onto* F. *If* E *is semi-Suslin and* F *is the locally convex hull of convex-Baire spaces, then* f *is open.*

Proof. According to 1.(19), $f^{-1}(0)$ is a closed subspace of E. By (2) $E/f^{-1}(0)$ is a semi-Suslin space. If h is the canonical surjection from E onto $E/f^{-1}(0)$, let g be the linear mapping from $E/f^{-1}(0)$ onto F such that $f = g \circ h$. According to 1.(14), g has closed graph and therefore g^{-1} has closed graph. We apply (19) to obtain that g^{-1} is continuous. Consequently g is open.

(24) *Let E be a semi-Suslin locally convex space. If E is convex-Baire, then E is complete.*

Proof. Suppose E non-complete. Take a vector x in $\hat{E} \sim E$. Let F the subspace of $\hat{E}$ generated by $E \cup \{x\}$. If z is any point of F, then $z = ax+y$, $a \in K$, $y \in E$. Set $Tz = y$. T is a linear mapping from F onto E. Since E is dense in F there is a net

(25) $\{y_i : i \in I, \geq\}$

in E which converges to x in F. The net

$$\{Ty_i : i \in I, \geq\}$$

coincides with (25) and therefore does not converge to $Tx = 0$ in E. Thus T is not continuous.

Let $\{z_i : i \in I, \geq\}$ be a net in F converging to the origin such that $\{Tz_i : i \in I \geq\}$ converges to the element y in E. We set $a_i x = z_i - Tz_i$. Then

$$\lim \{a_i x : i \in I \geq\} = \lim \{z_i - Tz_i : i \in I, \geq\} = - y$$

and consequently $y = 0$. Therefore the graph of T is closed. Since F is convex-Baire, we apply (19) to obtain that T is continuous which is a contradiction.

(26) *Let E be a semi-Suslin locally convex space. If E is convex-Baire, then E is a Fréchet space.*

Proof. X, T and $B_{m_1, m_2,\dots,m_p}$ have the same meaning as in the proof of (13). We set $M_{m_1, m_2,\dots,m_p}$ to denote the convex hull of $T(B_{m_1, m_2,\dots, m_p})$. We apply §1,6.(3) to obtain a sequence of positive integers (n_p) such that $M_{n_1, n_2,\dots,n_p}$ is a convex-Baire subset of E, $p = 1,2,\dots$

For every positive integer p, take an element z_p in $M_{n_1, n_2,\dots,n_p}$ and an open absolutely convex neighbourhood of the origin U_p in E such that

$$z_p + U_p \subset \bar{M}_{n_1, n_2, \ldots, n_p}$$

Suppose that

(27) $\{\frac{1}{2^p} U_p : p = 1,2,\ldots\}$

is not a fundamental system of neighbourhoods of the origin in E. We find a closed absolutely convex neighbourhood of the origin V in E with

$$\frac{1}{2} U_p \not\subset V, \; p = 1,2,\ldots$$

Then $U_p \sim 2^p V$ is a non-void open subset of E. Since

$$(M_{n_1, n_2, \ldots, n_p} - z_p) \cap U_p$$

is dense in U_p there is a point x_p in $M_{n_1, n_2, \ldots, n_p}$ such that $x_p - z_p$ belongs to $U_p \sim 2^p V$. Then

(28) $\frac{1}{2^p} (x_p - z_p) \notin V, \; p = 1,2,\ldots$

For every positive integer j there is a finite subset

$$A_j = \{x_{1,j}, x_{2,j}, \ldots, x_{n(j), j}\}$$

in $T(B_{n_1, n_2, \ldots, n_j})$ such that x_j belongs to the convex hull of A_j. Take a point $z_{p, j}$ in $B_{n_1, n_2, \ldots, n_j}$ such that

$$x_{p, j} \in Tz_{p, j}, \; p = 1,2,\ldots, n(j)$$

The sequence

$$z_{1,1}, z_{2,1}, \ldots, z_{n(1),1}, \ldots, z_{1,j}, z_{2,j}, \ldots, z_{n(j), j} \ldots$$

is a Cauchy sequence in (X, d) and therefore converges to a point in X. Then there is an element A in $S(E)$ such that

$$x_{p, j} \in A, \; p = 1,2,\ldots, n(j) \text{ and } j = 1,2,\ldots$$

and, since A is convex,

$$x_j \in A, \; j = 1,2,\ldots$$

Then $\Sigma \frac{1}{2^j} x_j$ converges to a point x in E. Analogously the series $\Sigma \frac{1}{2^j} z_j$ converges to a point z in E. Consequently the sequence $(\frac{1}{2^j}(x_j - z_j))$ converges to the origin in E which is in contradiction with (28). Therefore (27) is a fundamental system of neighbourhoods of the origin in E. The conclusion follows having in mind (24).

§ 5. ORDERED CONVEX-BAIRE SPACES AND SUPRABARRELLED SPACES

1. ORDERED CONVEX-BAIRE SPACES. Let E be a locally convex space. E is ordered convex-Baire if and only if, given any increasing sequence of rare closed convex subsets of E, their union has void interior. It is obvious that every convex-Baire space is ordered convex-Baire and every ordered convex-Baire space is barrelled. According to §3, 1(4) every barrelled space whose completion is Baire is ordered convex-Baire.

(1) *Let* E *be a locally convex space.* E *is ordered convex-Baire if and only if given any increasing sequence* (A_n) *of closed convex subsets of* E *covering* E, *there is a positive integer* p *such that* A_p *has non-void interior.*

Proof. Suppose E ordered convex-Baire and (A_n) any increasing sequence of closed convex subsets of E covering E. Since E has non-void interior there is a positive integer p such that A_p has non-void interior. Reciprocally, let (B_n) be an increasing sequence of closed convex subsets of E whose union has an interior point z. Let U be a neighbourhood of the origin in E such that z + U is contained in $\bigcup\{B_n : n = 1,2,\ldots\}$. We set

$$A_n = n\,(B_n - z) + z, \; n = 1,2,..$$

Then (A_n) is an increasing sequence of closed convex x subsets of E. If x is any point of E there is a positive integer r such that

$$\frac{1}{r}(x - z) \in \bigcup\{B_n : n = 1,2,\ldots\} - z = \bigcup\{B_n - z : n = 1,2,\ldots\}$$

since $\cup\{B_n : n = 1,2,\ldots\} - z$ is a neighbourhood of the origin in E. Therefore there is an integer q larger than r with

$$\frac{1}{r}(x - z) \in B_q - z$$

Then

$$x - z \in r(B_q - z) \subset q(B_q - z)$$

and consequently x belongs to A_q. Thus $\cup\{A_n : n = 1,2,\ldots\}$ coincides with E and therefore there is a positive integer p such that A_p has non-void interior. Then B_p has non-void interior. The proof is complete.

(2) *Let* E *be a locally convex space. Let* F *be a dense subspace of* E. *If* F *is ordered convex-Baire, then* E *is ordered convex-Baire.*

Proof. Let (A_n) be an increasing sequence of closed convex subsets of E covering E. Then $(A_n \cap F)$ is an increasing sequence of closed convex subsets of F covering F. Since F is ordered convex-Baire there is a positive integer p such that $A_p \cap F$ has non-void interior in F. Then A_p has non-void interior in E.

(3) *Let* E *be an ordered convex-Baire space. Let* F *be a closed subspace of* E. *Then* E/F *is ordered convex-Baire.*

Proof. Let (A_n) be an increasing sequence of closed convex subsets of E/F covering E/F. Let T be the canonical mapping from E onto E/F. $(T^{-1}(A_n))$ is an increasing sequence of closed convex subsets of E covering E. Since E is ordered convex-Baire there is a positive integer p such that $T^{-1}(A_p)$ has an interior point in E. Consequently A_p has non-void interior in F and the conclusion follows.

(4) *Let* E *be an ordered convex-Baire space. If* F *is a countable codimensional subspace of* E, *then* F *is ordered convex-Baire.*

Proof. Let G be the closure of F in E. Reasoning as we did in the proof of §3,2.(8), G is isomorphic to a separeted quotient of E and therefore G is ordered convex-Baire. Consequently it is enough to carry the proof supposing F dense in E. Let (A_n) be an increasing sequence of closed convex subsets of F covering F. According to §3,2.(9) F is barrelled and therefore every bounded set of $F'[\sigma(F', F)]$ is equicontinuous in F. We apply §3,1.(3) to obtain that $\hat{F}$ coincides with $\cup\{n\hat{A}_n : n = 1,2,\ldots,\}$. Since $\hat{F}$ coincides with $\hat{E}$, we apply (2) to obtain that $\hat{F}$ is ordered Baire-convex. Thus there is a positive integer p such that $p\hat{A}_p$ has non-void interior in $\hat{F}$. Since $p\hat{A}_p \cap F$ coincides with pA_p it follows that A_p has non-void interior in F and the conclusion follows.

(5) *If* $\{E_i : i \in I\}$ *is a family of ordered convex-Baire spaces, then* $E = \Pi\{E_i : i \in I\}$ *is ordered convex-Baire.*

Proof. For every subset H of I, E(H) has the same meaning as in §2, Section 1. Let (A_n) be an increasing sequence of closed convex subsets of E covering E. Without loss of generality we suppose that the origin of E is in A_n, n = 1,2,... We apply §2,.1(15) to obtain a finite subset J of I and a positive integer p such that A_p contains $E(I \sim J)$. We set

$$B_n = A_{p+n} \cap (-A_{p+n}),\ n = 1,2,\ldots$$

If x is any point of E there are positive integers q and r with

$$x \in A_{p+q},\ -x \in A_{p+r}$$

Then

$$x, -x \in A_{p+q+r}$$

and therefore x belongs to B_{q+r}. Consequently $\{B_n : n = 1,2,\ldots\}$ covers E. Obviously B_n contains $E(I \sim J)$, n = 1,2,... If J is the void set, clearly B_n coincides with E, n = 1,2,..., and therefore A_{p+n} has non-void interior, n = 1,2,... If J is distinct from the void set, let m be its cardinal number. Since J is finite there is a positive integer s such that $B_s \cap E(\{i\})$ has non-void interior in $E(\{i\})$ for every i in J. Since B_s is symmetric, $B_s \cap E(\{i\})$ is a neighbourhood of the origin in E for every i in J. We find an absolutely convex neighbourhood of the origin U_i in $E(\{i\})$ contained in B_s, $i \in J$. Then

$$\Pi\{\frac{1}{m} U_i : i \in J\} \times \Pi\{E_i : i \in I \sim J\}$$

is a neighbourhood of the origin in E contained in B_s. Then A_{p+s} has non-void interior in E. The proof is complete.

A locally convex space is said to be a Baire-like space if given any increasing sequence (A_n) of closed absolutely convex subsets of E covering E there is a positive integer p such that A_p is a neighbourhood of the origin in E.

(6) *Let* E *be a locally convex space.* E *is Baire-like if and only if* E *is ordered convex-Baire.*

Proof. It E is ordered convex-Baire, then E is Baire-like. Suppose now that E is Baire-like. Let (A_n) be an increasing sequence of closed convex subsets of E covering E. Without loss of generality we can suppose that the origin of E belongs to A_n, n = 1,2,... Let B_n be the closed convex hull of the union of all absolutely convex subsets of A_n. Then (B_n) is an increasing sequence of closed absolutely convex subsets of E. We shall see that this sequence covers E. If E is real and if x is any point of E we find a positive number p such that x and -x belong to A_p. Therefore the absolutely convex set $\{hx: -1 \leq h \leq 1\}$ is contained in A_p and thus x belongs to B_p. If E is complex and if z is any point of E, we find a positive integer q such that 2z, -2z, 2iz, -2iz are in A_q. Let a + bi a complex number, a and b being real numbers,with $|a + bi| \leq 1$.Then $|a| \leq 1$ and $|b| \leq 1$ and therefore 2az and 2biz to A_q; thus

$$(a + bi) z \in \frac{1}{2} A_q + \frac{1}{2} A_q = A_q,$$

from where it follows that the absolutely convex set $\{hz : h$ complex $|h| \leq 1\}$ is contained in Aq and therefore z belongs to B_q. In any case there is a positive integer r such that B_r is a neighbourhood of the origin in E and thus A_r has non-void interior in E. The proof is complete.

Results (2), (3), (4) and (5) can be found in SAXON [1] where different proofs are presented.

2. SUPRABARRELLED SPACES. A locally convex space E is suprabarrelled if and only if given any increasing sequence (E_n) of subspaces of E covering E there exists a positive integer p such that E_p is dense in E and barrelled.

(1) *If E is suprabarrelled, then E is ordered convex-Baire.*

Proof. Let (A_n) be an increasing sequence of closed absolutely convex subsets of E covering E. Let G_n be the linear hull of A_n endowed with the topology induced by the topology of E, n = 1,2,... Since E is suprabarrelled there is a positive integer p such thtat G_p is dense in E and barrelled. Since A_p is a barrel in G_p, clearly A_p is a neighbourhood of the origin in G_p. But G_p is dense in E and A_p is closed in E, therefore G_p coincides with E. We apply (6) to obtain the conclusion.

(2) *Let E be a suprabarrelled space. If F is a closed subspace of E, then E/F is suprabarrelled.*

Proof. Let G_n be an increasing sequence of subspaces of E/F covering E/F. Let T be the canonical mapping from E onto E/F. Then $(T^{-1}(G_n))$ is an increasing sequence of subspaces of E covering E. Consequently there is a positive integer p such that $T^{-1}(G_p)$ is dense in E and barrelled. Then G_p is dense in E/F. Since F is contained in $T^{-1}(G_p)$, G_p is isomorphic to $T^{-1}(G_p)/F$ and therefore G_p is barrelled. The proof is complete.

(3) *Let E be a locally convex space. Let F be a dense subspace of E. If F is suprabarrelled, then E is suprabarrelled.*

Proof. Let (E_n) be an increasing sequence of subspaces of E covering E. Then $(E_n \cap F)$ is an increasing sequence of subspaces of F covering F. Since F is suprabarrelled there is a positive integer p such that $E_p \cap F$ is dense in F and barrelled. Then $E_p \cap F$ is dense in E_p and therefore E_p is dense in E and barrelled.

(4) *Let E be a suprabarrelled space. Let F be a countable codimensional subspace of E. Then F is suprabarrelled.*

Proof. Let (F_n) be an increasing sequence of subspaces of F covering F. By (1), E is ordered convex-Baire and, by 1.(4), F is ordered convex-Baire. Let G_n be the closure of F_n in F, n = 1,2,... Then (G_n) is an increasing sequence of closed absolutely convex subsets of F covering F. Therefore there is a positive integer p such that G_p is a neighbourhood of the origin in F. Consequently G_p coincides with F. Therefore F_p is dense in F. Let G be an algebraic complement of F in E. Then $(F_{p+n} + G)$ is an increasing sequence of subspaces of E covering E and therefore there is a positive integer q such that $F_{p+q} + G$ is barrelled. Since F_{p+q} is a countable codimensional subspace of $F_{p+q} + G$ then F_{p+q} is barrelled. On the other hand, F_{p+q} contains F_p and therefore is dense in F. The proof is complete.

Let E be the topological product $\Pi\{E_i : i \in I\}$, E_i being a suprabarrelled space, $i \in I$. For every i in I, we denote also by E_i the subspace of E of all those elements $\{x_j : j \in I\}$ with $x_j = 0$ if j is distinct from i. Let E_0 be the subspace of E of all those elements with vanishing coordinates save a countable number of them. We take an increasing sequence (F_n) of subspaces of E_0 covering E_0. Let U_n be a barrel in F_n. We write V_n to denote the closure of U_n in E_0. We set G_n for the linear hull of V_n.

(5) *There is a positive integer* p *such that* G_p *contains* E_i *for all* i *in* I.

Proof. We suppose the property not true. Given a positive integer n_1 we take an index i_1 in I such that G_{n_1} does not contain E_{i_1}. Suppose we have selected the positive integers $n_1, n_2, \ldots, n_q$ and the indices $i_1, i_2, \ldots, i_q$ in I. Since E_{i_q} is suprabarrelled and since

$$\cup \{F_n \cap E_{i_q} : n = 1,2,\ldots\} = E_{i_q}$$

there is an integer $n_{q+1} > n_q$ such that $F_{n_{q+1}} \cap E_{i_q}$ is dense in E_{i_q} and barrelled and therefore $V_{n_{q+1}} \cap E_{i_q}$ is a neighbourhood of the origin in E_{i_q}, from where it follows that $G_{n_{q+1}}$ contains E_{i_q}. We select an index i_{q+1} in I such that $G_{n_{q+1}}$ does contain $E_{i_{q+1}}$.

For every positive integer r we take in E_{i_r} an one-dimensional subspace L_r not contained in G_{i_r}. Since the elements of the sequence (i_p) are pairwise distinct, the subspace L of E_o of all those elements $\{x_i : i \in I\}$ with

$$x_{i_r} \in L_r,\ x_i = 0,\ i \neq i_r,\ r = 1,2,\ldots$$

is isomorphic to $\Pi\{L_r : r = 1,2,\ldots\}$ and therefore is Baire. On the other hand, the family

$$\{m\ V_q : m,\ q = 1,2,\ldots\}$$

of closed absolutely convex subsets of E_o covers L and therefore there is a positive integer s such that $V_{n_s} \cap L$ is a neighbourhood of the origin in L; thus G_{n_s} contains L and that is a contradiction.

(6) *If* I *is countable, then* E *is suprabarrelled.*

Proof. First we suppose that I coincides with N. Then $E_o = E$. Now suppose that E is not suprabarrelled. Then we can take the sequences (F_n) and (U_n) such that U_n is not a neighbourhood of the origin in F_n, $n = 1, 2,\ldots$ Consequently V_n is not a neighbourhood of the origin in E_n, $n = 1,2, \ldots$ According to (5), there is a positive integer p such that G_p contains E_n, $n = 1,2,\ldots$ Since the product of barrelled spaces is barrelled (cf. KOTHE [1], Chapter Six, §27, Section 1) we have that E is barrelled and therefore, given an integer $n_1 > p$, there is a point x_1 in E such that x_1 is not in G_{n_1}. Proceeding by recurrence suppose that we have obtained the integers $n_1 < n_2 < \ldots < n_q$ and the points $x_1, x_2, \ldots, x_q$ of E. We set

$$H_q = \{\{x_n : n = 1,2,\ldots\} \in E : x_n = 0,\ n > q\},$$

$$k_q = \{\{x_n : n = 1,2,\ldots\} \in E : x_n = 0,\ n < q\}$$

and we take $n_{q+1} > n_q$ such that $G_{n_{q+1}}$ contains H_q. Therefore there is a point x_{q+1} in K_{q+1} which is not in $G_{n_{q+1}}$.

The set $A = \{x_1, x_2, \ldots, x_n, \ldots\}$ has finite projections in E_q, $q=1,2,\ldots$, and therefore its closed absolutely convex hull B is compact and thus E_B is a Banach space. The family

$$\{m V_q : m = 1,2,\ldots; q = p+1, p+2,\ldots\}$$

covers E and also covers E_B. Then there is an integer $r > p$ such that $V_{n_r} \cap E_B$ is a neighbourhood of the origin in E_B from where it follows that G_{n_r} contains A and that is a contradiction.

If I is finite, we set $E_n = \{0\}$, $n = 1,2,\ldots$, which are clearly suprabarrelled spaces and therefore $F = E \times \Pi \{E_n : n = 1,2,\ldots\}$ is suprabarrelled. Since F is isomorphic to E, E is suprabarrelled.

(7) *The topological product of suprabarrelled spaces is suprabarrelled.*

Proof. Let E be the topological product $\Pi\{E_i : i \in I\}$ defined before. According to (3) it is enough to show that E_o is suprabarrelled. Suppose that E_o is not suprabarrelled. Take the sequence (F_n) and (U_n) such that U_n is not a neighbourhood of the origin in F_n, $n = 1,2,\ldots$ Then V_n is not a neighbourhood of the origin in E_o, $n = 1,2,\ldots$

Let U be a barrel in E_o and suppose that U does not absorb the bounded subsets of E_o. Then a certain bounded sequence (z_p) in E_o is not absorbed by U. Obviously there is a countable set J of I such that the coordinates of z_p corresponding to indices in $I \sim J$ are zero, $p = 1,2,\ldots$ Let E(J) be the subspace of E

$$\{\{x_i : i \in I\} : x_i = 0, i \in I \sim J\}.$$

Then E(J) is isomorphic to $\Pi\{E_i : i \in J\}$ and therefore barrelled; thus $U \cap E(J)$ is a neighbourhood of the origin in E(J). Since the sequence (z_p) is in E(J), U absorbs this sequence and that is a contradiction. Therefore U absorbs the bounded subsets of E_o. Let W be the closure of U in E. If $\{y_i : i \in I\}$ is a point of E, let P be the subsets of all those elements $\{u_i : i \in I\}$ of E_o such that u_i is zero save a finite number of indices and such that u_i coincides with y_i when u_i is distinct of zero. Then P is a bounded subset of E_0 and $\{y_i : i \in I\}$ belongs to the closure of P in E. Since U absorbs P, it follows that W absorbs $\{y_i : i \in I\}$. Thus W is a barrel in E. Since E is barrelled, $U = W \cap E_0$ is a neighbourhood of the origin in E_0 and therefore E_0 is barrelled.

The barrelledness of E_0 implies that G_n is distinct from E_0, n = 1, 2,... We take x_n in $E_0 \sim G_n$, n = 1,2,... There is a countable subset H of I such that the coordinates of z_n corresponding to indices not belonging to H are zero, n = 1,2,... Then $\{x_1, x_2,...,x_n,...\}$ is contained in the subspace E(H) of E defined by

$$\{\{y_i : i \in I\} : y_i = 0, \ i \in I \sim H\}.$$

Since E(H) is isomorphic to $\Pi\{E_i : i \in H\}$, we apply (6) to obtain that E(H) is suprabarrelled. Then there is an integer q such that $F_q \cap E(H)$ is dense in E(H) and barrelled, from where it follows that G_q contains E(H) and therefore G_q contains $\{x_1, x_2,...,x_n,...\}$. This is a contradiction and the proof is complete.

The results of this section can be seen in VALDIVIA [10]. The following results can also be found in the aforementioned article: *If* E *is an infinite dimensional separable Fréchet space there is a dense subspace* F *of* E *satisfying the following conditions*: a) F *is suprabarrelled;* b) F *is not an inductive limit of unordered Baire-like spaces.*

An example of a suprabarrelled space which is not convex-Baire can be seen in Chapter Two, §5, Section 2. In Chapter Two, §5, Section 1, examples of ordered convex-Baire spaces which are not suprabarrelled are given.

§ 6. OTHER RESULTS ON THE CLOSED GRAPH THEOREM

1. GENERAL RESULTS ON THE CLOSED GRAPH THEOREM. If $\mathcal{E}$ is a class of locally convex spaces we denote by $\mathcal{E}_r$ the class of all locally convex spaces such that if E belongs to $\mathcal{E}$, F belongs to $\mathcal{E}_r$ and if f is a linear mapping with closed graph from E into F, then f is continuous.

(1) *Let $\mathcal{E}$ be a class of locally convex spaces. Let* F *be an element of* $\mathcal{E}_r$. *If* T *is a Hausdorff topology on* F *coarser than the original topology such that* F [T] *is a locally convex space, then* F [T] *belongs to* $\mathcal{E}_r$.

Proof. Let E be any element of $\mathcal{E}$ and let f be a linear mapping with closed graph from E into F [T]. Then the graph of f is closed in E x F and therefore f : E ⟶ F is continuous. Consequently, f : E ⟶ F [T] is continuous and the conclusion follows.

(2) *Let* $\mathcal{E}$ *be a class of locally convex spaces. Let* E *be an element of* $\mathcal{E}_r$. *If* F *is a closed subspace of* E, *then* F *belongs to* $\mathcal{E}_r$.

Proof. Let G be any element of $\mathcal{E}$ and let f be a linear mapping with closed graph from G into F. Since F is closed in E, f : G ⟶ E has closed graph and therefore it is continuous. The conclusion follows.

(3) *Let* F *and* G *be two subspaces of a locally convex space* E *such that* $F + G = E$ *and* $F \cap G = \{0\}$. *If* F *is closed and* T *is the projection from* E *onto* G *along* F, *then* T *has closed graph.*

Proof. Let $\{ x_i : i \in I, \geq \}$, be a net in E converging to the origin such that the net $\{ Tx_i : i \in I, \geq \}$ converges to z in G. Then the net

$$\{ x_i - T x_i : i \in I, \geq \}$$

is in F and converges to -z in E. Since F is closed it follows that -z belongs to F. Then z belongs to $F \cap G$ and consequently $z = 0$. We apply now § 4. 1. (6) to obtain that the graph of T is closed.

(4) *Let* E *and* F *be a locally convex spaces. Let* G *be a subspace of* E *of finite codimension. If* f : G ⟶ F *is a linear mapping with closed graph, then there is a linear mapping* g : E ⟶ F *with closed graph and whose*

restriction to G *coincides with* f.

Proof. Obviously it is enough to carry the proof in the case of G being an hyperplane of E. Then we suppose that G has codimension one in E. We take a vector x in E $\sim$ G. If z belongs to E, we write z = ax + y, a $\in$ K, y $\in$ G, and we set Tz = y. If G is closed in E, then T is a continuous mapping from E onto G and the mapping g = f o T has closed graph, according to § 4, 1.(9). On the other hand, it is obvious that the restriction of g to G coincides with f.

Now we suppose that G is dense in E. According to § 4, 1.(21), there exists a Hausdorff topology T on F coarser than the original topology such that F[T] is a locally convex space and f : G $\longrightarrow$ F [T] is continuous. Let H be the completion of F [T]. The mapping f can be extended to a continuous linear mapping h from E into H. If h(E) is contained in F, we set h = g. Then g has closed graph in E x F and the restriction of g to G coincides with f. If h(E) is distinct from F and u is any element of h(E), we write u = ah(x) + v, a $\in$ K, v $\in$ F, and we set Su = v. S is a linear mapping from the subspace h(E) of H onto F. If L is the linear hull of h(x), it follows that L and F[T] are subspaces of h(E) such that L $\cap$ F = {0}, L + F = h(E) and L is closed in h(E). Since S is the projection from h(E) onto F [T] along L we apply (3) to obtain that the graph of S is closed. We set g = S o h. According to §4,1.(9) g has closed graph in E x F[T] and therefore the graph of g is closed in E x F. Obviously, the restriction of g to G coincides with f. This completes the proof.

(5) *Let* E *be a class of locally convex spaces. If* E *belongs to* E_r *and* F *is a product of one-dimensional locally convex-spaces, then* E x F *belongs to* E_r.

Proof. We set G = E x F and we suppose that G does not belong to E_r. Then there exists an element H in E and a linear mapping f : H $\longrightarrow$ G with closed graph which is not continuous. According to §4,1.(21) there is a Hausdorff topology T on G coarser than the original topology such that G [T] is a locally convex space and f: H $\longrightarrow$ G [T] is continuous. Since F is a product of one-dimensional locally convex spaces, then T coincides on F with the topology of F. We set U for the restriction of T to E. Since F is complete we apply (3) to obtain that the projection T from G[T] onto E[U] along F has closed graph. According to §4, 1.(9), T o f is a mapping from

H into $E[\mathcal{U}]$ with closed graph and therefore $T \circ f: H \longrightarrow E$ has closed graph then $T \circ f : H \longrightarrow E$ is continuous. Since f is not continuous there is a net $\{x_i : i \in I, \geqslant\}$ in H converging to the origin such that

$$(6) \qquad \{f(x_i) : i \in I, \geqslant\}$$

does not converge to the origin in G. Now suppose that the net

$$(7) \qquad \{T \circ f(x_i) : i \in I, \geqslant\}$$

converges to the origin in E. Then (7) converges to the origin in $E[\mathcal{U}]$ and, since (6) converges to the origin in $G[\mathcal{T}]$ it follows that

$$\{f(x_i) - T \circ f(x_i) : i \in I \geqslant\}$$

converges to the origin in F. Consequently (6) converges to the origin in G and that is a contradiction. Therefore $T \circ f$ is not continuous and the conclusion follows.

We say that a class of locally convex spaces $\mathcal{E}$ is normal if it satisfies the following condition: if E belongs to $\mathcal{E}$, every subspace of $\hat{E}$ containing E belongs to $\mathcal{E}$.

(8) *Let $\mathcal{E}$ be a normal class of locally convex spaces. If F belongs to $\mathcal{E} \cap \mathcal{E}_r$, then F is complete.*

Proof. Let f be the identity on F. Suppose F non-complete. Take a vector x in $\hat{F} \sim F$. Let G be the subspace of $\hat{F}$, linear hull of $F \cup \{x\}$. According to (4), f can be extended to a linear mapping $g: G \longrightarrow F$ with closed graph. Since G belongs to $\mathcal{E}$ it follows that g is continuous. If

$$(9) \qquad \{x_i : i \in I, \geqslant\}$$

is a net in F converging to x in G, then the net

$$\{g(x_i) : i \in I, \geqslant\} = \{x_i : i \in I, \geqslant\}$$

converges to g(x) in F. Then (9) has two different limits in G. Thus F is complete.

(10) *Let $\mathcal{E}$ be a normal class of locally convex spaces such that if E belongs to $\mathcal{E}$ every closed hyperplane of E belongs to $\mathcal{E}$. Let H be a non-complete element of $\mathcal{E}$. If F belongs to $\mathcal{E}_r$ and if f is a continuous linear mapping from H onto F, there is a Hausdorff topology $\mathcal{T}$ on H strictly coar-*

ser than the original topology such that $H[T]$ *belongs to* E *and* $f : H[T] \longrightarrow F$ *is continuous.*

Proof. Take a vector x in $\widehat{H} \sim H$. Let G be the subspace of $\widehat{H}$ generated by $H \cup \{x\}$. According to (4), f can be extended to a linear mapping g with closed graph from G onto F. Since G belongs to E, it follows that g is continuous. Since f is an onto mapping there is an element y in H with $f(y) = g(x)$. If L is the linear hull of $y - x$, then $g(L) = \{0\}$. Consequently there is a continuous linear mapping k from G/L onto F such that if h is the canonical mapping from G onto G/L, then g coincides with $k \circ h$. Since G/L is isomorphic to a closed hyperplane of G, it follows that G/L belongs to E. If h_1 is the restriction of h to H it follows that h_1 is a continuous linear injective mapping from H onto G/L and since H is dense in G, h_1^{-1} is not continuous. Let $\{U_i : i \in I\}$ be a fundamental system of neighbourhoods of the origin in G/L. Then $\{h_1^{-1}(U_i) : i \in I\}$ is a fundamental system of neighbourhoods of the origin in H for a locally convex topology T strictly coarser than the original topology. Since $H[T]$ is isomorphic to G/L, we have that $H[T]$ belongs to E. Obviously $k \circ h_1 : H[T] \longrightarrow F$ is continuous Since f coincides with $k \circ h_1$ the conclusion follows.

(11) *Let* E *be a normal class of locally convex spaces. Let* G *be a dense subspace of a locally convex space* F. *Let* H *be an element of* E *and let* $f : G \longrightarrow H$ *be a linear mapping with closed graph in* $F \times H$. *If* G *belongs to* E, *then* G *coincides with* F.

Proof. Suppose G distinct from F. Take a vector x in $F \sim G$. Let L be the subspace of F generated by $G \cup \{x\}$. Since the graph of f is closed in $G \times H$, we apply (4) to obtain a linear mapping g with closed graph from L into H such that g coincides with f in G. Since L belongs to E, g is continuous. Let $\{x_i : i \in I \geqslant\}$ a net in G converging to x in L. Then the net $\{g(x_i) : i \in I \geqslant\}$ converges to $g(x)$ in H. In $F \times H$, the point $(x, g(x))$ does not belong to the graph $G(f)$ of f. On the other hand, the net

$$\{(x_i, g(x_i)) : i \in I \geqslant\}$$

belongs to $G(f)$ and converges to $(x, g(x))$ in $F \times H$. Since $G(f)$ is closed in $F \times H$, $(x, g(x))$ belongs to $G(f)$. This is a contradiction and therefore G coincides with F.

(12) *Let* $\mathcal{E}$ *be a normal class of locally convex space. Let* F *and* G *be elements of* $\mathcal{E}$ *and* $\mathcal{E}_r$, *respectively. If* f *is a continuous linear mapping from* F *into* G, *then* f *can be extended to a continuous linear mapping* g *from* $\widehat{F}$ *into* G.

Proof. $f : F \longrightarrow \widehat{G}$ can be extended to a continuous linear mapping g from $\widehat{F}$ into $\widehat{G}$. Clearly it is enough to show that $g(\widehat{F})$ is contained in G. Let x be any point of $\widehat{F}$. Let L be the subspace of $\widehat{F}$ generated by $F \cup \{x\}$. According to (4), there is a linear mapping $h : L \longrightarrow G$ with closed graph which coincides with f in F. Let k be the restriction of f to L. The mappings h and k are continuous from L into $\widehat{G}$ and coincide on a dense subspace F of L. Therefore they coincide on L and thus $f(x) = k(x) = h(x) \in G$. The proof is complete.

(13) *Let* $\mathcal{E}$ *be a normal class of metrizable locally convex spaces. Let* F *be a locally convex hull of elements of* $\mathcal{E}$. *If* F *belongs to* $\mathcal{E}_r$, *then* F *is ultrabornological.*

Proof. Let $\{E_i : i \in I\}$ a family of elements of $\mathcal{E}$. Suppose the existence of a linear mapping A_i from E_i into F, $i \in I$, such that the topology of F is the finest locally convex topology for which all the mappings A_i, $i \in I$, are continuous. According to (12), A_i can be extended to a continuous linear mapping B_i from $\widehat{E}_i$ into F. Clearly the topology of F is the finest locally convex topology on F for which all the mappings B_i are continuous, $i \in I$. Therefore F can be represented as a separated quotient of $\oplus\{\widehat{E}_i : i \in I\}$ (cf. KÖTHE [1], Chapter Four, §19, Section 1) and thus F is ultrabornological.

(14) *Let* $\mathcal{E}$ *be a class of locally convex spaces. Let* F *and* G *be subspaces of an element* E *of* $\mathcal{E}$ *such that* $F + G = E$, $F \cap G = \{0\}$. *If* F *is closed and if* G *belongs to* $\mathcal{E}_r$, *then* E *is the topological direct sun of* F *and* G.

Proof. Let f be the projection from E onto G along F. By (3), f has closed graph and therefore f is continuous from where the conclusion follows.

(15) *Let* $\mathcal{E}$ *be a class of locally convex spaces. Let* F *and* G *be closed subspaces of an element* E *of* $\mathcal{E} \cap \mathcal{E}_r$ *such that* $F + G = E$, $F \cap G = \{0\}$. *Then* E *is the topological direct sum of* F *and* G.

Proof. Since G is closed in E, G belongs to $\mathcal{E}_r$. We apply (14) to ob tain the conclusion.

If $\mathcal{E}$ is a class of locally convex spaces, we denote by $\mathcal{E}_s$ the class of locally convex spaces such that E belongs to $\mathcal{E}_s$ if and only if, given a sequence (E_n) of subspaces of E covering E, there is a positive integer p such that E_p is dense in E and belongs to $\mathcal{E}$.

(16) *Let $\mathcal{E}$ be a normal class of locally convex spaces. Let (G_n) be a sequence of subspaces of a locally convex space* E *covering* G *such that* G_n *belongs to* $\mathcal{E}_r$, n = 1,2,... *If* F *belongs to* $\mathcal{E}_s$ *and if* f *is a linear mapping with closed graph from* F *into* G, *then* f *is continuous and there is a positive integer* p *such that* f(F) *is contained in* G_p.

Proof. $(f^{-1}(G_n))$ is a sequence of subspaces of F covering F and therefore there is a positive integer p such that $f^{-1}(G_p)$ is dense in F and belongs to $\mathcal{E}$. It is immediate that the restriction f_p of f to $f^{-1}(G_p)$ has closed graph in F x G_p. We apply (11) to obtain that $f^{-1}(G_p)$ coincides with F. Finally, since F belongs to $\mathcal{E}$ and G_p is in $\mathcal{E}_r$, $f : F \longrightarrow G_p$ is continuous and therefore $f : F \longrightarrow G$ is continuous.

If $\mathcal{E}$ is a class of locally convex spaces, we denote by $\mathcal{E}_t$ the class of locally convex spaces such that E belongs to $\mathcal{E}_t$ if and only if, given an increasing sequence (E_n) of subspaces of E covering E, there is a positive integer p such that E_p is dense in E and belongs to $\mathcal{E}$.

(17) *Let $\mathcal{E}$ be a normal class of locally convex spaces. Let (G_n) be an increasing sequence of subspaces of a locally convex space* G *covering* G *such that* G_n *belongs to* $\mathcal{E}_r$, n = 1,2,... *If* F *belongs to* $\mathcal{E}_t$ *and if* f *is a linear mapping with closed graph from* F *into* G, *then* f *is continuous and there is a positive integer* p *such that* f(F) *is contained in* G_p.

Proof. See the proof of (16).

A class $\mathcal{E}$ of locally convex spaces is said to be maximal for the closed graph theorem if, given any locally convex space E which is not in $\mathcal{E}$, there is a non-continuous linear mapping $f : E \longrightarrow F$ with closed graph, being F an element of $\mathcal{E}_r$.

(18) *Let $\mathcal{E}$ be a maximal class of locally convex spaces for the closed*

theorem. If E *belongs to* $\mathcal{E}$, *then every finite codimensional subspace of* E *belongs also to* $\mathcal{E}$.

Proof. Let F be a finite codimensional subspace of E. Let G be any element of $\mathcal{E}_r$ and $f : F \longrightarrow G$ be a linear mapping with closed graph. According to (4), f can be extended to a linear mapping $g : E \longrightarrow G$ with closed graph. Since E belongs to $\mathcal{E}$ and G is in $\mathcal{E}_r$, g is continuous and therefore also f. Consequently F belongs to $\mathcal{E}$.

In VALDIVIA [7] the following result is included : if E is an ultrabornological space whose topology is not the strongest locally convex topology there is an hyperplane of E which is not ultrabornological. According to (18), if $\mathcal{E}$ is the class of all ultrabornological spaces, then $\mathcal{E}$ is not maximal for the closed graph theorem.

We say that a class $\mathcal{E}$ of locally convex spaces is regular if the following conditions are satisfied:

a) If E is an one-dimensional space, then E belongs to $\mathcal{E}$;

b) if F belongs to $\mathcal{E}$, every separated quotient of F belongs to $\mathcal{E}$;

c) if E_i belongs to $\mathcal{E}$, $i \in I$, then $\oplus\{E_i : i \in I\}$ belongs to $\mathcal{E}$.

Let $\mathcal{E}$ be a regular class of locally convex spaces. Let E be a locally convex space. Let

(19) $\{\mathcal{T}_i : i \in I\}$

be the family of all locally convex topologies on E, finer than the original topology, such that $E[\mathcal{T}_i]$ belongs to $\mathcal{E}$, $i \in I$. This family is non-void since the strongest locally convex topology on E belongs to (19), according to a) and c). Let $\mathcal{U}$ be the topology on E such that $E[\mathcal{U}]$ is the locally convex hull of the family of locally convex spaces $\{E[\mathcal{T}_i] : i \in I\}$. According to b) and c), $E[\mathcal{T}]$ belongs to $\mathcal{E}$. Moreover $\mathcal{U}$ is the coarset topology of (19). We call $E[\mathcal{U}]$ the associated space to E of class $\mathcal{E}$.

(20) *Let* $\mathcal{E}$ *be a class of locally convex spaces. If* $\mathcal{E}$ *is maximal for the closed graph theorem, then* $\mathcal{E}$ *is regular.*

Proof. Let F be any element of $\mathcal{E}_r$. If E is an one-dimensional, locally convex space every linear mapping from E into F is continuous and therefore belongs to $\mathcal{E}$. If G is an element of $\mathcal{E}$ and if H is a closed subspace of G, let h be the canonical mapping from G onto G/H. Let f : G/H ⟶ F be a linear mapping with closed graph. By §4, 1.(9), f o h: G ⟶ F has closed graph. Consequently f o h is continuous and therefore f is continuous from where it follows that G/H belongs to $\mathcal{E}$. Finally, let $\{E_i : i \in I\}$ a family of elements of $\mathcal{E}$ and let g be a linear mapping with closed graph from $\oplus\{E_i : i \in I\}$ into F. For every i of I, let g_i be the restriction of g to E_i. Then g_i is linear and has closed graph; therefore g_i is continuous. Consequently g is continuous and therefore $\oplus\{E_i : i \in I\}$ belongs to $\mathcal{E}$. Thus, conditions a), b) and c) are satisfied.

(21) *Let* $\mathcal{E}$ *be a normal and regular class of locally convex spaces. Then, if* F *belongs to* $\mathcal{E}_r$, *its associated space* $F[\mathcal{U}]$ *of class* $\mathcal{E}$ *is complete.*

Proof. If E is any element of $\mathcal{E}$, every closed hyperplane of H of E is isomorphic to a separated quotient of E and therefore H belongs to $\mathcal{E}$. Let f : $F[\mathcal{U}]$ ⟶ F be the canonical injection. Suppose $F[\mathcal{U}]$ non-complete. We apply (10) obtain a Hausdorff topology $\mathcal{T}$ on F, strictly coarser than $\mathcal{U}$, such that F $[\mathcal{T}]$ belongs to $\mathcal{E}$ and f : F $[\mathcal{T}]$ ⟶ F is continuous. This is clearly a contradiction.

Let $\mathcal{E}$ be a class of locally convex space satisfying the following two conditions:

1) Every separated quotient of an element of $\mathcal{E}$ belongs to $\mathcal{E}$;

2) the topological product of two elements of $\mathcal{E}$ is in $\mathcal{E}$.

We need the class $\mathcal{E}$ only to prove the following result:

(22) *Let* E *be an element of* $\mathcal{E}$. *Let* F *be a locally convex space. Let* f : E ⟶ F *be a continuous linear mapping. If* $\mathcal{T}$ *is a topology on* F, *finer than the original topology, such that* $F[\mathcal{T}]$ *belongs to* $\mathcal{E}$ *and* f : E ⟶ F $[\mathcal{T}]$ *is not continuous, then there is a topology* $\mathcal{U}$ *on* F, *finer than the original topology but strictly coarser than* $\mathcal{T}$ *such that* $F[\mathcal{U}]$ *belongs to* $\mathcal{E}$.

Proof. We set G = E x F $[\mathcal{T}]$. Let g be the mapping from G into $F[\mathcal{T}]$ defined by

$$g(x, y) = f(x) + y, \quad x \in E, \; y \in F.$$

Obviously g is linear and continuous and therefore $g^{-1}(0)$ is closed in G. Let $\{U_i : i \in I\}$ be the family of all absolutely convex subsets of F such that $g^{-1}(U_i)$ is a neighbourhood of the origin in G. Then $\{U_i : i \in I\}$ is a fundamental system of neighbourhoods of the origin in F for a locally convex topology $\mathcal{U}$ finer than the original topology. Obviously $F[\mathcal{U}]$ is isomorphic to $G/g^{-1}(0)$ and therefore belongs to $\mathcal{E}$. If h is the restriction of $g: E \times F[\mathcal{T}] \longrightarrow F[\mathcal{U}]$ to $\{0\} \times F[\mathcal{T}]$, then h is continuous. If z belongs to F, then $g(0, z)=z$ and therefore $\mathcal{U}$ is coarser than $\mathcal{T}$.

Let T be the mapping from E into $E \times E[\mathcal{T}]$ such that $Tx = (x, 0)$, $x \in E$. Then T is continuous. The mapping $g \circ T : E \longrightarrow F[\mathcal{T}]$ coincides with f and therefore is not continuous. Consequently $g : E \times F[\mathcal{T}] \longrightarrow F[\mathcal{T}]$ is not continuous and thus $\mathcal{U}$ is strictly coarser than $\mathcal{T}$. The proof is complete.

(23) *Let $\mathcal{E}$ be a regular class of locally convex spaces. Let E be an element of $\mathcal{E}$. Let F be a locally convex space and let f be a continuous linear mapping from E into F. If $F[\mathcal{T}]$ is the associated space to F of class $\mathcal{E}$, then $f : E \longrightarrow F[\mathcal{T}]$ is continuous.*

Proof. According to (22), if $f : E \longrightarrow F[\mathcal{T}]$ is not continuous there is a topology $\mathcal{U}$ on F, finer than the original topology and strictly coarser than $\mathcal{T}$ such that $F[\mathcal{U}]$ belongs to $\mathcal{E}$. That is a contradiction.

(24) *Let E and F be locally convex spaces. Let f be a linear mapping with closed graph from E into F. Let G be a dense subspace of E. If the restriction h of f to G is continuous and F is complete, then f is continuous.*

Proof. Since F is complete, h can be extended to a continuous linear mapping $g : E \longrightarrow F$. Let z be a point of E. Take a net $\{z_i : i \in I, \geq\}$ in G converging to z in E. Since h is continuous, we have that $\{f(z_i) : i \in I \geq\}$ is a Cauchy net in F and therefore converges to a point x in F. Since the graph of f is closed, x coincides with f(z). Then

$$g(z) = \lim \{g(z_i) : i \in I, \geq\} = \lim \{f(z_i) : i \in I\} = f(z)$$

and the conclusion follows.

(25) *Let $\mathcal{E}$ be a maximal class of locally convex spaces for the closed graph theorem. If for some element E of $\mathcal{E}$ there is a subspace G of $\hat{E}$ containg E which does not belong to $\mathcal{E}$, then there is an element F in $\mathcal{E}_r$ whose associated space $F[\mathcal{T}]$ of class $\mathcal{E}$ is not complete.*

Proof. Since G does not belong to $\mathcal{E}$, there is a non-continuous linear mapping f with closed graph from G into a space F of $\mathcal{E}_r$. Let F[T] be the associated space to F of class $\mathcal{E}$. Suppose F[T] complete. If g denotes the restriction of f to E, g has closed graph and therefore g is continuous. By (23), g : E ⟶ F[T] is continuous and since f : G ⟶ F[T] has closed graph, we apply (24) to obtain that f is continuous and that is a contradiction.

(26) *Let* $\mathcal{E}$ *be a regular class of locally convex spaces. Let* F *be an element of* $\mathcal{E}_r$. *If* U *is a Hausdorff topology on* F, *coarser than the original topology, such that* F[U] *is a locally-convex space, then the associated spaces to* F[U] *and* F *of class* $\mathcal{E}$ *coincide.*

Proof. Let F[T] be the associated space to F[U] of class $\mathcal{E}$. Let f be the identity mapping from F[T] into F[U]. The linear mapping f is a continuous and f : F[T] ⟶ F has closed graph. Therefore f : F[T] ⟶ F is continuous and thus T is finer than the topology of F. consequently F[T] is the associated space to F of class $\mathcal{E}$.

(27) *Let* $\mathcal{E}$ *be a regular class of locally convex spaces. Let* F *be a locally convex space satisfying the following condition: if* U *is any Hausdorff topology on* F, *coarser than the original topology, such that* F[U] *is a locally convex space, then the associated spaces to* F[U] *and* F *of class* $\mathcal{E}$ *coincide. Then* F *belongs to* $\mathcal{E}_r$.

Proof. Suppose that F does not belongs to $\mathcal{E}_r$. Then there is a non-continuous linear mapping f with closed graph from a space E of $\mathcal{E}$ into F. According to §4, 1.(21), we can find a Hausdorff topology T on F, coarser than the original topology, such that F[T] is a locally convex space and f : E ⟶ F[T] is continuous. Let F[V] be the associated space to F[T] of class $\mathcal{E}$. By (23), f: E ⟶ F[V] is continuous and, since F[V] is the associated space to F of class $\mathcal{E}$, it follows that f is continuous and this is a contradiction.

(28) *Let* $\mathcal{E}$ *be a regular class of locally convex spaces. If* F *is a locally convex space which does not belong to* $\mathcal{E}_r$, *there is a non-continuous injective linear mapping from a space* E *of* $\mathcal{E}$ *onto* F *with closed graph.*

Proof. According to (27), there is a Hausdorff topology U on F, coarser than the original topology, such that F[U] is a locally convex space such

such that $F[\mathcal{U}]$, which is the associated space to $F[\mathcal{U}]$ of class $\mathcal{E}$, is different from the associated space to F of class $\mathcal{E}$. We set $E = F[\mathcal{T}]$ and f for the identity on F. Then $f : E \longrightarrow F[\mathcal{U}]$ is continuous and therefore $f : E \longrightarrow F$ has closed graph. Obviously $f : E \longrightarrow F$ is not continuous.

Given a class $\mathcal{E}$ of locally convex spaces, we set $\mathcal{E}_0$ to denote the subclass of $\mathcal{E}_r$ such that E belongs to $\mathcal{E}_0$ if and only if every separated quotient of E belongs to $\mathcal{E}_r$.

(29) *Given a class* $\mathcal{E}$ *of locally convex spaces, let* E *and* F *be elements of* $\mathcal{E}$ *and* $\mathcal{E}_0$ *respectively. Let* f *be a linear mapping from a subspace* G *of* F *onto* E. *If the graph of* f *is closed in* $F \times E$, *then* f *is open.*

Proof. The graph of f is obviously closed in $G \times E$ and therefore $f^{-1}(0)$ is a closed subspace of G. Let H be the closure of $f^{-1}(0)$ in F. Suppose that H is not contained in G. Let z be a point of H which is not in $f^{-1}(0)$. We find a net

$$\{z_i : i \in I, \geq\}$$

in $f^{-1}(o)$ converging to z in F. Then

$$\{(z_i, f(z_i)) : i \in I, \geq\}$$

is a net in the graph $G(f)$ of f converging to $(z, 0)$ in $F \times E$. Therefore $(z, 0)$ belongs to $G(f)$ and thus z belongs to G which is a contradiction. Therefore $f^{-1}(0)$ is closed in F. Let h be the canonical mapping from F onto $F/f^{-1}(0)$ and let k be the restriction of h to G. It is obvious that k is the canonical mapping from G onto $G/f^{-1}(0)$. Let $g : G/f^{-1}(0) \longrightarrow F$ be the linear onto mapping such that $g \circ k$ coincides with f. It is not difficult to check that g has closed graph in $(F/f^{-1}(0)) \times E$. Therefore $g^{-1} : E \longrightarrow F/f^{-1}(0)$ is a linear mapping with closed graph. Since E belongs to $\mathcal{E}$ and $F/f^{-1}(0)$ is in $\mathcal{E}_r$, it follows that g^{-1} is continuous and thus g is open. Then f is open.

(30) *Let* $\mathcal{E}$ *be a class of locally convex spaces. If* F *is a locally convex space which is not in* $\mathcal{E}_0$, *there is an element* E *of* $\mathcal{E}$ *and a linear onto mapping* $f : E \longrightarrow F$ *with closed graph which is not open.*

Proof. Since F is not in $\mathcal{E}_0$, there is a closed subspace H of F such that F/H does not belong to $\mathcal{E}_0$. According to (28), there is a space E in $\mathcal{E}$ and a non-continuous injective onto linear mapping g with closed graph defined on E with values in F/H. If h is the canonical mapping from F onto F/H, it is enough to consider $f = g^{-1} \circ h$ to obtain the conclusion.

(31) *Let* $\mathcal{E}$ *be a normal class of locally convex spaces. Let* F *be an element of* $\mathcal{E}_0$. *Let* f *be a linear mapping with closed graph from* F *into* E, E *being a locally convex space. If* f(F) *is dense in* E *and belongs to* $\mathcal{E}$, *then* f(F) *coincides with* E.

Proof. Let h be the canonical mapping from F into $F/f^{-1}(0)$. Let g be the mapping from $F/f^{-1}(0)$ into E such that $g \circ h$ coincides with f. Then $g^{-1} : f(F) \longrightarrow F/f^{-1}(0)$ is linear and has closed graph in $E \times (F/f^{-1}(0))$. Since E belongs to $\mathcal{E}$ and $E/f^{-1}(0)$ is in $\mathcal{E}_0$, we apply (11) to obtain the conclusion.

(32) *Let* $\mathcal{E}$ *be a class of locally convex spaces. Let* E *be an element of* $\mathcal{E}_0$ *If* T *is a Hausdorff topology on* E, *coarser than the original topology such that* E[T] *is a locally convex space, then* E[T] *belongs to* $\mathcal{E}_0$.

Proof. Let F be any element of $\mathcal{E}$. Let f be a linear mapping with closed graph from E[T] onto F. Then $f : E \longrightarrow F$ has closed graph and is open by (29). Therefore $f : E[T] \longrightarrow F$ is open. Apply (30) to obtain that E[T] belongs to $\mathcal{E}_0$.

(33) *Let* $\mathcal{E}$ *be a class of locally convex spaces. Let* E *be an element of* $\mathcal{E}_0$. *If* G *is a closed subspace of* E, *then* G *belongs to* $\mathcal{E}_0$.

Proof Let F be any element of $\mathcal{E}$. Let f be a linear mapping with closed graph from G onto F. Since G is closed in E, f has closed graph in $E \times F$. According to (29), f is open. The conclusion follows applying (30).

(34) *Let* $\mathcal{E}$ *be a class of locally convex spaces. If* E *belongs to* $\mathcal{E}_0$ *and if* F *is a topological product of one-dimensional locally convex spaces, then* $E \times F$ *belongs to* $\mathcal{E}_0$.

Proof. We set $G = E \times F$. Let H be a closed subspace of G. Let T be the canonical mapping from G onto G/H. The subspace T(F) of G/H is isomorphic to a product of one-dimensional locally convex spaces and therefore has a topological complement L in G/H (cf. BOURBAKI [1], Chapter IV,

§1, Ex. 13). By (32), the subspace T(E) of G/H belongs to $\mathcal{E}_0$. If f is the projection from G/H onto L along T(F), then f is continuous and, since T (E) + T(F) coincides with G/H, it follows that f(T(E)) coincides with L. Then L is the continuous image of T(E) by the restriction of f to T(E). Consequently L belongs to $\mathcal{E}_0$. Finally, since G/H is isomorphic to the product L x T(F), we apply (5) to reach the conclusion.

(35) *Let* E *and* F *be locally convex spaces. Let* f *be a linear mapping from* E *into* F. *If there is a fundamental system of closed absolutely convex neighbourhoods* $\{U_i : i \in I\}$ *of the origin in* F *such that* $f^{-1}(U_i)$ *is closed in* E *for every* i *in* I, *then the graph of* f *is closed.*

Proof. If G is the closure of f(E) in F it is enough to prove that the graph of f is closed in E x G. We set

$$V_i = U_i \cap G, \ i \in I$$

Consider f as a mapping from E into G and let $g : G'[\sigma(G', G)] \longrightarrow E^*[\sigma(E^*, E)]$ be the transposed to f. If u belongs to G', there is an index j in I such that u belongs to W_j, W_j being the polar set of V_j in G'. Since g is continuous, $g(W_j)$ is a compact subset of $E^*[\sigma(E^*, E)]$. Let P_j and Q_j be the polar sets of $f^{-1}(V_j)$ in E' and E* respectively. Since $f^{-1}(V_j)$ is a closed absolutely convex subset of E, then Q_j is the closure of P_j in $E^*[\sigma(E^*, E)]$. We have that $g(W_j)$ coincides with Q_j and therefore there is a net

$$\{v_h : h \in H, \geq\}$$

in P_j which $\sigma(E^*, E)$- converges to g(u). Since f(E) is dense in G, then g is injective and therefore $g^{-1}(P_j)$ is contained in W_j and therefore the net

$$\{g^{-1}(v_h) : h \in H, \geq\}$$

has an adherent point v in the compact subset W_j of $G'[\sigma(G', G)]$. Since g is continuous we have that g(v) coincides with g(u) and consequently v coincides with u. Then $g^{-1}(E')$ is dense in $G'[\sigma(G', G)]$. Accordingly, $\sigma(G, g^{-1}(E')) = \mathcal{T}$ is a Hausdorff topology on G. Since $f : E \longrightarrow G[\mathcal{T}]$ is continuous we have that $f : E \longrightarrow G$ has closed graph.

Result (4) is due to IYAHEN [1]. (5) and (34) can be found in EBERHARDT [1]. Results (10), (11), (12), (13), (16), (17), (18) and (31) appear here for the first time. (8), (21), (24), (25) and (35) can be seen in VALDIVIA [11]. (23), (26) and (27), which generalize results due to KŌMURA [1], can be found in EBERHARDT [1].

For other results of general type on the closed graph theorem we refer to EBERHARDT [1], VALDIVIA[11] and POWELL [1].

2. BARRELLED SPACES AND THE CLOSED GRAPH THEOREM. A locally convex space E is a Γ_r-space if given any quasi-complete subspace G of $E^*[\sigma(E^*, E)]$ such that $G \cap E'$ is dense in $E'[\sigma(E', E)]$, then G contains E'.

Let E be a locally convex space. Let $\mathcal{E}$ be the class of the barrelled spaces. We denote by E^t the barrelled space associated to E, i.e., the space associated to E of class $\mathcal{E}$.

(1) *Let* E *be a* Γ_r*-space. If* $\mathcal{E}$ *is the class consisting of all barrelled spaces, then* E *belongs to* $\mathcal{E}_r$.

Proof. Let $\mathcal{T}$ be a Hausdorff topology on E, coarser than the original topology, such that $E[\mathcal{T}]$ is a locally convex space. The topological dual G of $E[\mathcal{T}]^t$, endowed with the weak topology is a quasi-complete subspace of $E^*[\sigma(E^*, E)]$. The topological dual of $E[\mathcal{T}]$ is a dense subspace H of $E'[\sigma(E', E)]$. Obviously G contains H and therefore $G \cap E'$ is dense in $E'[\sigma(E', E)]$. Then G contains E' and therefore $E[\mathcal{T}]^t$ coincides with E^t. We apply (27) to reach the conclusion.

(2) *Let* $f: E \longrightarrow F$ *be a linear mapping with closed graph,* E *and* F *being barrelled and* Γ_r*-spaces respectively. Then* f *is continuous.*

Proof. It is a straightforward consequence of (1).

Results (3) and (4) are particular cases of 1.(21) and 1.(11) respectively.

(3) *If* E *is a* Γ_r*-space, then* E^t *is complete.*

(4) *Let* E *be a locally convex space. Let* G *be a dense barrelled subspace of* E. *Let* $f: G \longrightarrow F$ *be a linear mapping,* F *being a* Γ_r*-space. If* f *has closed graph in* E x F, *then* G *coincides with* E.

(5) *Let* E *be an unordered Baire-like space. Let* (E_m) *be a sequence of*

subspaces of E *covering* E. *Then there is a positive integer* p *such that* E_p *is dense in* E *and unordered Baire-like.*

Proof. See proof of §2, 2.(5).

(6) *Let* (F_n) *be a sequence of subspaces of a locally convex space* F *covering* F *such that* F_n *is a* Γ_r*-space,* n = 1,2,... *Let* f *be a linar mapping with closed graph from* E *into* F, E *being an unordered Baire-like space. Then there is a positive integer* p *such that* f(E) *is contained in* F_p *and* $f : E \longrightarrow F_p{}^t$ *is continuous.*

Proof. It follows easily from (5), 1.(16) and 1.(23).

(7) *Let* (F_n) *be an increasing sequence ob subspaces of a locally convex space* F *covering* F *such that* F_n *is a* Γ_r*-space,* n = 1,2,... *Let* f *be a linear mapping with closed graph from* E *into* F, E *being a suprabarrelled space. Then there is a positive integer* p *such that* f(E) *is contained in* F_p *and* $f : E \longrightarrow F_p{}^t$ *is continuous.*

Proof. It follows easily from 1.(17) and 1.(23).

(8) *If* E *is the class of all barrelled spaces and if* E *belongs to* E_r *then* E *is* Γ_r*-space.*

Proof. Let G be a quasi-complete subspace of $E^*[\sigma(E^*, E)]$ such that $G \cap E'$ is dense in $E'[\sigma(E', E)]$. Clearly $E[\mu(E, G)]$ is barrelled. Let f be the identity mapping on E. Then $f : E[\sigma(E, E')] \longrightarrow E$ has closed graph and therefore $f : E[\mu(E, E')] \longrightarrow E$ has also closed graph and therefore is continuous. Then G contains E' and the conclusion follows.

(9) *Let* F *be a locally convex space. If* F *is not a* Γ_r*-space, there is a barrelled space* E *and a non-continuous injective linear mapping with closed graph defined on* E *onto* F.

Proof. It is an immediate consequence from (8) and 1.(28).

Let E and F be locally convex spaces. A linear mapping f from E into F is called nearly continuous if, for every neighbourhood of the origin U in F, the closure of $f^{-1}(U)$ in E is a neighbourhood of the origin in E.

(10) *Let* F *be a locally convex space.* F *is a* Γ_r*-space if and only if given any linear mapping* $f : E \longrightarrow F$, E *being an arbitrary locally convex spaces, such that* $f : E \longrightarrow F^t$ *is nearly continuous, then* f *is continuous.*

Proof. First we suppose F a Γ_r-space. Let $\mathcal{U}$ be the strongest locally convex topology on E. Let G be the topological dual of F^t endowed with the weak topology. Let $g : G \longrightarrow E^*[\sigma(E^*, E)]$ be the transposed mapping of $f : E[\mathcal{U}] \longrightarrow F^t$. We denote by H the subspace $g^{-1}(E')$ of G. Let A be a bounded closed absolutely convex subset of H and let $\bar{A}$ be the closure of A in G. If A° is the polar set of A in F^t, A° is a barrel in F^t and therefore a neighbourhood of the origin in F^t. Since the graph of f is closed we apply §4. 1.(21) to obtain that $H \cap F'$ is dense in $F'[\sigma(F', F)]$. Since $f : E \longrightarrow F[\sigma(F, H)]$ is continuous, $f^{-1}(A^\circ)$ is closed in E and therefore $f^{-1}(A^\circ)$ is a neighbourhood of the origin in E. If $g(A)^\circ$ is the polar set of g(A) in E we have that $g(A)^\circ$ coincides with $f^{-1}(A^\circ)$ and therefore g(A) is a relatively compact subset of $E'[\sigma(E', E)]$. Now take a point u of $\bar{A}$ and let

$$\{u_i : i \in I, \geq\}$$

a net in A converging to u in G. Then

$$\{g(u_i) : i \in I, \geq\}$$

is a Cauchy net in $E'[\sigma(E', E)]$ contained in g(A) and therefore converges to a point v in $E'[\sigma(E', E)]$. Consequently g(u) coincides with v and thus u belongs to A. Then $A = \bar{A}$, from where it follows that H is quasi-complete. Since $H \cap F'$ is dense in $F'[\sigma(F', F)]$, we have that H contains F'. Consequently f is weakly continuous. Finally, if U is a closed absolutely convex neighbourhood of the origin in F, then $f^{-1}(U)$ is closed in E and therefore a neighbourhood of the origin. Thus f is continuous.

Now we suppose that F is not a Γ_r-space. According to (9) there is a non-continuous linear mapping with closed graph $f : E \longrightarrow F$, E being a barrelled space. Obviously f is nearly continuous. The conclusion follows.

Let E be a locally convex space E is B_r-complete if given any dense subspace G of $E'[\sigma(E', E)]$ which meets every equicontinuous closed subset of $E'[\sigma(E', E)]$ in a closed set, then G coincides with E'.

(11) *If* E *is a* B_r-*complete space then* E *is a* Γ_r-*space.*

Proof. Let G be a quasi-complete subspace of $E^*[\sigma(E^*, E)]$ such that $G \cap E'$ is dense in $E'[\sigma(E', E)]$. Let A be an equicontinuous closed subset of $E'[\sigma(E', E)]$. Then $(G \cap E') \cap A = G \cap A$ is closed and bounded in G and therefore compact. Consequently $G \cap E'$ coincides with E' and the conclusion follows.

(12) *If* E *is a barrelled* Γ_r*-space, then* E *is* B_r*-complete.*

Proof. Let G be a dense subspace of $E'[\sigma(E', E)]$ which intersects every closed equicontinuous subset of $E'[\sigma(E', E)]$ in a closed set. Since E is barrelled, G is a quasi-complete subset of $E^*[\sigma(E^*, E)]$ such that $G \cap E'$ is dense in $E'[\sigma(E', E)]$. Then G coincides with E' and the conclusion follows.

A locally convex space E is a Γ-space if given any quasi-complete subspace of $E^*[\sigma(E^*, E)]$, then $G \cap E'$ is closed in $E'[\sigma(E', E)]$.

(13) *Let* E *be a* Γ*-space. If* $\mathcal{E}$ *is the class of all barrelled spaces, then* E *belongs to* $\mathcal{E}_0$.

Proof. Let F be a closed subspace of E. Let H be the subspace of E* orthogonal to F. We set $L = H \cap E'$. We can identify H and L with the algebraic and topological dual of E/F respectively. Let G be a quasi-complete subspace of $E^*[\sigma(E^*, E)]$ such that $G \cap L$ is dense in $L[\sigma(L, E/F)]$. Then G is a quasi-complete subspace of $E^*[\sigma(E^*, E)]$ such that the closure of $G \cap E'$ in $E'[\sigma(E', E)]$ coincides with L. Since E is a Γ-space, we have that $G \cap L$ coincides with L and thus E/F is a Γ_r-space. It follows from (1) that E/F belongs to $\mathcal{E}$. The conclusion is now obvious.

(14) *Let* E *be a* Γ*-space. Let* $f : G \longrightarrow F$ *be an onto linear mapping,* G *being a subspace of* E *and* F *being barrelled. If the graph of* f *is closed in* E x F, *then* f *is open.*

Proof. It follows easily from (13) and 1.(29)

(15) *Let* E *be a* Γ*-space. Let* G *be a dense barrelled subspace of a locally convex space* F. *If* f *is a linear mapping from* E *onto* G *with closed graph in* E x F, *then* f(E) *coincides with* F.

Proof. It is a particular case of 1.(31).

(16) *If* $\mathcal{E}$ *is the class of all barrelled spaces and if* E *belongs to* $\mathcal{E}_0$, *then* E *is a* Γ*-space.*

Proof. Let G be a quasi-complete subspace of $E^*[\sigma(E^*, E)]$. Let H be the closure of $G \cap H'$ in $E'[\sigma(E', E)]$. Let F be the subspace of E orthogonal to H. Then F is closed in E. Since E/F belongs to E_r, we have that E/F is a Γ_r-space. The algebraic and topological dual of E/F can be identified with H and $L = H \cap E'$ respectively. It is obvious that $G \cap H$ is quasi-complete in $H[\sigma(H, E/F)]$ and $G \cap L$ is dense in $L[\sigma(L, E/F)]$ from where it follows that G contains L. Then $G \cap E'$ is closed in $E'[\sigma(E', E)]$ and the conclusion follows.

(17) *Let* E *be a locally convex space. If* E *is not a* Γ-*space, there is a non-open onto linear mapping* $f : E \longrightarrow F$ *with closed graph,* F *being a barrelled space.*

Proof. It follows from (16) and 1.(30).

Let E and F be locally convex spaces. Let f be a linear mapping from E onto F. f is nearly open if for every neighbourhood of the origin U in F the closure of f(U) in F is a neighbourhood of the origin in F.

(18) *Let* E *be a locally convex space.* E *is a* Γ-*space if and only if given any onto linear mapping* $f : E \longrightarrow F$ *with closed graph,* F *being an arbitrary locally convex space, such that* $f : E^t \longrightarrow F$ *is nearly open, then* f *is open.*

Proof. First suppose E is a Γ-space. Since the graph of f is closed, we have that $f^{-1}(0)$ is closed in E. Let h be the canonical mapping from E onto $E/f^{-1}(0)$. Let g be the injective linear mapping from $E/F^{-1}(0)$ onto F such that f coincides with $g \circ h$. Then g has closed graph. Since $f : E^t \longrightarrow F$ is nearly open, it follows that $g : E^t/f^{-1}(0) \longrightarrow F$ is nearly open. Since $E^t/f^{-1}(0)$ is barrelled, its topology is finer than the topology of $(E/f^{-1}(0))^t$ and therefore $g : (E/f^{-1}(0))^t \longrightarrow F$ is nearly open. Then $g^{-1} : F \longrightarrow (E/f^{-1}(0))^t$ is nearly continuous and continuous by (8). Consequently f is open.

Now suppose that E is not a Γ-space. We apply (17) to obtain a barrelled space F and a non-open onto mapping $f : E \longrightarrow F$ with closed graph. Obviously $f : E^t \longrightarrow F$ is nearly open. The proof is complete.

Let E be a locally convex space. E is B-complete if, given any subspace G of $E'[\sigma(E', E)]$ intersecting every closed equicontinuous subset of $E'[\sigma(E', E)]$ in a closed set, then G is closed. Obviously every B-complete space is B_r-complete.

(19) *Let* E *be a* B-*complete space. Then* E *is a* Γ-*space.*

Proof. Let G be a quasi-complete subspace of $E^*[\sigma(E^*, E)]$. Let A be a closed equicontinuous subset of $E'[\sigma(E', E)]$. Then $(G \cap E') \cap A = G \cap A$ is closed and bounded in G and therefore compact. Consequently $G \cap E'$ is closed in $E'[\sigma(E', E)]$ and the conclusion follows.

(20) *If* E *is a barrelled* Γ-*space, then* E *is* B-*complete.*

Proof. Let G be a subspace of $E'[\sigma(E', E)]$ which intersects every closed equicontinuous subset of $E'[\sigma(E', E)]$ in a closed set. Since E is barrelled, G is a quasi-complete subspace of $E^*[\sigma(E^*, E)]$ such that $G \cap E' = G$. Therefore G is closed and the conclusion follows.

(21) *Let* (F_n) *be a sequence of subspaces of a locally convex space* F *covering* F *such that there is a topology* T_n *on* F_n *finer than the original topology such that* $F_n[T_n]$ *is a Fréchet space,* n = 1,2,... *Let* f *be a linear mapping with closed graph from* E *into* F, E *being a locally convex Baire space. Then there is a positive integer* p *such that* f(E) *is contained in* F_p *and* $f : E \longrightarrow F_p[T_p]$ *is continuous.*

Proof. By using Krein-Smulian's theorem (cf. HORVÁTH [1], Chapter, §10, p. 246) every Fréchet space is B-complete. Consequently F_n is a Γ_r-space and $F_n[T_n]$ is the associated barrelled space to F_n, n = 1,2,... It is enough to apply (6) to reach the conclusion.

(22) *Let* E *be a barrelled* B-*complete space. Let* $f : E \longrightarrow F$ *be a linear mapping,* F *being a Fréchet space. If* $g : F' \longrightarrow E'$ *is the transposed mapping of* f *and if* g(F') *is closed in* $E'[\sigma(E', E)]$, *then* f *is an homomorphism and* f(E) *is closed in* F.

Proof. Since g(F') is closed in $E'[\sigma(E', E)]$ we have that $f : E[\sigma(E, E')] \longrightarrow F[\sigma(F, F')]$ is an homomorphism (cf. KÖTHE [2], Chapter Seven, §32, Section 3). The subspace f(E) of F is metrizable and therefore has its Mackey topology; then $E/f^{-1}(0)$ is isomorphic to f(E). If $\mathcal{E}$ is the class of all barrelled spaces, then $E/f^{-1}(0)$ belongs to $\mathcal{E} \cap \mathcal{E}_r$ and, according to 1.(8), $E/f^{-1}(0)$ is complete. Consequently f(E) is closed in F.

(23) *Let* E *be a barrelled* B-*complete space. Let* $f : E \longrightarrow F$ *be a linear*

mapping, F *being a Fréchet space. If* $g : F' \longrightarrow E'$ *is the transposed mapping of* f *and if* g *is injective and* $g(F')$ *is closed in* $E'[\sigma(E', E)]$, *then* f *is an homomorphism from* E *onto* F.

Proof. Since g is injective, then $f(E)$ is dense in F. We apply (22) to reach the conclusion.

(24) *Let* (F_n) *be an increasing sequence of subspaces of a locally convex space* F *covering* F. *Let* T_n *be a topology on* F_n, *coarser than the original topology, such that* $F_n[T_n]$ *is a Banach space and* T_n *is finer than* T_{n+1}, $n = 1,2,\ldots$ *Let* E *be an ordered convex-Baire space. Let* $f : E \longrightarrow F$ *be a linear mapping with closed graph. Then there is a positive integer* p *such that* $f(E)$ *is contained in* F_p *and* $f : E \longrightarrow F_p[T_p]$ *is continuous.*

Proof. For every positive integer n, let A_n be a bounded absolutely convex neighbourhood of the origin in $F_n[T_n]$ such that A_n is contained in A_{n+1}. Let (B_n) be the closure of $f^{-1}(A_n)$ in E. The increasing sequence $(n B_n)$ of closed absolutely convex subset of E covers E and therefore there is a positive integer p such that $p B_p$ is a neighbourhood of the origin in E. If E_p is the linear hull of $f^{-1}(A_p)$ and if f_p is the restriction of f to E_p, then $f_p : E_p \longrightarrow F_p$ has closed graph and $f_p : E_p \longrightarrow F_p[T_p]$ is nearly continuous. Since F_p is a Γ_r-space and $F_p[T_p]$ coincide with $F_p{}^t$, we apply (8) to obtain that $f_p : E_p \longrightarrow F_p[T_p]$ is continuous. Let x be a point of E. Let $\{x_i : i \in I, \geq\}$ a net in F_p converging to x. Then

$$\{f(x_i) : i \in I, \geq\} = \{f_p(x_i) : i \in I, \geq\}$$

is a Cauchy net in $E_p[T_p]$ and therefore converges to a point z. Then z coincides with $f(x)$ and therefore E_p coincides with E. The proof is complete.

(25) *Let* E *be a locally convex space. If* E *is not barrelled, there is a Banach space* F *and a non-continuous linear mapping* $f : E \longrightarrow F$ *with closed graph.*

Proof. Let U be a barrel in E which is not a neighbourhood of the origin. Let p be be the gauge of U. We set

$$F = \{x \in E : p(x) = 0\}$$

Then F is a closed subspace of E. Let f be the canonical mapping from E onto E/F. Since F is contained in U we have that

$$\{\frac{1}{n} f(U) : n = 1,2,...\}$$

is a fundamental system of closed neighbourhoods of the origin in the linear space E/F for a locally convex topology E/F. Its completion G is a Banach space. Let V be the closure of f(U) in G. We have that

$$\{\frac{1}{n} V : n = 1,2,...\}$$

is a fundamental system of closed neighbourhoods of the origin in G. Now we consider f as a mapping from E into G. Then

$$f^{-1}(\frac{1}{n} V) = f^{-1}(\frac{1}{n} f(U)) = \frac{1}{n} U, n = 1,2,...$$

and, applying 1.(36), the graph of f is closed. Since $U = f^{-1}(V)$ is not a neighbourhood of the origin in E, then f is not continuous. The proof is complete.

(26) *The class of all barrelled spaces is maximal for the closed graph theorem.*

Proof. Since Every Banach space is a Γ_r-space we apply (25) to reach the conclusion.

The definition of Γ_r-space and Γ-space are taken from VALDIVIA [12], where results (2), (9), (14) and (17) are included but proven in a different way. The same theorems are also included in ADASCH [1]. Results (2) and (9) can be found in VALDIVIA [23]. Result (3) can be found in EBERHARDT [2] and ADASCH [2]. Weaker results than (6) can be seen in A.P. ROBERTSON and W. ROBERTSON [2], VALDIVIA [11] and TODD and SAXON [1]. Result (7) is taken from VALDIVIA [10].

The concept of nearly continuous and nearly open mapping is due to PTÁK [1], and it is useful to characterize the B_r-complete and B-complete spaces respectively (cf. KÖTHE [2], Chapter Seven, §34, Section 6). In a similar way we characterize the Γ_r-spaces and Γ-spaces. Result (21) is due to GROTHENDIECK [1]. Theorem (24) can be seen in SAXON [1] and (25) in MAHOWALD [1].

The following articles contain results relating the duality theory with the closed graph theorem: KALTON [1], MARQUINA [1], PERSSON [1], Mac INTOSH [1] and VALDIVIA [24].

§ 7. FINITELY ADDITIVE BOUNDED MEASURES

1. PROPERTIES OF FINITELY ADDITIVE BOUNDED MEASURES. Given a set X, let $\mathcal{A}$ be a σ-algebra on X. We set R^+ to denote the non-negative real numbers. If A belongs to $\mathcal{A}$, we set $\mathcal{F}(A)$ to denote the family of all finite partitions $\{A_1, A_2,\dots,A_n\}$ of A with A_j in $\mathcal{A}$, $j = 1,2,\dots,n$.

A K-valued finitely additive measure λ on $\mathcal{A}$ is a mapping from $\mathcal{A}$ in the field K satisfying:

1) $\lambda(\emptyset) = 0$;
2) if A and B are disjoint elements of $\mathcal{A}$, then $\lambda(A \cup B) = \lambda(A) + \lambda(B)$.

In this section we use the term "finitely additive measure" with the meaning" K-valued finitely additive measure on $\mathcal{A}$". If λ takes only real values we say that λ is real and if λ takes only non-negative real values we say that λ is positive. If K is the field of the complex numbers, then we can write

$$\lambda(A) = \lambda_1(A) + i\,\lambda_2(A),\ A \in \mathcal{A}.$$

i being the imaginary unity and $\lambda_1(A)$ and $\lambda_2(A)$ real numbers. Clearly λ_1 and λ_2 are real finitely additive measures. We say that λ is bounded if the following condition is satisfied:

3) There is a positive integer h such that

$|\lambda(A)| < h$ for every A in $\mathcal{A}$.

Now suppose that λ is a finitely additive measure. For every A of $\mathcal{A}$ we set

$$|\lambda|(A) = \sup\{\sum_{j=1}^{n} |\lambda(A_j)| : \{A_1, A_2,\dots,A_n\} \in \mathcal{F}(A)\}.$$

Then $|\lambda|$ is a function defined on $\mathcal{A}$ and valued in $R^+ \cup \{\infty\}$. We set $\|\lambda\| = |\lambda|(X)$

(1) *If* A *and* B *are disjoint elements of* $\mathcal{A}$, *then* $|\lambda|\ (A \cup B) = |\lambda|\ (A) + |\lambda|\ (B)$.

Proof. If $\{E_1, E_2, \ldots, E_n\}$ and $\{F_1, F_2, \ldots, F_m\}$ are elements of $F(A)$ and $F(B)$ respectively, then

$$\{E_1, E_2, \ldots, E_n, F_1, F_2, \ldots, F_m\} \in F(A \cup B)$$

and therefore

$$\sum_{h=1}^{n} |\lambda(E_h)| + \sum_{k=1}^{m} |\lambda(F_k)| \leq |\lambda|\ (A \cup B)$$

from where it follows

(2) $$|\lambda|\ (A) + |\lambda|\ (B) \leq |\lambda|\ (A \cup B)$$

On the other hand, if $\{G_1, G_2, \ldots, G_n\}$ belongs to $F(A \cup B)$ we have that

$$\{G_1 \cap A, G_2 \cap A, \ldots, G_n \cap A\} \in F(A),$$

$$\{G_1 \cap B, G_2 \cap B, \ldots, G_n \cap B\} \in F(B)$$

and therefore

$$\sum_{j=1}^{n} |\lambda(G_j)| \leq \sum_{j=1}^{n} |\lambda(G_j \cap A)| + \sum_{j=1}^{n} |\lambda(G_j \cap B)|$$

$$\leq |\lambda|\ (A) + |\lambda|\ (B)$$

and accordingly

(3) $$|\lambda|\ (A \cup B) \leq |\lambda|\ (A) + |\lambda|\ (B)$$

The conclusion follows applying (2) and (3).

(4) *Let* (λ_n) *be a sequence of finitely additive measures. If*

(5) $$\sup\ \{|\lambda_n|\ (X) : n = 1,2,\ldots\} = \infty$$

and

(6) $$\sup\ \{|\lambda_n(A)| : n = 1,2,\ldots\} < \infty$$

for every A *of* $\mathcal{A}$, *there is a sequence* (A_n) *of pairwise disjoint elements of* $\mathcal{A}$ *and sequence* (n_p) *of positive integers such that*

(7) $$|\lambda_{n_p}(A_{p+1})| \geq \sum_{j=1}^{p} |\lambda_{n_j}(A_j)| + p - 1, \quad p = 1,2,\ldots$$

Proof. We set $\lambda_n = \alpha_n + i\,\beta_n$ with α_n and β_n real and

$$H = \sup\{|\alpha_n|(X) : n = 1,2,\ldots\}$$

If H is infinite we write $\mu_n = \alpha_n$. If H is finite it follows from

$$|\lambda_n(A)| \leq |\alpha_n(A)| + |\beta_n(A)|, \quad A \in \mathcal{A},$$

that

$$\sup\{|\beta_n(X)| : n = 1,2,\ldots\} = \infty;$$

then we set $\mu_n = \beta_n$.

We write $n_1 = 1$, $A_1 = A_2 = \emptyset$, $B_1 = B_2 = X$. Suppose that for a positive integer p we have found elements $A_1, A_2, \ldots, A_{p+1}, B_1, B_2, \ldots, B_{p+1}$ of $\mathcal{A}$ and positive integers $n_1 < n_2 < \ldots < n_p$ such that

(8) $$A_{p+1} \cap B_{p+1} = \emptyset, \quad A_{p+1} \subset B_p,$$

(9) $$|\mu_{n_p}(A_{p+1})| \geq \sum_{j=1}^{p} |\lambda_{n_p}(A_j)| + p - 1,$$

(10) $$|\mu_{n_p}(B_{p+1})| \geq \sum_{j=1}^{p} |\lambda_{n_p}(A_j)| + p - 1,$$

(11) $$\sup\{|\mu_n|(B_{p+1}) : n = 1,2,\ldots\} = \infty.$$

Obviously (8), (9), (10) and (11) are verified for p = 1. We set

$$h = \sup\{\sum_{j=1}^{p+1} |\lambda_n(A_j)| + p + |\lambda_n(B_{p+1})| : n = 1,2,\ldots\}$$

According to (6), h is finite and therefore there is an integer n_{p+1} such that

$$|\mu_{n_{p+1}}|(B_{p+1}) > 2h.$$

Take $\{M_1, M_2, \ldots, M_q\}$ in $\mathcal{F}(B_{p+1})$ with

$$\sum_{j=1}^{q} |\mu_{n_{p+1}}(M_j)| > 2h.$$

We set

$$P = \cup \{M_j : \mu_{n_{p+1}}(M_j) \geq 0, 1 \leq j \leq q\}$$

$$Q = \cup \{M_j : \mu_{n_{p+1}}(M_j) < 0, 1 \leq j \leq q\}.$$

Then

$$\mu_{n_{p+1}}(P) - \mu_{n_{p+1}}(Q) > 2h$$

from where it follows that at least one of the inequalities

$$\mu_{n_{p+1}}(P) > h, \ -\mu_{n_{p+1}}(Q) > h$$

is true. If $\mu_{n_{p+1}}(P) > h$, then

$$|\mu_{n_{p+1}}(Q)| = |\mu_{n_{p+1}}(B_{p+1}) - \mu_{n_{p+1}}(P)|$$

$$\geq \mu_{n_{p+1}}(P) - |\lambda_{n_{p+1}}(B_{p+1})| > h - |\lambda_{n_{p+1}}(B_{p+1})|$$

$$\geq \sum_{j=1}^{p+1} |\lambda_{n_{p+1}}(A_j)| + p.$$

If $-\mu_{n_{p+1}}(Q) > h$ it follows analogously

$$\mu_{n_{p+1}}(P) \geq \sum_{j=1}^{p+1} |\lambda_{n_{p+1}}(A_j)| + p$$

On the other hand,

$$|\mu_n|(P) + |\mu_n|(Q) = |\mu_n|(B_{p+1}), \ n = 1,2,\ldots$$

and consequently one of the inequalities

$$\sup\{|\mu_n|(P) : n = 1,2,\ldots\} = \infty,$$

$$\sup\{|\mu_n|(Q) : n = 1,2,\ldots\} = \infty$$

is true. Therefore we have proved that there is an element $\{A_{p+2}, B_{p+2}\}$

in $F(B_{p+1})$ such that (8), (9), (10) and (11) are verified taking p+1 instead of p.

It is obvious that the elements of the sequence (A_n) are pairwise disjoint. Finally

$$|\lambda n_p(A_{p+1})| \geq |\mu n_p(A_{p+1})| \geq \sum_{j=1}^{p} |\lambda n_p(A_j)| + p-1,\ p = 1,2,\ldots$$

(12) *A finitely additive measure* λ *is bounded if and only if for every sequence* (A_n) *of pairwise disjoint elements of* A *the series* $\Sigma|\lambda(A_n)|$ *is convergent.*

Proof. Suppose that λ is bounded and the existence of a sequence (A_n) of pairwise disjoint elements of A such that $\Sigma|\lambda_n(A_n)|$ is divergent. We set $\lambda = \lambda_1 + i\lambda_2$ with λ_1 and λ_2 real. Then

$$\Sigma|\lambda(A_n)| \leq \Sigma|\lambda_1(A_n)| + \Sigma|\lambda_2(A_n)|.$$

If $\Sigma|\lambda_1(A_n)|$ is divergent we set $\lambda_1 = \mu$. If $\Sigma|\lambda_1(A_n)|$ is convergent it follows that $\Sigma|\lambda_2(A_n)|$ is divergent and we set $\lambda_2 = \mu$. We write

$$P = \{n \in N : \mu(A_n) \geq 0\},\ Q = \{n \in N : \mu(A_n) < 0\}.$$

If $\Sigma\{\mu(A_n) : n \in P\}$ is divergent we set P = H. If the former series is convergent, then $\Sigma\{\mu(A_n) : n \in Q\}$ is divergent and we set Q = H. Given a positive number h we can find a finite subset L of H such that

$$|\Sigma\{\mu(A_n) : n \in L\}| > h.$$

Then

$$|\lambda(\cup\{A_n : n \in L\})| \geq |\mu(\cup\{A_n : n \in L\})|$$

$$= |\Sigma\{\mu(A_n) : n \in L\}| > h$$

and that is a contradiction.

If λ is not bounded we have

$$|\lambda(A)| \leq |\lambda|(A) \leq |\lambda|(X)$$

for every A of $\mathcal{A}$ and therefore $|\lambda|(X) = \infty$. We apply (4) for $\lambda_n = \lambda$, n = 1,2,..., to obtain a sequence (A_n) of pairwise disjoint elements of $\mathcal{A}$ such that $\Sigma|\lambda_n(A_n)|$ is divergent.

(13) *If a finitely additive measure* λ *is bounded, then* $|\lambda|$ *is a bounded finitely additive measure.*

Proof. Obviously $|\lambda|(\emptyset) = 0$. If $|\lambda|(X)$ is infinite we apply (4) for $\lambda_n = \lambda$, n = 1,2,..., to obtain a sequence (A_n) of pairwise disjoint elements of $\mathcal{A}$ such that $\Sigma|\lambda_n(A_n)|$ is divergent. Now we apply (12) to obtain that λ is not bounded. Therefore $|\lambda|(X) < \infty$. The conclusion follows from (1).

We denote by $B(\mathcal{A})$ the set of all bounded finitely additive measures. If λ and μ belong to $B(\mathcal{A})$ and if h is in K we set

$$(\lambda + \mu)(A) = \lambda(A) + \mu(A), \quad (h\lambda)(A) = h\lambda(A)$$

for every A of $\mathcal{A}$. Clearly $\lambda+\mu$ and $h\lambda$ belongs to $B(\mathcal{A})$. In what follows we suppose $B(\mathcal{A})$ endowed with the linear structure defined by the former operations. If λ is real we set $\lambda^+ = |\lambda|$ and $\lambda^- = |\lambda| - \lambda$. Then λ^- is positive and $\lambda = \lambda^+ - \lambda^-$. Consequently result (14) follows.

(14) *If* λ *is a bounded real finitely additive measure, then* λ *can be written as the difference of two bounded positive finitely additive measures.*

(15) $||.||$ *is a norm on* $B(\mathcal{A})$.

Proof. Consider $\lambda, \mu \in B(\mathcal{A})$ and $h \in K$. We have that

$$||h\lambda|| = \sup \{ \sum_{j=1}^{n} |h\lambda(A_j)| : \{A_1, A_2, \ldots, A_n\} \in \mathcal{F}(X)\}$$

$$= |h| \sup \{ \sum_{j=1}^{n} |\lambda(A_j)| : \{A_1, A_2, \ldots, A_n\} \in \mathcal{F}(X)\}$$

$$= |h| \cdot ||\lambda||,$$

$$||\lambda + \mu|| = \sup \{ \sum_{j=1}^{n} |\lambda(A_j) + \mu(A_j)| :$$

$\{A_1, A_2,\ldots,A_n\} \in F(X)\}$

$$\leq \sup \{\sum_{j=1}^{n} |\lambda (A_j)| : \{A_1, A_2,\ldots,A_n\} \in F(X)\}$$

$$+ \sup \{ \sum_{j=1}^{n} | \mu (A_j)| : \{A_1, A_2,\ldots,A_n\} \in F(X) \}$$

$$= || \lambda || + || \mu ||.$$

In what follows we suppose $B(A)$ endowed with the norm $||\cdot||$.

(16) $B(A)$ *is a Banach space.*

Proof. Let (λ_n) be a Cauchy sequence in $B(A)$. There is a positive number h such that $||\lambda_n|| < h$, $n = 1,2,\ldots$ If A is any element of A we have that

$$|\lambda_n (A) - \lambda_m (A)| \leq |\lambda_n - \lambda_m| (A) \leq ||\lambda_n - \lambda_m||$$

and therefore $(\lambda_n(A))$ is a Cauchy sequence in K and therefore converges to a number $\lambda(A)$. It is obvious that λ is a finitely additive measure. On the other hand,

$$|\lambda (A)| \leq |\lambda (A) - \lambda_1(A)| + |\lambda_1 (A)| \leq ||\lambda - \lambda_1|| + ||\lambda_1|| < 3 h$$

and therefore λ belongs to $B(A)$. Finally, given any positive number ε, we find a positive integer p such that

$$||\lambda_n - \lambda_m|| < \varepsilon,\ n,\ m > p$$

If $\{M_1, M_2,\ldots, M_q\}$ belongs to $F(X)$, we have that

$$\sum_{j=1}^{q} |(\lambda_m - \lambda_n) (M_j)| \leq ||\lambda_m - \lambda_n|| < \varepsilon,\ n, m > p,$$

and therefore

$$\sum_{j=1}^{q} |(\lambda_m-\lambda) (M_j)| \leq \varepsilon,\ m > p$$

and thus

$$||\lambda_m - \lambda|| \leq \varepsilon,\ m > p.$$

Consequently (λ_n) converges to and the conclusion follows.

(17) *Let* (μ_n) *be a sequence in* B(A). *Let* A_n *be a sequence of pairwise disjoint elements of* A. *Given* $\varepsilon > 0$ *there is a subsequence* (A_{n_p}) *Of* (A_n) *such that*

$$|\mu_{n_p}| (\cup \{A_{n_q} : q = p + 1, p + 2, \ldots \}) < \varepsilon, \ p = 1,2,\ldots$$

Proof. Let $\{N_1, N_2, \ldots, N_r, \ldots\}$ be a partition of N such that N_r is infinite, $r = 1,2,\ldots$ We set $n_1 = 1$. Since $|\mu_{n_1}|$ belongs to B(A), the series

$$\sum_r |\mu_{n_1}| (\cup \{A_n : n \in N_r\})$$

is convergent and therefore there is a positive integer p such that

$$|\mu_{n_1}| (\cup \{A_n : n \in N_p\}) < \varepsilon.$$

The sequence $(A_n : n \in N_p)$ can be written in the form $(A_{(1, n)})$ where $(1, n)$ belongs to N_p and $(1, n) < (1, n+1)$, $n = 1,2,\ldots$ Proceeding by recurrence we suppose that for a positive integer r we have obtained the subsequence $(A_{(r, n)})$ of (A_n). We set $n_{r+1} = (r, r)$. Since $\mu_{n_{r+1}}$ belongs to B (A), the series

$$\sum_s |\mu_{n_{r+1}}| (\cup \{A_{(r, n)} : n \in N_s\})$$

is convergent and therefore we find a positive integer q such that

$$|\mu_{n_{r+1}}| (\cup \{A_{(r, n)} : n \in N_q\}) < \varepsilon.$$

We write the sequence $(A_{(r, n)} : n \in N_q)$ in the form $(A_{(r+1,n)})$ such that, if $r_1 < r_2 < \ldots < r_h < \ldots$ are all the elements of N_q, then $(r + 1, h) = (r, r_h)$, $h = 1,2,\ldots$

Then (A_{n_r}) is a subsequence of (A_n) suchthat

$$|\mu_{n_q}| (\cup \{A_{n_q} : q = p + 1, p + 2,\ldots\})$$

$$\leq |\mu_{n_p}| (\cup \{A_{(p, n)} : n = 1,2,\ldots\}) < \varepsilon, \ p = 1,2,\ldots$$

(18) *Let* H *be a subset of* B(A). H *is bounded if and only if*

(19) $\sup \{|\lambda(A)| : \lambda \in H\} < \infty$

for every A *of* $\mathcal{A}$.

Proof. If H is bounded there is $h > 0$ such that $||\lambda|| < h$, $\lambda \in H$. Then, if A belongs to $\mathcal{A}$ we have that

$$|\lambda(A)| \leq |\lambda|(A) \leq ||\lambda|| < h$$

and therefore (19) follows.

Now suppose that (19) is verified and that H is not bounded. We find a sequence (λ_n) in H such that $\lim ||\lambda_n|| = \infty$. Apply (4) to obtain a sequence (A_n) of pairwise disjoint elements of $\mathcal{A}$ and a sequence (n_p) of positive integers such that (7) is verified. We set $\mu_1 = \lambda_1$ and $\mu_{p+1} = \lambda_{n_p}$, $p = 1, 2, \ldots$

Then

$$|\mu_{p+1}(A_{p+1})| \geq \sum_{j=1}^{p} |\mu_{p+1}(A_j)| + p - 1, \; p = 1,2,\ldots$$

According to (17), we can find a subsequence (A_{m_p}) of (A_n) such that

$$|\mu_{m_p}| (\cup \{A_{m_q} : q = p+1, p+2, \ldots\}) < \varepsilon, \; p = 1,2,$$

Then, if we set

$$B = \cup \{A_{m_q} : q = 1,2,\ldots\}$$

it follows that

$$|\mu_{m_p}(B)| \geq |\mu_{m_p}(\cup \{A_{m_q} : q = 1,2,\ldots,p\})|$$

$$-|\mu_{m_p}(\cup \{A_{m_q} : q = p+1, p+2,\ldots\})|$$

$$\geq |\sum_{q=1}^{p} \mu_{m_p}(A_{m_q})| - 1 \geq |\mu_{m_p}(A_{m_p})|$$

$$- \sum_{q=1}^{p-1} |\mu_{m_p}(A_{m_q})| - 1 \geq |\mu_{m_p}(A_{m_p})|$$

$$- \sum_{q=1}^{m_p - 1} |\mu_{m_p}(A_q)| - 1 \geq m_p - 3, \; p = 2,3,\ldots$$

and therefore

$$\sup\{|\mu_n(B)| : n = 1,2,\ldots\} = \infty$$

which is in contradiction with (19).

Result (18) is due to DIEUDONNÉ for the case X = N and A the family of all the parts of N.

2. THE SPACES $\ell_0^\infty(X, A)$ and $\ell^\infty(X, A)$. Let X be a set and let A be a σ-algebra on X. For every A of A we set e_A to denote the characteristic function of A, i.e., e_A is the function defined in X which takes the value one in every point of A and zero in every point of $X \sim A$. Let $\ell_0^\infty(X, A)$ be the linear space over K generated by $\{e_A : A \in A\}$. If f belongs to $\ell_0^\infty(X, A)$ we set

$$||f|| = \sup\{|f(x)| : x \in X\}.$$

We suppose $\ell_0^\infty(X, A)$ endowed with the norm $||.||$. $\ell^\infty(X, A)$ is the completion of $\ell_0^\infty(X, A)$ and therefore a Banach space. We set $M(X, A)$ to denote the conjugate space of $\ell^\infty(X, A)$. We shall use $||.||$ to denote the norm on $\ell^\infty(X, A)$ and also on $M(X, A)$. The conjugate space of $\ell_0^\infty(X, A)$ can be identified with $M(X, A)$.

If A is the σ-algebra of all the parts $P(X)$ of X we write $\ell_0^\infty(X)$ and $\ell^\infty(X)$ instead of $\ell_0^\infty(X, P(X))$ and $\ell^\infty(X, P(X))$ respectively. When X = N and $A = P(N)$ we write ℓ_0^∞ and ℓ^∞.

The norm in the Banach space B(A) is denote by $||.||$.

If u is an element of $M(X, A)$ and if A belongs to A we set $T_u(A) = \langle e_A, u\rangle$. If $A = \emptyset$ it is obvious that $T_u(A) = 0$. If A and B are disjoint elements of A, then

$$T_u(A \cup B) = \langle e_{A \cup B}, u\rangle = \langle e_A, u\rangle + \langle e_B, u\rangle$$

$$= T_u(A) + T_u(B)$$

and therefore T_u is a finitely additive measure on A. Since $||e_A|| = 1$, $A \in A$, $A \neq \emptyset$, there is a positive integer h such that

$$|T_u(A)| = |\langle e_A, u\rangle| \leq h\,||u||$$

for every A of $\mathcal{A}$ and therefore Tu belongs to $B(\mathcal{A})$.

(1) $T : M(X, \mathcal{A}) \longrightarrow B(\mathcal{A})$ *is linear*

Proof. Take u and v in $M(X, \mathcal{A})$ and k in K. If A belongs to $\mathcal{A}$, we have that

$$T(u+v)(A) = \langle e_A, u + v\rangle = \langle e_A, u\rangle + \langle e_A, v\rangle$$

$$= Tu(A) + Tv(A),$$

$$T(ku)(A) = \langle e_A, ku\rangle = k \langle e_A, u\rangle = k\, Tu$$

and therefore T is linear.

(2) T *is a norm-preseving isomorphism from* $M(X, \mathcal{A})$ *onto* $B(\mathcal{A})$.

Proof. Let λ be any element of $B(\mathcal{A})$. If f belongs to $\ell_0^\infty(X, \mathcal{A})$, it is not difficult to check

(3) $$f = h_1 e_{A_1} + h_2 e_{A_2} + \dots + h_n e_{A_n},$$

$A_1, A_2, \dots, A_n$ being pairwise disjoint elements of $\mathcal{A}$ and $h_j \in K$, $j = 1,2, \dots, n$. W set

(4) $$\langle f, w\rangle = h_1 \lambda(A_1) + h_2 \lambda(A_2) + \dots + h_n \lambda(A_n).$$

Clearly the definition of w does not depend on the representation of f. The linearity of w is easy to check. We have that

$$\sup\{|\langle f, w\rangle| : ||f|| \leq 1\}$$

$$= \sup \{\sum_{j=1}^{n} |h_j \langle e_{A_j}, w\rangle| : ||f|| \leq 1\}$$

$$\leq \sup \{\sum_{j=1}^{n} |\lambda(A_j)| : ||f|| \leq 1\} \leq ||\lambda||.$$

Therefore w belongs to $M(X, \mathcal{A})$, $||w|| \leq ||\lambda||$ and

$$Tw(A) = \langle e_A, w\rangle = \lambda(A),\ A \in \mathcal{A},$$

and thus $Tw = \lambda$.

Finally, given $\varepsilon > 0$ there is an element $\{B_1, B_2, \dots, B_n\}$ in $\mathcal{F}(X)$ such that

$$\sum_{j=1}^{n} |\lambda(B_j)| > ||\lambda|| - \varepsilon$$

Let k_j be an element of K with $|k_j| = 1$ and $k_j <e_{B_j}, w> = |<e_{B_j}, w>|$, j = 1, 2,...,n. Then

$$||\sum_{j=1}^{n} k_j \cdot e_{B_j}|| \leq 1$$

and therefore

$$|w| \geq |<\sum_{j=1}^{n} \lambda(B_j), w>| = \sum_{j=1}^{n} |<e_{B_j}, w>|$$

$$= \sum_{j=1}^{n} |\lambda(B_j)| \geq ||\lambda|| - \varepsilon$$

from where it follows that $||w|| \leq ||\lambda||$. Thus $||w|| = ||\lambda||$ and the conclusion follows.

According to (2), M (X, A) can be identified with B(A) by means of the mapping T.

An element u of M (X, A) is said to be real if <f, u> is real for every real f of $\ell^\infty(X, A)$. If <f, u> is non-negative for every $f \geq 0$ of $\ell^\infty(X, A)$, we say that u is a positive linear form. It is obvious that T maps the real elements of M(X, A) in real elements of B(A) and also the positive elements of M (X, A) in positive elements of B(A).

If u is a real element of M(X, A) we apply 1.(14) to obtain positive elements λ and μ in B(A) such that $Tu = \lambda - \mu$. Then $T^{-1}\lambda$ and $T^{-1}\mu$ are positive. Therefore

(5) *Every real element of* M(X, A) *can be written as the difference of two positive elements.*

(6) $\ell_0^\infty(X, A)$ *is barrelled.*

Proof. Since every bounded set of M(X, A) is equicontinuous on $\ell^\infty(X, A)$ we have to wee that if H is a $\sigma(M(X, A), \ell_0^\infty(X, A))$-bounded subset of M(X, A), then H is bounded in M(X, A). Given any A of A we have that

$$\sup\{|<e_A, u>| : u \in H\} < \infty.$$

Since $<e_A, u>$ coincides with $T_u(A)$, we have that

$$\sup\{|Tu(A)| : u \in T(H)\} < \infty.$$

We apply 1. (18) to obtain that

$$\sup \{||Tu|| : u \in T(H)\} < \infty$$

and, since $||Tu|| = || u ||$, we have that

$$\sup \{|| u || : u \in H \} < \infty$$

and the proof is complete.

It is possible to show that the space $\ell_0^\infty(X, A)$ is suprabarrelled (VALDIVIA, [25]).

§ 8. WEAKLY REALCOMPACT LOCALLY CONVEX SPACES

1. A THEOREM OF CORSON. Let X be a completely regular Hausdorff topological space. Let βX be the Stone - Cech compactification of X. The realcompactification G of X is a subspace of βX containing X and verifying the following conditions:

a) If f is a continuous real function on X, there is a continuous real function on G whose restriction to X coincides with f;

b) If x is a point of $\beta X \sim G$, there is a continuous real function on X which does not admit a continuous extension to the subspace $X \cup \{x\}$ of βX.

If X coincides with G, X is said to be real compact. The following result can be seen in GILLMAN and JERISON [1], Chapter 8

(1) *If X is a Lindelöf space, then X is realcompact.*

(2) *If X is metrizable and separable, then every subspace of X is realcompact.*

(3) *If X is realcompact, then every closed subspace of X is realcompact.*

(4) *Let U be a topology on X finer that the original topology such that X [U] is completely regular. If X is metrizable and separable, then X [U] is realcompact.*

(5) *Let* $\{X_i : i \in I\}$ *be a family of completely regular Hausdorff topological spaces. If* X_i *is realcompact,* $i \in I$, *then* $\Pi\{X_i : i \in I\}$ *is realcompact.*

Let E be a locally convex space. Let $\{E_i : i \in I\}$ be the family of all separable closed subspaces of $E[\sigma(E', E)]$. We denote by E_η the subspace of E'^* such that u is in E_η if and only if the restriction of u to every E_i is continuous, $i \in I$. If $\mathcal{T}$ is the topology on E' such that $E'[\mathcal{T}]$ is the inductive limit of the family $\{E_i : i \in I\}$, it is obvious that E_η coincides with the topological dual of $E'[\mathcal{T}]$.

(6) *If* E *coincides with* E_η, *then* $E[\sigma(E, E')]$ *is realcompact.*

Proof. Let G_i be the topological dual of E_i endowed with the weak topology. There is a closed subspace H of $\oplus\{E_i : i \in I\}$ such that $\oplus\{E_i : i \in I\}/H$ is isomorphic to $E'[\mathcal{T}]$. The topological dual of $\oplus\{E_i : i \in I\}$ can be identified with $\Pi\{G_i : i \in I\}$ and therefore $E[\sigma(E, E')]$ is isomorphic to the closed subspace of $\Pi\{G_i : i \in I\}$ orthogonal to H. Let H_i be the linear hull of a countable dense subset of E_i. Then $G_i[\sigma(G_i, H_i)]$ is metrizable and separable and, according to (4), G_i is realcompact. We apply (5) and (3) to obtain that $E[\sigma(E,E')]$ is realcompact.

Let $\mathcal{F}$ be the family of all continuous real functions on $E[\sigma(E, E')]$. For every f of $\mathcal{F}$ we have a pseudometric d_f on E such that

$$d_f(x, y) = |f(x) - f(y)|, \quad x, y \in E.$$

The family of pseudometrics $\{d_f : f \in \mathcal{F}\}$ defines an uniformity $\mathcal{U}$ on $E[\sigma(E, E')]$ compatible with its topology. The topological space $E[\sigma(E, E')]$ is realcompact if and only if the uniform space $(E, \mathcal{U})$ is complete (cf. GILLMAN and JERISON [1], Chapter 15, 15.13).

(7) *If* $E[\sigma(E, E')]$ *is realcompact, then* E *coincides with* E_η.

Proof. Let u be an element of E_η. For every i of I the restriction u_i of u to E_i is continuous and therefore we apply Hahn-Banach's theorem to obtain an element v_i in E which coincides with u_i in E_i. If i and k are in I we set $v_i \geq v_k$ if F_i contains F_k. Then

(8) $\{v_i : i \in I, \geq\}$

is a net in E converging obviously to u in $E_\eta[\sigma(E_\eta, E')]$.

Let $\{w_j : j \in J\}$ be an algebraic basis of E'. Let z_j be the element of E'* such that $<z_j, w_j> = 1$, $<z_j, w_k> = 0$, $k \neq j$, $k, j \in J$. Let H_j be the subspace of $E'^*[\sigma(E'^*, E')]$ generated by z_j, $j \in J$. Then $E'^*[\sigma(E'^*, E')]$ can be identified in the usual way with $H = \Pi\{H_j : j \in J\}$, i.e., if $(y_j : j \in J)$ belongs to H and $w = \Sigma\{a_j w_j : j \in J\}$, $a_j \in K$, $j \in J$, $a_j = 0$ save for a finite number of indices j, then

$$<(y_j : j \in J), w> = \Sigma\, a_j < y_j, w_j>.$$

If S is a subset of I, we set H(S) to denote the subspace of H:

$$\{(y_j : j \in J)\}: y_j = 0, j \in J \sim S\}.$$

Let g be a continuous real function on $E[\sigma(E, E')]$. Let Q be the set of all rational numbers. For every r of Q we set

$$A_r = \{x \in E : g(x) < r\},$$

$$B_r = \{x \in E : g(x) > r\}.$$

We find open subsets M_r and N_r in H such that $M_r \cap E = A_r$ and $N_r \cap E = B_r$. Since E is dense in H, we have that $M_r \cap N_r = \emptyset$. Since H_j is metrizable and separable, $j \in J$, we apply §1, 2.(6) to obtain sequences (P_n^r) and (Q_n^r) of pairwise disjoint cylinders in H such that

$$\cup\{P_n^r : n = 1,2,..\} \text{ and } \cup\{Q_n^r : n = 1,2,...\}$$

are dense subsets of M_r and N_r respectively. We find a countable subset L_r in J such that there are open subsets C_n^r and D_n^r in $H(L_r)$ with

$$P_n^r = C_n^r \times \Pi\{H_j : j \in J \sim L_r\},$$

$$Q_n^r = D_n^r \times \Pi\{H_j : j \in J \sim L_r\}$$

for every positive integer n. We set $L = \cup\{L_r : r \in Q\}$. We determine an index k in I such that the closed linear hull of $\{w_j : j \in L\}$ in $E'[\sigma(E', E)]$ coincides with E_k. Let i and h be elements of I with $i \geq k$, $h \geq k$.

Suppose that $g(v_i) < g(v_h)$. We find a rational number r such that $g(v_i) < r < g(v_h)$. We set

$$C^r = \cup \{C_n^r : n = 1,2,\ldots\},$$

$$D^r = \cup \{D_n^r : n = 1,2,\ldots\}.$$

Then C^r and D^r are disjoint open subsets of $H(L_r)$. Let C_r and D_r be the projection on $H(L_r)$ of M_r and N_r respectively and we suppose that $C_r \cap D_r \neq \emptyset$. Then

$$((C_r \cap D_r) \times \Pi\{H_j : j \in J \sim L\}) \cap M_r$$

is a non-void open subset of M_r disjoint with $\cup \{P_n^r : n = 1,2,\ldots\}$. That is a contradiction and therefore $C_r \cap D_r = \emptyset$. We write v_i and v_h as elements of H, i.e.,

$$v_i = (v_i^j : j \in J),\ v_h = (v_h^j : j \in J).$$

Since v_i is in M_r, v_h is in N_r and C_r is disjoint with D_r there is an index m in L_r such that v_i^m is distinct from v_h^m. Then

$$\langle u_k, w_m\rangle = \langle v_i, w_m\rangle = \langle (v_i^j, j \in J), w_m\rangle$$

$$= \langle v_i^m, w_m\rangle \neq \langle v_h^m, w_m\rangle = \langle (v_h^j : j \in J), w_m\rangle$$

$$= \langle v_h, w_m\rangle = \langle u_k, w_m\rangle$$

and that is a contradiction. Therefore $g(v_i) = g(v_h)$ and thus

$$d_g(v_i, v_h) = |g(v_i) - g(v_h)| = 0,\ i, h \in I,\ i, h \geq k,$$

from where it follows that (8) is a Cauchy net in $(E, \mathcal{U})$. Since $E[\sigma(E, E')]$ is realcompact (8) converges in this space to an element v. It is obvious that u coincides with v. Therefore E coincides with E_η.

Let $\mathcal{A}$ be the family of all equicontinuous subset of $E'[\sigma(E', E)]$ such that A belongs to $\mathcal{A}$ if and only if A is contained in a separable subspace of $E'[\sigma(E', E)]$. The polar sets in E of the elements of $\mathcal{A}$ are a fundamental system of neighbourhoods of the origin for a topology $\eta(E, E')$

compatible with the dual pair <E, E'>.

(9) *If* $E[\eta (E, E')]$ *is complete, then* $E[\sigma (E, E')]$ *is realcompact.*

Proof. Let u be an element of E_η. Let A be a subset of A. If F is a closed separable subspace of $E'[\sigma (E', E)]$ containing A, the restriction of u to F is continuous and therefore the restriction of u to A is continuous. We apply Pták-Collins' theorem (cf. KÖTHE [1], Chapter Four, §21, Section 9) to obtain that u belongs to E and the conclusion follows.

(10) *Let* E *be a* B-*complete space.* $E[\sigma (E, E')]$ *is realcompact if and only if* $E[\eta (E, E')]$ *is complete.*

Proof. If $E[\eta (E, E')]$ is complete it is enough to apply (9). Now suppose $E[\sigma (E, E')]$ realcompact. Then E coincides with E_η. Let u be a linear form on E' continuous on every element of A. We set $H = \{z \in E' : \langle u, z\rangle = 0\}$. Let F be a separable closed subspace of $E'[\sigma (E', E)]$. Let B be a closed absolutely convex equicontinuous subset of $E'[\sigma (E', E)]$. Then $B \cap F$ belongs to A and therefore $(F \cap H) \cap B$ is closed. Since E is B-complete, $F \cap H$ is closed and therefore the restriction of u to F is continuous. Consequently u belongs to E_η and thus u belongs to E. The conclusion follows applying Pták-Collins' theorem.

Results (6) and (7) for Banach spaces in a slight different form are due to CORSON [1].

2. A CERTAIN CLASS OF SUBSETS OF POSITIVE INTEGERS. In this section we shall construct a family of parts of N which will be needed later.

Let Ω be the first non-countable ordinal. We write A_0 to denote the set of the odd positive integers. Fix an ordinal α, $0 < \alpha < \Omega$ and suppose we have constructed $A_\beta \subset N$ for every ordinal number β with $0 \leq \beta < \alpha$ verifying the following conditions: $N \sim A_\beta$ is infinite and $A_{\beta_1} \sim A_{\beta_2}$ is finite and $A_{\beta_2} \sim A_{\beta_1}$ is infinite if $0 < \beta_1 < \beta_2 < \alpha$. If there is an ordinal γ such that $\alpha = \gamma + 1$ we take an infinite subset H in $N \sim A_\gamma$ such that $(N \sim A_\gamma) \sim H$ is infinite and we set $A_\alpha = A_\gamma \cup H$. Then $N \sim A_\alpha$ is infinite and, if $0 \leq \beta < \alpha$, we have that $A_\alpha \sim A_\beta$ is infinite and $A_\beta \sim A_\alpha$ is finite. Now we suppose α a limit ordinal. We order the ordinals γ, with $\gamma < \alpha$, as a sequence (γ_n). We set $n_1 = 1$. Suppose the positive integers $n_1, n_2, \ldots, n_k$ already constructed. n_{k+1} is the first integer larger than n_j, $j = 1,2,\ldots,k$, for which

$$\gamma_{n_{k+1}} > \gamma_n,\ 1 \leqslant n \leqslant n_k.$$

Obviously we can take p_1 in $A_{\gamma_{n_1}}$ and p_j in $A_{\gamma_{n_j}}$ such that

$$p_j \notin A_{\gamma_n},\ n < n_j,\ j = 2,3,\ldots$$

We set

$$A_\alpha = N \sim \{p_1, p_3,\ldots,p_{2n-1},\ldots\}$$

Then $N \sim A_\alpha$ is infinite. If $0 \leqslant \beta < \alpha$ we find a positive integer h such that $\beta < \gamma_h$. Since

$$p_{h+j} \notin A_{\gamma_h},\ j = 1,2,\ldots,$$

setting

$$B_h = N \sim \{p_{h+1}, p_{h+2},\ldots,p_{h+n},\ldots\},$$

it follows that B_h contains A_{γ_h} and since $A_\alpha \sim B_h$ is infinite, we have that $A_\alpha \sim A_{\gamma_h}$ is infinite and thus $A_\alpha \sim A_\beta$ is infinite. On the other hand,

$$A_{\gamma_h} \sim A_\alpha = A_{\gamma_h} \cap \{p_1, p_3,\ldots,p_{2n+1},\ldots\}$$

from where it follows that $A_\beta \sim A_\alpha$ is finite. The following result has been proven:

(1) *If Ω is the first non-countable ordinal there is $A_\alpha \subset N$ for every $\alpha<\Omega$ such that $A_\alpha \sim A_\beta$ is finite and $A_\beta \sim A_\alpha$ is infinite if $0 \leqslant \alpha < \beta < \Omega$.*

An equivalent statement of result (1) can be found in RUDIN [1].

3. THE SPACE ℓ^∞/c_0. Let M be the conjugate of the Banach space ℓ^∞. The subspace of ℓ^∞ of all the K-valued sequences which are convergent to zero in the space c_0. We set E to denote the Banach space ℓ^∞/c_0. If L is the subspace of M orthogonal to c_0, we identify L with E' in the usual way.

Let Ω be the first non-countable ordinal. For every $\alpha < \Omega$ we denote by A_α the subset of the positive integers constructed in 2.(1). Let x_α be the characteristic function of A_α defined on N. Then

$$\{x_\alpha : \alpha < \Omega, \geqslant\}$$

is a net in ℓ^∞.

(1) *Let* u *be a linear real form on* ℓ^∞ *belonging to* L. *Then* u *can be written as the difference of two positive elements of* L.

Proof. In §7, Section 2 we take X = N and A for the family of all parts of N. M coincides with M(X, A). Let T : M(X, A) ⟶ B(X, A) be the mapping defined in that section. We set $\lambda = Tu$. Let A be a finite subset of N and let B be a subset of A. Then B is finite and therefore

$$\lambda\,(B) = \langle e_B, u\rangle = 0$$

from where it follows $|\lambda|\,(A) = 0$. We set $\lambda^+ = |\lambda|$ and $\lambda^- = |\lambda| - \lambda$. Then λ is the difference of the positive elements λ^+ and λ^- of B (X, A). We write

$$u_1 = T^{-1}(\lambda^+),\ u_2 = T^{-1}(\lambda^-)$$

Then $u = u_1 - u_2$ and u_2 beings positive elements of M. If A is a finite subset of N we have that

$$\langle e_A, u_1\rangle = \lambda^+(A) = 0,$$

from where it follows that u_1 vanishes in c_0. Consequently u_1 and u_2 belong to L.

(2) *If* u *is an element of* L, *there is an ordinal* $\beta < \Omega$ *such that* $u(x_\beta) = u(x_\alpha)$ *for* $\beta \leqslant \alpha < \Omega$.

Proof. First we suppose that u is a positive linear form. If $\beta < \gamma < \Omega$, we have that $A_\beta \sim A_\gamma$ is a finites set $A_{\beta\gamma}$. We set $x_{\beta\gamma}$ to denote the element of ℓ^∞ characteristic function of $A_{\beta\gamma}$. Then $x_{\beta\gamma}$ belongs to c_0 and, since $x_{\beta\gamma} + x_\gamma - x_\beta$ is a positive element of ℓ^∞, it follows that

$$u(x_\beta) \leqslant u(x_{\beta\gamma} + x_\gamma) = u(x_{\beta\gamma}) + u\,(x_\gamma) = u\,(x_\gamma). \tag{3}$$

If the net $\{u(x_\alpha) : \alpha < \Omega, \geqslant\}$ is not bounded, we use (3) to find a sequence of ordinals

$$\beta_1 < \beta_2 < \ldots < \beta_n < \ldots < \Omega$$

such that

$$u(x_{\beta n}) > n,\ n = 1,2,\ldots$$

If β is an ordinal with $\beta_n < \beta < \Omega$, $n = 1,2,\ldots$, we have that

$$n < u(\beta_n) < u(\beta),\ n = 1,2,\ldots,$$

which is a contradiction. Therefore there is a real number h such that

$$h = \sup \{u(x_\alpha) : \alpha < \Omega\}.$$

According to (3), we can find a sequence of ordinals

$$\gamma_1 < \gamma_2 < \ldots < \gamma_n < \ldots < \Omega$$

such that

$$u(x\gamma_n) > h - \frac{1}{n},\ n = 1,2,\ldots$$

If β is an ordinal with $\gamma_n < \beta < \Omega$, $n = 1,2,\ldots$, and $\beta \leq \alpha < \Omega$ it follows that

$$u(x_\alpha) \geqslant u(x_\beta) \geqslant u(x_{\gamma_n}) > h - \frac{1}{n},\ n = 1,2,\ldots,$$

and therefore $u(x_\alpha) = h$. The conclusion follows.

Suppose now that u is real. According to (1) we write $u = u_1 - u_2$ with u_1 and u_2 positive and belonging to L. Than an ordinal β can be found such that

$$u_1(x_\alpha) = u_1(x_\beta),\ u_2(x_\alpha) = u_2(x_\beta),\ \beta \leq \alpha < \Omega,$$

and consequently

$$u(x_\alpha) = u_1(x_\alpha) - u_2(x_\alpha) = u_1(x_\beta) - u_2(x_\beta) = u(x_\beta),\ \beta \leq \alpha < \Omega$$

Finally, if u is complex, we write $u = u_1 + i\,u_2$ with real u_1 and u_2. We apply the former construction to reach our conclusion.

For every u in L, the net $\{u(x_\alpha) : \alpha < \Omega \geqslant\}$ has a limit f(u) in K according to (2).

(4) f *is a linear form on* L.

Proof. Take u and v in L and h in K, We have that

$$f(u + v) = \lim \{(u + v)(x_\alpha) : \alpha < \Omega, \geqslant\}$$

$$= \lim \{u(x_\alpha) : \alpha < \Omega \geqslant\} + \lim v(x_\alpha) : \alpha < \Omega, \geqslant\}$$

$$= f(u) + f(v),$$

$$f(hu) = \lim \{(hu)(x_\alpha) : \alpha < \Omega, \geqslant\}$$

$$= h \lim \{u(x_\alpha) : \alpha < \Omega, \geqslant\} = h\, f(u)$$

and the conclusion follows.

(5) *If* F *is a separable closed subspace of* $L[\sigma(L, E)]$, *then the restriction of* f *to* F *is continuous,*

Proof. Let A be a countable dense subset of F. We suppose A endowed with the topology induced by $\sigma(L, E)$. For every n in A, we find an ordinal $\alpha_n < \Omega$ with $u(x_{\alpha_n}) = u(x_\alpha)$, $\alpha_n \leqslant \alpha < \Omega$. Since A is countable there is an ordinal $\beta < \Omega$ such that $\alpha_n < \beta < \Omega$ u ∈ A. Consequently

$$u(x_\beta) = u(x_\alpha),\ \beta \leqslant \alpha < \Omega,\ u \in A.$$

Therefore the net $\{u(x_\alpha) : \alpha < \Omega \ \geqslant\}$ converges to f(u) uniformly on A. For every $\alpha < \Omega$, x_α is a continuous linear form on F and thus x_α is uniformly continuous on A for the canonical uniformity. Then f is uniformly continuous on A for the canonical uniformity. Accordingly, there is a continuous extension g of f to F. If x is any point of F we have that $A \cup \{x\}$ is countable and therefore the restriction of f to $A \cup \{x\}$ is continuous. Since A is dense in $A \cup \{x\}$ and since f and g coincide on A, it follows that f(x) = g(x). Thus the restriction of f to F is continuous.

For every ordinal $\alpha < \Omega$, let U_α be the family of all subsets U of N such that $(A_{\alpha+1} \sim A_\alpha) \sim U$ is finite. Since $A_{\alpha+1} \sim A_\alpha$ is infinite, it follows that $\mathcal{U}_\alpha$ is a filter in N. Let $\mathcal{V}_\alpha$ be an ultrafilter in N finer than $\mathcal{U}_\alpha$, If A is any subset of N we set

$$\lambda_\alpha(A) = 0 \text{ if } A \notin \mathcal{V}_\alpha,\ \lambda_\alpha(A) = 1 \text{ if } A \in \mathcal{V}_\alpha$$

Since $\emptyset \notin \mathcal{V}_\alpha$, it follows that $\lambda_\alpha(\emptyset) = 0$. Let A and B be any disjoint subsets of N. If $A \cup B \notin \mathcal{V}_\alpha$, then A and B are not in $\mathcal{V}_\alpha$ and therefore

$$0 = \lambda_\alpha(A \cup B) = \lambda_\alpha(A) + \lambda_\alpha(B),$$

If $A \cup B$ belongs to $\mathcal{V}_\alpha$ it is obvious that either A or B is in $\mathcal{V}_\alpha$. Consequently

$$1 = \lambda_o(A \cup B) = \lambda_\alpha(A) + \lambda_\alpha(B).$$

Therefore λ_α is a positive finitely additive measure defined on all the parts of N. Recalling §7, Section 2, there is an element w_α in M such that

$$\lambda_\alpha(A) = \langle e_A, w_\alpha \rangle \ , \ A \subset N.$$

On the other hand, if A is a finite subset of N, $N \sim A$ belongs to $\mathcal{U}_\alpha$ and thus $A \notin \mathcal{V}_\alpha$ from where it follows $\lambda_\alpha(A) = 0$. Therefore w_α is in L.

If $||.||$ denotes the norm on M we have that

$$||w_\alpha|| = |\lambda_\alpha|(N) = \lambda_\alpha(N) = 1.$$

Consequently $\{w_\alpha : \alpha \in \Omega, \geqslant\}$ is an equicontinuous net in L and therefore has an adherent point w in $L[\sigma(L, E)]$.

If $\alpha < \beta < \Omega$, then

$$(A_{\beta+1} \sim A_\beta) \sim (N \sim A_\alpha)$$

$$= A_{\beta+1} \cap (N \sim A_\beta) \cap A_\alpha \subset A_\alpha \sim A_\beta.$$

Since $A_\alpha \sim A_\beta$ is finite, $N \sim A_\alpha$ belongs to $\mathcal{U}_\beta$ and thus A_α does not belong to $\mathcal{V}_\beta$. Then

$$w_\beta(x_\alpha) = \langle x_\alpha, w_\beta \rangle = \langle e_{A_\alpha}, w_\beta \rangle = \lambda_\beta(A_\alpha) = 0$$

and consequently

$$\lim \ \{w_\beta(x_\alpha) : \beta < \Omega \ \geqslant\} = 0.$$

Then

(6) $$f(w) = \lim \{w(x_\alpha) : \alpha < \Omega \ \geqslant\}$$

$$= \lim \ \{\lim \{w_\beta(x_\alpha) : \beta < \Omega \ \geqslant\} : \alpha < \Omega, \geqslant \} = 0$$

If $\beta < \alpha < \Omega$, then

$$(A_{\beta+1} \sim A_\beta) \sim A_\alpha \subset A_{\beta+1} \sim A_\alpha$$

and, since $A_{\beta+1} \sim A_\alpha$ is finite, A_α belongs to $\mathcal{V}_\beta$ and thus

$$w_\beta(x_\alpha) = \langle x_\alpha, w_\beta \rangle = \langle e_{A\alpha}, w_\beta \rangle = \lambda_\beta \ (A_\alpha) = 1.$$

Then

$$\lim \{w_\beta(x_\alpha) : \alpha < \Omega \ \geqslant\} = 1$$

If we suppose f in ℓ^∞/c_0 we have that

$$f(w) = \lim w_\alpha (f) : \alpha < \Omega, \geq \}$$

$$= \lim \{\lim \ w_\beta (x_\alpha) : \alpha < \Omega \geq \} : \beta < \Omega \geq \} = 1$$

which is in contradiction with (6). Then we have proved the following result :

(7) f *does not belongs to* ℓ^∞/c_0.

(8) ℓ^∞/c_0, *endowed with the weak topology is not realcompact.*

Proof. Let E_η be the space introduced in Section 1. According to (5) and (7), E_η is distinct from E and, applying 1. (7), we obtain that $E\,\sigma[\,(E, E')\,]$ is not realcompact.

Result (8) is due to CORSON [1].

§ 9. INDUCTIVE LIMITS OF SEQUENCES OF LOCALLY CONVEX SPACES

1. GENERALIZED INDUCTIVE LIMITS. Let E be a linear space over K. Let (E_n) be an increasing sequence of subspaces of E covering E. For every positive integer n, let A_n be an absorbing an absolutely convex subset in E_n and let $\mathcal{T}_n$ be a topology on E_n such that $E_n[\mathcal{T}_n]$ is a locally convex space Let $\mathcal{U}_n$ be the topology on A_n induced by $\mathcal{T}_n$. Suppose $2A_n$ contained in A_{n+1} and $\mathcal{U}_{n+1}$ inducing on A_n a topology coarser than $\mathcal{U}_n$.

We consider the family $\mathcal{U}$ of all absolutely convex subsets U of E such that $U \cap A_n$ is a neighbourhood of the origin in $A_n[\,_n]$, n = 1,2, ...

(1) *$\mathcal{U}$ is a fundamental system of neighbourhood of the origin for a linear topology $\mathcal{T}$ on E.*

Proof. Let U and V be elements of $\mathcal{U}$. Given a positive integer n we find neighbourhoods of the origin U_n and V_n in $E_n[\mathcal{T}_n]$ such that

$$U_n \cap A_n \subset U, \qquad V_n \cap A_n \subset V.$$

Then $U_n \cap V_n \cap A_n$ is a neighbourhood of the origin in A_n $[\mathcal{U}_n]$ contained in $U \cap V$. Consequently $U \cap V$ belongs to $\mathcal{U}$. On the other hand, $U \cap A_{n+1}$ is a neighbourhood of the origin in A_{n+1} $[\mathcal{U}_{n+1}]$ and therefore there is an absolutely convex neighbourhood of the origin W_{n+1} in E_{n+1} $[\mathcal{T}_{n+1}]$ with $W_{n+1} \cap A_{n+1} \subset U$. The set $(\frac{1}{2} W_{n+1}) \cap A_n$ is a neighbourhood of the origin in A_n for the topology $\mathcal{V}_n$ induced on A_n by $\mathcal{U}_{n+1}$. Since $\mathcal{V}_n$ is coarser than $\mathcal{U}_n$, we have that $(\frac{1}{2} W_n) \cap A_n$ is a neighbourhood of the origin in A_n $[\mathcal{U}_n]$. Moreover

$$(\tfrac{1}{2} W_{n+1}) \cap A_n = \tfrac{1}{2} (W_{n+1} \cap (2 A_n))$$

$$\subset \tfrac{1}{2} (W_{n+1} \cap A_{n+1}) \subset \tfrac{1}{2} U$$

and therefore $\frac{1}{2} U$ belongs to $\mathcal{U}$. Since U is absorbing in E, the conclusion follows.

Generally the topological space E $[\mathcal{T}]$ is not Hausdorff. We call E $[\mathcal{T}]$ the generalized inductive limit of the sequence of pairs (E_n $[\mathcal{T}_n]$, A_n) n = 1,2,... If A_n coincides with E_n, then $E[\mathcal{T}]$ is the inductive limit of the sequence (E_n $[\mathcal{T}_n]$) of locally convex spaces. In this case, if E_n $[\mathcal{T}_n]$ is Fréchet, n = 1,2,..., then $E[\mathcal{T}]$ is an (LF)-space; if $E_n[\mathcal{T}_n]$ is Banach, n = 1,2,..., then $E[\mathcal{T}]$ is an (LB)-space and if E_n $[\mathcal{T}_n]$ is normed, n = 1,2,..., then $E[\mathcal{T}]$ is an (LN)-space.

(2) *For every positive integer* n, *the restriction of* $\mathcal{T}$ *to* A_n *is coarser than* $\mathcal{U}_n$.

Proof. Let

(3) $\{x_i : i \in I \geq\}$

be a net in A_n $[\mathcal{U}_n]$ converging to x. Let U be a neighbourhood of the origin in E $[\mathcal{T}]$. The net (3) converges to x in $A_{n+1}[\mathcal{U}_{n+1}]$ and, since $2A_n$ is contained in A_{n+1}, it follows that

$$\{x_i - x : i \in I, \geq\}$$

converges to the origin in $A_{n+1}[\mathcal{U}_{n+1}]$. Since $U \cap A_{n+1}$ is a neighbourhood of the origin in $A_{n+1}[\mathcal{U}_{n+1}]$ there is an index j in I with

$$x_i - x \in U, \ i \geq j$$

and therefore

$$x_i \in x + U, \ i \geq j$$

from where it follows that (3) converges to x in $E[\mathcal{T}]$. The conclusion follows

(4) *Let* f *be a linear mapping from* $E[\mathcal{T}]$ *into a locally convex space* F. f *is continuous if and only if its restriction* f_n *to* $A_n[\mathcal{U}_n]$ *is continuous for* n = 1,2,...

Proof. If f is continuous, then $f_n : A_n[\mathcal{W}_n] \to F$ is continuous, n = 1,2,..., $\mathcal{W}_n$ being the restriction of $\mathcal{T}$ to A_n. Since $\mathcal{W}_n$ is coarser than $\mathcal{U}_n$, then $f_n : A_n[\mathcal{U}_n] \to F$ is continuous.

Now suppose that $f_n : A_n[\mathcal{U}_n] \to F$ is continuous, n = 1,2,... Let V be an absolutely convex neighbourhood of the origin in F. Then $f^{-1}(V)$ is an absolutely convex subset of E. Since $f^{-1}(V) \cap A_n$ coincides with $f_n^{-1}(V)$ it follows that $f^{-1}(V)$ is a neighbourhood of the origin in $E[\mathcal{T}]$. The conclusion follows.

If we fix a positive integer n, let $\mathcal{S}_n$ be the topology on E_n such that $E_n[\mathcal{S}_n]$ is the generalized inductive limit of the sequence $(E_n[\mathcal{T}_n], m A_n)$ m = 1,2,... Since $\mathcal{S}_n$ is clearly finer than $\mathcal{T}_n$, it follows that $E_n[\mathcal{S}_n]$ is a Hausdorff space.

(5) *For every positive integer* n, $\mathcal{S}_{n+1}$ *induces on* E_n *a topology coarser than* $\mathcal{S}_n$.

Proof. Let V be an absolutely convex neighbourhood of the origin in $E_{n+1}[\mathcal{S}_{n+1}]$ and suppose that $V \cap E_n$ is not a neighbourhood of the origin in $E_n[\mathcal{S}_n]$. Then there is a positive integer p such that $V \cap (p A_n)$ is not a neighbourhood of the origin in pA_n for the topology induced by $\mathcal{T}_n$. Consequently there is a net $\{x_i : i \in I, \geq\}$ in $p\, A_n$ converging to the origin in $E_n\ [T_n]$ and x_i is not in V for every i in I. Therefore the net $\{\frac{1}{p}\ x_i : i \in I, \geq\}$ converges to the origin in $A_n\ [U_n]$ and $\frac{1}{p}\ x_i$ is not in $\frac{1}{p}\ V$ for every i of I. The set $\frac{1}{p}\ V$ is a neighbourhood of the origin in $E_{n+1}\ [S_{n+1}]$ and therefore $(\frac{1}{p}\ V) \cap A_{n+1}$ is a neighbourhood of the origin in $A_{n+1}\ [U_{n+1}]$ and since U_{n+1} induces a topology on A_n coarset than U_n, we have that $(\frac{1}{p} V) \cap A_n$ is a neighbourhood of the origin in $A_n\ [U_n]$ and this there is an index j in I such that

$$\frac{1}{p}\ x_i \in (\frac{1}{p}\ V) \cap A_n \subset \frac{1}{p}\ V,\ i \geq j,$$

which is a contradiction. The conclusion follows.

According to (5) we are able to define a topology S on E such that $E[S]$ is the inductive limit of the sequence $(E_n\ [S_n])$.

(6) $E\ [T]$ *coincides with* $E[S]$.

Proof. Let U be an absolutely convex neighbourhood of the origin in $E\ [T]$. Fix the positive integers n and m. Since $2\ A_p$ is contained in A_{p+1}, $p = 1,2,\dots$, there is a positive integer q such that $m\ A_n$ is contained in A_{n+q}. According to (5) the topology on E_n induced by S_{n+q} is coarser than S_n and therefore $U \cap A_{n+q}$ which is a neighbourhood of the origin in $A_{n+q}\ [U_{n+q}]$, intersects $m\ A_n$ in a neighbourhood of the origin for the topology induced in this set by S_n. Consequently $U \cap E_n$ is a neighbourhood of the origin in $E_n\ [S_n]$. Thus S is coarser than T.

Let V be an absolutely convex neighbourhood of the origin in $E\ [S]$. Given a positive integer n, $V \cap E_n$ is a neighbourhood of the origin in $E_n\ [S_n]$ and thus $V \cap A_n$ is a neighbourhood of the origin in $A_n\ [U_n]$. Thus S is finer than T and therefore S coincides with T.

All the results of this section can be found in GARLING [1] except (5) and (6).

2. GENERALIZED STRICT INDUCTIVE LIMITS. We shall use the same notations of Section 1. In what follows we suppose that the restriction of $\mathcal{U}_{n+1}$ to A_n coincides with $\mathcal{U}_n$, n= 1,2,... Then we say that $E[\mathcal{T}]$ is the generalized strict inductive limit of sequence $(E_n[\mathcal{T}_n], A_n)$, n = 1,2,... In particular, if E_n coincides with A_n, n = 1,2,..., $E[\mathcal{T}]$ is the strict inductive limit of the sequence $(E_n[\mathcal{T}_n])$.

(1) *Let n be a positive integer. Let W be a neighbourhood of the origin in $A_n[\mathcal{U}_n]$. Then there is a neighbourhood of the origin U in $E[\mathcal{T}]$ such that $U \cap A_n$ is contained in W.*

Proof. If $n_1 < n_2 < \ldots < n_p < \ldots$ is a sequence of positive integers it is not difficult to check that $E[\mathcal{T}]$ is the generalized strict inductive limit of the sequence $(E_{n_p}[\mathcal{T}_{n_p}], A_{n_p})$, p = 1,2,... On the other hand,

$$4\, A_m \subset 2\, A_{m+1} \subset A_{m+2},\ m = 1,2,\ldots$$

Consequently we can perform the proof for n = 1 and supposing $3\, A_m$ contained in A_{m+1}, for m = 1, 2,..., without loss of generality.

Given a positive integer m let

(2) $\{x_i : i \in I, \geqslant\}$

be a net in e A_m. If (2) converges to x in $3\, A_m$ for the topology induced by $\mathcal{T}_m$, then the net

(3) $\{\frac{1}{3}\, x_i : i \in I, \geqslant\}$

converges to $\frac{1}{3}$ x in $A_m[\mathcal{U}_m]$ and consequently in $A_{m+1}[\mathcal{U}_{m+1}]$; thus (2) converges to x in $A_{m+1}[\mathcal{U}_{m+1}]$.

Now we suppose that (2) converges to x in $A_{m+1}[\mathcal{U}_m]$. Then (3) converges to $\frac{1}{3}$ x in A_m for the topology induced by $\mathcal{U}_{m+1}$ which coincides with $\mathcal{U}_m$ and therefore (2) converges to x in $3A_m$ for the topology induced by $\mathcal{T}_m$.

We conclude that $\mathcal{T}_m$ and $\mathcal{T}_{m+1}$ coincide on $3\, A_m$.

We find an absolutely convex neighbourhood of the origin V_1 in $E_1[\mathcal{T}_1]$

such that $V_1 \cap A_1$ is contained in W. Since $(\frac{1}{2} V_1) \cap (3 A_1)$ is a neighbourhood of the origin for the topology induced by T_1 and since T_1 coincides with T_2 on $3 A_1$, there is an absolutely convex neighbourhood of the origin V_2 in $E_2 [T_2]$ with

$$V_2 \cap (3 A_1) \subset (\frac{1}{2} V_1) \cap (3 A_1)$$

Proceeding by recurrence suppose that, for an integer $m > 1$, we have found the absolutely convex neighbourhood of the origin V_m in $E_m [T_m]$. Since $(\frac{1}{2} V_m) \cap (3A_m)$ is a neighbourhood of the origin in $3A_m$ for the topology induced by T_m and since T_m coincides with T_{m+1} on $3 A_m$, there is an absolutely convex neighbourhood of the origin V_{m+1} an $E_{m+1} [T_{m+1}]$ with

$$V_{m+1} \cap (3 A_m) \subset (\frac{1}{2} V_m) \cap (3 A_m).$$

We set U to denote the absolutely convex hull of

$$\cup \{(\frac{1}{2} V_m) \cap A_m : m = 1,2,\dots\}.$$

Since $U \cap A_m$ contains $(\frac{1}{2} V_m) \cap A_m$, which is a neighbourhood of the origin in $A_m [U_m]$, $m = 1,2,\dots$, we have that U is a neighbourhood of the origin in $E [T]$.

Let z be any element of $U \cap A_1$. Then z can be written in the form

$$\sum_{j=1}^{q} a_j z_j, \ a_j \in K, \ |a_j| \leq 1, \ z_j \in (\frac{1}{2} V_j) \cap A_j,$$

$$j = 1,2,\dots,q, \ q > 2.$$

We set

$$y_r = \sum_{j=r}^{q} a_j z_j, \ r = 2,3,\dots,q.$$

Then

$$y_r = z - \sum_{j=1}^{r-1} a_j z_j.$$

If $r = 2$, then

$$y_2 \in A_1 + A_1 = 2 A_1.$$

If $r > 2$, then

$$y_r \in \frac{1}{3^{r-2}} A_{r-1} + \sum_{j=1}^{r} \frac{1}{3^{j-2}} A_{r-1}$$

$$\subset \frac{1}{3} A_{r-1} + \left(\sum_{j=0}^{\infty} \frac{1}{3^j}\right) A_{r-1} = \frac{1}{3} A_{r-1} + \frac{3}{2} A_{r-1} \subset A_{r-1}$$

Since y_q coincides with $a_q z_q$, then

$$y_q \in V_q \cap (2A_{q-1}) \subset (\tfrac{1}{2} V_{q-1}) \cap (2A_{q-1})$$

Proceeding by recurrence, suppose that, for an integer r with $2 < r < q$ we know that

$$y_r \in (\tfrac{1}{2} V_{r-1}) \cap (2A_{q-1}).$$

Then

$$y_{r-1} = y_r + a_{r-1} z_{r-1} \in \frac{1}{2} V_{r-1} + \frac{1}{2} V_{r-1} = V_{r-1}$$

and thus

$$y_{r-1} \in V_{r-1} \cap (2 A_{r-2}) \subset (\tfrac{1}{2} V_{r-2}) \cap (2 A_{r-2}).$$

Consequently

$$z = y_2 + a_1 z_1 \in (\tfrac{1}{2} V_1) \cap (2A_1) + \tfrac{1}{2} V_1 \subset V_1$$

and therefore z belongs to $V_1 \cap A_1$. Then $U \cap A_1$ is contained in W.

(4) *T is a Hausdorff topology*

Proof. Let x be a point of E distinct from the origin. We find a positive integer p such that x belongs to E_p. Let h be a real number with $0 < h < 1$ such that hx belongs to A_p. Since T_p is a Hausdorff topology, there is a neighbourhood of the origin V in $E_p [T_p]$ such that hx is not in V. We apply (3). to obtain an absolutely convex neighbourhood of the origin U in $E[T]$ such that $U \cap A_p$ is contained in V. Then hx is not in U and therefore x is not in U. The conclusion follows.

For the next result we consider two locally convex topologies $\mathcal{V}$ and $\mathcal{W}$ on a linear space F. Let A be an absolutely convex subset of F. Let $\{U_i : i \in I\}$ be a fundamental system of neighbourhoods of the origin for the topology $\mathcal{M}$ induced by $\mathcal{V}$ in A and also for the topology $\mathcal{N}$ induced by $\mathcal{W}$ in A.

(5) *The topologies* $\mathcal{M}$ *and* $\mathcal{N}$ *coincide.*

Proof. Let

(6) $\{x_j : j \in J, \geq\}$

be a net in $A[\mathcal{M}]$ converging to x. Then

(7) $\{\frac{1}{2}(x_j - x) : j \in J, \geq\}$

is a not in $A[\mathcal{M}]$ converging to the origin. Consequently (7) converges to the origin in $A[\mathcal{N}]$ and therefore (6) converges to x in $A[\mathcal{N}]$. The conclusion follows.

(8) *For every positive integer* n, $\mathcal{T}$ *coincides with* $\mathcal{T}_n$ *in* A_n.

Proof. By 1.(2), $\mathcal{T}$ induces on A_n a topology coarser than $\mathcal{U}_n$. According to (3), there is a fundamental system of neighbourhood of the origin in A_n common to $\mathcal{U}_n$ and to the topology induced by $\mathcal{T}$. We apply (5) to reach the conclusion.

The results of this section can be found in GARLING [1].

3. GENERALIZED HYPERSTRICT INDUCTIVE LIMITS. The same notations of Section 1 are used here. In this section we suppose that the restriction of $\mathcal{U}_{n+1}$ to A_n coincides with $\mathcal{U}_n$ and that A_n is closed in $A_{n+1}[\mathcal{U}_{n+1}]$. We say that $E[\mathcal{T}]$ is the generalized hyperstrict inductive limit of the sequence $(E_n[\mathcal{T}_n], A_n)$, n = 1,2,... In particular, if E_n coincides with A_n, n = 1,2,.., $E[\mathcal{T}]$ is the hyperstrict inductive limit of the sequence $(E_n[\mathcal{T}_n])$ of locally convex spaces.

(1) *For every positive integer* n, A_n *is closed in* $E[\mathcal{T}]$.

Proof. Let x be an adherent point of A_n in $E[T]$. Let

(2) $\{x_i : i \in I, \geq\}$

be a net in A_n converging to x in $E[T]$. Since $2A_p$ is contained in A_{p+1}, $p = 1, 2, \ldots$, we have that

$$E = \cup \{A_p : p = 1,2,\ldots\}$$

and therefore there is a positive integer q such that x belongs to A_{n+q}. Since A_{n+q-1} is closed in $A_{n+q}[T_{n+q}]$ and T coincides with U_{n+q} in A_{n+q} and the net (2) is contained in A_{n+q-1}. If $q > 1$, we proceed as we did before to obtain that x belongs to A_{n+q-2}. Repeating this argument we obtain that x belongs to A_n.

We write G to denote the completion of $E[T]$. Let B_n be the closure of A_n to G, $n = 1,2,\ldots$

(3) *The sequence* (B_n) *covers* G.

Proof. We suppose the property not true. Let z be a point of $G \sim \cup \{B_n : n = 1, 2,\ldots\}$. We apply Hahn-Banach theorem to find an element u_n in G' with

$$|\langle y, u_n\rangle| < 1, \; y \in B_n, \; \langle z, u_n\rangle = 1, \; n = 1,2,\ldots$$

If v_n is the restriction of u_n to E, let M be the set of all the elements of the sequence (v_n). Let U be the polar set of M in E. Since

$$|\langle x, v_{n+p}\rangle| < 1, \; x \in A_n, \; p = 1,2,\ldots,$$

we have that $U \cap A_n$ is a neighbourhood of the origin in $A_n[U_n]$ and therefore U is a neighbourhood of the origin in $E[T]$. Then the sequence (v_n) is T-equicontinuous and, since G is the completion of $E[T]$, (u_n) is an equicontinuous sequence of G'. Let u be an adherent point of (u_n) in $G'[\sigma(G', G)]$. If we take any point x in E, let m be a positive integer such that x belongs to A_m. Since $2^m A_m$ is contained in A_{2m+1} it follows that mx is in A_{2m+r}, $r = 1,2,\ldots$, and therefore

$$|<mx, u_{2m+r}>| < 1, \quad r = 1,2,\ldots,$$

i.e.

$$|<x, u_{2m+r}>| < \frac{1}{m}, \quad r = 1,2,\ldots$$

Consequently $<x, u> = 0$. On the other hand,

$$<z, u> = \lim <z, u_m> = 1.$$

Then u is a non-vanishing continuous linear form on G which vanishes in the dense subset E of G and this is a contradiction.

(4) *If* A_n *is a complete subset of* $E[T]$, $n = 1,2,\ldots$, *then* $E[T]$ *is complete.*

Proof. Since B_n coincides with A_n, we apply (3) to reach the conclusion.

(5) *If* B *is a bounded subset of* $E[T]$ *there is a positive integer* q *such that* B *is contained in* A_q .

Proof. Let D be the closed absolutely convex hull of B in G. Since G is complete, then G_D is a Banach space. $(B_n \cap G_D)$ is a sequence of closed absolutely convex subset of C_D covering G_D. Therefore there is a positive integer q such that $B_q \cap G_D$ is a neighbourhood of the origin in G_D. Take $0 < h < 1$ such that hD is contained in B_q. Let m be a positive integer with $h2^m > 1$. Then

$$D \subset h^{-1}B_q \subset 2^m B_q \subset B_{q+m+1}.$$

According to (1), $B_n \cap E$ coincides with A_n, $n = 1,2,\ldots$, and thus B is contained in A_{q+m+1}.

Results (1) and (5) can be seen in GARLING [1]. Result (4) is due to RAÍKOV [2]. The proofs presented here are different from the original ones.

4. PROPERTIES OF THE WEAK TOPOLOGIES ON CERTAIN (LN)-SPACES. Let $G\ [T]$ be

a locally convex space. Let B be an absolutely convex subset of G and let u be a linear form on G. We set $\mathcal{U}$ to denote the topology on B induced by $\mathcal{T}$.

(1) *The restriction of* u *to* B $[\mathcal{U}]$ *is continuous if and only if* $u^{-1}(0) \cap B$ *is closed in* $B[\mathcal{U}]$.

Proof. If the restriction of u to $B[\mathcal{U}]$ is continuous it is immediate that $u^{-1}(0) \cap B$ is closed in B $[\mathcal{U}]$. Now we suppose that the restriction of u to $B[\mathcal{U}]$ is not continuous. Suppose $u^{-1}(0) \cap (2B)$ closed in 2B for the topology induced by $\mathcal{T}$. Since u does not vanish on B and this set is balanced, given $\varepsilon > 0$ there is a point z in B with $0 < |u(z)| < \varepsilon$. We find and absolutely convex neighbourhood of the origin U in $G[\mathcal{T}]$ such that

$$(2) \qquad (z + U) \cap u^{-1}(0) \cap (2B) = \emptyset.$$

Suppose the existence of a point y in $U \cap B$ such that $|u(y)| > \varepsilon$. We set

$$- t = \frac{u(z)}{u(y)} y$$

Then t belongs to $U \cap B$ and therefore

$$z + t \in (z + U) \cap u^{-1}(0) \cap 2B$$

which is in contradiction with (2). Consequently

$$|u(x)| \leq \varepsilon, \ x \in U \cap B,$$

and thus the restriction of u to $B[\mathcal{U}]$ is continuous in the origin. On the other hand, there is a net $\{y_j : j \in J, \geq\}$ in $B[\mathcal{U}]$ converging to y_0 such that the net $\{u(y_j) : j \in J, \geq\}$ does not converge to $u(y_0)$. Then the net

$$\{\frac{1}{2}(y_j - y_0) : j \in J, \geq\}$$

converges to the origin in $B[\mathcal{U}]$ and

$$\{u(\frac{1}{2}(y_j - y_0) \ j \in J, \geq\}$$

does not converge to cero which is a contradiction. Consequently $u^{-1}(0) \cap (2B)$ is not closed in 2B for the topology induced by 2 $\mathcal{T}$.

There is a point x in $2B \sim (u^{-1}(0) \cap (2B))$ and a net $\{x_i : i \in I, \geq\}$

in $u^{-1}(0) \cap (2B)$ converging to x in G $[T]$. Then $\frac{1}{2}x$ is in $u^{-1}(0) \cap B$ and the net $\{\frac{1}{2} x_i : i \in I, \geq\}$ is contained in $u^{-1}(0) \cap B$ and converges to $\frac{1}{2}x$ in B $[U]$. Consequently $u^{-1}(0) \cap B$ is not closed in B $[U]$..

Let F be a locally convex space. Let (U_m) be a sequence of locally convex topologies on F finer than the original topology. For every positive integer n suppose U_n finer than U_{n+1} and set F'_n for the topological dual of F $[U_n]$.

(3) *Let A be an absolutely convex subset of F satisfying the following conditions.*

a) *A is U_n-metrizable, $n = 1,2,\ldots$;*

b) *given any sequence (x_n) in A converging to the origin in F there is a positive integer p such that (x_n) converges to the origin in $F[\sigma(F, F'_p)]$.*

Then there is a positive integer q and a neighbourhood of the origin U in F $[U_1]$ such that $\sigma(F, F'_q)$ and $\sigma(F, F'_{q+n})$ coincide on $A \cap U$, $n = 1,2,\ldots$

Proof. Let (U_r) be a sequence of absolutely convex neighbourhoods of the origin in F $[U_1]$ such that $(A \cap U_r)$ is a decreasing fundamental system of neighbourhoods of the origin in A for the topology induced by U_1.

Suppose the property not true. We set $n_1 = 1$. Proceeding by recurrence we suppose the positive integers $n_1, n_2, \ldots, n_p$ already constructed. Then there is an integer $n_{p+1} > n_p$ such that $\sigma(F, F'_{n_{p+1}})$ does not coincide with $\sigma(F, F'_{n+1})$ on $A \cap U_p$. We find an element u in F'_{n_p} whose restriction to $A \cap U_p$ is not continuous for the topology U induced by $\sigma(F, F'_{n_{p+1}})$ on $A \cap U_p$. According to (1), $u^{-1}(0) \cap A \cap U_p$ is not closed in $(A \cap U_p)$ $[U]$. The closed convex sets in F coincide for the topologies $U_{n_{p+1}}$ and $\sigma(F, F'_{n_{p+1}})$ and, since A is $U_{n_{p+1}}$-metrizable, there is a sequence (x_m) in $u^{-1}(0) \cap A \cap U_p$ which $U_{n_{p+1}}$-converges to a point x of $A \cap U_p \sim (u^{-1}(0) \cap A \cap U_p)$. The sequence $(\frac{1}{2}(x_m - x))$ $U_{n_{p+1}}$-converges to the origin in $A \cap U_p$ and does not converge to the origin in $A \cap U_p$ for the topology $\sigma(F, F'_{np})$.

We set z_{pm}, m = 1,2,..., instead of $(\frac{1}{2}(x_m - x))$. We shall see that the sequence

(4) $$z_{11}, z_{12}, z_{21}, \ldots, z_{1m}, z_{2(m-1)}, \ldots, z_{m1}, \ldots$$

of A converges to the origin in F. Indeed, let W be an neighbourhood of the origin in F. Then $W \cap A$ is a neighbourhood of the origin in A for the topology induced by $\mathcal{U}_1$ and therefore there is a positive integer q such that $U_q \cap A$ is contained in W and thus

$$z_{pm} \in W, \quad p = q + 1, q + 2, \ldots$$

On the other hand, $W \cap A$ is a neighbourhood of the origin in A for the topology induced by $\mathcal{U}_{n_{q+1}}$ and therefore there is a positive integer r such that

$$z_{pm} \in W, \quad j = 1,2,\ldots,q; \quad m = r + 1, \quad r + 2, \ldots$$

Consequently

$$z_{pm} \in W \quad \text{for} \quad p + m > q + r$$

and thus (4) converges to the origin in F. Then there is a positive integer s such that (4) converges to the origin in $F[\sigma(F,F'_s)]$. Consequently the subsequence z_{sm}, m = 1,2,..., of (4) converges to the origin in $F[\sigma(F,F'_{n_{s+1}})]$ which is a contradiction.

Let $E[\mathcal{T}]$ be a locally convex space. Let (E_n) be an increasing sequence of subspaces of E covering E. For every positive integer n let $\mathcal{T}_n$ be a locally convex topology of E_n finer than the original topology such that $\mathcal{T}_{n+1}$ induces on E_n a topology coarser than $\mathcal{T}_n$. We suppose that $E[\mathcal{T}]$ is the inductive limit of the sequence $(E_n[\mathcal{T}_n])$ of locally convex spaces. We set E' for the topological dual of $E[\mathcal{T}]$ and G_n for the topological dual of $E_n[\mathcal{T}_n]$.

Following RETAKH [1] we say that the inductive limit $E[\mathcal{T}]$ of the sequence $(E_n[\mathcal{T}_n])$ has the property M_0 if, for every positive integer m, there is an absolutely convex neighbourhood of the origin U_m in $E_m[\mathcal{T}_m]$ with U_m contained in U_{m+1} verifying the following condition:

(5) $\forall n\ \exists j > n\ \forall k > j\ \forall f \in G_j\ \forall \varepsilon > 0$

$\exists g \in G_k\ \forall x \in U_n : |f(x) - g(x)| < \varepsilon$

From (5) it follows that

$\forall n\ \exists j > n\ \forall k > j\ \forall f \in G_j\ \forall \varepsilon > 0$

$\exists g \in G_k\ \forall x \in U_n : |f(x) - g(x)|\ \ \frac{\varepsilon}{2^m}$

and consequently

(6) $\forall n\ \exists j > n\ \forall k > j\ \forall f \in G_j \forall\ \varepsilon > 0$

$\exists g \in G_k\ \forall x\ \in 2^n\ U_n : |f(x) - g(x)| < \varepsilon$

Since $2(2^n U_n)$ is contained in $2^{n+1}U_{n+1}$ it follows from (6) that in the definition of the property M_o we can take $2\ U_m$ contained in U_{m+1}, a fact we suppose in what follows.

It is not difficult to show that the condition (5) is equivalent to the following condition (DE WILDE [3]):

(7) $\forall n \exists\ j > n\ \forall k > j : \sigma(E_j,\ G_j)$ coincides

with $\sigma(E_k,\ G_k)$ on U_n.

(8) *If the sequence* (U_m) *verifies condition* (7),*given a positive integer* n *there exists an integer* $h > n$ *such that* $\sigma(E_h,\ G_h)$ *coincides with* $\sigma(E,\ E')$ *on* U_n.

Proof. We set $n_1 = 1$. Proceeding by recurrence we suppose the positive integers $n_1, n_2, \ldots, n_p$ already constructed. We find an integer $n_{p+1} > n_p$ such that $\sigma(E_{n_{p+1}}, G_{n_{p+1}})$ coincides with $\sigma(E_{n_{p+r}}, G_{n_{p+r}})$ on U_{n_p}, $r = 2,3, \ldots$

For every positive integer p we write $F_p = E_{n_p}$, $V_p = U_{n_p}$ and $\mathcal{V}_p$ to denote the topology induced by $\sigma(E_{n_{p+1}}, G_{n_{p+1}})$ on F_p. Then $\mathcal{V}_p$ coincides with $\mathcal{V}_{p+1}$ on V_p. We have that $2\ V_p$ is contained in V_{p+1}. Let $\mathcal{V}$ be the locally convex topology on E such that $E[\mathcal{V}]$ is the generalized strict inductive limit of the sequence.

$$(F_n\ [\mathcal{V}_n],\ V_n),\ n = 1,2,\ldots,$$

According to 2.(8), $\mathcal{V}_p$ coincides with $\mathcal{V}$ in V_p.

Let u be a continuous linear form on $E[\mathcal{V}]$. Then $u^{-1}(0) \cap V_p$ is closed in V_p for the topology induced by $\mathcal{V}_p$ and thus $u^{-1}(0) \cap V_p$ is closed in V_p for the topology induce by $\mathcal{T}_{n_p}$. Since V_p is a neighbourhood of the origin in $F_p\ [\mathcal{T}_{n_p}]$ it follows that the restriction of u to this space is continuous. Consequently u belongs to E'. If v is any element of E', its restriction to $F_{p+1}\ [\sigma(F_{p+1}, G_{n_{p+1}})]$ is continuous and therefore the restriction of v to V_p is continuous for the topology induced by $\mathcal{V}_p$. We apply now 1.(4) to obtain that v belongs to the topological dual of $E\ [\mathcal{V}]$. Then $\mathcal{V}$ is compatible with the dual pair <E, E'> and therefore the restriction of $\mathcal{V}$ to V_p is finer than the restriction of $\sigma(E, E')$ to V_p.

Let W be a neighbourhood of the origin of the origin in $E[\mathcal{V}]$. $W \cap V_p$ is a neighbourhood of the origin in V_p for the topology induced by $\mathcal{V}_p$ and therefore there is a finite set $A = \{u_1, u_2, \ldots, u_q\}$ in $G_{n_{p+1}}$ such that, if $A°$ is the polar set of A in $E_{n_{p+1}}$, $A° \cap V_p$ is contained in $W \cap V_p$. First we suppose that K is the field of the complex numbers. If j is an integer with $1 \leq j \leq q$ we set $u_j = v_j + i\, w_j$, v_j and w_j being real continuous linear forms on $E_{n_{p+1}}\ [\mathcal{T}_{n_{p+1}}]$. We set

$$P_j = \{x \in V_p : v_j(x) \geq \tfrac{1}{2}\},\ Q_j = \{x \in V_p : w_j(x) \geq \tfrac{1}{2}\}$$

Let A_j and B_j be the closures of P_j and Q_j in $E[\mathcal{V}]$ respectively. The origin of E is neither in A_j nor in B_j and therefore we can find real continuous linear forms z_j and t_j on $E[\mathcal{V}]$ with

$$z_j(x) > 1 \text{ for } x \in A_j,\ t_j(x) > 1 \text{ for } x \in B_j$$

We set M to denote the polar set in E of $\{x_1, y_1, \ldots, x_q, y_q\}$ with

$$x_j(\cdot) = z_j(\cdot) - iz_j(i\cdot),\quad y_j(\cdot) = t_j(\cdot) - it_j(i)\quad j = 1,2,\ldots,q$$

Then M is a neighbourhood of the origin in $E[\sigma\ (E, E')]$. Now we take any point x in $V_p \sim W$. There is an integer h with $1 \leq h \leq q$ such that $|u_h(x) > 1|$. Then, at least one of the following inequalities

$$|v_h(x)| > \frac{1}{2}, \ |w_h(x)| > \frac{1}{2}$$

is true and therefore x or -x belongs to $A_h \cup B_h$. Consequently max $\{|z_h(x)|\ |t_h(x)|\} > 1$ and therefore $|x_h(x)| > 1$ or $|y_n(x)| > 1$ and thus x is not in M. Thus $M \cap A_p$ is contained in $W \cap A_p$. If K is the field of the real numbers we proceed as we did before taking $w_j = t_j = 0$, $j = 1,2,\ldots,q$, to obtain that $M \cap A_p$ is contained in $W \cap A_p$. Therefore the restriction of V to V_p is coarser than the restriction of $\sigma(E, E')$ to V_p. Consequently V coincides with $\sigma(E, E')$ on V_p.

Finally, given a positive integer n, we find a positive integer p with $n > n_p$. We take $h = n_{p+1}$. Then $\sigma(E_h, G_h)$ and $\sigma(E, E')$ coincide on U_{n_p} and therefore on U_n.

Result (9) follows easily.

(9) *The inductive limit* $E[T]$ *of the sequence* $(E_n\ [T_n])$ *has the property* M_0 *if and only if, for every positive integer* m, *there is an absolutely convex neighbourhood of the origin* U_m *in* $E_m[T_m]$ *with* U_m *contained in* U_{m+1} *such that, given any positive integer* n, *there is an integer* $h > n$ *such that* $\sigma(E, E')$ *and* $\sigma(E_h, G_h)$ *coincide on* U.

Given a locally convex space H, we say that a sequence (x_n) in H is weak-locally convergent to x if there is a bounded closed absolutely convex subset B of H such that $x_n, x \in B$, $n = 1,2,\ldots$, and (x_n) converges to x for the weak topology of H_B.

(10) *Let* E *be an* (LN)*-space. Then the two following conditions are equivalent:*

(1) *If* (x_n) *is any sequence in* E *converging to the origin, then* (x_n) *is weak-locally convergent to the origin;*

(2) *if* D *is any bounded subset of* E, *there is a bounded closed absolutely convex subset* B *of* E *such that* $D \subset B$ *and* $\sigma(E, E')$ *coincide on* D *with the weak topology of* E_B.

Proof. Suppose that E is the inductive limit of the sequence of normed spaces (E_n $[T_n]$). We take in E_1 $[T_1]$ a bounded absolutely convex neighbourhood of the origin U_1. Proceeding by recurrence suppose that, for a positive integer p, the bounded absolutely convex neighbourhood of the origin U_p in E_p $[T_p]$ is already constructed. Since U_p is bounded in E_{p+1} $[T_{p+1}]$ we find a bounded absolutely convex neighbourhood of the origin U_{p+1} in E_{p+1} $[T_{p+1}]$ such that $2U_p$ is contained in U_{p+1}.

Let V_n be the closure of U_n in E. If U_n° is the polar set of U_n in E' we have that

$$\{U_n^{\circ} : n = 1,2,\ldots\}$$

is a fundamental system of neighbourhood of the origin in E' for a topology $\mathcal{W}$ such that E'$[\mathcal{W}]$ is a Fréchet space. If P is any bounded subset of E, then P is $\mathcal{W}$-equicontinuous and therefore there is a positive integer q such that P is contained in V_q. On the other hand, E is clearly the inductive limit of (E_{V_n}). Consequently, by changing E_n $[T_n]$ for E_{V_n} if necessary, we can suppose that for every bounded subset of E there is a certain U_n containing it.

Given a positive integer n we set F = E_n and we suppose F endowed with the topology induced by the topology of E. We write A = U_n. Let $\mathcal{U}_m$ be the topology induced on F by T_{n+m-1}, m = 1,2,... Then A is $\mathcal{U}_m$-metrizable, m = 1,2,..., and, if 1) holds, given any sequence (x_n) in A converging to the origin in F, there is a positive integer p such that (x_n) converges to the origin in F$[\sigma(F, F'_n)]$, F'_p being the topological dual of F $[\mathcal{U}_p]$. We apply (3) to obtain a positive integer q and a neighbourhood of the origin U in F $[\mathcal{U}_1]$ such that $\sigma(F, F'_q)$ and $\sigma(F, F'_{q+m})$ coincide on A ∩ U for m = 1,2,... And since A ∩ U absorbs A, $\sigma(F, F'_q)$ and $\sigma(F, F'_{q+m})$ coincide on A, m = 1,2,... Consequently $\sigma(E_{n+q-1}, G_{n+q-1})$ coincides with $\sigma(E_{n+q+m}, G_{n+q+m})$ on U_n, m = 1,2,... Therefore condition (7) is satisfied and thus property M_0 holds. Given any bounded subset D of E we find a posi tive integer n such that D is contained in U_n. By (8), there is an integer $h > n$ such that $\sigma(E, E')$ and $\sigma(E_h, G_h)$ coincide on U_n. Consequently, taking $B = U_h$, we have that D is contained in B and $\sigma(E, E')$, we have that D is contained in B and $\sigma(E, E')$ coincides with the weak topology of E_B on D. Therefore 2) follows from 1). Obviously 1) follows from 2).

In the inductive limit $E[T]$ of the sequence of locally convex spaces $(E_n [T_n])$ we say that a bounded subset A of $E[T]$ is regular if there is a positive integer n such that A is contained in E_n and T_n-bounded, Analogously, we say that a bounded sequence (x_n) of $E[T]$ is regular if its elements constitue a regular set.

We say that $E[T]$ is sequentially quasi-retractive if given any sequence (x_p) in $E[T]$ converging to the origin and regular there is a positive integer n such that (x_p) is contained in E_n and converges to the origin for the weak topology of E_n.

We say that $E[T]$ is boundedly quasi-retractive if given any bounded regular subset A of $E[T]$ there is a positive integer n such that A is contained in E_n and $\sigma(E, E')$ coincides on A with the weak topology of $E_n [T_n]$.

(11) *If* E *is the inductive limit of the sequence of normed spaces* $(E_n [T_n])$ *the following conditions are equivalent:*

i) E *is sequentially quasi-retractive;*

ii) E *has property* M_o;

iii) E *is boundedly quasi-retractive.*

Proof. We set G_n to denote the topological dual of $E_n [T_n]$, n = 1, 2,... For every positive integer m we take a bounded absolutely convex neighbourhood of the origin U_m in $E_m [T_m]$ such that $2U_m$ is contained in U_{m+1}. If i) is satisfied, given any positive integer n we proceed as we did in the proof of (10) taking $A = U_n$ to obtain an integer $h > n$ such that $\sigma(E_h, G_h)$ coincides with $\sigma(E, E')$ on U_n. Therefore E has property M_o. Now suppose that ii) is satisfied. For every positive integer m, we find an absolutely convex neighbourhood of the origin V_m in E_m $[T_m]$ such that, given any positive integer n, there is an integer $h > n$ such that $\sigma(E_h, G_h)$ coincides with $\sigma(E, E')$ on V_n. Given a bounded regular subset A of E, we take n such that A is a bounded subset of E_n $[T_n]$. We find $k > 0$ such that kA is contained in V_n. Then $\sigma(E_h, G_h)$ and $\sigma(E, E')$ coincide on kA, consequently coincide on A. Thus iii) is satisfied. Clearly i) follows from iii).

Except (1) the results of this section appear here for the first time.

5. SOME PROPERTIES OF THE (LN)-SPACES. Let $E[T]$ be a locally convex space which is the inductive limit of the sequence of locally convex spaces (E_n $[T_n]$). Following RETAKH [1], we say that $E[T]$ has property M if, for every positive integer m, there is an absolutely convex neighbourhood of the origin U_m in E_m $[T_m]$, with U_m contained in U_{m+1}, verifying the following condition:

(1) $\forall n\ \exists j > n\ \forall k > j$: T_j coincides with T_k on U_n.

It is easy to check that (U_m) can be taken such that $2U_m$ is contained in U_{m+1}, $m = 1,2,\ldots$

(2) *If the sequence* (U_m) *verifies condition* (1), *given a positive integer* n *there is an integer* $k > n$ *such that* T_h *coincides with* T *on* U_n.

Proof. We set $n_1 = 1$. Proceeding by recurrence suppose the positive integers $n_1, n_2, \ldots, n_p$ already constructed. We find an integer $n_{p+1} > n_p$ such that $T_{n_{p+1}}$ coincides with $T_{n_{p+r}}$ on U_{n_p}, $r = 2,3,\ldots$

For every positive integer p, we write $F_p = E_{n_p}$, $V_p = U_{n_p}$ and V_p for the topology induced on F_p by $T_{n_{p+1}}$. Then V_p coincides with V_{p+1} on V_p. Clearly $2V_p$ is contained in V_{p+1}. Let V be the locally convex topology on E such that $E[V]$ is the generalized strict inductive limit of the sequence $(F_n\ [V_n], V_n)$, $n = 1,2,\ldots$ According to 2.(8), V_p coincides with V on V_p.

Let U be an absolutely convex neighbourhood of the origin in $E[T]$. Then $U \cap V_p$ is a neighbourhood of the origin in V_p for the topology induced by V_p. Since V_p is coarser than T_{n_p} and since V_p is a neighbourhood of the origin in $E_{n_p}[T_{n_p}]$ it follows that $U \cap V_p$ is a neighbourhood of the origin in $E_{n_p}[T_{n_p}]$. Consequently U is a neighbourhood of the origin in $E[T]$. Now suppose that V is an absolutely convex neighbourhood of the origin in $E[T]$. Then $V \cap E_{n_{p+1}}$ is a neighbourhood of the origin in $E_{n_{p+1}}[T_{n_{p+1}}]$ and therefore $V \cap V_p$ is a neighbourhood of the origin in V_p for the topology induced by V_p. Consequently, V is a neighbourhood of the origin in $E[V]$.

Then V coincides with T. Finally, given a positive integer n we find a positive integer p such that $n < n_p$. We take $h = n_{p+1}$. Then T_h and T coincide on U_{n_p} and therefore U_n.

Now result (3) follows easily

(3) *The inductive limit* $E[T]$ *of the sequence* $E_n[T_n]$ *has property* M *if and only if, for every positive integer* m, *there is an absolutely convex neighbourhood of the origin* U_m *with* U_m *contained in* U_{m+1} *such that, given any positive integer* n, *there is an integer* $h > n$ *such that* T_h *coincides with* T *on* U_n.

Fix a positive integer n and take a balanced bounded subset B of $E[T]$ contained in $E_n[T_n]$. We set U and U_m to denote the topologies on B induced by T and T_m respectively, $m = n+1, n+2, \ldots$

(4) *If, given any sequence* (x_m) *in* B-B *which converges to the origin in* $E[T]$, *there is a positive integer* p *such that* (x_m) *converges to the origin in* $E_p[T_p]$, *then there is an integer* $q > n$ *such that* $B[U]$ *and* $B[U_q]$ *have the same convergent sequences.*

Proof. Suppose the property not true. We set $n_1 = n$. Proceeding by recurrence suppose the positive integers $n_1, n_2, \ldots, n_p$ already constructed. Then there is an integer $n_{p+1} > n_p + n$ and a sequence (x_m) which converges to x in $B[U]$ but does not converge to x in $B[U_{n_{p+1}}]$. The sequence $(\frac{1}{p}(x_m - x))$ of B - B converges to the origin fot the topology induced by $T_{n_{p+1}}$

We set z_{pm}, m = 1,.2,..., instead of $(\frac{1}{p}(x_m - x))$. We shall show that the sequence

(5) $z_{11}, z_{12}, z_{21}, \ldots, z_{1m}, z_{2(m-1)}, \ldots, z_{m1}, \ldots$

converges to the origin in B-B for the topology induced by T. Indeed let W be a neighbourhood of the origin in $E[T]$. Since B-B is a bounded subset of $E[T]$, there is a positive integer q such that

$$\frac{1}{q}(B-B) \subset W$$

and therefore

$$z_{pm} \in W, \ p = q+1, \ q+2, \ldots$$

On the other hand, $W \cap (B-B)$ is a neighbourhood of the origin in B-B for the topology induced by $\mathcal{U}_{n_{q+1}}$ and therefore there is a positive integer r such that

$$z_{pm} \in W, \ j = 1,2,\ldots, \ q; \ m = r+1, \ r+2, \ldots$$

Consequently

$$z_{pm} \in W \quad \text{for } p+m > q+r$$

and therefore (5) converges to the origin in $E[T]$. Then there is a positive integer a such that (5) converges to the origin in $E_s\ [T_s]$. Consequently the subsequence z_{sm}, m = 1,2,..., of (5) converges to the origin in $E_{n_{s+1}}[T_{n_{s+1}}]$ which is a contradiction.

Let H be a locally convex space. A sequence (x_m) converges to x in the sense of Mackey, or (x_m) is locally convergent to x, if there is a bounded closed absolutely convex subset B of H such that $x_m \in B$, m = 1,2, ..., and (x_m) converges to x in H_B. The space H satisfies the Mackey condition if every sequence of H which converges to the origin, is convergent to the origin in the sense of Mackey. H satisfies the strict Mackey condition if, given any bounded subset A of H, there is a bounded closed absolutely convex subset B of H such that A is contained in B and the topology of H_B coincides on A with the topology on H. It is obvious that if H satisfies the strict Mackey condition, then H satisfies the Mackey condition.

(6) *Let* E *be an* (LN)*-space. If* E *satisfies the Mackey condition, then* E *satisfies the strict Mackey condition.*

Proof. Proceeding as we did in the proof of 4.(10), E can be represented as the inductive limit of a sequence of normed spaces ($E_n\ [T_n]$) such that there is an absolutely convex bounded neighbourhood of the origin U_m in $E_m\ [T_m]$ such that 2 U_m is contained in U_{m+1} and given any bounded subset A of E there is a positive integer q such that A is contained in U_q.

Given a positive integer n, we set B = U_n. Then B is a balanced bounded subset of E contained in $E_n\ [T_n]$. Let (x_n) be a sequence in B-B converging to the origin in E. Since E satisfies the Mackey condition, there is a bounded closed absolutely convex subset D of E such that B-B is contained in D and (x_m) converges to the origin in E_D. We find a positive integer p such that D is contained in U_p. Then (x_m) converges to the origin in $E_p[T_p]$. We apply (4) to obtain an integer q > n such that B has the same convergent sequences for the topologies induced by T_q and by topology of E. Since E_n $[T_n]$ is metrizable, n = 1,2,..., Then T_{q+r} and T_q coincide on U_n = B. Consequently E has property M.

Given any bounded subset A of E we find a positive integer n such that A is contained in U_n. By (2), there is an integer h > n such that the topology of E coincides with T_h on U_n. Consequently the topology of E coincides with the topology of E_{U_n} on A. The conclusion follows.

We say that the inductive limite $E[T]$ of the sequence ($E_n\ [T_n]$) of locally convex spaces is sequentially semi-retractive if, given any sequence (x_p) in $E[T]$ converging to the origin and regular, there is a positive integer n such that (x_p) is in $E_n\ [T_n]$ and converges to the origin in $E_n\ [T_n]$.

We say that $E[T]$ is boundedly semi-retractive if given any bounded regular subset A of E, there is a positive integer n such that A is contained in E_n and T_n coincides with T on A.

(7) *If* $E\ [T]$ *is the inductive limit of the sequence of normed spaces* (E_n $[T_n]$), *the following conditions are equivalent:*

a) $E\,[T]$ *is sequentially semi-retractive;*

b) $E\,[T]$ *has property* M;

c) $E\,[T]$ *is boundedly semi-retractive.*

Proof. For every positive integer m, we take a bounded absolutely convex neighbourhood of the origin U_m in $E_m\,[T_m]$ such that $2U_m$ is contained in U_{m+1}. If a) holds, we proceed as we did in the proof of (6) to obtain, for any positive integer n, an integer $h > n$ such thtat T_h coincides with T on U_n. Consequently $E\,[T]$ has property M. Now we suppose that b) holds. We every positive integer m we find an absolutely convex neighbourhood of the origin V_m in $E_m\,[T_m]$ such that, given any positive integer n, there is an integer $h > n$ such thtat T_h coincides with T on V_n. Given a bounded regular subset A of E we take n such that A is bounded in $E_n\,[T_n]$ and we find $k > 0$ such that kA is contained in V_n. Then T_h and T coincide on kA and accordingly coincide on A. Finally a) follows from c).

The inductive limit $E[T]$ of the sequence $(E_n\,[T_n])$ is sequentially retractive if given any sequence (x_m) in $E[T]$ converging to the origin there is a positive integer n such that (x_m) is contained in $E_n\,[T_n]$ and converges to the origin in this space (FLORET [1]).

$E[T]$ is sequentially compact-regular if given any sequentially compact A of $E[T]$ there is a positive integer n such that A is contained in $E_n\,[T_n]$ and A is sequentially compact in this space (NEUS [1]).

$E[T]$ is compact-regular if given any compact subset A of $E[T]$ there is a positive integer n such that A is contained in $E_n\,[T_n]$ and A is compact in this space (BIERSTEDT and MEISE [1]).

$E[T]$ is boundedly retractive if given any bounded subset A of $E[T]$ there is a positive integer n such that A is contained in E_n and T coincides with T_n on A (BIERSTEDT and MEISE [1]).

$E[T]$ is strongly boundedly retractive if given a positive integer n there is an integer $h > n$ such that for every bounded subset A of $E[T]$ contained in $E_n\,[T_n]$ the topologies T and T_h coincide on A (BIERSTEDT and MEISE [1]),

(8) *If* $E[T]$ *is sequentially retractive, then* $E[T]$ *is sequentially compact regular.*

Proof. Let A be a sequentially compact subset of $E[T]$ Suppose A is not in E_m, m = 1,2,... For every positive integer m, take x_m in $A \sim E_m$. We extract a subsequence (z_m) from the sequence (x_m) which is convergent in $E[T]$ to a point z of A. Then $(z_m - z)$ converges to the origin in $E[T]$ and is not in any E_q. That is a contradiction.

We find a positive integer p such that A is contained in E_p. Let B be the balanced hull of A. Then B is sequentially compact and therefore bounded. Every sequence of B-B which converges to the origin in $E[T]$, T_q-converges to the origin for some positive integer q. Therefore we can apply (4) to obtain a positive integer $n > p$ such that B has the same convergent sequences for T and T_n. Thus B is sequentially compact in $E_n[T_n]$. The proof is complete.

(9) *If* $E\,T$ *is the inductive limit of the sequence of normed spaces* $((E_n)([T_n]))$, *the following conditions are equivalent:*

1) $E[T]$ *is sequentially retractive;*
2) $E[T]$ *is sequentially compact-regular;*
3) $E[T]$ *is compact-regular;*
4) $E[T]$ *is boundedly retractive;*
5) $E[T]$ *is strongly boundedly retractive.*

Proof. If 1) is satisfied, then 2) follows from (8). Now suppose that 2) holds. Given a compact subset A of $E[T]$, then A is sequentially compact and therefore there is a positive integer p such that A is contained in $E_p[T_p]$ and sequentially compact in this space. Therefore A is T_p-compact and 3) follows.

If 3) holds, let A be a bounded subset of $E[T]$ and suppose A not contained in E_n, n = 1,2,..., We select x_m in $A \sim E_m$, m = 1,2,... Then $\{0, x_1, \frac{1}{2}x_2, \dots, \frac{1}{m}x_m, \dots\}$ is a compact subset of $E[T]$ which is not contained in E_n, n = 1,2,..., and this is a contradiction. Therefore there is a positive integer p such that A is contained in E_p. Let B the balanced hull of A. Then B is bounded in $E[T]$. Let (z_m) be a sequence in B-B T-convergent to the origin. Since

(10) $\{0, z_1, z_2, \ldots, z_n, \ldots\}$

is T-compact, there is an integer $r > p$ such that (10) is T_r-compact and thus (z_m) T_r-converges to the origin. We apply (4) to obtain an integer $n > p$ such that B has the same convergent sequences for the topologies T_n and T. If (a_m) is any sequence of elements of K converging to zero with $|a_m| \leq 1$, $m = 1,2,\ldots$, and if (y_m) is a sequence in B, the T-boundedness of B implies that $(a_m y_m)$ T-converges to the origin and thus T_n-converges to the origin from where it follows that B is T_n-bounded. Consequently A is regular. Then $E[T]$ is sequentially semi-retractive and every bounded subset of $E[T]$ is regular. We apply (7) to obtain that $E[T]$ is boundedly semi-restrictive from where 4) follows.

Suppose 4) true. Let U_1 be a bounded absolutely convex neighbourhood of the origin in $E[T_1]$. We set $n_1 = 1$. Proceeding by recurrence suppose that a positive integer n_p and a bounded absolutely convex neighbourhood of the origin U_{n_p} in $E_{n_p}[T_{n_p}]$. We find an integer $n_{p+1} > n_p$ such that $T_{n_{p+1}}$ coincides with T on U_{n_p}. Let $U_{n_{p+1}}$ be a bounded absolutely convex neighbourhood of the origin in $E_{n_{p+1}}[T_{n_{p+1}}]$ which contains U_{n_p}.

Given a positive integer n, we find an integer q such that $n < n_q$. If D is a bounded subset of $E_n[T_n]$ there is $k > 0$ such that kD is contained in U_n and therefore $T_{n_{q+1}}$ and T coincide on kD and thus on D. Consequently 5) follows from 4). Clearly 1) is implied by 5).

If the inductive limit $E[T]$ of the sequence of normed spaces $(E_n[T_n])$ is sequentially retractive then it is obviously sequentially semi-retractive and, according (7), $E[T]$ has property M. On the other hand, $E[T]$ can be property M and be not sequentially retractive as the following example shows: Let E be an infinite dimensional Banach space. Take a subspace F of E of infinite countable codimension. Let $\{x_1, x_2, \ldots, x_m, \ldots\}$ be a cobasis of F in E. We set F_n to denote the subspace of E, linear hull of $F \cup \{x_1,$

$x_2,\ldots,x_n\}$. Then E coincides with $\cup \{F_n : n = 1,2,\ldots\}$, and since E is a Baire space there is a positive integer p such that F_p is dense in E. We set $E_n = F_{n+p}$, n = 1,2,... Then E_1 is dense in E, E_n is contained in E_{n+1}, E_n is distinct from E_{n+1}, n = 1,2,..., and E is the inductive limit of the sequence of normed spaces (E_n). Let B be the closed unit ball of E. We set $U_n = B \cap E_n$, n = 1,2,..., Then, for every positive integer m, U_m is absolutely convex neighbourhood of the origin in E_m, U_m is contained in U_{m+1} and the topology of E_{m+1} coincides with the topology of E on U_m. Therefore the inductive limit E of (E_n) has property M. On the other hand, take a point x_m in $E_{m+1} \sim E_m$ and determine an increasing sequence (n_p) of positive integers such that $(\frac{1}{n_p} x_p)$ converges to the origin in E. The sequence $(\frac{1}{n_p} x_p)$ is not contained in E_n, n = 1,2,..., and therefore E is not sequentially retractive.

Results (2), (3) and (7) appear here for the first time. Result (4), (6), (8) and (9) can be found in NEUS [1]. The proofs of (6) and (9) in NEUS [1] are distinct from the proofs given here.

CHAPTER TWO
SEQUENCE SPACES

In this Chapter a detailed exposition of the perfect, echelon and co-echelon spaces of G. KÖTHE is included. Examples of sequence spaces which answer several questions on aspects of the general theory of locally convex spaces are given. An example of a Banach space which is an hyperplace of its strong bidual due to R.C. James inspires the end of the Chapter where a construction of some vector-valued sequence spaces can be found.

§ 1. SCALAR SEQUENCE SPACES

1. SCALAR SEQUENCE SPACES. A sequence space λ is a linear space of sequences $x = (x_m) = (x_1, x_2, \ldots, x_m, \ldots)$ in K. If

$$y = (y_m) \in \lambda, \quad z = (z_m) \in \lambda, \quad h \in K,$$

then

$$y + z = (y_m + z_m), \quad h\, y = (h\, y_m).$$

ω is the space of all the sequences. ϕ is the subspace of ω of all sequences $x = (x_m)$ which have finitely many non-zero x_m. In what follows we shall consider sequence spaces containing ϕ.

To every sequence space λ we associate the sequence space $\lambda^{\times} = \lambda^{\alpha}$ of all those sequences $u = (u_m)$ such that

$$\Sigma\, |x_m\, u_m| < \infty$$

for every $(x_m) \in \lambda$ and we call $\lambda^{\times}$ the α - dual of λ.

We consider the dual pair $<\lambda,\lambda^{\times}>$ with the bilinear form

$$<x, u> = \Sigma x_m u_m, \; x = (x_m) \in \lambda, \quad u = (u_m) \in \lambda^{\times}$$

It is obvious that λ is contained in $(\lambda^{\times})^{\times} = \lambda^{\times\times}$. λ is perfect if $\lambda = \lambda^{\times\times}$. λ is normal if $(h_m x_m) \in \lambda$, $|h_m| \leq 1$, when $(x_m) \in \lambda$.

$\lambda^{\times}$ is always a perfect space. The spaces ℓ^p, $1 \leq p \leq \infty$, are also perfect (cf. KÖTHE [1], Chapter Six, § 30, Section 1).

The space c_0 of all the sequences converging to zero is normal. On the other hand, its α- dual is ℓ^1 and the α - dual of ℓ^1 is ℓ^{∞}. Thus c_0 is not perfect.

The sequence $z = (z_m)$, $z_m=1$, $m=1,2,\ldots$ is an element of ℓ^{∞}_0 but the sequence $(\frac{1}{m})$ is not an element of ℓ^{∞}_0 and thus ℓ^{∞}_0 is not normal

(1) *ϕ is total in* $\lambda[\sigma(\lambda,\lambda^{\times})]$.

Proof. Let e_m be the sequence with nulle coordinates but the m-- th coordinate which is 1, m = 1,2, ..., 2. If $u = (u_m) \in \lambda$ is not nulle there is a positive integer p such that $u_p \neq 0$. But $e_p \in \phi$ and

$$<e_p, u> = u_p \neq 0,$$

and the conclusion follows.

2. SUBSETS IN A SEQUENCE SPACE. Given a scalar sequence $u=(u_m)$ we set

$$u^n = \{v = (v_m) \in \omega: |v_m| \leq |u_m|, \quad m = 1,2,\ldots\}$$

and we call it the normal hull of $\{u\}$. Given a subset A in ω we define the normal hull A^n of A as $A^n = \cup\{u^n : u \in A\}$. If $A = A^n$ we say that A is normal.

Given a sequence space λ, a bounded set A in $\lambda[(\lambda,\lambda^{\times})]$ and an element $u = (u_m) \in \lambda^{\times}$ we set

$$P_A(u) = \sup\{|\Sigma x_m u_m| : (x_m) \in A\},$$

$$P_{(A)}(u) = \sup\{\Sigma |x_m u_m| : (x_m) \in A\}.$$

The polar set A° of A in $\lambda^{\times}$ coincides with $\{v \in \lambda^{\times} : p_A(v) \leq 1\}$.

Analogously, if B is a bounded set in $\lambda^{\times}[\sigma(\lambda^{\times},\lambda)]$ and if $x = (x_m) \in \lambda$, we set

$$p_B(x) = \sup \{|\Sigma x_m u_m| : (u_m) \in B\}$$

$$p_{(B)}(x) = \sup \{\Sigma |x_m u_m| : (u_m) \in B\}.$$

The polar set B^o of B in λ coincides with

$$\{z \in \lambda : p_B(z) < 1\}.$$

Let A be a family of normal absolutely convex closed and bounded sets in $\lambda^{\times}[\sigma(\lambda^{\times},\lambda)]$ covering $\lambda^{\times}$ and such that the following conditions hold: a) If $A, B \in A$ there is $C \in A$ such that $A \cup B \subset C$. b) If $h \in K$ and if $A \in A$ then $hA \in A$. The family

$$\{A^o : A \in A\}$$

is a fundamental system of 0-neighbourhoods in λ for a locally convex topology T on λ.

Let B be a family of normal absolutely convex closed and bounded sets in $[\sigma(\lambda,\lambda^{\times})]$ covering λ and such that properties a) and b) are verified taking A as B. The family

$$\{B^o : B \in B\}$$

is a fundamental system of 0-neighbourhoods in $\lambda^{\times}$ for a locally convex topology U on $\lambda^{\times}$.

(1) *The system of seminorms*

(2) $$\{p_{(A)} : A \in A\}$$

defines the topology T *on* λ.

Proof. It is obvious that

$$\{x \in \lambda : p_{(A)}(x) \leq 1\} \subset A^o \quad \text{if} \quad A \in A$$

On the other hand, if $z = (z_m)$ belongs to A^o and if $u = (u_m)$ belongs to A there are $h_m \in K$ with $|h_m| = 1$ and

$$z_m h_m u_m = |z_m u_m|, \quad m = 1,2,\ldots$$

Since A is normal, $(h_m u_m) \in A$ and thus

$$\left|\sum_{m=1}^{\infty} z_m h_m u_m\right| = \sum_{m=1} |z_m u_m| \leq 1.$$

Thus

$$A^o \subset \{x \in \lambda : p_{(A)}(x) \leq 1\}$$

and then

$$A^o = \{x \in \lambda : p_{(A)}(x) \leq 1\}$$

and the topology T is defined by the system of seminorms (2).

Analogously follows:

(3) *The system of seminorms*

$$\{p_{(B)} : B \in \mathcal{B}\}$$

defines the topology U *on* $\lambda^\times$.

(4) *If* M *is a bounded set in* $\lambda[T]$ *then* $M^n \cap \lambda$ *is bounded.*

Proof. Given $A \in \mathcal{A}$ there is a positive number $h > 0$ such that

$$p_{(A)}(x) \leq h, \text{ for all } x \in M.$$

If $y = (y_m)$ belongs to $M^n \cap \lambda$ there is $x = (x_m) \in M$ such that

$$|y_m| \leq |x_m|, \quad m = 1,2,\ldots$$

Since $p_{(A)}(y) \leq p_{(A)}(x) \leq h$ the conclusion follows.

Analogously follows :

(5) *If* M *is a bounded set* $\lambda^\times [U]$ *then* M^n *is bounded.*

If a,b are elements of K we write

$$(a;b) = \frac{a}{b} \text{ if } b \neq 0, \ (a;b) = 0 \text{ if } b = 0$$

We set $|a;b|$ instead of $|(a;b)|$.

(6) *If* A *is a normal set of* ω *then its absolutely convex hull* B *is normal.*

Proof. Let $z = (z_m)$ be an element of B^n. There are elements

$$x^j = (x_m^{(j)}) \in A \text{ and } h_j \in K, \quad j = 1,2,..,q, \quad \text{with } \sum_{j=1}^{q} |h_j| \leq 1,$$

such that

$$|z_m| \leq \left|\sum_{j=1}^{q} h_j x_m^{(j)}\right|, \quad m = 1,2,\ldots$$

We find $h_m^{(j)} \in K$ such that

$$h_m^{(j)} h_j x_m^{(j)} = |h_j x_m^{(j)}| \quad \text{and} \quad |h_m^{(j)}| = 1$$

Setting

$$k_m = (h_m^{(j)} z_m ; \sum_{j=1} | h_j x_m^{(j)}|)$$

it follows that

$$y^j = (k_m x_m^{(j)}) \in A$$

since A is normal and $|k_m| < 1$. Thus

$$\sum_{j=1}^{q} h_j y^j = \sum_{j=1}^{q} (h_j (h_m^{(j)} z_m ; \sum_{j=1}^{q} |h_j x_m^{(j)}|) x_m^{(j)}) = (z_m) \in B,$$

and therefore $B = B^n$.

(7) *If* M *is a normal subset in* λ *then its closure* P *in* $\lambda[T]$ *is normal.*

Proof. Let $z = (z_m)$ be an element of P^n and let $x = (x_m)$ be an element of P such that

$$|z_m| \leq |x_m|, \quad m = 1,2,...$$

Let

$$\{x^j = (x_m^{(j)}) : j \in J, > \}$$

be a net in M T-converging to x. Setting

$$z^j = ((z_m ; x_m) x_m^{(j)})$$

it follows that the net

(8) $\{ z^j : j \in J, > \}$

is contained in M since

$$| (z_m ; x_m) x_m^{(j)}| < | x_m^{(j)} |, \quad j \in J, \quad m = 1,2,...$$

On the other hand, if $A \in \mathcal{A}$ and $j \in J$

$$p_{(A)}(z^j - z) = \sup \{ \Sigma |(z_m^{(j)} - z_m) u_m| : (u_m) \in A \}$$

$$= \sup \{ \Sigma |(z_m ; x_m)(x_m^{(j)} - x_m) u_m | : (u_m) \in A \}$$

$$\leq \sup \{ \Sigma |(x_m^{(j)} - x_m) u_m | : (u_m) \in A \} = p_{(A)}(x^j - x),$$

and thus the net (8) T-converges to z and $z \in P$. Thus $P = P^n$

Analogously, follows:

(9) *If* M *is a normal set in* $\lambda^{\times}$ *then its closure in* $\lambda^{\times}[U]$ *is normal.*

3. THE NORMAL TOPOLOGY OF A SEQUENCE SPACE. We have the following result:

(1) *If* λ *is a sequence space and if* $u = (u_m) \in \lambda^{\times}$ *then* u^n *is a* $\sigma(\lambda^{\times},\lambda)$ *-compact absolutely convex subset of* $\lambda^{\times}$.

Proof. If $v = (v_m)$ and $w = (w_m)$ are elements of u^n and if, h, $k \in K$ with $|h| + |k| \leq 1$ then

$$|hv_m + kw_m| \leq (|h| + |k|)\,|u_m| \leq |u_m|, \qquad m = 1,2,\ldots,$$

and thus u^n is contained in $\lambda^{\times}$ and it is absolutely convex. Let

$$(2) \qquad \{ v^j = (v_m^{(j)}) : j \in J, \geqslant \}$$

be a net in u^n. Then $|v_m^{(j)}| \leq |u_m|$, $j \in J$, $m = 1,2,\ldots,$ and therefore a subnet of (2), which we denote by (2) again, can be extracted such that

$$\lim \ \{v_m^{(j)} : j \in J, \geqslant\} = v_m \in K, \qquad m = 1,2, \ldots$$

We set $v = (v_m)$. Clearly $|v_m| \leq |u_m|$, $m = 1,2,\ldots,$ and therefore v belongs to u^n. On the other hand, given $\varepsilon > 0$ and $x = (x_m) \in \lambda$ there is a positive integer q such that

$$\sum_{m=q+1}^{\infty} |x_m u_m| < \frac{\varepsilon}{4}$$

we determine an index $i \in J$ such that

$$|x_m (v_m^{(j)} - v_m)| < \frac{\varepsilon}{2q}, \quad j \in J, \quad j \geqslant i, \quad m = 1,2,\ldots,q.$$

If $j \in J$, $j \geqslant i$, it follows that

$$|\Sigma\, x_m (v_m^{(j)} - v_m)| \leq \sum_{m=1}^{q} |x_m(v_m^{(j)} - v_m)| + \sum_{m=q+1}^{\infty} |x_m v_m^{(j)}|$$

$$+ \sum_{m=q+1}^{\infty} |x_m v_m| < \frac{\varepsilon}{2} + 2 \sum_{m=q+1}^{\infty} |x_m u_m| < \varepsilon$$

and thus the net (2) converges to v in $\lambda^{\times}[\sigma(\lambda^{\times},\lambda)]$ and therefore u^n is $\sigma(\lambda^{\times}, \lambda)$ - compact.

(3) *The family* $N = \{u^n : u \in \lambda^\times\}$ *satisfies the following conditions:*

a) $\cup\{A : A \in N\} = \lambda^\times$;

b) *if* $A, B \in N$ *there is* $C \in N$ *such that* $A \cup B \subset C$

c) *if* $h \in K$ *and if* $A \in N$ *then* $h A \in N$.

Proof. If $u \in \lambda^\times$ then $u \in u^n$ and thus N covers $\lambda^\times$. On the other hand, if $u = (u_m) \in \lambda^\times$, $v = (v_m) \in \lambda^\times$ and $h \in K$ it follows that $w = (|u_m|+|v_m|) \in \lambda^\times$, $u^n \cup v^n \subset w^n$ and $hu^n = (hu)^n$.

Since N satisfies the conditions imposed on family A of the former section, the polar sets in λ of the elements of family N are a fundamental system of 0-neighbourhoods in λ for a locally convex topology $\nu(\lambda,\lambda^\times)$. According to (1), $\nu(\lambda,\lambda^\times)$ is compatible with the dual pair $\langle\lambda,\lambda^\times\rangle$. The topology $\nu(\lambda,\lambda^\times)$ is the normal topology of the sequence space λ and it is defined by the system of seminorms

$$\{P_{(u^n)} : u \in \lambda^\times\}.$$

If $x = (x_m)$, $y = (y_m)$ are in ω and if $y \geqslant 0$, i.e., $y_m \geqslant 0$, $m = 1, 2, \ldots$, we set

$$P_y(x) = \Sigma|x_m|\, y_m.$$

If $u = (u_m) \in \lambda^\times$, we set $v = (|u_m|)$ and $\bar{p}_{(u^n)} = p_v$ and, thus, $\nu(\lambda,\lambda^\times)$ can be defined by the system of seminorms

$$\{p_u : u \geqslant 0,\ u \in \lambda^\times\}.$$

If λ is normal and if x is in λ, then x^n is included in λ. According to (1), x^n is $\sigma(\lambda^{\times\times},\lambda^\times)$-compact and thus $\sigma(\lambda,\lambda^\times)$-compact. If M denotes the family $\{z^n : z \in \lambda\}$, we can prove (as we did for family N) that M satisfies the properties enjoyed by family B of the former section. Therefore, the topology $\nu(\lambda^\times,\lambda)$ on $\lambda^\times$ of the uniform convergence on each element of family M is compatible with the dual pair $\langle\lambda,\lambda^\times\rangle$ and it can be described by the system of seminorms

$$\{p_{(x^n)} : x \in \lambda\}$$

and also by

$$\{p_x : x \geqslant 0,\ x \in \lambda\}.$$

(4) *If* M *is a bounded set in* $\lambda[\sigma(\lambda,\lambda^\times)]$, *then* $M^n \cap \lambda$ *is* $\sigma(\lambda,\lambda^\times)$*-bounded.*

Proof. Since $\nu(\lambda,\lambda^\times)$ is compatible with the dual pair $\langle\lambda,\lambda^\times\rangle$, then

M is $\upsilon(\lambda,\lambda^{\times})$-bounded and $M^n \cap \lambda$ is $\upsilon(\lambda,\lambda^{\times})$-bounded by 2. (4). The conclusion follows.

Analogously follows:

(5) *If* λ *is normal and if* M *is a bounded set in* $\lambda^{\times}[\sigma(\lambda^{\times},\lambda)]$, *then* M^n *is* $\sigma(\lambda^{\times},\lambda)$ *-bounded.*

Results (6) and (7) are obvious consequences of 2. (7) and 2. (9) respectively.

(6) *If* M *is a normal subset of* λ, *then its closure in* $\lambda[\upsilon(\lambda,\lambda^{\times})]$ *is normal.*

(7) *If* λ *is normal and if* M *is a normal subset of* $\lambda^{\times}$, *then the closure of* M *in* $\lambda^{\times}[\upsilon(\lambda^{\times},\lambda)]$ *is normal.*

If $n_1, n_2, n_m, \ldots,$ is a reordering of the positive integers, let $T : \omega \to \omega$ be the bijection $Tx = (x_{n_m})$ with $x = (x_m) \in \omega$. Given a sequence space λ_1, we set $\lambda_2 = T(\lambda_1)$. It is clear that $\lambda^{\times}_2$ coincides with $T(\lambda^{\times}_1)$. If $z = (z_m)$ belongs to λ_1 and $u = (u_m)$ belongs to $\lambda^{\times}_1$ it follows that

$$< z, u > = \Sigma z_m u_m = \Sigma z_{n_m} u_{n_m} = < T x, T u >$$

and therefore T establishes and isomorphism between $\lambda_1[\sigma(\lambda_1,\lambda^{\times})]$ and $\lambda_2[\sigma(\lambda_2,\lambda^{\times}_2)]$, $\lambda_1[\upsilon(\lambda_1,\lambda^{\times}_1)]$ and $\lambda_2[\upsilon(\lambda_2,\lambda^{\times}_2)]$ and $\lambda_1[\mu(\lambda_1,\lambda^{\times}_1)]$ and $\lambda_2[\mu(\lambda_2,\lambda^{\times}_2)]$. Accordingly, for the study of some topological properties in sequence spaces, the order in which the coordinates are placed is not fundamental and, thus, index sets different from the positive integers can be used For instance, λ can be a space of double sequences $x = (x_{pq})$ containing the space ϕ of all double sequences whose elements are all zero but a finite number of them. Then its α- dual $\lambda^{\times}$ is formed by all double sequences (u_{pq}) such that $\Sigma \mid x_{pq} u_{pq} \mid < \infty$ being the linear form of the dual pair $< \lambda\ \lambda^{\times} >$ defined by $\Sigma x_{pq} u_{pq}$.

4. PROPERTIES OF THE α - DUAL OF A SEQUENCE SPACE. We have the following result:

(1) *If the sequence space* λ *is normal, then* $\lambda^{\times}[\upsilon(\lambda^{\times},\lambda)]$ *is complete.*

Proof. Let

(2) $\{u^j = (u_m^{(j)}) : j \in J, \geq\}$

be a $\nu(\lambda^\times,\lambda)$-Cauchy net in $\lambda^\times$. Then (2) is a $\sigma(\lambda^\times,\phi)$-Cauchy net and therefore

$$\lim \{<e_m,u^j> : j \in J, \geq\} = \lim \{u_m^{(j)} : j \in J, \geq\} = u_m \in K, m = 1,2,..$$

If $x = (x_m)$ belongs to λ, $x \geq 0$, and if $\varepsilon > 0$, and index $i \in J$ can be selected such that

$$p_x (u^h-u^k) = \Sigma x_m |u_m^{(h)}-u_m^{(k)}| \leq \varepsilon, \ h, k \in J, h, k \geq i.$$

Given any positive integer r we have that

$$\sum_{m=1}^{r} x_m |u_m^{(h)}-u_m^{(k)}| < \varepsilon$$

and thus

$$\sum_{m=1}^{r} x_m |u_m^{(h)}-u_m| \leq \varepsilon$$

and so

(3) $\Sigma x_m |u_m^{(h)}-u_m| \leq \varepsilon$

From (3) it follows that if we set $u = (u_m)$ then

$$u^h-u \in \lambda^\times$$

and, therefore

$$u^h- (u^h-u) = u \in \lambda^\times$$

and the net (2) converges to u in $\lambda^\times[\nu(\lambda^\times,\lambda)]$.

The following result follows from (1):

(4) *If λ is normal then $\lambda^\times[\mu(\lambda^\times,\lambda)]$ is complete.*

Result (9) is an extension of result-(12) of SCHUR. In order to prove (5) we shall use the "sliding hump" method due to LEBESGUE-TOEPLITZ.

(5) *Let λ be a normal sequence space. If*

(6) $u^r = (u_m^{(r)}), \ r = 1,2,...,$

is a sequence in $\lambda^\times[\sigma(\lambda^\times,\lambda)]$ converging to the origin, then (6) converges to the origin in $\lambda^\times[\nu(\lambda^\times,\lambda)]$.

Proof. If (6) does not converge to the origin in $\lambda^\times[\nu(\lambda^\times,\lambda)]$ we can find $\varepsilon > 0$, $x = (x_m) \in \lambda$, $x \geq 0$, and a subsequence of (6), which we denote by (6), such that

$$p_x(u^r) = \Sigma x_m |u_m^{(r)}| > 4\varepsilon, \ r = 1,2,\ldots$$

By recurrence, sequences of integers

$$1 = r_1 < r_2 < \ldots < r_p < \ldots$$

$$1 = m_1 < m_2 < \ldots < m_p < \ldots$$

can be constructed in the following way: we find m_2 such that

$$\sum_{m=1}^{m_2-1} x_m |u_m^{(1)}| > 3\varepsilon, \ \sum_{m=m_2}^{\infty} x_m |u_m^{(1)}| < \varepsilon.$$

Suppose we have obtained r_p, m_p and m_{p+1}, for an integer $p \geqslant 1$, and such that

(7) $$\sum_{m=1}^{m_p-1} x_m|u_m^{(r_p)}| < \varepsilon, \ \sum_{m=m_p}^{m_{p+1}-1} x_m|u_m^{(r_p)}| > 3\varepsilon, \ \sum_{m=m_{p+1}}^{\infty} x_m|u_m^{(r_p)}| < \varepsilon,$$

Since (6) converges to the origin for $\sigma(\lambda^\times,\lambda)$ and, therefore, coordinatewise, we determine $r_{p+1} > r_p$ such that

$$\sum_{m=1}^{m_{p+1}-1} x_m |u_m^{(r_{p+1})}| < \varepsilon.$$

Then

$$\sum_{m=m_{p+1}}^{\infty} x_m|u_m^{(r_{p+1})}| = \sum_{m=1}^{\infty} x_m|u_m^{(r_{p+1})}| - \sum_{m=1}^{m_{p+1}-1} x_m|u_m^{(r_{p+1})}| > 4\varepsilon-\varepsilon = 3\varepsilon$$

and, therefore, we find $m_{p+2} > m_{p+1}$ such that

$$\sum_{m=m_{p+1}}^{m_{p+2}-1} x_m|u_m^{(r_{p+1})}| > 3\varepsilon, \ \sum_{m=m_{p+2}}^{\infty} x_m |u_m^{(r_{p+1})}| < \varepsilon.$$

We consider now the sequence (h_m) in K such that

$$|h_m| = 1, \ m=1,2,\ldots, \ h_m u_m^{(r_p)} = |u_m^{(r_p)}|, \ m_p \leqslant m < m_{p+1}, \ 1=1,2,\ldots$$

Since λ is normal, the element $z = (h_m x_m)$ belongs to λ and, accordingly,

(8) $$\lim_p \langle z, u^{r_p}\rangle = 0$$

On the other hand, for $p = 1,2,\ldots$, it follows from (7) that

$$|\langle z,u^{r_p}\rangle| = |\Sigma h_m x_m u_m^{(r_p)}| \geqslant |\sum_{m=m_p}^{m_{p+1}-1} h_m x_m u_m^{(r_p)}$$

$$- \sum_{m=1}^{m_p-1} x_m |u_m^{(r_p)}| - \sum_{m=m_{p+1}}^{\infty} x_m |u_m^{(r_p)}| \geqslant \sum_{m=m_p}^{m_{p+1}-1} x_m|u_m^{(r_p)}| - 2\varepsilon > \varepsilon,$$

which is in contradiction with (8).

(9) *Let* λ *be a normal sequence space. If*

(10) $$u^r = (u_m^{(r)}),\ r = 1,2,\ldots,$$

is a Cauchy sequence in $\lambda^\times[\sigma(\lambda^\times,\lambda)]$, *then* (10) *converges in* $\lambda^\times[\nu(\lambda^\times,\lambda)]$.

Proof. According to (1), the space $\lambda^\times[\nu(\lambda^\times,\lambda)]$ is complete and, thus, it is enough to prove that (10) is a Cauchy sequence in $\lambda^\times[\nu(\lambda^\times,\lambda)]$. Let us suppose that (10) is not a $\nu(\lambda^\times,\lambda)$- Cauchy sequence. There is $\varepsilon>0$, $x = (x_m) \in \lambda$, $x \geq 0$, and a sequence

$$r_1 < r_2 < \ldots < r_p < \ldots$$

of positive integer sucht that

(11) $$\Sigma x_m \,|u_m^{(r_p)} - u_m^{(r_{p+1})}| > \varepsilon,\ p = 1,2,\ldots$$

It is obvious that the sequence

$$u^{r_p} - u^{r_{p+1}},\ p = 1,2,\ldots$$

converges to the origin in $\lambda^\times[\sigma(\lambda_\times,\lambda)]$ and, according to (11), it does not converge to the origin in $\lambda^\times[\nu(\lambda^\times,\lambda)]$, which is in contradiction with (5).

(12) *If*

(13) $$u^r = (u_m^{(r)}),\ r = 1,2,\ldots$$

is a Cauchy sequence in ℓ^1 $[\sigma(\ell^1,\ell^\infty)]$, *Then* (13) *converges in the Banach space* ℓ^1.

Proof. Taking $\lambda = \ell^\infty$ it follows that $\lambda^\times = \ell^1$. If e is the sequence 1,1,...,1,... the closed unit ball of ℓ^∞ is e^n and therefore $\nu(\ell^1,\ell^\infty)$ is the norm topology of ℓ^1 and the conclusion follows from (9) observing that ℓ^∞ is normal.

From (9) it follows also

(14) *If* λ *is a normal sequence space, then* $\lambda^\times[\sigma(\lambda^\times,\lambda)]$ *is sequentially complete.*

(15) *If* λ *is a normal sequence space and if* A *is a compact set in* $\lambda^\times[\sigma(\lambda^\times,\lambda)]$, *then* A *is compact in* $\lambda^\times[\nu(\lambda^\times,\lambda)]$.

Proof. A is $\sigma(\lambda^\times,\lambda)$-metrizable since $\sigma(\lambda^\times,\phi)$ is a metrizable topology. If

$$T: A[\sigma(\lambda^{\times}, x)] \longrightarrow A[\nu(\lambda^{\times}, \lambda)]$$

is the canonical injection it follows, according to (9), that every convergent sequence of $A[\sigma(\lambda^{\times},\lambda)]$ is transformed by T in a convergent sequence in $A[\nu(\lambda^{\times},\lambda)]$ and thus T is continuous and, therefore, and homomorphism.

(16) *If* λ *is a normal sequence space and if* x *belongs to* λ, *then* $\sigma(\lambda,\lambda^{\times})$ *and* $\mu(\lambda,\lambda^{\times})$ *coincide on* x^n.

Proof. x^n is an equicontinuous set in λ for the topology $\nu(\lambda^{\times},\lambda)$ and, therefore, $\sigma(\lambda,\lambda^{\times})$ and the topology of the uniform convergence on the absolutey convex and compact sets of $\lambda^{\times}[\nu(\lambda^{\times},\lambda)]$ coincide on x^n. The conclusion follows from (15).

If $x = (x_m)$ is an element of ω we call section of order p of x, and we represent it by $x(p)$, to the sequence (z_m) such that $z_r = x_r$ $r = 1,2,\ldots, p$, $z_r = o$, $r > p$.

The following three results can be found also in KÖTHE's book [1].

(17) *Given the sequence space* λ, $\lambda^{\times\times}$ *is the sequential completion of* $\lambda[\nu(\lambda,\lambda^{\times})]$.

Proof. The topologies $\nu(\lambda^{\times\times},\lambda^{\times})$ and $\nu(\lambda,\lambda^{\times})$ coincide on λ. $\lambda^{\times\times}[\nu(\lambda^{\times\times},\lambda^{\times})]$ is complete, according to (1). The sequential completion E of $\lambda[\nu(\lambda,\lambda^{\times})]$ is the minimal sequentially complete subspace of $\lambda^{\times\times}[\nu(\lambda^{\times\times},\lambda^{\times})]$ containing λ. If $x \in \lambda^{\times\times}$, then

$$x, x(m) \in x^n, \quad m = 1,2,\ldots$$

and, since x^n is $\sigma(\lambda^{\times\times},\lambda^{\times})$-compact, we apply (15) to obtain that x^n is $\nu(\lambda^{\times\times},\lambda^{\times})$-compact. Since the sequence $x(m)$, $m = 1,2,\ldots$, of λ converges to x for the topology $\sigma(\lambda^{\times\times},\phi)$ it also converges to x for $\nu(\lambda^{\times\times},\lambda^{\times})$ and, therefore, $E = \lambda^{\times\times}$. It also follows that $\lambda^{\times\times}[\nu(\lambda^{\times\times},\lambda^{\times})]$ is the completion of $\lambda[\nu(\lambda,\lambda^{\times})]$.

(18) *The sequence space* λ *is perfect if and only if* $\lambda[\sigma(\lambda,\lambda^{\times})]$ *is sequentially complete.*

Proof. If λ is perfect, then $\lambda = \lambda^{\times\times}$ and (14) implies the sequential completeness of $\lambda[\sigma(\lambda,\lambda^{\times})]$. If λ is not perfect, then $\lambda[\sigma(\lambda,\lambda^{\times})]$ is not sequentially complete according go (17).

(19) *The sequence space* λ *is perfect if and only if* λ *is* $\nu(\lambda,\lambda^{\times})$*-complete.*

Proof. If λ is perfect then $\lambda = \lambda^{\times\times}$ and, according to (1),

$\lambda[\nu(\lambda,\lambda^{\times})]$ is complete. If λ is not perfect we apply (17) to obtain that $\lambda[\nu(\lambda,\lambda^{\times})]$ is not complete.

In order to prove theorems (20) and (25) we consider the normal sequence space λ and a family $\mathcal{B}$ of normal absolutely convex compact sets in $\lambda[\sigma(\lambda,\lambda^{\times})]$ covering λ and such that a) if $A, B \in \mathcal{B}$ there is $C \in \mathcal{B}$ such that $A \cup B \subset C$; b) if $h \in K$ and if $A \in \mathcal{B}$ then $hA \in \mathcal{B}$. We denote by $\mathcal{U}$ the topology on $\lambda^{\times}$ of the uniform convergence on each element of $\mathcal{B}$.

(20) *Let* A *be a bounded set in* $\lambda^{\times}[\mathcal{U}]$. A *is relatively compact if and only if for every* $B \in \mathcal{B}$ *the decreasing sequence.*

(21) $$\sup\{\sum_{m=r}^{\infty} |x_m u_m| : (x_m) \in B, (u_m) \in A\}, \ r = 1,2,\ldots,$$

converges to zero.

Proof. Let us suppose that the sequence (21) converges to zero. Let

(22) $$\{u^j = (u_m^{(j)}) : j \in J, \geq\}$$

be a net in A. Since A is $\sigma(\lambda^{\times},\phi)$-bounded, we can find a subnet of (22), which we denote again by (22), such that

(23) $$\lim \{u_m^{(j)} : j \in J, \geq\} = u_m \in K, \ m = 1,2,\ldots,$$

Given $\varepsilon > 0$ and $B \in \mathcal{B}$ we can find a positive integer p such that for every $x = (x_m) \in B$ and for every $j \in J$

$$\sum_{m=p+1}^{\infty} |x_m u_m^{(j)}| < \frac{\varepsilon}{3} .$$

Then, if s is any positive integer, we have that

$$\sum_{m=p+1}^{s} |x_m u_m^{(j)}| < \frac{\varepsilon}{3}.$$

and thus

$$\sum_{m=p+1}^{s} |x_m u_m| \leq \frac{\varepsilon}{3}$$

and so

$$\sum_{m=p+1}^{\infty} |x_m u_m| < \frac{\varepsilon}{3} ,$$

implying that $u = (u_m)$ belongs to $\lambda^{\times}$, since $\mathcal{B}$ covers λ. Since B is $\sigma(\lambda,\phi)$-bounded, we have that $\sup\{|x_r| : x = (x_m) \in B, \ r = 1,2,\ldots,p\} = h < \infty$

According to (23), an index $i \in J$ can be selected such that

$$|h(u_m^{(j)} - u_m)| < \frac{\varepsilon}{3p}, \quad j \geq i,$$
$$m = 1,2, \ldots, p.$$

For every $x = (x_m) \in B$ and $j \geq i$ it follows that

$$| \langle x, u^j - u \rangle | = | \sum_{n=1}^{\infty} x_m (u_m^{(j)} - u_m)|$$
$$\leq \sum_{m=1}^{p} |h(u_m^{(j)} - u_m)| + \sum_{m=p+1}^{\infty} |x_m u_m^{(j)}|$$
$$+ \sum_{m=p+1}^{\infty} |x_m u_m| < \frac{\varepsilon}{3} + \frac{\varepsilon}{3} + \frac{\varepsilon}{3} = \varepsilon ,$$

and therefore (22) converges to u in $\lambda^{\times}[\,\mathcal{U}\,]$ and thus A is $\mathcal{U}$ -relatively compact.

We suppose now that (21) converges to $k > 0$. We take a sequence in A.

(24) $\quad u^r = (u_m^{(r)}, \qquad r = 1,2,\ldots,$

and a sequence of positive integers

$$1 = m_1 < m_2 < \ldots < m_r < \ldots$$

such that

$$\sup \{ \sum_{m=m_r}^{\infty} |x_m u_m^{(r)}| : (x_m) \in B\} > \frac{k}{2}.$$

If A is $\mathcal{U}$ - relatively compact, then A is $\sigma(\lambda^{\times},\lambda)$ - relatively compact and, thus, A is $\mathcal{U}$ - metrizable. Then a subsequences of (24), which we denote by (24) again, can be extracted $\mathcal{U}$ - converging to $u = (u_m) \in \lambda^{\times}$. Since $\sigma(\lambda^{\times},\phi)$ and $\mu(\lambda^{\times},\lambda^{\times\times})$ coincide on u^n, it follows that $u(r)$, $r = 1,2, \ldots,$ $\mu(\lambda^{\times},\lambda^{\times\times})$ - converges to u and, therefore, there is a positive integer q such that

$$\sup \sum_{m=m_q}^{\infty} |x_m u_m| : (x_n) \in B\} < \frac{k}{8}$$

The subset of K

$$\{x_r : r = 1, 2, \ldots, m_q, \ (x_m) \in B\}$$

is bounded and, thus, there is a positive integer $s > q$ such that for $r \geqslant s$

$$\sup \{\sum_{m=1}^{m_q - 1} |x_m (u_m^{(r)} - u_m)| : (x_m) \in B\} <$$

$$< \frac{k}{8}, \quad p_{(B)}(u^r - u) < \frac{k}{4}$$

Then

$$\frac{k}{4} > p_{(B)}(u^s - u) = \sup \{\sum_{m=1}^{\infty} |x_m(u_m^{(s)} - u_m)| : (x_m) \in B\}$$

$$\geqslant \sup \{\sum_{m=m_q}^{\infty} |x_m(u_m^{(s)} - u_m)| : (x_m) \in B\}$$

$$\sup \{\sum_{m=m_s}^{\infty} |x_m u_m^{(s)}| : (x_m) \in B\}$$

$$- \sup \{\sum_{m=m_s}^{\infty} |x_m u_m| : (x_m) \in B\}$$

$$- \frac{k}{8} \geqslant \frac{k}{2} - \frac{k}{8} - \frac{k}{8} = \frac{k}{4},$$

which is a contradiction. Thus A is $\mathcal{U}$- relatively compact.

If A is a subset of ω, we denote by A^{an} the minimal absolutely convex and normal subset of ω containing A which, according to 2. (6), coincides with the absolutely convex hull of A^n. We call A^{an} the normal absolutely convex hull of A.

(25) *Let* A *be a relatively compact set in* $\lambda^{\times}[\mathcal{U}]$. *Then* A^{an} *is relatively compact in* $\lambda^{\times}[\mathcal{U}]$.

Proof. Given $\varepsilon > 0$ and $B \in \mathcal{B}$ we apply (20) to obtain a positive integer q such that for $r \geqslant q$.

$$\sup \{ \sum_{m=r}^{\infty} |x_m u_m| : (x_m) \in B, \quad (u_m) \in A\} < \varepsilon$$

If $v = (v_m) \in A^{an}$, there are

$$u^j = (u_m^{(j)}) \in A, \qquad v^j = (v_m^{(j)}) \in \lambda^\times, \qquad h_j \in K,$$

$$j = 1, 2, \ldots, s,$$

such that

$$\sum_{j=1}^{s} |h_j| \leq 1, \quad |v_m^{(j)}| \leq |u_m^{(j)}|,$$

$$m = 1,2, \ldots, \qquad v = \sum_{j=1}^{s} h_j v^j.$$

If $r \geq q$ and if $x = (x_m)$ belongs to B, it follows that

$$\sum_{m=r}^{\infty} |x_m v_m| = \sum_{m=r}^{\infty} |x_m \sum_{j=1}^{s} h_j v_m^{(j)}|$$

$$\leq \sum_{j=1}^{s} |h_j| \sum_{m=r}^{\infty} |x_m u_m^{(j)}| < \varepsilon$$

and thus

$$\{\sup \sum_{m=r}^{\infty} |x_m v_m| : (x_m) \in B, \quad (v_m) \in A^{an}\} < \varepsilon$$

and, accordingly to (20) again, we have that A^{an} is relatively com - pact in $\lambda^\times[\mathcal{U}]$

Accordingly to (15), the results (26) and (27) are particular cases of (20) and (25), respectively.

(26) *Let* λ *be a normal sequence space and let* A *be a bounded set in* $\lambda^\times[\sigma(\lambda^\times,\lambda)]$. A *is* $\sigma(\lambda^\times,\lambda)$*-relatively compact if and only if for every* $x = (x_m) \in \lambda$, $x \geq 0$, *the sequence*

$$\sup \{ \sum_{m=r}^{\infty} x_m |u_m| : (u_m) \in A\}, \qquad r = 1,2,\ldots$$

converges to zero.

(27) *Let λ be a normal sequence space and let A be a relatively compact set in $\lambda^\times[\sigma(\lambda^\times,\lambda)]$. Then A^{an} is $\sigma(\lambda^\times,\lambda)$ - relatively compact.*

If λ is a normal sequence space and if A is relatively compact set in $\lambda^\times$ $[\sigma(\lambda^\times,\lambda)]$, then the closure B of A^{an} in $\lambda^\times[\sigma(\lambda^\times,\lambda)]$ is compact normal and absolutely convex (see 3.(7)) and, thus, if we call M to the family of all normal absolutely convex and compact sets in $\lambda^\times[\sigma(\lambda^\times,\lambda)]$ we have that

(28) *If λ is normal, then the topology $\mu(\lambda,\lambda^\times)$ can be described by the system of seminorms*

$$\{p_{(A)} : A \in M\}$$

(29) *If A is a compact set in $\lambda^\times[\,U\,]$, then A^n is compact.*

Proof. According to (25), A^n is U- relatively compact. It is enough to see that A^n is $\sigma(\lambda^\times,\lambda)$ - sequentially closed. Let

(30) $$u^r = (u_m^{(r)}), \qquad r = 1,2, \ldots,$$

be a sequence in A^n $\sigma(\lambda^\times,\lambda)$ - converging to $u = (u_m)$. For every positive integer r we can find $v^r = (v_m^{(r)}) \in A$ such that

(31) $$|u_m^{(r)}| \leq |v_m^{(r)}| \qquad m = 1,2, \ldots$$

We extract a subsequence from (30), which we denote by (30) again, such that $\lim_r v_m^{(r)} = v_m$, $m = 1,2,\ldots$ The element $v = (v_m)$ belongs to A and according to (31) $|u_m| \leq |v_m|$, $m = 1,2,\ldots$, and thus $u \in A^n$.

Result (32) is easy consequence of (29).

(32) *If λ is a normal sequence space, then the normal hull of every compact set in $\lambda^\times[\sigma(\lambda^\times,\lambda)]$ is compact.*

5. PRECOMPACT SETS IN SEQUENCE SPACES. Given a sequence space λ, let A be a family of normal absolutely convex closed and $\beta(\lambda^\times,\lambda)$- bounded sets in $\lambda^\times[(\lambda^\times,\lambda)]$ covering $\lambda^\times$ and such that a) if A, B $\in$ A, there is C $\in$ A such that $A \cup B \subset C$; b) if $h \in K$ and if $A \in$ A, then $hA \in$ A.

(1) *Every element of A is $\sigma(\lambda,\lambda^{\times\times})$- bounded.*

Proof. Given $B \in \mathcal{A}$, let $x = (x_m)$ be an element of $\lambda^{\times\times}$.$\{x(r)$: $r = 1,2,\ldots\}$ is a bounded set in $\lambda[\sigma(\lambda, \lambda^{\times})]$ and, therefore, there is $h>0$ such that

$$p_{(B)}(x(r)) \leq h, \; r = 1,2,\ldots,$$

If $u = (u_m)$ is any element of B it follows that

$$\sum_{m=1}^{r} |x_m u_m| \leq h, \; m = 1,2,\ldots,$$

and thus

$$\sum_{m=1}^{\infty} |x_m u_m| \leq h.$$

Accordingly, x is bounded on B and, thus, B is $\sigma(\lambda^{\times},\lambda^{\times\times})$-bounded.

Because of (1), the topology $\mathcal{U}$ on $\lambda^{\times\times}$ of the uniform convergence on the elements of $\mathcal{A}$ is a locally convex topology. Let $\mathcal{T}$ be the topology on λ induced by the topology $\mathcal{U}$.

(2) *The closure of in* $\lambda^{\times\times}[\mathcal{U}]$ *with the topology induced by* $\mathcal{U}$ *is the completion* E *of* $\lambda[\mathcal{T}]$.

Proof. If u belongs to $\lambda^{\times}$ there is $B \in \mathcal{A}$ such that u belongs to B and, therefore, $u^n \subset B$ which implies that $\mathcal{U}$ is finer than $\nu(\lambda^{\times\times},\lambda^{\times})$. Since $\lambda^{\times\times}[\nu(\lambda^{\times\times},\lambda^{\times})]$ is complete it follows that $\lambda^{\times\times}[\mathcal{U}]$ is complete and the conclusion follows.

In what follows on this section, we shall suppose that the family $\mathcal{A}$ is constitued by normal absolutely convex closed and compact sets in $\lambda^{\times}[\sigma(\lambda^{\times},\lambda)]$ covering $\lambda^{\times}$ and satisfying properties a) and b). Observe that if $B \in \mathcal{A}$ then $\lambda_B^{\times}$ is a Banach space and therefore B is $\beta(\lambda^{\times},\lambda)$-bounded.

In the last section we have obtained some properties of the compact sets in sequence spaces. More results of this type will be given now.

(3) *If* $x = (x_m)$ *belongs to the completion* E *of* $\lambda[\mathcal{T}]$, *then the sequence of the sections of* x.

(4) $\qquad x(r), \; r = 1,2,\ldots$

$\mathcal{U}$-converges to x.

Proof. Suppose that (4) does not converge to x in $\lambda^{\times\times}[\mathcal{U}]$. Then there is $B \in \mathcal{A}$,$\varepsilon > 0$ and a sequence $u^r=(u_m^{(r)})$,$r=1,2,\ldots$, in B such that

$$\sum_{m=r+1}^{\infty} |x_m u_m^{(r)}| \geqslant \varepsilon, \quad r = 1,2,\ldots$$

If

(5) $v^r = (v_m^{(r)}),\ r = 1,2,\ldots,$ is the sequence in B such that $v_m^{(r)} = 0,\ m \leqslant r,\ |v_m^{(r)}| = |u_m^{(r)}|,\ v_m^{(r)} x_m = |v_m^{(r)} x_m|\ ,\ m > r,$

then (5) $\sigma(\lambda,\phi)$-converges to the origin and, since B is $\sigma(\lambda^\times,\lambda)$-compact, the sequence (5) converges to the origin in $\lambda^\times[\sigma(\lambda^\times,\lambda)]$. On the othe hand, x is in E and, thus, x is $\sigma(\lambda^\times,\lambda)$- continuous on B. Thus

$$0 = \lim |< x,v^r > | = \lim |< x-x(r),v^r >= \lim \sum_{m=r+1}^{\infty} |x_m u_m^{(r)}| = 0$$

which is a contradiction.

(6) *The completion* E *of* $\lambda[T]$ *is a normal sequence space.*

Proof. If $x = (x_m)$ is an element of E, let $y = (y_m)$ be an element of x^n. Given $\varepsilon > 0$ and $B \in A$ we apply the former result to obtain a positive integer r such that $p_{(B)}(x-x(p) \leqslant \varepsilon\ , p \geqslant r$. For those values of p we have that if $u = (u_m)$ is in B

$$\sum_{m=p+1}^{\infty} |y_m u_m| \leqslant \sum_{m=p+1}^{\infty} |x_m u_m| \leqslant p_{(B)}(x-x(p)) \leqslant \varepsilon,$$

and, therefore, y (r), r = 1,2,.., converges to y in $\lambda^{\times\times}[U]$ and, thus, y belongs to E.

(7) *If* A *is compact set in* $\lambda[T]$ *then for every* $B \in A$ *the sequence.*

(8) $$\sup\{\sum_{m=r}^{\infty} |x_m u_m| : (x_m) \in A,\ (u_m)\ \ B\},\quad r = 1,2,\ldots,$$

converges to zero.

Proof. Suppose that (8) converges to $k > 0$. We can find $B \in A$ and a sequence $x^r = x_m^{(r)} \in A,\ r = 1,2,\ldots,$ such that

$$\sup\{\sum_{m=r}^{\infty} |x_m^{(r)} u_m| : (u_m) \in B\} > \frac{k}{2},\quad r = 1,2,\ldots$$

A subsequence of (x^r), which we denote by (x^r) again, can be extracted converging to $x = (x_m)$ in A for the topology T. According to (3) and since (x^r) T-converges to x, there is a positive integer s such that

$$p_{(B)}(x-x^s) < \frac{k}{4}\ ,\quad p_{(B)}(x-x(s-1)) < \frac{k}{4}$$

Then

$$\frac{k}{2} \leq \sup \{ \sum_{m=s}^{\infty} |x_m^{(s)} u_m| : (u_m) \in B\}$$

$$\leq \sup \{ \sum_{m=s}^{\infty} |x_m^{(s)} - x_m) u_m| : (u_m) \in B\} + \sup \sum_{m=s}^{\infty} |x_m u_m| : (u_m) \in B\}$$

$$\leq p_{(B)} (x - x^s) + p_{(B)} (x - x(s-1) < \frac{k}{2},$$

which is a contradiction.

(9) *Let* A *be a bounded set in* λ [*T*] *such that for every* B $\in$ A *the sequence* (8) *converges to zero. If* A *is* $\sigma(\lambda^{\times\times}, \lambda^{\times})$ *closed then* A *is compact.*

Proof. If we consider A as a subset of $\lambda^{\times\times}[\sigma(\lambda^{\times\times}, \lambda^{\times})]$, we apply 4.(26) to obtain that A is $\sigma(\lambda^{\times\times}, \lambda^{\times})$-compact. Given a sequence $x^r = (x_m^{(r)})$ $r = 1,2,\ldots$, in A we can extract a subsequence, which we denote by (x^r) again, such that $\sigma(\lambda, \lambda^{\times})$- converges to $x = (x_m)$ in A.

Given $\varepsilon > 0$ we can find a positive integer s such that

$$\sup \{ \sum_{m=s+1}^{\infty} |z_m u_m| : (z_m) \in A, (u_m) \in B\} < \frac{\varepsilon}{4}.$$

Since (x^r) $\sigma(\lambda, \lambda^{\times})$-converges to x we can find a positive integer p such that

$$\sup \{ \sum_{m=1}^{s} |(x_m^{(r)} - x_m) u_m| : (u_m) \in B\} < \frac{\varepsilon}{2}, \ r \geq p.$$

Then, if $r \geq p$ it follows that

$$p_{(B)}(x - x^r) = \sup \{ \sum_{m=1}^{\infty} |(x_m - x_m^{(r)}) u_m| : (u_m) \in B\}$$

$$\leq \sup\{ \sum_{m=1}^{s} |(x_m - x_m^{(r)} u_m| : (u_m) \in B\}$$

$$+ \sup \{ \sum_{m=s+1}^{\infty} |x_m u_m| : (u_m) \in B\}$$

$$+ \sup \{ \sum_{m=s+1}^{\infty} |x_m^{(r)} u_m : (u_m) \in B\} < \frac{\varepsilon}{2} + \frac{\varepsilon}{4} + \frac{\varepsilon}{4} = \varepsilon$$

Then (x^r) *T*-converges to x and, thus, A is compact.

(10) *If* A *is a compact set in* $\lambda[T]$ *and if* λ *is normal, then* A^n *is compact.*

Proof. A is $\sigma(\lambda^{\times\times}, \lambda^{\times})$-compact and according to 4.(32). A^n is

σ $(\lambda^{\times\times},\lambda^{\times})$ - compact and, thud it is σ $(\lambda^{\times\times},\lambda^{\times})$ - closed. If $B \in A$ we set

$$a_r = \sup \{ \sum_{m=r}^{\infty} |x_m u_m| \ : (x_m) \in A, \ (u_m) \in B \}$$

$$b_r = \sup \{ \sum_{m=r}^{\infty} |x_m u_m| \ : (x_m) \in A^n, \ (u_m) \in B \}$$

It is obvious that $a_r = b_r$, $r = 1,2,\ldots$ Since A is compact, we apply (7) to obtain that (a_r) converges to zero. According to 2. (4), A^n is bounded and, thus, we apply (9) to obtain that A^n is compact.

(12) *Let A be a precompact set in* λ $[T]$. *Then* A^n λ *is precompact.*

Proof. Let M be the closure of A in a the completion of λ $[T]$. According to (6), E is normal and, since every B in A is σ $(\lambda^{\times},E)$ - compact we apply (10) to obtain that M^n is compact. Then $A^n \cap \lambda$ is precompact since it is contained in M^n.

(13) *Let A be a compact set in* λ $[\sigma(\lambda,\lambda^{\times})]$. *If* λ *is normal, then* A^n *is compact.*

Proof. We take $T = \nu(\lambda,\lambda^{\times})$ and we apply 4. (15) and (10).

(14) *Let A be a compact set in* $\lambda[\mu(\lambda,\lambda^{\times})]$. *If* λ *is normal then* A^n *is compact.*

Proof. We take $T = \mu(\lambda,\lambda^{\times})$ and we apply 4. (28) and (10).

6. QUASIBARRELLED SEQUENCE SPACES. We have the following results :

(1) *Let* λ *be a sequence space and let A be a normal closed set in* $\lambda^{\times}$ $[\mu(\lambda^{\times},\lambda^{\times\times})]$. *Then A is* σ $(\lambda^{\times},\phi)$- *closed.*

Proof. If $u = (u_m)$ belongs to $\lambda^{\times}$ and if u is in the closure of A in $\lambda^{\times}[\sigma(\lambda^{\times},\phi)]$, there is a net in A

$$\{ u^j = (u_m^{(j)}) : j \in J, \geqslant \}$$

such that $\lim \{u_m^{(j)} : j \in J, \geqslant \} = u_m$, $m = 1,2,\ldots$

Given a positive integer p we set

$$E_p = \{v(p) : v \in \lambda^{\times}\}.$$

E_p is a subspace of $\lambda^{\times}$ of dimension p and, thus, $\sigma(\lambda^{\times},\phi)$ and $\mu(\lambda^{\times},\lambda^{\times\times})$ coincide on E_p and, therefore, $E_p \cap A$ is $\sigma(\lambda^{\times},\phi)$-closed. Since A is normal, the net

$$\{u^j(p) : j \in J, \geq\}$$

is in $E_p \cap A$ and $\sigma(\lambda^{\times},\phi)$-converges to $u(p)$. Then the sequence $u(p)$, $p=1,2,\ldots$ is contained in A.

Finally, the sequence $(u(p))$ converges to u in the $\sigma(\lambda^{\times},\lambda^{\times\times})$- compact set u^n and, according to 4.(16), it converges to u for the topology $\mu(\lambda^{\times},\lambda^{\times\times})$ and, thus, $u \in A$.

(2) *Let* λ *be a sequence space and let* A *be a normal closed and bounded set in* $\lambda^{\times}[\mu(\lambda^{\times},\lambda^{\times\times})]$. *Then* A *is* $\sigma(\lambda^{\times},\phi)$*-compact.*

Proof. ϕ is normal and ω is its α-dual. The topology $\nu(\omega,\phi)$ coincides with $\sigma(\omega,\phi)$ and, according to 4.(1), $\omega[\sigma(\omega,\phi)]$ is complete. The closure B of A in $\omega[\sigma(\omega,\phi)]$ is compact. Take $u = (u_m)$ in B and, having in mind the proof of (1), the sequence $u(p)$, $p = 1,2,\ldots$, is in A. Let $x=(x_m)$ be an element of $\lambda^{\times\times}$. If

$$h_m \in K, \ |h_m| = 1, \ h_m x_m u_m = |x_m u_m|, \ m = 1,2,\ldots,$$

the element $z = (h_m x_m)$ belongs to $\lambda^{\times\times}$. On the other hand, since A is $\mu(\lambda^{\times},\lambda^{\times\times})$-bounded, there is a scalar $h > 0$ such that

$$|\langle z,u(p)\rangle| = \sum_{m=1}^{p} h_m x_m u_m = \sum_{m=1}^{p} |x_m u_m| \leq h, \ p = 1,2,\ldots,$$

and therefore

$$\sum_{m=1}^{\infty} |x_m u_m| \leq h$$

Thus $u \in \lambda^{\times}$. We apply (1) to obtain that A is $\sigma(\lambda^{\times},\phi)$-closed and, therefore, u belongs to A which concludes the proof.

(3) *Let* Q *be a bounded set in* $\lambda^{\times}[\beta(\lambda^{\times},\phi)]$, *being* λ *a sequence space. Then* Q *is* $\sigma(\lambda^{\times},\lambda^{\times\times})$*-bounded.*

Proof. Let $x = (x_m)$ be an element of $\lambda^{\times\times}$. Since $(x(m))$ $\sigma(\lambda^{\times\times},\lambda^{\times})$-converges to x, the set $\{x(1),x(2),\ldots\}$ is bounded in $\phi[\sigma(\phi,\lambda^{\times})]$ and thus x is a linear form on $\lambda^{\times}$ bounded in Q.

If A denotes the family of all normal absolutely convex closed and bounded sets in $\lambda^{\times}[\sigma(\lambda^{\times},\lambda^{\times\times})]$, being λ a sequence space, it follows that A is a fundamental system of bounded sets in $\lambda^{\times}[\sigma(\lambda^{\beta},\lambda^{\times\times})]$. Indeed, given P a $\sigma(\lambda^{\times},\lambda^{\times\times})$-bounded set it is $\nu(\lambda^{\times},\lambda^{\times\times})$-bounded and according to 2.(4) its normal hull M is $\nu(\lambda^{\times},\lambda^{\times\times})$-bounded and thus M is $\sigma(\lambda^{\times},\lambda^{\times\times})$-bounded.

Let M be the family of all absolutely convex bounded sets in $\lambda^{\times}[\beta(\lambda^{\times},\phi)]$. Since every element of A is $\sigma(\lambda^{\times},\phi)$-compact according to (2), then A is contained in M. On the other hand, if B is a bounded set in $\lambda^{\times}[\beta(\lambda^{\times},\phi)]$, then B is $\sigma(\lambda^{\times},\lambda^{\times\times})$-bounded according to (3). Therefore A is a fundamental system of bounded sets in $\lambda^{\times}[\beta(\lambda^{\times},\phi)]$.

Now results (4) and (5) follow easily.

(4) *If λ is a sequence space, then $\beta(\lambda^{\times\times},\lambda^{\times})$ and $\mu(\phi,\lambda^{\times})$ coincides on ϕ.*

(5) *If λ is a sequence space, then $\phi[\mu(\phi,\lambda^{\times})]$ is quasibarrelled.*

We denote by λ_r the closure of ϕ in $\lambda[\beta(\lambda^{\times\times},\lambda^{\times})]$ endowed with the topology $\mu(\lambda_r,\lambda^{\times})$.

(6) *If the sequence space λ_1 is contained in λ_r, then $\lambda_1[\mu(\lambda_1,\lambda^{\times})]$ is quasibarrelled.*

Proof. According to 5.(2), λ_1 is contained in the completion λ_r of $\phi[\mu(\phi,\lambda^{\times})]$. Then $\phi[\mu(\phi,\lambda^{\times})]$ is a dense subspace of $\lambda_1[\mu(\lambda_1,\lambda^{\times})]$ and the conclusion follows from (5).

(7) *If a sequence space λ_1 verifies*

$$\lambda_1 \subset \lambda,\ \lambda_1 \not\subset \lambda_r,$$

then $\lambda[\mu(\lambda_1,\lambda^{\times})]$ is not quasibarrelled.

Proof. If x is a vector of λ_1 which is not in the completion λ_r of $\phi[\mu(\phi,\lambda^{\times})]$, according to PTAK-COLLINS' theorem (cf. KÖTHE [1], Chapter Four, §21, Section 9) there is a subset A in $\lambda^{\times}[\sigma(\lambda^{\times},\phi)]$ which is absolutely convex and compact on which x is not continuous. Since x belongs to λ_1 it follows that A is not $\sigma(\lambda^{\times},\lambda_1)$-compact. On the other hand, A is bounded in $\lambda^{\times}[\beta(\lambda^{\times},\lambda)]$ and thus $\lambda[\mu(\lambda_1,\lambda^{\times})]$ is not quasibarrelled.

$\lambda_r[\mu(\lambda_r,\lambda^{\times})]$ is then the largest quasibarrelled space of all those sequence spaces λ_1 contained in a given λ such that $\lambda_1[\mu(\lambda_1,\lambda^{\times})]$ is quasibarrelled.

(8) *Given a sequence space* λ *the following conditions are equivalent:*

a) $\lambda[\mu(\lambda,\lambda^{\times})]$ *is quasibarrelled;*

b) $\mu(\lambda,\lambda^{\times})$ *and* $\beta(\lambda^{\times\times},\lambda^{\times})$ *coincide on* λ;

c) *if* B *is a precompact set in* $\lambda[\beta(\lambda^{\times\times},\lambda^{\times})]$, *then* $B^n \cap \lambda$ *is precompact;*

d) *for every* $x \in \lambda$ *the sequence* $(x(r))$ *converges to* x *in* $\lambda[\beta(\lambda^{\times\times},\lambda^{\times})]$

Proof. Let A be an absolutely convex and $\sigma(\lambda^{\times},\lambda)$-compact set in $\lambda^{\times}$. Then A is $\sigma(\lambda^{\times},\phi)$-compact and, according to (4), A is $\sigma(\lambda^{\times},\lambda^{\times\times})$-bounded. Thus $\mu(\lambda,\lambda^{\times})$ is coarser than $\beta(\lambda^{\times\times},\lambda^{\times})$ on λ. On the other hand, if B is $\sigma(\lambda^{\times},\lambda^{\times\times})$-bounded, then B is $\beta(\lambda^{\times},\lambda)$-bounded since $\lambda^{\times}[\mu(\lambda^{\times},\lambda^{\times\times})]$ is complete. Thus $\beta^*(\lambda,\lambda^{\times})$ is finer than $\beta(\lambda^{\times\times},\lambda^{\times})$ on λ. Therefore b) can be deduced from a).

If b) is verified, $\mu(\lambda,\lambda^{\times})$ is the topology on λ of the uniform convergence on every normal absolutely convex compact set in $\lambda^{\times}[\sigma(\lambda^{\times},\lambda)]$. We apply 5.(12) to obtain that $B^n \cap \lambda$ is precompact in $\lambda[\mu(\lambda,\lambda^{\times})] = \lambda[\beta(\lambda^{\times\times},\lambda^{\times})]$.

If c) is verified and if x belongs to λ, then $x^n \cap \lambda$ is precompact in $\lambda[\beta(\lambda^{\times\times},\lambda^{\times})]$ since $\{x\}$ is $\beta(\lambda^{\times\times},\lambda^{\times})$-compact. Since $x(r)$ belongs to $x^n \cap \lambda$ $r = 1,2,\ldots$, and $(x(r))$ $\sigma(\lambda,\phi)$-converges to x, then $(x(r))$ converges to x for the topology $\beta(\lambda^{\times\times},\lambda^{\times})$.

Finally, if d) holds λ is contained in λ_r and then a) follows.

(9) *Given a normal sequence space* λ *the following conditions are equivalents:*

a) $\lambda[\mu(\lambda,\lambda^{\times})]$ *is barrelled;*

b) $\mu(\lambda,\lambda^{\times})$ *coincides with* $\beta(\lambda,\lambda^{\times})$ *in* λ;

c) *if* B *is a compact set in* $\lambda[\beta(\lambda,\lambda^{\times})]$, *then* B^n *is compact;*

d) *for every* x *in* λ, $(x(r))$ *converges to* x *in* $\lambda[\beta(\lambda,\lambda^{\times})]$;

e) $\lambda[\beta(\lambda,\lambda^{\times})]$ *is sequentially separable;*

f) $\lambda[\beta(\lambda,\lambda^{\times})]$ *is separable.*

Proof. If λ is normal, then $\lambda^{\times}[\mu(\lambda^{\times},\lambda)]$ is complete and, thus, if λ is quasibarrelled then it is barrelled. On λ the topologies $\beta(\lambda^{\times\times},\lambda^{\times})$ and $\beta(\lambda,\lambda^{\times})$ coincide.

According to (8), b) follows from a).

If b) holds, we apply 5.(10) instead of 5.(12) in the proof of the former result to obtain c).

d) follows from c) as in (8).

Obviously, e) follows from d).

f) is a trivial consequence from e).

Finally, let F be the topological dual of $\lambda[\beta(\lambda,\lambda^{\times})]$ and let B be an absolutely convex closed and bounded set in $\lambda^{\times}[\sigma(\lambda^{\times},\lambda)]$. The closure D of B in F $[\sigma(F,\lambda)]$ is $\sigma(F,\lambda)$-compact and this if f) holds, then D is $\sigma(F,\lambda)$-metrizable. Let u be an element of D and let (u^r) be a sequence in B $\sigma(F,\lambda)$-converging to u. Since λ is normal, the space $\lambda^{\times}[\sigma(\lambda^{\times},\lambda)]$ is a sequentially complete and thus u belongs to $\lambda^{\times}$ and a) follows.

Compare result (9) with the analogous theorem given by KÖTHE [1] for perfect spaces.

The space λ_r has been introduced by T. and Y. KOMURA [1].

7. BOUNDED LINEAR FORMS ON A SEQUENCE SPACE. In this section, λ is a perfect sequence space. A linear form on λ is bounded if it is bounded on every bounded set in $\lambda[\sigma(\lambda,\lambda^{\times})]$.

(1) *Let* T *be a linear form on* λ *which is bounded on sets* $x^n, x \in \lambda$. *Then* T *is bounded.*

Proof. Suppose T is not bounded. There is a normal absolutely convex closed and bounded set A in $\lambda[\sigma(\lambda,\lambda^{\times})]$ such that T is not bounded on A. We find a sequence $x^r = (x_m^{(r)})$, r = 1,2,..., in A such that

(2) $$|T(x^r)| \geq 2^{2r}, \quad r = 1,2,\ldots$$

We set

$$y_m^{(r)} = |x_m^{(r)}|, \quad y^r = (y_m^{(r)}), \quad r = 1,2,\ldots$$

Since A is normal, y^r belongs to A. On the other hand, $\lambda[\mu(\lambda,\lambda^{\times})]$ is complete and thus λ_A is a Banach space in which the series

$$\Sigma\, 2^{-r} y^r$$

converges to an element z. The vectors

$$u^r = 2^{-r}x^r, \quad r = 1,2,\ldots,$$

are in z^n and therefore the series

$$\Sigma 2^{-r} u^r$$

converges to u in the Banach space λ_{z^n}. Since T is bounded in z^n it follows that

$$T(u) = T\,(\Sigma 2^{-r}u^r) = \Sigma T(2^{-r}u^r)$$

ando so

$$\lim T\,(2^{-r}u^r) = 0$$

On the other hand,

$$|T(2^{-r}u^r)| = |T(2^{-2r}x^r)| = 2^{-2r}\,|T(x^r)| \geq 1$$

According to (2). That is a contradiction.

We consider now a normal sequence space λ_1 contained in λ and endowed with the topology $\sigma(\lambda_1,\lambda^{\times})$. The most important result in this section is that every bounded linear form on λ_1 can be extended to a bounded linear form on λ. We set

$$\Delta = \{x \in \lambda : x \geq 0\}$$

Given elements x and y in λ we write

$$x \leq y \text{ if } y - x \geq 0.$$

We suppose first that λ is real and T is a bounded linear form on λ_1. We define on Δ a real-valued function S such that

$$S(z) = \sup\,\{T(x) : 0 \leq x \leq z,\ x \in \lambda_1\},\ z \in \Delta.$$

(3) *The following property holds:*

$$S(u) : S(v) \leq S(u+v),\ u,\ v \in \Delta.$$

Proof. $S(u) + S(v) = \sup\,\{T(x) : 0 \leq x \leq u,\ x \in \lambda_1\}$

$$+ \sup\,\{T(y) : 0 \leq y \leq v,\ y \in \lambda_1\}$$

$$= \sup\,\{T(x) + T(y) : 0 \leq x \leq u,\ 0 \leq y \leq v,\ x,\ y \in \lambda_1\}$$

$$= \sup\,\{T(x+y) : 0 \leq x \leq u,\ 0 \leq y \leq v,\ x,y \in \lambda_1\}$$

$$\leq \sup\,\{T(x+y) : 0 \leq x+y \leq u+v,\ x+y \in \lambda_1\} = S\,(u+v).$$

(4) *The following property holds:*

$$S(u+v) \leq S(u) + S(v),\ u,v, \in \Delta.$$

Proof. Given $\varepsilon > 0$ we find $x \in \lambda_1$, $0 \leq x \leq u+v$, such that

$$S(u+v) < T(x) + \varepsilon.$$

If z is the minimum of the sequences x and u, we set w = x-z. Since λ_1 is normal, then z belongs to λ_1 and thus w belongs to λ_1. We have

$$0 \leq z \leq u, \ 0 \leq w \leq v.$$

Therefore

$$S(u+v) < T(x)+\varepsilon = T(w+z)+\varepsilon = T(w) + T(z) +\varepsilon$$
$$\leq S(u) + S(v) + \varepsilon$$

and thus

$$S(u+v) \leq S(u) + S(v)$$

(5) *If* $h \geq 0$ *and* $u \in \Delta$, *then*

$$S(hu) = hS(u).$$

Proof. $S(hu) = \sup \{T(x) : 0 \leq x \leq hu, \ x \in \lambda_1\}$

$$= \sup \{T(hy) : 0 \leq y \leq u, \quad y \in \lambda_1\}$$
$$= h \sup \{T(y): 0 \leq y \leq u, \ y \in \lambda_1 \ = hS(u)$$

Every element $u = (u_m) \in \lambda$ can be written

$$u = u^1 - u^2, \ u^1, u^2 \in \Delta,$$

f.i.,

$$u_m^{(1)} = u_m \text{ if } u_m \geq 0, \ u_m^{(1)} = 0 \text{ if } u_m < 0,$$

$$u_m^{(2)} = -u_m \text{ if } u_m \leq 0, \ u_m^{(2)} = 0 \text{ if } u_m > 0,$$

$$u^1 = (u_m^{(1)}) \text{ and } u^2 = (u_m^{(2)}).$$

We define S on u as

$$S(u) = S(u^1) - S(u^2).$$

If we have another representation of u:

$$u = v^1 - v^2, \ v^1, \quad v^2 \in \Delta$$

then

$$u^1 + v^2 = u^2 + v^1$$

and, according to (3) and (4),

$$S(\breve{u}^1) + S(v^2) = S(u^2) + S(v^1),$$

and therefore

$$S(u^1) - S(u^2) = S(v^1) - S(v^2)$$

Then, the definition of S(u) does not depend of the particular choice of the representation of u.

(6) S *is a linear form on* λ.

Proof. Let u and v be elements of λ and let h be any real number. We set

$$u = u^1-u^2,\ v = v^1-v^2,\ u^1,u^2,v^1,v^2 \in \Delta$$

Then

$$S(u+v) = S(u^1+v^1-u^2-v^2) = S\ (u^1+v^1)$$

$$- S(u^2+v^2) = S(u^1) + S(v^1) - S(u^2) - S(v^2)$$

$$= S\ (u^1-u^2) + S(v^1-v^2) = S(u) + S(v).$$

If h is larger or equal than zero,

$$S(hu) = S(hu^1-hu^2) = S(hu^1) - S(hu^2)$$

$$= h\ S(u^1) - h\ S(u^2) = h\ S(u^1-u^2) = h\ S(u).$$

If h is negative,

$$S(hu) = S((-h)u^2-(-h)u^1) = (-h)\ S\ (u^2) - S(u^1)$$

$$= h\ S(u^1) - hS(u^2) = hS(u^1-u^2) = hS(u).$$

(7) S *is a bounded linear form on* λ.

Proof. According to (1) it is enough to see that if $z = (z_m)$ is in Δ , then S is bounded in z^n. If $u = (u_m)$ belongs to z^n, we set

$$v_m = u_m \text{ if } u_m \geq 0,\ v_m = 0 \text{ if } u_m \leq 0,$$

$$w_m = -u_m \text{ if } u_m \leq 0,\ w_m = 0 \text{ if } u_m \geq 0,$$

We set $v = (v_m)$ and $w = (w_m)$. It follows that $u = v-w$, $0 \leq v \leq z$, $0 \leq w \leq z$ and thus

$$|S(u)| = |S(v)-S(w)| \leq |S(v)| + |S(w)|$$

$$= S(v) + S(w) \leq 2\ S(z)$$

and we are done.

For every $x \in \lambda_1$ we set

$$U(x) = S(x) - T(x).$$

U is a bounded linear form on λ_1. We apply to U the same method used to deduce S from T to obtain a bounded linear form V on λ.

(8) *There is a bounded linear form* T_1 *on* λ *which coincides with* T *on* λ_1.

Proof. Setting $T_1 = S - V$ it follows that T_1 is linear and bounded on λ. T_1 coincides with T on λ_1. Indeed, if $v \in \lambda_1$, $v \geq 0$, we have that

$$S(v) = \sup \{T(x) : 0 \leq x \leqslant v, \; x \in \lambda_1\} \geqslant T(v)$$

and therefore

$$U(v) = S(v) - T(v) \geqslant 0,$$

i.e. U is larger or equal than zero on every positive element of λ_1. If $0 \leqslant x \leqslant v$, $x \in \lambda_1$, we have that $v-x \in \lambda_1$, $v-x \geqslant 0$, and thus

$$U(v-x) \geqslant 0, \text{ i.e., } U(v) \geqslant U(x)$$

Then

$$V(v) = \sup \{ U(x) : 0 \geqslant x \leqslant v, \; x \in \lambda_1 \} = U(v),$$

i.e., V coincides with U on the positive elements of λ_1 and therefore V and U coincide on λ_1. For every $x \in \lambda_1$,

$$T_1(x) = S(x) - V(x) = S(x) - U(x) = T(x).$$

We suppose now λ complex. Every element of λ can be written as $(x_m + i\, y_m)$ with x_m and y_m real, m= 1,2,..,

Let $Z: \to \omega$ defined as

$$Z(x) = (x_1, y_1, \ldots, x_m, y_m, \ldots)$$

being $x = (x_m + i y_m)$. We set $Z(\lambda_1) = \Lambda_1$ and $Z(\lambda) = \Lambda$. It is obvious that Λ_1 is a normal real sequence space.

(9) Z *is a real topological isomorphism of* $\lambda[\nu(\lambda, \lambda^\times)]$ *onto* $\Lambda[\nu(\Lambda, \Lambda^\times)]$.

Proof. It is obvious that Z is a real algebraic isomorphism of λ onto Λ. Let

$$\{x^j = (x_m^{(j)} + i y_m^{(j)}) : j \in J, \geq\}$$

be a net in $\lambda[\nu(\lambda, \lambda^\times)]$ converging to zero. Let $(u_1, v_1, \ldots, u_m, v_m, \ldots)$ be an element of $\Lambda^\times$. If $(x_m + i y_m) \in \lambda$, then $ix = (-y_m + i x_m) \in \lambda$ and therefore

$$(x_1, y_1, \ldots, x_m, y_m, \ldots) \text{ and } (-y_1, x_1, \ldots, -y_m, x_m, \ldots)$$

belong to Λ and so

$$\Sigma(|x_m u_m| + |y_m v_m|) < \infty, \quad \Sigma(|y_m u_m| + |x_m v_m|) < \infty.$$

Let u be $(u_m + iv_m)$. We have that

$$\Sigma|(x_m + iy_m)(u_m + iv_m)| \leqslant \Sigma\ (|x_m u_m| + |y_m v_m| + |y_m u_m| + |x_m v_m|) < \infty$$

and therefore u belongs to $\lambda^\times$. Then

$$\lim \{\Sigma(|x_m^{(j)} u_m| + |y_m^{(j)} v_m|) : j \in J, \geqslant\}$$

$$\leqslant \lim \{\Sigma|x_m^{(j)} + iy_m^{(j)}| . |u_m + iv_m| : j \in J, \geqslant\} = 0$$

and thus Z is ontinuous of $\lambda[\mu\ (\lambda,\lambda^\times)]$ onto $\Lambda[\mu\ (\Lambda,\Lambda^\times)\ ]$.

We show now that Z^{-1} is continuous of $\Lambda[\nu\ (\Lambda,\Lambda^\times)]$ onto $\lambda[\nu\ (\lambda,\lambda^\times)]$. Let $u = (u_m + iv_m)$ be and element of $\lambda^\times$ and let

$$\{x^j = (x_m^{(j)} + iy_m^{(j)}) : j \in J, \geqslant\}$$

be a net in λ such that $\{Z(x^j) : j \in J, \geqslant\}$ converges to the origin in Λ. If $(x_1, y_1, \ldots, x_m, y_m, \ldots)$ belongs to $\lambda^\times$, then

$$\Sigma(|x_m u_m| + |y_m v_m|) \leqslant \Sigma|x_m + iy_m| . |u_m + iv_m| < \infty$$

and therefore $(u_1, v_1, \ldots, u_m, v_m, \ldots)$ belongs to $\Lambda^\times$. Since $iu = (-v_m + iu_m)$ belongs to $\lambda^\times$ it follows that $(-v_1, u_1, \ldots, -v_m, u_m, \ldots)$ is in $\Lambda^\times$. Then

$$\lim \{\Sigma|x_m^{(j)} + iy_m^{(j)}| . |u_m + iv_m| : j \in J \geqslant\}$$

$$\leqslant \lim\ \{\Sigma(|x_m^{(j)} u_m| + |y_m^{(j)} v_m|) : j \in J, \geqslant\}$$

$$+ \lim\ \{\Sigma(|x_m^{(j)} v_m| + |y_m^{(j)} u_m| : j \in J, \geqslant\} = 0$$

which concludes the proof.

(10) *Λ is a perfect sequence space.*

Proof. If

(11) $\{x^j : j \in J, \geqslant\}$

is a Cauchy net in $\Lambda[\nu(\Lambda,\Lambda^{\times})]$, it follows that

$$\{Z^{-1}(x^j) : j \in J, \geq\}$$

is a Cauchy net in $\lambda[\nu(\lambda,\lambda^{\times})]$ and, according to 4.(19), this net converges in $\lambda[\nu(\lambda,\lambda^{\times})]$ to an element u. Then (11) converges to Z(u) in $\Lambda[\nu(\Lambda,\Lambda^{\times})]$ and we apply again 4.(19) to obtain that Λ is perfect.

Let T be a bounded linear form on λ_1. For every $x \in \lambda_1$ we set

(12) $\quad T(x) = X(x)+i\,Y(x)$

with X(x),Y(x) real numbers. If we consider λ as a linear space over the field of the real numbers, then X is a bounded linear form on $\lambda_1[\nu(\lambda_1,\lambda^{\times})]$ and thus $X\circ Z^{-1}$ is a bounded linear form on $\Lambda[\nu(\Lambda_1,\Lambda^{\times})]$. Then $X\circ Z^{-1}$ is bounded on $\Lambda_1[\sigma(\Lambda_1,\Lambda)]$ According to (8) we obtain a bounded linear form W on $\Lambda[\nu(\Lambda,\Lambda^*)]$ coinciding with $X\circ Z^{-1}$ on Λ_1. Then $W\circ Z$ is a bounded real linear form on λ. We set

$$T_1(x) = (W\circ Z)(x) = i(W\circ Z)(ix),\ x \in \lambda.$$

(13) T_1 *is a bounded linear form on* λ *coinciding with* T *on* λ_1.

Proof. It is easy to see that T_1 is a linear form on the complex space λ. Since W_0Z is bounded on the real space λ, T_1 is bounded on the complex space λ. If we write in (12) ix instead of x, it follows that

$$iT(x) = T(ix) = X(ix) + iY(ix) = iX(x) - Y(x)$$

and therefore

$$X(ix) = -Y(x)$$

and thus

$$T(x) = X(x) - iX(ix)$$

Then, if x belongs to λ_1,

$$T_1(x) = (W\circ Z)(x) - i(W\circ Z)(ix)$$

$$= X(x) - iX(ix) = T(x)$$

and the result is proven.

Thus, in the real or complex case we have the following result:

(14) *If T is a bounded linear form on* λ_1 *there is a bounded linear form on* λ *coinciding with T on* λ_1.

8. BORNOLOGICAL SEQUENCE SPACES. Given a sequence space λ and an element $x \in \lambda$ we write ϕ_x to denote the normed space $\phi_{x^n \cap \phi}$ with $x^{n} \cap \phi$ as closed unit ball.

(1) *The space* $\phi[\mu(\phi,\lambda^{\times})]$ *is the locally convex hull of the family of normed spaces.*

(2) $\{ \phi_x : x \in \lambda\}$.

Proof. Let T be a linear form on ϕ which is continuous on every ϕ_x, $x \in \lambda$. Given any $z = (z_m) \in \lambda$ we find $h_m \in K$, $|h_m| = 1$, with $h_m z_m T(e_m) = |z_m T(e_m)|$, $m = 1,2, \ldots$ The vectors $\sum_{m=1}^{r} h_m z_m e_m$, $r = 1,2, \ldots$, are in the closed unit ball of ϕ_z and therefore there is a positive number M such that $T(\sum_{m=1}^{r} h_m z_m e_m) = \sum_{m=1}^{r} h_m z_m T(e_m) = \sum_{m=1}^{r} |z_m T(e_m)| \leq M$,

and consequently $\sum |z_m T(e_m)| \leq M$ and thus $(T(e_m)) \in \lambda^{\times}$. Since T coincides with $(T(e_m))$ on ϕ it follows that T is continuous on $\phi[\mu(\phi,\lambda^{\times})]$. Consequently, $\phi[\mu(\phi,\lambda^{\times})]$ is the locally convex hull of the family of normed spaces. (2).

(3) *If* λ *is a normal sequence space, then* $\phi[\mu(\lambda,\lambda^{\times})]$ *is the inductive limit of the family* (2).

Proof. (2) is ordered by inclusion. The conclusion follows from (1).

A straigforward conclusion of (1) is the following:

(4) *The space* $\phi[\mu(\phi,\lambda^{\times})]$ *is bornological.*

We denote by $\rho(\lambda,\lambda^{\times})$ the associated bornological topology to $\sigma(\lambda,\lambda^{\times})$ Let λ_b the closure of ϕ in $\lambda[\rho(\lambda,\lambda^{\times})]$. Since $\rho(\lambda,\lambda^{\times})$ is finer than the topology on λ induced by $\beta(\lambda^{\times\times},\lambda^{\times})$ and since λ_r is the completion of $\phi[\beta(\lambda^{\times\times},\lambda^{\times})] = \phi[\mu(\phi,\lambda^{\times})]$, we have that λ_b is contained in λ_r. On the other hand, every bounded set of $\phi[\mu(\phi,\lambda^{\times})]$ is $\rho(\lambda,\lambda^{\times})$-bounded and, since $\phi[\mu(\lambda,\lambda^{\times})]$ is bornological, we have that $\phi(\lambda,\lambda^{\times})$ coincides with $\mu(\lambda_1,\lambda^{\times})$ on λ_1 for every sequence space λ_1 contained in λ_b.

(5) *If* λ_1 *is a normal sequence space contained in* λ_b *then* $\lambda_1[\mu(\lambda_1,\lambda^{\times})]$ *is bornological.*

Proof. Let T be a bounded linear form on $\lambda_1[\mu(\lambda_1,\lambda^{\times})]$. We apply

7. (14) to cbtain a bounded linear form T_1 on $\lambda^{\times\times}[\sigma(\lambda^{\times\times},\lambda^{\times})]$ coinciding with T on λ_1. The restriction to $\lambda[\rho(\lambda,\lambda^{\times})]$ of T_1 is continuous and therefore T is continuous on $\lambda_1[\rho(\lambda_1,\lambda^{\times})] = \lambda_1[\mu(\lambda_1,\lambda^{\times})]$.

(6). *If λ_1 is a sequence space contained in λ and not contained in λ_b, then $\lambda[\mu(\lambda_1,\lambda^{\times})]$ is not bornological*

Proof. It is easy.

Given a sequence space λ such that λ_b is normal and given the family of all bornological sequence spaces $\lambda_1[\mu(\lambda_1,\lambda^{\times})]$ whit $\lambda_1 \subset \lambda_b$, there is a largest member in this family, namely $\lambda_b[\mu(\lambda_b,\lambda^{\times})]$.

The idea of using the space λ_b to study bornological spaces is due to T. and Y. KOMURA [1].

9. ULTRABORNOLOGICAL SEQUENCE SPACES. Let λ be a normal sequence space If x is any element of λ we set $\phi(x)$ to denote the closure of $\phi \cap \lambda_{x^n}$ in the Banach space λ_{x^n}. We suppose $\phi(x)$ endowed with the topology induced by λ_{x^n}. We set $\phi^1 = \cup \{ \phi(x) : x \in \lambda\}$ and we suppose ϕ^1 endowed with the topology $\mu(\phi^1,\lambda^{\times})$. Since ϕ^1 is contained in λ_b, $\mu(\phi^1,\lambda^{\times})$ coincides with $\beta(\lambda^{\times\times},\lambda^{\times})$ on ϕ^1.

(1) *The sequence space ϕ^1 is normal.*

Proof. It is easy.

(2) *The sequence space ϕ^1 is the inductive limit of the family of Banach spaces*

(3) $\{ \phi(x) : \in \lambda \}$.

Proof. Since λ is normal the family (3) is ordered by inclusion. On the other hand, let f be a linear form on ϕ^1 which is continuous on every $\phi(x)$, $x \in \lambda$, and let g be its restriction to ϕ. By 8. (1) g can be extended to a continuous linear form X on ϕ^1. If $u \in \phi^1$ there is $x \in \lambda$ such that $u \in \phi(x)$. Since f and X are continuous on $\phi(x)$ coincide on ϕ, we have that $f(u) = X(u)$ and, therefore, $f = X$. If T denotes the inductive limit topology of the family (3), we have seen that the topological dual of $\phi^1[T]$ coincides with the topological dual of ϕ^1. On the other hand, ϕ^1 and $\phi^1[T]$ have their own Mackey topologies and, consequently, $\phi^1 = \phi^1[T]$

10. SCHWARTZ SEQUENCE SPACES. A locally convex space E is a Schwartz space if given any equicontinuous setA of E' there is an absolutely convex

closed subset B in E' $[\sigma(E',E)]$ which is a equicontinuous set containing A and such that A is precompact in the Banach space E'_B.

Some properties of Schwartz spaces are contained in GROTHENDIECK [2] HORVATH [1] and JARCHOW [2].

Let λ be a sequence space and let v be an element of $\lambda^\times$. We denote by $||.||_v$ the norm of the space $\lambda^\times_{v^n}$. If $w \in \lambda^\times_{v^n}$, we have that

$$||w||_v = \inf\{h : h > 0, w \in h v^n\} = \inf\{h, h > 0, |w_m| \leq$$

$$\leq h |v_m|, m = 1,2,\dots\} = \sup\{|w_m; v_m| : m = 1,2,\dots\}$$

(1) *If the sequence space* λ*, endowed with the topology* $\nu(\lambda,\lambda^\times)$*, is Schwartz, given* $u = (u_m) \in \lambda^\times$*, there is* $v = (v_m) \in \lambda^\times$ *such that*

$$|u_m| \leq |v_m|, \ m = 1,2,\dots, \text{ and } \lim (u_m; v_m) = 0.$$

Proof. The set u^n is $\nu(\lambda, \lambda^\times)$-equicontinuous and therefore there is $v = (v_m) \in \lambda^\times$ such that $u^n \subset v^n$ and such that u^n is precompact in $\lambda^\times_{v^n}$. If $u_m^{(r)} = u_m$, $m = r$, $u_m^{(r)} = 0$, $m \neq r$, the sequence

(2) $\qquad (u_m^{(r)})$, $r = 1,2,\dots,$

is contained in v^n and converges to the origin for the topology $\sigma(\lambda^\times,\phi)$. Thus (2) converges to the origin in $\lambda^\times_{v^n}$ and therefore

$$\lim ||(u_m^{(r)})|| = \lim \sup\{|u_m^{(r)}; v_m| : m = 1,2,\dots\}$$

$$= \lim |u_r; v_r| = 0$$

and consequently $\lim (u_m; v_m) = 0$.

(3) *If given any* $u = (u_m) \in \lambda^\times$ *there is* $v = (v_m) \in \lambda^\times$ *with*

$$|u_m| \leq |v_m| \quad m = 1,2,\dots, \text{ and } \lim (u_m; v_m) = 0$$

then $\lambda[\nu(\lambda,\lambda^\times)]$ *is a Schwartz space*

Proof. Let u and v be vector of $\lambda^\times$ satisfying both conditions. Clearly $u^n \subset v^n$. Let

(4) $\qquad w^r = (w_m^{(r)})$, $r = 1,2,\dots,$

be a sequence in u^n. Since u^n is compact and metrizable for the topology $\sigma(\lambda^\times,\lambda)$ a subsequence of (4), which we denote by (4) again can be extracted converging to $w = (w_m) \in u^n$ for the topology $\sigma(\lambda^\times,\phi)$. Given any $\varepsilon > 0$

we find a positive integer q such that

$$|u_m;v_m| < \frac{\varepsilon}{3},\ m \geq q,$$

and therefore

$$|w_m;v_m| < \frac{\varepsilon}{3},\ |w_m^{(r)};\ v_m| < \frac{\varepsilon}{3},\ m > q,\ r = 1,2,\ldots$$

We can find a positive integer s with

$$|w_m^{(r)} - w_m;v_m| < \frac{\varepsilon}{3},\ r \geq s,\ m = 1,2,\ldots,q.$$

Consequently, if $r \geq s$ we have that

$$||w_m^{(r)} - w_m||_v = \sup\{|w_m^{(r)} - w_m;\ v_m| : m = 1,2,\ldots\}$$

$$\leq \sup\{|w_m^{(r)} - w_m;v_m| : m = 1,2,\ldots,q\} + \sup\{|w_m^{(r)};v_m| : m = q+1,q+2,\ldots\}$$

$$+ \sup\{|w_m;v_m| : m = q+1,q+2,\ldots\} < \frac{\varepsilon}{3} + \frac{\varepsilon}{3} + \frac{\varepsilon}{3} = \varepsilon,$$

which completes the proof.

Results (5) and (6) can be obtained analogously as results (1) and (3) respectively.

(5) *If* λ *is a normal sequence space and if* $\lambda^\times[\nu(\lambda^\times,\lambda)]$ *is a Schwartz space, then given* $x = (x_m) \in \lambda$ *there is* $y = (y_m) \in \lambda$ *such that*

$$|x_m| \leq (y_m),\ m = 1,2,\ldots,\ \text{and}\ \lim\ (x_m;y_m) = 0$$

(6) *If* λ *is a normal sequence space and if given any* $x = (x_m) \in \lambda$ *there is* $y = (y_m) \in \lambda$ *with*

$$|x_m| \leq |y_m|,\ m = 1,2,\ldots,\ \textit{and}\ \lim\ (x_m;y_m) = 0,$$

then $\lambda^\times[\nu(\lambda^\times,\lambda)]$ *is a Schwartz space.*

Result (5) and (6) can be found in KÖTHE [1] for echelon spaces. More results an Schwartz sequence spaces can be seen in RUCKLE and SWART [1].

11. NUCLEAR SEQUENCE SPACES. Let E and F be normed spaces. We denote by U the closed unit ball of E and by U° its polar set in E'. We represent by $|.|$ the gauge of U° and by $||.||$ the norm in F. A mapping $T: E \longrightarrow F$ is nuclear if there are sequences (u_m), (y_m) in E' and F respectively such that

$$\Sigma|u_m| \cdot ||y_m|| < \infty \text{ and } T(x) = \Sigma\langle x,u_m\rangle\, y_m,\ x \in E.$$

A locally convex space G is nuclear if given in $G'[\sigma(G',G)]$ an equicontinuous absolutely convex closed subset A there is a subset B in $G'[\sigma(G',G)]$,

$B \supset A$, equicontinuous absolutely convex and closed, such that the canonical injection $J : G'_A \to G'_B$ is nuclear.

(1) *If a sequence space* λ *is nuclear for the topology* $\nu(\lambda,\lambda^\times)$, *then given* $(u_m) \in \lambda^\times$ *there is* $(v_m) \in \lambda^\times$ *such that*

$$|u_m| < |v_m|, \; m = 1,2,\ldots, \textit{ and } \; \Sigma|u_m;v_m| < \infty.$$

Proof. If $u = (u_m) \in \lambda^\times$ then u^n is absolutely convex closed subset of $\lambda^\times[\sigma(\lambda^\times,\lambda)]$ which is $\upsilon(\lambda,\lambda^\times)$-equicontinuous. We find in $\lambda^\times$ a subset A, which is $\nu(\lambda,\lambda^\times)$-equicontinuous absolutely convex and $\sigma(\lambda^\times,\lambda)$-closed such that $u^n \subset A$ and the canonical injection $J : \lambda^\times{}_{u^n} \to \lambda^\times{}_A$ is nuclear. Let $v \in \lambda^\times$ be such that $A \subset v^n$. There are sequences (z^r) in $(\lambda^\times{}_{u^n})'$ and (y^r) in $\lambda^\times{}_A$ such that

(2) $\qquad \Sigma|z^r|.||y^r|| < \infty, \; J(x) = \Sigma <x,z^r> y^r, x \in \lambda^\times{}_{u^n},$

$|.|$ being the gauge of the polar set in $(\lambda^\times{}_{u^n})'$ of u^n and $||.||$ being the norm in $\lambda^\times{}_A$.

Give positive integers r and m we find $a_{mr} \in K$, $|a_{mr}| = 1$, such that

$$a_{mr} <u_m e_m, z^r> = |<u_m e_m z^r>|$$

For every positive integer s

$$\sum_{m=1}^{s} a_{mr} u_m e_m \in u^n$$

and thus

$$|z^r| \geq |< \sum_{m=1}^{s} a_{mr} u_m e_m, z^r>| = \sum_{m=1}^{s} |<u_m e_m, z^r>|$$

and consequently

$$\sum_m |<u_m e_m, z^r>| \leq |z^r$$

Since A is contained in v^n it follows that

$$||y^r|| \geq \sup \{|y_m^{(r)};v_m| : m = 1,2,\ldots\}$$

Setting in (2) $x = u_m e_m$ we have that

$$u_m = \sum_r <u_m e_m, z^r> y_m^{(r)}$$

and accordingly

$$|u_m;v_m| = |\sum_r <u_m e_m, z^r> \ y_m^{(r)};v_m| \leq \sum_r |<u_m e_m, z^r>|\cdot|y_m^{(r)};v_m|$$

$$\leq \sum |<u_m e_m, z^r>|\cdot|| y^r ||,$$

from where it follows

$$\sum |u_m;v_m| < \sum_r ||y^r|| \sum_m |<u_m e_m, z^r>| \leq \sum |z^r|.|| y^r|| < \infty,$$

which completes the proof.

(3) *If given any* $(u_m) \in \lambda^\times$ *there is* $(v_m) \in \lambda^\times$ *with*
(4) $|u_m| \leq |v_m|$, m= 1,2,..., *and* $\Sigma|u_m;v_m| < \infty$, *then* $\lambda[\upsilon(\lambda,\lambda^\times)]$ *is a nuclear space.*

Proof. Let A be a subset of $\lambda^\times[\sigma(\lambda^\times\lambda)]$ $\upsilon(\lambda,\lambda^\times)$-equicontinuous absolutely convex and closed. We take $u = (u_m) \in \lambda^\times$ such that $A \subset u^n$. We find $v = (v_m) \in \lambda$ satisfiyng (4). Let $J : \lambda^\times_A \longrightarrow \lambda^\times_{vn}$ be the canonical injection. $|\cdot|$ denotes the gauge of the polar set in $(\lambda^\times_A)'$ of A and $||\cdot||$ denotes the norm in $\lambda^\times_{vn}$.

Let r be a positive integer. The vector $d^r = (d_m^{(r)})$ of λ such that $d_r^{(r)} = 0$ if $v_r = 0$, $d_r^{(r)} = \frac{1}{v_r}$ if $v_r \neq 0$, $d_m^{(r)} = 0$ if $m \neq r$, is a continuous linear form on $\lambda^\times[\sigma(\lambda^\times,\lambda)]$. Let x^r be the restriction of d^r to $\lambda^\times_A$. We have that

$$|x^r| = \sup\{|<w,x^r>| : w = (w_m) \in A\}$$

$$= \sup\{|w_r;v_r| ; w = (w_m) \in A\} \leq |u_r;v_r|.$$

Setting $y^r = v^r e^r \in \lambda^\times_{vn}$, if $x = (x_m)$ is any vector of $\lambda^\times_A$ we have that

$$J(x) = x = (x_m) = \Sigma (x_r;v_r)v_r e_r = \Sigma <x,x^r> y^r,$$

$$\Sigma |x^r| \cdot ||y^r|| \leq \Sigma |x^r| \leq \Sigma |u_r;v_r| < \infty$$

and therefore J is nuclear. The proof is complete.

(5) *If* $\lambda[\upsilon(\lambda,\lambda^\times)]$ *is nuclear then it is a Schwartz space.*

Proof. If $\lambda[\upsilon(\lambda,\lambda^\times)]$ is nuclear, then (1) holds and 10. (3) is satisfied and $\lambda[\upsilon(\lambda,\lambda^\times)]$ is Schwartz.

Results (6), (7) and (8) are obtained analogously as (1), (3) and (5) respectively.

(6) *If* λ *is a normal space and if* $\lambda^\times$ $[\upsilon\,(\lambda^\times,\lambda)]$ *is a nuclear space, then given* $(x_m) \in \lambda$ *there is* $(y_m) \in \lambda$ *such that*

$$|x_m| \leq |y_m|, \quad m = 1,2, \ldots, \quad \text{and} \quad \Sigma\,|x_m ; y_m| < \infty$$

(7) *If* λ *is a normal sequence space and if given any* $(x_m) \in \lambda$ *there is* $(y_m) \in \lambda$ *with*

$$|x_m| < |y_m|, \quad m \,.\, 1,2, \ldots, \text{and} \quad \Sigma\,|x_m ; y_m| < \infty$$

then $\lambda^\times[\,\upsilon(\lambda^\times,\lambda)\,]$ *is a nuclear space.*

(8) *If* λ *is a normal sequence space and if* $\lambda^\times[\,\upsilon\,(\lambda^\times,\lambda)]$ *is nuclear then* $\lambda^\times[\,\upsilon(\lambda^\times,\lambda)]$ *is a Schwartz space.*

Results (6) and (7) are due to PIETSCH [2] and to GROTHENDIECK [1] (cf. PIETSCH [1], Chapter 6, 6.1).

§ 2. ECHELON AND CO – ECHELON SPACES

1. ECHELON AND CO-ECHELON SPACES. Let

(1) $\alpha_r = (a_m^{(r)})$, $r = 1,2,\ldots,$ be a sequence of elements of ω satisfying

1. $\alpha_{r+1} \geq \alpha_r \geq 0$, $r = 1,2, \ldots;$
2. For every positive integer m there is a positive integer r such that $a_m^{(r)} > 0$.

We set $\lambda = \{(x_m) \in \omega : \sum_m |x_m|\; a_m^{(r)} < \infty\,, \quad r = 1,2,\ldots\}$.

Let λ_1 be the set of all elements of ω such that $(u_m) \in \lambda_1$ if and only if there is a positive integer r and $h > 0$, depending on the sequence (u_m), with

$$|u_m| \leq h\, a_m^{(r)}, \qquad m = 1,2,\ldots,$$

or, what is the same, λ_1 is the union of all sets which are scalar multi_ ples of the normal hull α^n_r of α_r, $r = 1,2,\ldots$

From condition 2. follows that $e_m \in \lambda_1$, m = 1,2,..., From condition 1. we obtain that the sum of two elements of λ_1 is in λ_1. Now result (2) is obvious.

(2) *λ_1 is a normal sequence space.*

(3) *λ is the α-dual of λ_1.*

Proof. Since $\alpha_r \in \lambda_1$ for every positive integer r, we have that if (x_m) belongs to $\lambda^\times_1$.

$$\sum_m |x_m \, a_m^{(r)}| < \infty, \; r = 1,2,\ldots,$$

and therefore (x_m) belongs to λ. Thus $\lambda^\times_1 \subset \lambda$. On the other hand, if $(y_m) \in \lambda$ and $(u_m) \in \lambda_1$, we can find $h > 0$ and a positive integer r such that

$$|u_m| \leq h a_m^{(r)}, \; m = 1,2,\ldots$$

Then

$$\Sigma |y_m u_m| \leq h \, \Sigma |y_m| \, a_m^{(r)} < \infty$$

and consequently (y_m) belongs to $\lambda^\times_1$. λ coincides with $\lambda^\times_1$.

Result (4) is now obvious

(4) *λ is a perfect space.*

Since λ_1 is a normal space we apply §1,4.(1) to obtain that $\lambda[\nu(\lambda,\lambda_1)]$ is complete. In what follows we shall suppose λ endowed with the topology $\nu(\lambda,\lambda_1)$. It is obvious that

$$\{r\, \alpha_r^n : r = 1,2,\ldots\}$$

is a fundamental system of $\nu(\lambda,\lambda_1)$-equicontinuous absolutely convex and $\sigma(\lambda_1,\lambda)$-closed subsets of λ_1. Therefore λ is a Fréchet space and its topology can be defined by the system of seminorms $||.||_r$, r = 1,2,..., such that if $x = (x_m) \in \lambda$, then

$$||x||_r = \Sigma |x_m| \, a_m^{(r)}, \; r = 1,2,\ldots$$

(5) *The sequence space λ_1 is perfect.*

Proof. Let (u^s) be a Cauchy sequence in $\lambda_1[\sigma(\lambda_1,\lambda)]$. Since λ is a Fréchet space and

$$M = \{u^s : s = 1,2,\ldots\}$$

is a bounded subset of $\lambda_1[\sigma(\lambda_1,\lambda)]$, then M is a $\nu(\lambda,\lambda_1)$-equicontinuous and therefore there is a positive integer r such that $M \subset r\,\alpha_r^n$. The set $r\,\alpha_r^n$ is $\sigma(\lambda_1,\lambda)$-compact and consequently (u^s) $\sigma(\lambda_1,\lambda)$-converges to an element of $r\ \alpha_r^n$. We conclude that $\lambda_1[\sigma(\lambda_1,\lambda)]$ is sequentially complete and applying §1,4.(18) λ_1 is perfect.

From (5) we deduce that λ_1 coincides with $\lambda^\times$. In what follows we shall use $\lambda^\times$ instead of λ_1 and we shall suppose that $\lambda^\times$ is provided with the strong topology $\beta(\lambda^\times,\lambda)$.

We say that λ is an echelon space defined by the system (1) of steps. $\lambda^\times$ is a co-echelon space defined by the steps (1).

We shall give sometimes the steps (1) as double sequences, i.e.,

$$\alpha_r = (a_{ij}^{(r)}),\ r = 1,2,\dots$$

Then λ is a space of double sequences:

$$\lambda = \{(x_{ij}) : \Sigma|x_{ij}|\ a_{ij}^{(r)} < \infty,\ r = 1,2,\dots\}.$$

Let E_1 be the topological dual of $\lambda^\times$ and let E_2 be the space of all bounded linear forms on $\lambda^\times$. Since this last space is complete it follows that E_2 coincides with the topological dual of the associated ultrabornological space G to $\lambda^\times$. We identify λ with a subspace of E_1. Let H_1 and H_2 be the subspaces of E_1 and E_2 respectively orthogonal to the subspace ϕ of $\lambda^\times$. We denote by ψ the closure of ϕ in $\lambda^\times$

(6) H_2 *is a topological complement of* λ *in* $E_2[\beta(E_2,\lambda^\times)]$

Proof. Since λ and H_2 are closed subspaces of the Fréchet space E_2 $[\beta(E_2,\lambda^\times)]$ it is enough to see that H_2 is an algebraic complement of λ in E_2. The subspace ϕ of $\lambda^\times$ is dense in $\lambda^\times[\sigma(\lambda^\times,\lambda)]$ and therefore the zero vector is the only common element to λ and H_2. We take now a vector T in E_2 T is bounded on every bounded subset of the subspace ϕ of $\lambda^\times$ and, according the proof of §1,8.(1), the sequence $(T(e_m))$ is an element of the α-dual λ of $\lambda^\times$. Since ϕ is the linear hull of $\{e_m : m = 1,2,\dots\}$ we have that

$$T = T - (T(e_m)) + (T(e_m)),\ T - (T(e_m)) \in H_2,\ (T(e_m)) \in \lambda,$$

which completes the proof.

(7) H_1 *is the topological complement of* λ *in* $E_1[\beta(E_1,\lambda^\times)]$.

Proof. It is an obvious consequence of the equality

$$H_1 = H_2 \cap E_1.$$

(8) *If* A *is a* $\sigma(E_2,\lambda^\times)$*-bounded subset of* H_2 *there is a* $\sigma(E_1\text{-}\lambda^\times)$*-bounded subset* B *of* H_1 *with closure in* $E_2[\sigma(E_2,\lambda^\times)]$ *containing* A.

Proof. Let M be the polar set of A in $\lambda^{\times}$. M is absolutely convex and bornivorous in $\lambda^{\times}$. Applying a result of GROTHENDIECK (cf. KÖTHE [1], Chapter Six, §29, Section 4) there is a barrel U in $\lambda^{\times}$ contained in M. If P is the polar set of U in E_1 it is obvious that A is contained in the closure of P in $E_2[\sigma(E_2,\lambda^{\times})]$. Let P_1 be the projection of P in λ along H_1 and let P_2 be the projection of P in H_1 along λ. According to (7), P_1 and P_2 are $\sigma(E_1,\lambda^{\times})$-bounded. For every u of A a take a net in P

(9) $\{u^i : i \in I, \geq\}$

converging to u in $E_2[\sigma(E_2,\lambda^{\times})]$. We set

$$u^i = u_1^i + u_2^i,\ u_1^i \in P_1,\ u_2^i \in P_2.$$

Since P_1 is relatively compact in $E_1[\sigma(E_1,\lambda^{\times})]$ we take the net (9) with

$$\{u_1^i : i \in I, \geq\}$$

converging to an element u_1 in $E_1[\sigma(E_1,\lambda^{\times})]$. Then the net

$$\{u_2^i : i \in I, \geq\}$$

converges in $E_2[\sigma(E_2,\lambda^{\times})]$ to an element u_2 of H_2. It is obvious that $u = u_1+u_2$ from where it follows that $u_1 \in H_1$. Consequently the set

$$L = \{u_1 : u \in A\}$$

is in H_1 and it is $\sigma(E_1,\lambda^{\times})$-bounded.

If we set $B = L+P_2$, B is a bounded set in $E_1[\sigma(E_1,\lambda^{\times})]$ contained in H_1 satisfying the hypothesis. Indeed, if we take $u \in A$, the net

$$\{u_1 + u_1^i : i \in I, \geq\}$$

is in B and its limit in $E_2[\sigma(E_2,\lambda^{\times})]$ coincides with $u_1+u_2 = u$. The proof is complete.

(10) *ψ coincides with the closure of ϕ in* G.

Proof. The closure of ϕ in G coincides with the subspace S of $\lambda^{\times}$ orthogonal to H_2. According to (8), H_1 is $\sigma(E_2,\lambda^{\times})$-dense in H_2 and therefore the subspace ψ orthogonal to H_1 in $\lambda^{\times}$ coincides with S.

(11) *ψ is a normal sequence space.*

Proof. The family of all normal bounded closed absolutely convex subsets of λ is a fundamental system of bounded sets of λ. On the other hand, ϕ is a normal subset of $\lambda^{\times}$. We apply now §1,2.(7) to obtain that ψ is normal.

(12) *If* $\lambda^\times/\psi$ *is barrelled, then it is bornological.*

Proof. It is an obvious consequence of result (8).

(13) ψ *considered as topological subspace of* $\lambda^\times$ *is an* (LB)-*space.*

Proof. We apply results (10), (11) and §1,8.(5) to obtain that ψ is bornological for the topology $\mu(\psi,\lambda)$ which coincides on ψ with the topology induced by $\beta(\lambda^\times,\lambda)$. Since $\lambda^\times$ is complete ψ is complete and

$$\{r\,\alpha_r^n \cap \psi : r = 1,2,\ldots\}$$

is a fundamental system of complete absolutely convex bounded sets of ψ. The proof is complete.

(14) λ *is the strong dual of a complete* (LB)-*space.*

Proof. The topology $\beta(\lambda^\times,\lambda)$ coincides with $\mu(\phi,\lambda)$ on the subspace ϕ of $\lambda^\times$. Accordingly λ can be identified with the topological dual of the (LB)-space ψ. The topology $\beta(\lambda,\psi)$ is coarser than the topology of λ. Since $\lambda[\beta(\lambda,\psi)]$ is a Fréchet space we apply the closed graph theorem to the identity from $\lambda[\beta(\lambda,\psi)]$ into λ to obtain that λ coincides with $\lambda[\beta(\lambda,\psi)]$. The proof is complete.

2. REFLEXIVE ECHELON SPACES. In this section we shall suppose that λ is an echelon space defined by the system of steps 1.(1).

(1) *If* λ *is reflexive then* λ *is a Montel space.*

Proof. If A is a compact subset of $\lambda[\sigma(\lambda,\lambda^\times)]$ we have that A is compact for the topology $\nu(\lambda,\lambda^\times)$, according to §1,4.(15). This topology coincides with the topology of λ. Thus if λ is reflexive, λ is a Montel space.

(2) *If* λ *is not a Montel space there is a positive integer* r *and a sequence of positive integers*

$$n_1 < n_2 < \ldots < n_m < \ldots$$

such that

$$\inf\{(a_{n_m}^{(r)};\, a_{n_m}^{(k)}) : m = 1,2,\ldots\} > 0,\; k = r, r+1,\ldots$$

Proof. If λ is not a Montel space it is not reflexive and therefore, applying §1,4.(26), there is a bounded set A in λ and an element $u = (u_m)$ in $\lambda^\times$, $u \geq 0$, such that the sequence

$$\sup\{\sum_{m=s} |x_m|u_m : (x_m) \in A\},\; s = 1,2,\ldots,$$

does not converge to zero. According to the definition of $\lambda^\times$ there is a positive integer r such that u belongs to a scalar multiple of α_r^n. Consequently the sequence

$$(3)\qquad \sup\{\sum_{m=s}^{\infty} |x_m|\, a_m^{(r)} : (x_m) \in A\},\ s = 1,2,\ldots,$$

does not converges to zero.

Since A is bounded in λ we have that

$$\sup\{\sum_m |x_m|\, a_m^{(k)} : (x_m) \in A\} < M_k < \infty,\ k = 1,2,\ldots$$

From (3) we see that there is $h>0$ verifying

$$(4)\qquad \sup\{\sum_{m=s}^{\infty} |x_m|\, a_m^{(r)} : (x_m) \in A\} < h,\ s = 1,2,\ldots$$

Given positive integers s and k, $k \geq r$, we set N_{sk} to denote the subset of all elements m of $\{s+1,s+2,\ldots\}$ with

$$(a_m^{(r)};\ a_m^{(k)}) \leq \frac{h}{2^k M_k}$$

Applying (4) we obtain an element $(y_m) \in A$ with

$$\sum_{m=s+1}^{\infty} |y_m|\, a_m^{(r)} > h$$

Then

$$\sum_{m\in N_{sk}} |y_m|\, a_m^{(r)} = \sum_{m\in N_{sk}} |y_m|\, a_m^{(k)}\, (a_m^{(r)};a_m^{(k)})$$

$$\leq \frac{h}{2^k M_k} \sum_{m\in N_{sk}} |y_m|\, a_m^{(k)} \leq \frac{h}{2^k M_k} \sum |y_m|\, a_m^{(k)} < \frac{h}{2^k}$$

We shall prove now the existence of an integer $m(s) > s$ verifying

$$(5)\qquad (a_{m(s)}^{(r)};\ a_{m(s)}^{(k)}) > \frac{h}{2^k M_k},\ k = r, r+1,\ldots$$

Indeed, if (5) is not true we have

$$\{s+1,s+2,\ldots\} = \bigcup_{k=r}^{\infty} N_{sk}$$

and consequently

$$h < \sum_{m=s+1}^{\infty} |y_m|\, a_m^{(r)} \leq \sum_{k=r}^{\infty} \sum_{m\in N_{sk}} |y_m|\, a_m^{(r)} < \sum_{k=r}^{\infty} < \frac{h}{2^k} \leq h,$$

which is a contradiction.

We set $m(1) = n_1$, $m(n_s) = n_{s+1}$, $s=1,2,\ldots$ Then

$$\inf\{(a_{n_m}^{(r)};a_{n_m}^{(k)}) : m = 1,2,\ldots\} \geq \frac{h}{2^k M_k} > 0, \; k=r,r+1,\ldots$$

(6) *Given the echelon space* λ, *if there is a positive integer* r *and a sequence of positive integers*

$$n_1 < n_2 < \ldots < n_m < \ldots$$

such that

$$\inf\{(a_{n_m}^{(r)};a_{n_m}^{(k)}); m = 1,2,\ldots\} = m_k > 0, \; k = r, r+1,\ldots$$

then λ *is not a Montel space.*

Proof. Let E be the sectional subspace of λ such that the vector (z_m) of λ belongs to E if and only if $z_m = 0$, $m \neq n_s$, $s = 1,2,\ldots$ For every $(a_m) \in \ell^1$ we set $X((a_m))$ for the sequence (x_s) such that

$$x_s = 0 \text{ if } s \neq n_m, \; x_{n_m} = \frac{1}{a_{n_m}^{(r)}} a_m, \; m = 1,2,\ldots$$

For every positive integer $k \geq r$ we have

$$(7) \qquad \|X((a_m))\|_k = \sum_m |x_{n_m}| \, a_{n_m}^{(k)} = \sum_m \frac{|a_m|}{(a_{n_m}^{(r)};a_{n_m}^{(k)})} \leq \frac{1}{m_k} \sum |a_m|$$

from where it follows that X is a mapping from ℓ^1 in E. Obviously X is linear and injective and according to (7) X is continuous.

We take now any vector (y_m) of E. If

$$b_m = a_{n_m}^{(r)} y_{n_m}, \; m = 1,2,\ldots,$$

it follows that

$$\sum |b_m| = \sum_m |y_{n_m}| \, a_{n_m}^{(r)} = \sum_m |y_m| \, a_m^{(r)} = \|(y_m)\|_r$$

and therefore $(b_m) \in \ell^1$, $X((b_m)) = (y_m)$ and the mapping X^{-1} from E into ℓ^1 is continuous.

E is isomorphic to ℓ^1 and, since this last space is not reflexive, then λ is not a Montel space.

(8) *The echelon space* λ *is not a Montel space if and only if has a sectional subspace isomorphic to* ℓ^1.

Proof. If λ is not a Montel space the conditions of (2) are verified. We apply now the proof of (6) to obtain that λ has a sectional subspace isomorphic to ℓ^1. Obviously λ is not a Montel space if λ has a sectional subspace isomorphic to ℓ^1.

We represent now by F a separable Fréchet space, $F \neq \{0\}$. Let

$$p_1 < p_2 < \ldots < p_m < \ldots$$

be a fundamental system of continuous seminorms on F such that p_1 is not identically nulle. We set

$$H = \{x \in F : p_1(x) = 0\}.$$

Since H is a closed subspace of F, $F \neq H$, there is a sequence (x_m) in $F \sim H$ such that

$$\{x_1, x_2, \ldots, x_m, \ldots\}$$

is a dense subset of F. We consider the family of sequences

$$(9) \qquad \beta_r = (p_r(x_m)),\ r = 1,2,\ldots$$

We have that $\beta_{r+1} \geqslant \beta_r$, $r = 1,2,\ldots$, and $p_r(x_m) \neq 0$, $r,m = 1,2,\ldots$, and thus (9) is a system for an echelon space μ.

For every $(a_m) \in \mu$ we set

$$f((a_m)) = \Sigma a_m x_m.$$

(10) f *is a continuous linear mapping from* μ *into* F.

Proof. Given the seminorm p_r and the element (a_m) we have that

$$\sum_m |a_m|\ p_r(x_m) < \infty$$

and therefore the series

$$\Sigma a_m x_m$$

is convergent in F and therefore f is well defined. Obviously f is linear. On the other hand,

$$p_r(f((a_m))) = p_r\ (\Sigma a_m x_m) \leqslant \sum_m |a_m|\ p_x(x_m)$$

from where the continuity of f follows.

(11) *The mapping* f *is onto.*

Proof. Given an element x of F we find a positive integer n_1 such that

$$p_1(x - x_{n_1}) < \frac{1}{2}$$

We proceed by recurrence. Suppose we have constructed

$$x_{n_1}, x_{n_2}, \ldots, x_{n_r}, n_1 < n_2 < \ldots < n_r$$

and that

$$(12) \qquad p_r(x - x_{n_1} - \frac{1}{2} x_{n_2} - \ldots - \frac{1}{2^{r-1}} x_{n_r}) < \frac{1}{2^r} .$$

Given the vector of F

$$y_r = 2^r (x - x_{n_1} - \frac{1}{2} x_{n_2} - \ldots - \frac{1}{2^{r-1}} x_{n_r})$$

we determine a positive integer n_{r+1} larger than n_r such that

$$p_{r+1}(y_r - x_{n_{r+1}}) < \frac{1}{2} .$$

Then

$$p_{r+1}(x - x_{n_1} - \frac{1}{2} x_{n_2} - \ldots - \frac{1}{2^r} x_{n_{r+1}})$$

$$= p_{r+1}(\frac{1}{2^r} y_r - \frac{1}{2^r} x_{n_{r+1}}) = \frac{1}{2^r} p_{r+1}(y_r - x_{n_{r+1}}) < \frac{1}{2^{r+1}} .$$

Let (b_m) be the sequence of numbers defined in the following way:

$$b_m = 0 \text{ if } m \neq n_r, \; b_{n_r} = \frac{1}{2^{r-1}}, \; r = 1,2,\ldots$$

Given a positive integer k we have that if $r > k$

$$b_{n_r} p_k(x_{n_r}) = p_k(b_{n_r} x_{n_r}) \leq p_r(x - x_{n_1} - \frac{1}{2} x_{n_2} - \ldots - \frac{1}{2^{r-1}} x_{n_r})$$

$$+ p_{r-1}(x - x_{n_1} - \frac{1}{2} x_{n_2} - \ldots - \frac{1}{2^{r-2}} x_{n_{r-1}}) < \frac{1}{2^r} + \frac{1}{2^{r-1}} < \frac{1}{2^{r-2}}$$

and therefore

$$\sum_m |b_m| \, p_k(x_m) < \infty$$

from where it follows that (b_m) is an element of μ. Finally, having in mind (12) it follows that

$$f((b_m)) = \Sigma b_m x_m = \Sigma b_{n_r} x_{n_r} = x$$

which completes the proof.

(13) *The space* F *is isomorphic to a separated quotient of* μ.

Proof. Since $f: \mu \longrightarrow F$ is a continuous linear mapping and since μ and F are Fréchet spaces, the open mapping theorem shows that f is an homomorphism. Consequently F is isomorphic to $\mu / f^{-1}(o)$. The proof is complete.

Now we consider the echelon space Λ defined by the system of steps

$$\gamma_r = (a_{ij}^{(r)}),\ r = 1,2,\ldots.$$

such that

$$a_{ij}^{(r)} = j^r p_r(x_j),\ i < r,$$

$$a_{ij}^{(r)} = r^{r+i} p_r(x_j),\ i \geq r,$$

$$j = 1,2,\ldots$$

(14) *The echelon space* Λ *is a Montel space.*

Proof. If Λ is not a Montel space we apply result (2) to obtain a positive integer r and a sequence of pairs of positive integers

$$(i_m, j_m),\ m = 1,2,\ldots,$$

whose elements are different such that

(15) $$\inf\{(a_{i_m j_m}^{(r)}; a_{i_m j_m}^{(k)}) : m = 1,2,\ldots\} > 0,\ k = r, r+1,\ldots$$

If the sequence (i_m) is not bounded and if we fix an integer $k>r$ it follows that if $i_m \geq k$

$$(a_{i_m j_m}^{(r)}; a_{i_m j_m}^{(k)}) = (r^{r+i_m} p_r(x_{j_m}); k^{k+i_m} p_k(x_{j_m})) < (\frac{r}{k})^{i_m}$$

and therefore (15) does not hold. Thus the sequence (i_m) is bounded by a positive integer $h \geq r$ and therefore the sequence (j_m) is not bounded. Then

$$(a_{i_m j_m}^{(r)}; a_{i_m j_m}^{(h+2)}) \leq (a_{i_m j_m}^{(h+1)}; a_{i_m j_m}^{(h+2)})$$

$$= (j_m^{h+1} p_{h+1}(x_{j_m}); j_m^{h+2} p_{h+2}(x_{j_m})) < \frac{1}{j_m}$$

and again (15) does not hold and we arrive to a contradiction. Thus Λ is a Montel space.

For every (a_{ij}) in Λ we set

$$g((a_{ij})) = (\sum_i a_{ij})$$

(16) g: $\Lambda \to \mu$ *is a continuous linear mapping.*

Proof. Given a positive integer r we have

$$\sum_j |\sum_i a_{ij}| p_r(x_j) \leq \sum |a_{ij}|\ p_r(x_j) \leq \sum |a_{ij}|\ a_{ij}^{(r)} < \infty$$

if $(a_{ij}) \in \Lambda$. It follows that g is continuous and obviously linear.

For every $(u_m) \in \mu^{\times}$ we set

$$Y((u_m)) = (v_{ij})$$

with

$$v_{ij} = u_j, \; i,j = 1,2,\ldots$$

We know the existence of $h > 0$ and a positive integer r such that

$$|u_m| \leq hp_r(x_m), \; m = 1,2,\ldots$$

and therefore

$$|v_{ij}| \leq hp_r(x_j) \leq ha_{ij}^{(r)}, \; i,j = 1,2,\ldots,$$

and thus the double sequence (v_{ij}) belongs to $\Lambda^{\times}$. Consequently, $Y:\mu^{\times} \to \Lambda^{\times}$ is a linear continuous mapping.

(17) *The mapping Y is the transposed mapping of g.*

Proof. If $(a_{ij}) \in \Lambda$ and $(u_j) \in \mu^{\times}$ we have

$$\langle g(a_{ij}),(u_j)\rangle = \sum_j (\sum_i a_{ij})u_j = \Sigma a_{ij}u_j = \langle (a_{ij}),Y(u_j)\rangle$$

which completes the proof.

(18) *$Y(\mu^{\times})$ is sequentially closed in $\Lambda^{\times}[\sigma(\lambda^{\times},\lambda)]$.*

Proof. Let

$$(u_m^{(s)}), \; s = 1,2,\ldots,$$

be a sequence in $\mu^{\times}$ such that

$$Y((u_m^{(s)})), \; s = 1,2,\ldots,$$

converges to (w_{ij}) in $\Lambda^{\times}[\sigma(\Lambda^{\times},\Lambda)]$. Then

$$w_{ij} = w_{2j} = \ldots = w_{ij} = \ldots = u_j, \; j = 1,2,\ldots$$

We find $h > 0$ and a positive integer r such that

$$|w_{ij}| \leq ha_{ij}^{(r)}, \; i,j = 1,2,\ldots$$

Then

$$|u_j| = |w_{rj}| \leq ha_{rj}^{(r)} = hr^{2r}p_r(x_j), j=1,2,\ldots,$$

from where it follows that $(u_j) \in \mu^{\times}$. Finally it is obvious that

$$Y((u_j)) = (w_{ij}).$$

(19) *$Y(\mu^{\times})$ is closed in $\Lambda^{\times}[\sigma(\lambda^{\times},\lambda)]$.*

Proof. Let A be any compact subset of $\Lambda^{\times}[\sigma(\Lambda^{\times},\times)]$. Since Λ is separable we have that A is metrizable. According to (18), $A \cap Y(\mu^{\times})$ is sequentially $\sigma(\Lambda^{\times},\Lambda)$-closed and therefore $A \cap Y(\mu^{\times})$ is $\sigma(\Lambda^{\times},\Lambda)$-closed. We apply the theorem of KREIN-SMULIAN (cf. HORVATH [1], Chapter 3, §10) to obtain that $Y(\mu^{\times})$ is $\sigma(\Lambda^{\times},\Lambda)$-closed.

(20) *The echelon space μ is isomorphic to a quotient of the echelon space Λ.*

Proof. Since Λ and μ are Fréchet spaces and since Y is injective and its image is $\sigma(\Lambda^{\times},\Lambda)$-closed we have that $g:\Lambda\to\mu$ is an homomorphism. Thus μ is isomorphic to $\Lambda/g^{-1}(0)$.

(21) *If E is a Fréchet separable space, then E is isomorphic to a quotient of a Montel echelon space.*

Proof. If E = {0} we consider any Montel echelon space and its quotient by itself is isomorphic to E. If E = F ≠ {0} we apply result (13) to obtain μ such that

$$\mu/f^{-1}(0) \sim F.$$

On the other hand, according to (20), we have that

$$\Lambda/g^{-1}(o) \simeq \mu.$$

Then E is isomorphic to the quotient of the echelon space Λ by $(g \circ f)^{-1}(0)$.

Results (2), (6) and (8) are due to KÖTHE [1] and [3]. Theorem (21) can be found in VALDIVIA [26].

3. TOTALLY REFLEXIVE ECHELON SPACES. Following GROTHENDIECK [2] we say that a locally convex space E is totally reflexive if every separated quotient of E is reflexive.

(1) *Let E be a locally convex space satisfying*

1. *In $E'[\sigma(E'E)]$ if A is a bounded subset there is an equicontinuous closed absolutely convex subset $B \supset A$ such that the weak topology of E'_B and $\sigma(E',E)$ coincide on A.*
2. *Every separated quotient of E is complete.*

Then E is totally reflexive.

Proof. Let F be a closed subspace of E and let $F^{\perp}$ be the subspace of $E'[\sigma(E',E)]$ orthogonal to F. $F^{\perp}$ can be identified in the usual way with the topological dual of E/F. Let w be a linear form on $F^{\perp}$ bounded on the bounded subsets of $F^{\perp}$. If A is a bounded subset of $F^{\perp}$ we have, according to 1., that A is equicontinuous and that w is continuous on A for the topology induced by $\sigma(E',E)$. Then E/F is barrelled and we apply PTAK-COLLINS's theorem (cf. KÖTHE [1], Chapter Four, §21, Section 9) to obtain that w is in E/F. Consequently E/F is reflexive.

In this section we denote by λ an echelon space defined by the system of steps 1.(1).

(2) *The echelon space* λ *is a Schwartz space if and only if given a positive integer* r *there is an integer* k>r *such that*

(3) $$\lim (a_m^{(r)};a_m^{(k)}) = 0$$

Proof. Let (u_m) be an element of $\lambda^{\times}$. We find a positive integer r and h > 0 such that

$$|u_m| \leq ha_m^{(r)}, \qquad m = 1,2,...$$

We suppose the existence of an integer k >r such that condition (3) is satisfied. Then if

$$v_m = ha_m^{(k)}, \quad m = 1,2,...,$$

we have that

$$(v_m) \in \lambda^{\times}, \; |u_m| \leq |v_m|, \quad m = 1,2,...,$$

$$\lim (u_m,v_m) = 0$$

and, applying §1,10.(3), λ is a Schwartz space.

If λ is a Schwartz space, given the positive integer, r, we have that $(a_m^{(r)}) \in \lambda^{\times}$ and applying §1,10.(1) we find $(v_m) \in \lambda^{\times}$ such that

$$a_m^{(r)} \leq |v_m|, \; m = 1,2,..., \text{ and } \lim (a_m^{(r)};v_m) = 0$$

There is h > 0 and an integer k > r with

$$|v_m| \leq h\, a_m^{(k)}, \; m = 1,2,...,$$

from where

$$\lim (a_m^{(r)};a_m^{(k)}) = 0$$

which completes the proof.

We suppose now that λ is a Montel space which is not Schwartz. According to (2) there is a positive integer r such that, for every $k > r$, the sequence $(a_m^{(r)};a_m^{(k)})$, $m = 1, 2,\ldots$, does not converge to zero. Thus we can find a sequence of positive integers

$$q_1 < q_2 < \ldots < q_i < \ldots$$

such that

$$(a_{q_i}^{(r)};\ a_{q_i}^{(r+1)}) > b_1 > 0,\ i = 1,2,\ldots$$

According to 2.(6), we extract a subsequence (m_i) from (q_i) such that for a certain $k_1 > r+1$

$$\lim(a_{m_i}^{(r)};a_{m_i}^{(k_1)}) = 0.$$

We set

$$I_1 = \{m_1,m_2,\ldots,m_i,\ldots\}.$$

Proceeding by recurrence we suppose the pairwise disjoint infinite subsets of positive integers $I_1,I_2,\ldots,I_q$ constructed in such a way that if p is an integer with $1 \leq p \leq q$ and if

$$I_p = \{r_1,r_2,\ldots,r_i,\ldots\}$$

$$r_1 < r_2 < \ldots < r_i < \ldots,$$

there is an integer $k_p > r+p$ such that

$$\lim\ (a_{r_i}^{(r)};\ a_{r_i}^{(k_p)}) = 0$$

$$(a_{r_i}^{(r)};\ a_{r_i}^{(r+p)}) > b_p > 0,\ i = 1,2,\ldots$$

We set

$$H_q = \cup\ \{I_p : p = 1,2,\ldots,q\}$$

Ordering the elements of H_q in a sequence

$$n_1 < n_2 < \ldots < n_i < \ \ldots$$

we obtain for every integer $s > k_p$, $p = 1,2,\ldots,q$, that

(4) $$\lim\ (a_{n_i}^{(r)};a_{n_i}^{(s)}) = 0$$

Since the sequence $(a_m^{(r)};a_m^{(s)})$, $m = 1,2,\ldots$, does not converge to zero, it follows from (4) that there is in $N \sim H_q$ an infinity of elements

$$s_1 < s_2 < \ldots < s_i < \ldots$$

such that

$$(a_{s_i}^{(r)}; a_{s_i}^{(r+q+1)}) > b_{q+1} > 0, \ i = 1,2,\ldots$$

Since λ is a Montel space there is a subsequence (t_i) of (s_i) and an integer $k_{q+1} > r+q+1$ such that

$$\lim (a_{t_i}^{(r)}; a_{t_i}^{(k_{q+1})}) = 0$$

If I_{q+1} is the set $\{t_1,t_2,\ldots,t_i,\ldots\}$ we have constructed by recurrence the infinite family

$$\{I_1,I_2,\ldots,I_q,\ldots\}$$

with the properties indicated above. We represent the elements of I_p by

$$p1, p2, \ldots, pi \ldots$$

Then

$$(5) \qquad (a_{pi}^{(r)}; a_{pi}^{(r+p)}) > b_p > 0, \ i = 1,2,\ldots, \ p = 1,2,\ldots$$

Now we consider the echelon space Λ defined by the following system of steps:

$$\beta_s = (a_{ij}^{(s)}), \ s = r, r+1, \ldots$$

If μ_1 is the sectional subspace of all those elements (x_m) of λ such that $x_m = 0$, $m \in N \sim \bigcup_{q=1}^{\infty} I_q$, then Λ is isomorphic to μ_1. If μ_2 is the sectional subspace of λ of all those elements (x_m) of λ such that $x_m = 0$, $m \in \bigcup_{q=1}^{\infty} I_q$, then λ is the topological direct sum of μ_1 and μ_2. Consequently Λ is isomorphic to a quotient of λ.

Let Δ be the linear space over K of all double sequence (x_{ij}) satisfying

$$(6) \qquad |x_{ij}| \leqslant b a_{1j}^{(r)}, \ x_{ij} = \frac{a_{ij}^{(r+1)}}{a_{1j}^{(r)}} x_{1j},$$

$$j = 1,2,\ldots, \ i = 2,3,\ldots$$

b being a positive number depending of the sequence (x_{ij})

(7) *Δ is a linear subspace of $\Lambda^{\times}$.*

Proof. If the double sequence (x_{ij}) verifies (6) we have that, for $j = 1,2,\ldots,$

$$|x_{1j}| \leqslant b a_{1j}^{(r)} \leqslant b a_{1j}^{(r+1)}, \ |x_{ij}| = \frac{a_{ij}^{(r+1)}}{a_{1j}^{(r)}} |x_{1j}| \leqslant b a_{ij}^{(r+1)},$$

$$i = 2,3,\ldots$$

and therefore $(x_{ij}) \in \Lambda^{\times}$.

(8) *Λ is closed in $\Lambda^{\times}[\sigma(\Lambda^{\times},\Lambda)]$.*

Proof. We apply KREIN-SMULIAN's theorem and, since Λ is separable, it is enough to prove that Δ is sequentially closed in $\Lambda^{\times}[\sigma(\Lambda^{\times},\Lambda)]$. Let

(9) $\qquad (x_{ij}^{(k)})\ k = 1,2,\dots,$

be a sequence in Δ converging to (y_{ij}) in $\Lambda^{\times}[\sigma(\Lambda^{\times},\Lambda)]$. We can find a positive integer $m > 1$ and $h > 0$ such that

$$|x_{ij}^{(k)}| \leq h a_{ij}^{(r+m)},\ i,j,k = 1,2,\dots$$

According to (5) we have, for $k,j = 1,2,\dots$, that

$$|x_{mj}^{(k)}| \leq h a_{mj}^{(r+m)} = \frac{h}{(a_{mj}^{(r)};a_{mj}^{(r+m)})} a_{mj}^{(r)} \leq \frac{h a_{mj}^{(r)}}{b_m}$$

and therefore

$$\frac{h a_{mj}^{(r)}}{b_m} \geq |x_{mj}^{(k)}| = \frac{a_{mj}^{(r+1)}}{a_{1j}^{(r)}} |x_{ij}^{(k)}| \geq \frac{a_{mj}^{(r)}}{a_{1j}^{(r)}} |x_{1j}^{(k)}|$$

from where it follows

(10) $\qquad |x_{1j}^{(k)}| \leq \frac{h}{b_m} a_{1j}^{(r)}$

From (10) and from the coordinatewise convergence of the sequence (9) to (y_{ij}) it follows, for $j = 1,2,\dots$, that

$$|y_{1j}| \leq \frac{h}{b_m} a_{ij}^{(r)}, y_{ij} = \frac{a_{ij}^{(r+1)}}{a_{ij}^{(r)}} y_{1j},\ i = 2,3,\dots$$

and consequently $(y_{ij}) \in \Delta$.

(11) *The space Δ endowed with the topology induced by $\sigma(\Lambda^{\times},\Lambda)$ is isomorphic to $\ell^{\infty}[\sigma(\ell^{\infty},\ell^{1})]$.*

Proof. Let $f:\ell^{\infty} \to \Lambda^{\times}$ be the mapping defined by

$$f((z_j)) = (x_{ij}),\ (z_j) \in \ell^{\infty},$$

such that for $j = 1,2,\dots$

(12) $\qquad x_{1j} = a_{1j}^{(r)} z_j,\ x_{ij} = a_{ij}^{(r+1)} z_j,\ i = 2,3,\dots$

If $||.||$ denotes the norm in ℓ^∞ it follows that

$$|x_{1j}| \leq ||(z_m)||\ a_{1j}^{(r)},\ x_{ij} = \frac{a_{ij}^{(r+1)}}{a_{1j}^{(r)}}\ x_{1j},\ i = 2,3,\ldots$$

and therefore $f(\ell^\infty) \subset \Delta$. From (12) we know that f is injective. If $(u_{ij}) \in \Lambda$ then

$$\Sigma|u_{ij}|\ a_{ij}^{(r+1)} < \infty$$

and therefore

$$(\sum_i |u_{ij}|\ a_{ij}^{(r+1)}) \in \ell^1.$$

Also

$$|<(u_{ij}),f((z_j))>| = |\Sigma u_{ij} x_{ij}| \leq \sum_j(\sum_i |u_{ij}|\ a_{ij}^{(r+1)})|z_j|$$

from where it follows that

(13) $\quad f: \ell^\infty[\sigma(\ell^\infty,\ell^1)] \longrightarrow \Lambda^\times[\sigma(\Lambda^\times,\Lambda)]$

is continuous.

Given the element (x_{ij}) of Δ verifying (6), we set

$$z_j = \frac{x_{1j}}{a_{1j}^{(r)}},\ j = 1,2,\ldots$$

Then $(z_j) \in \ell^\infty$ and $f((z_j)) = (x_{ij})$. Therefore $f(\ell^\infty) = \Delta$ Finally, since Δ is closed in $\Lambda^\times[\sigma(\Lambda^\times,\Lambda)]$, (13) is an isomorphism onto.

(14) *If the echelon space λ is not a Schwartz space, then λ has a quotient isomorphic to* ℓ^1.

Proof. If λ is not a Montel space we apply 2.(8) to obtain that λ has a sectional subspace isomorphic to ℓ^1. Consequently λ has a quotient isomorphic to ℓ^1.

If λ is a Montel space which is not a Schwartz space we know the existence of a quotient of λ isomorphic to the space Λ described above. If L is the subspace of Λ orthogonal to Δ we have that Λ/L is isomorphic to ℓ^1 (see result (11)) and the conclusion follows.

(15) *The echelon space is totally reflexive if and only if λ is a Schwartz space.*

Proof. If λ is a Schwartz space, λ satisfies conditions 1. and 2. of the space E in (1) and therefore λ is totally reflexive.

If λ is not a Schwartz space we apply result (14) to obtain that λ is not totally reflexive.

The results above in this section can be found in VALDIVIA [11] and [12]. We finish the section with a characterization of nuclearity of echelon spaces.

(16) *The echelon space λ is nuclear if and only if given a positive integer r there is an integer $s>r$ such that*

(17) $$\sum_m (a_m^{(r)};a_m^{(s)}) < \infty$$

Proof. We suppose that λ is nuclear. Given a positive integer r we have that α_r belongs to $\lambda^\times$ and therefore, applying §1,11.(1), we find an element $(v_m) \in \lambda^\times$ such that

$$a_m^{(r)} \leq |v_m|,\ m=1,2,\ldots,\ \text{and}\ \sum_m |a_m^{(r)};v_m| < \infty$$

There is a positive integer $s > r$ and a positive real number h such that

$$|v_m| \leq h a_m^{(s)},\ m = 1,2,\ldots$$

Then

$$\sum_m (a_m^{(r)};\ a_m^{(s)}) \leq h \sum_m |a_m^{(r)},v_m| < \infty.$$

Reciprocally, given any element $(u_m) \in \lambda^\times$ we determine a positive integer r and a real number $h > 0$ such that

$$|u_m| < h a_m^{(r)},\ m = 1,2,\ldots$$

Let s be an integer larger than r satisfying (17). By taking

$$v_m = h a_m^{(s)},\ m = 1,2,\ldots,$$

We have that $(v_m) \in \lambda^\times$ and that §1,11.(3) is satisfied. Thus λ is nuclear.

4. QUASINORMABLE ECHELON SPACES. Following GROTHENDIECK [2] a locally convex space E is quasinormable if given an equicontinuous subset A of E' there is an equicontinuous closed absolutely convex subset of $E'[\sigma(E',E)]$ with $A \subset B$ and such that the strong topology $\beta(E',E)$ coincides on A with the topology of the Banach space E'_B. If E is a quasibarrelled space, E is quasinormable if and only if the strict Mackey condition is satisfied.

In this section λ is an echelon space defined by the system of steps

1.(1).

We set ϕ^1 to denote the subspace of $\lambda^\times[\beta(\lambda^\times,\lambda)]$ of all those vectors which are limit in the sense of Mackey of sequences contained in ϕ. Let ϕ_m be the closure of $\phi\cap\lambda^\times{}_{\alpha_m}^n$ in the Banach space $\lambda^\times{}_{\alpha_m}^n$, m = 1,2,... We suppose ϕ_m endowed with the topology induced by $\lambda^\times{}_{\alpha_m}^n$. $||\cdot||_m$ is the norm of $\lambda^\times{}_{\alpha_m}^n$.

(1) *ϕ^1 in an (LB)-space.*

Proof. If u belongs to $\lambda^\times$ we set $\phi(u)$ to denote the closure of $\phi\cap\lambda^\times{}_u^n$ in the Banach space $\lambda^\times{}_u n$. We suppose $\phi(u)$ provided with the topology induced by $\lambda^\times{}_u n$. We have that

$$\phi^1 = \cup\{\phi(u) : u \in \lambda^\times\}$$

We proved in §1, 9.(2) that ϕ^1 is the inductive limit of the family

$$\{\phi(u) : u \in \lambda^\times\}$$

Fixing $u \in \lambda^\times$ there is a positive integer r such that u belongs to a scalar multiple of $\alpha_r{}^n$ and consequently $\phi(u)$ is contained in $\phi(\alpha_r) = \phi_r$ and the canonical injection from $\phi(u)$ into ϕ_r is continuous. Then ϕ^1 is the inductive limit of the increasing sequence of Banach spaces

$$\{\phi_r : r = 1,2,\ldots\}.$$

A locally convex space E is said to be locally complete if it is complete in the sense of Mackey, i.e., if every sequence in E which is a Cauchy sequence in the sense of Mackey is Mackey-convergent.

(2) *If there is a positive integer* p *such that given any integer* $q > p$ *there is an integer* $r > q$ *and a subset* P *of* N *with*

(3) $$\inf\{(a_m^{(p)};a_m^{(q)}) : m \in P\}>0,\ \inf\{a_m^{(p)};a_m^{(r)}) : m \in P\}= 0,$$

then ϕ^1 is not locally complete.

Proof. By (3) P is infinite and therefore there is a sequence in P

$$m_1 < m_2 < \ldots < m_i < \ldots$$

with

(4) $$\inf\{(a_{m_i}^{(p)};a_{m_i}^{(q)}) : i = 1,2,\ldots\} = a > 0.$$

(5) $$\lim_i (a_{m_i}^{(p)};a_{m_i}^{(r)}) = 0.$$

We consider the vector $u = (u_m)$ of $\lambda^\times{}_{\alpha_p}n$ such that

$$u_m = 0,\ m \neq m_i,\ u_{m_i} = a_{m_i}^{(p)} \quad i = 1,2,\ldots$$

By (5) we have that

$$\lim_m (u_m;a_m^{(r)}) = 0$$

and therefore, given $\varepsilon > 0$, there is a positive integer k with

$$(u_m;a_m^{(r)}) < \varepsilon, \ m>k.$$

We have that the section u(m) of order m of u belongs to $\phi \cap \lambda^{\times}{}_{\alpha_r}n$, m = 1, 2,... If m > k it follows that

$$||u-u(m)||_r \leq \sup\{|(u_m;a_m^{(r)})|: m = k+1,k+2,...\} < \varepsilon$$

and thus u belongs to ϕ_r.

We take now a vector $v = (v_m)$ of $\phi \cap \lambda^{\times}{}_{\alpha_q}n$ There is a positive integer h such that $v_m = 0$ if m > k. Then, according to (4),

$$||u-v||_q = \sup\{|u_m-v_m;\ a_m^{(q)}| : m = 1,2,...\}$$

$$\geq \sup\{(u_m;a_m^{(q)}) : m = h+1,h+2,...\} \geq a$$

and therefore $u \notin \phi_q$. Then

$$\phi_r \cap \lambda^{\times}{}_{\alpha_p}n \neq \phi_q \cap \lambda^{\times}{}_{\alpha_p}n$$

and consequently the sequence

$$(\phi_m \cap \lambda^{\times}{}_{\alpha_p}n)$$

of closed subspace of $\lambda^{\times}{}_{\alpha_p}n$ is increasing and not stationary. We apply BAIRE's category theorem in the Banach space $\lambda^{\times}{}_{\alpha_p}n$ to obtain that

$$\cup\{\phi_m \cap \lambda^{\times}{}_{\alpha_p}n : m = 1,2,...\},$$

which coincides with ϕ^1, is a non-closed subspace of $\lambda^{\times}{}_{\alpha_p}n$ and therefore ϕ^1 is not locally complete.

(6) *Given any positive integer* p *suppose that there is an integer* q>p *with the following property: if* P *is any subset of* N *with*

(7) $\inf\{(a_m^{(p)};a_m^{(q)}) : m \in P\} > 0,$

then

(8) $\inf\{(a_m^{(p)};a_m^{(r)}) : m \in P\} > 0, \ r = q+1,q+2,...$

Then ϕ^1 *is complete.*

Proof. We suppose that ϕ^1 is not complete. We take a vector $u = (u_m)$ in the closure of ϕ in $\lambda^{\times}$ which is not in ϕ^1. Let p be a natural number with $u \in \lambda^{\times}_{\alpha_p n}$ We find an integer $q > p$ such that, if P is any subset of N satifying (7), then (8) is verified. Since $u \in \lambda^{\times}_{\alpha_q n}$ and $u \notin \phi_q$, the distance from u to ϕ_q in $\lambda^{\times}_{\alpha_q n}$ is larger than a positive number a. Given a positive integer k we have that $u(k)$ belongs to ϕ_q and therefore

$$||u-u(k)||_q = \sup \{|u_m ; a_m^{(q)}| : m = k+1, k+2, \ldots\} > a$$

and thus there is a sequence of positive integers $m_1 < m_2 < \ldots < m_i \ldots$ with

$$\inf \{|u_{m_i} ; a_{m_i}^{(q)}| : i = 1,2, \ldots\} \geqslant a.$$

We find a positive number b such that $|u_m| \leqslant b\, a_m^{(q)}$, $m = 1,2,\ldots$ Then

$$\inf \{(a_{m_i}^{(p)} ; a_{m_i})^{(q)} : i = 1,2,\ldots\} > \frac{1}{b} \inf \{|u_{m_i} : a_{m_i}^{(q)}| : i = 1,2, \ldots \} > \frac{a}{b} > 0.$$

We apply the conditions of the theorem to obtain

$$\inf \{(a_{m_i}^{(p)} ; a_{m_i})^{(r)}) : i = 1, 2, \ldots\} > a_r > 0, \quad r=q+1,q+2..$$

Let M be the subset of λ of all those sequences (x_m) such that $x_m = 0$, $m \neq m_i$, $i = 1,2,\ldots$, and $\Sigma|x_m|\, a_m^{(q)} \leqslant 1$.

If (z_m) belongs to M and if r is an integer larger than q we have that

$$\Sigma\, |z_m| a_m^{(r)} \leqslant \Sigma\, |z_{m_i}| \frac{a_{m_i}^{(p)}}{(a_{m_i}^{(p)} ; a_{m_i}^{(r)})} \leqslant \frac{1}{a_r} \Sigma\, |z_m| a_m^{(p)} \leqslant \frac{1}{a_r}$$

form where it follows that M is a bounded set of λ. Consequently there is an element $v = (v_m)$ of ϕ such that

(9) $\qquad \sup \{| \langle x, u-v \rangle | : x \in M\} < a.$

We find a positive integer j such that $v_{m_j} = 0$. The sequence (w_m) satisfying $w_m = 0$, $m \neq m_j$, $w_{m_j} = \dfrac{1}{a_{m_j}^{(q)}}$, is in M and therefore

$$\sup \{| \langle x, u-v \rangle : x \in M\} \geqslant | \langle w, u-v \rangle | = |u_{m_j} ; a_{m_j}^{(q)}| \geqslant a$$

which is in contradiction with (9). Thus ϕ^1 is complete.

(10) *If* ϕ^1 *is locally complete then* ϕ^1 *is complete.*

Proof. We suppose ϕ^1 is not complete. According to (6) there is a positive integer p such that if $q > p$ there is an integer $r > q$ and a subset P of N with

$$\inf \{(a_m^{(p)};a_m^{(q)}):m \in P\} > 0, \ \inf \{(a_m^{(p)};a_m^{(r)}) : m \in P\} = 0$$

We apply (2) to obtain that ϕ^1 is not locally complete.

(11) *If* ϕ^1 *is complete* λ *is quasinormable.*

Proof. Given a bounded set A in $\lambda^\times$ we can find a positive integer p such that A is a bounded set in $\lambda^\times{}_{\alpha_p}n$. According to (2) there is a positive $q > p$ such that, if P is any subset of N with

(12) $$\inf \{(a_m^{(p)};a_m^{(q)}) : m \in P\} > 0,$$

the following is verified:

(13) $$\inf \{(a_m^{(p)};a_m^{(r)}) : m \in P\} > 0, \ r = q+1,q+2,\ldots$$

We suppose that $\beta(\lambda^\times,\lambda)$ and the topology of $\lambda^\times{}_{\alpha_q}n$ does not coincide on A. We. can find a net

$$\{y^i : i \in I, \geqslant\}$$

in A converging to $y \in A$ for $\beta(\lambda^\times,\lambda)$ and not converging to y for the topology of $\lambda^\times{}_{\alpha_q}n$. Then we find a positive number c and a subnet

(14) $$\{x^j = (x_m^{(j)}) : j \in J, \geqslant\}$$

of

$$\{y^i - y : i \in I, \geqslant\}$$

such that

$$||(x_m^{(j)}||_q = \sup\{|x_m^{(j)};a_m^{(q)}| : m = 1,2,\ldots\} > c, \ j \in J.$$

Consequently, for every $j \in J$, we can find a positive integer m_j such that

$$|x_{m_j}^{(j)}; \ a_{m_j}^{(q)}| > c.$$

For a suitable positive number b we have that

$$|x_m^{(j)}| \leqslant b \, a_m^{(p)}, \ j \in J, \ m = 1,2,\ldots$$

andh therefore

$$(a_{m_j}^{(p)};a_{m_j}^{(q)}) \geqslant \frac{1}{b} |x_{m_j}^{(j)};a_{m_j}^{(q)}| > \frac{c}{b}, \ j \in J.$$

Taking

$$P = \{m_j : j \in J\}$$

(12) is verified and therefore (13) is verified; so

$$\inf \{(a_m^{(p)};a_m^{(r)}) : m \in P\} > a_r > 0, \ r = q+1, q+2, \ldots$$

Let M be the subset of all those elements (x_m) of λ such that $x_m = 0$, $m \notin P$, and

$$\Sigma |x_m| \, a_m^{(p)} \leq 1.$$

If (z_m) belongs to M and if r is an integer larger than q it follows that

$$\Sigma |z_m| \, a_m^{(r)} = \sum_{m \in P} |z_m| \frac{a_m^{(p)}}{(a_m^{(p)};a_m^{(r)})} < \frac{1}{a^r} \Sigma |z_m| \, a_m^{(p)} < \frac{1}{a^r}$$

from where it follows that M is a bounded set of λ. Thus, having in mind that the net (14) converges to the origin in $\lambda^\times$, there is an index $j_o \in J$ such that

(15) $$\sup \{|\langle x,x^j \rangle| : x \in M\} < c, \ j \geq j_o.$$

On the other hand, if we take $k \in P$, the sequence (w_m) verifying $w_m = 0$, $m \neq k$, $w_k = \frac{1}{a_k^{(q)}}$, belongs to M and accordingly, for $j \geq j_o$, we apply(15) to obtain

$$c > |\langle w,x^j \rangle| = |x_k^{(j)};a_k^{(q)}| > c$$

which is a contradiction. Thus $\beta(\lambda^\times,\lambda)$ and the topology of $\lambda^\times{}_{\alpha q}n$ coincide on A. Consequently λ is quasinormable.

(16) *If the echelon space λ is not quasinormable, then λ is the strong dual of a non-complete* (LB)*-space.*

Proof. According to the proof of 1.(14), λ is the strong dual of the subspace ϕ of $\lambda^\times$ as well as the strong dual of the subspace ψ of $\lambda^\times$. Since $\phi \subset \phi^1 \subset \psi$ it follows that λ is the strong dual of ϕ^1. Since λ is not quasinormable, we apply (11) to obtain that ϕ^1 is not complete. Finally ϕ^1 is an (LB)-space (see (1)). The proof is complete.

We suppose now that ϕ^1 is not complete. According to (6) we can find a positive integer p such that, for every integer $q > p$ there is an integer $r > q$ and a subset H of N such that

$$\inf \{(a_m^{(p)};a_m^{(q)}) : m \in H\} > 0, \ \inf \{(a_m^{(p)};a_m^{(r)}) : m \in H\} = 0$$

We set $q = q_1$, $r = q_2$. We determine an increasing sequence of positive integer

$$m_{11} < m_{12} < \ldots < m_{1i} < \ldots$$

with

$$\lim_i (a_{m_{1i}}^{(p)}; a_{m_{1i}}^{(q)}) = c_1 > 0, \quad \lim_i (a_{m_{1i}}^{(p)}; a_{m_{1i}}^{(q_2)}) = 0$$

Proceeding by recurrence we suppose constructed, for $s = 1,2,\dots,k$, the sequence

$$m_{s1} < m_{s2} < \dots < m_{si} < \dots$$

of positive integers and the numbers q_s, q_{s+1} of N satisfying

$$(17) \qquad \lim_i (a_{m_{si}}^{(p)}; a_{m_{si}}^{(q_s)}) = c_s > 0,$$

$$(18) \qquad \lim_i (a_{m_{si}}^{(p)}; a_{m_{si}}^{(q_{s+1})}) = 0$$

Starting with q_{k+1} we can find an integer $q_{k+2} > q_{k+1}$ and a subset P of N such that

$$\inf \{(a_m^{(p)}; a_m^{(q_{k+1})}) : m \in P\} > 0, \quad \inf \{(a_m^{(p)}; a_m^{(q_{k+2})}) : m \in P\} = 0$$

By (18) we can suppose P disjoint with

$$\{m_{si} : s = 1,2,\dots,k;\ i = 1,2,\dots\}.$$

Now we choose a sequence

$$m_{(k+1)1}, < , m_{(k+1)2}, < , m_{(k+1)i} < \dots$$

in P such that

$$\lim_i (a_{m_{(k+1)i}}^{(p)}; a_{m_{(k+1)i}}^{(q_{k+1})}) = c_{k+1} > 0$$

$$\lim_i (a_{m_{(k+1)i}}^{(p)}; a_{m_{(k+1)i}}^{(q_{k+2})}) = 0$$

Let Λ in the echelon space defined by the sistem of teps

$$\beta_k = (b_{ij}^{(k)}), \quad k = 1,2,\dots$$

where

$$b_{ij}^{(1)} = a_{m_{ij}}^{(p)},$$

$$b_{ij}^{(k)} = a_{m_{ij}}^{(q_{k+1})}, \quad k = 2,3,\dots, \qquad i,j = 1,2,\dots$$

Obvously, Λ and $\Lambda^{\times}$ are isomorphic to sectional subspaces of λ and $\lambda^{\times}$ respectively. In particular, if μ is the sectional subspace of $\lambda^{\times}$ of all those elements, (u_m) with $u_m = 0$ if $m \neq m_{ij}, i,j = 1,2,\dots$, we have that $\Lambda^{\times}$ is isomorphic to μ.

(19) *If ϕ^1 is not complete, then $\lambda^\times$ does not verify the Mackey convergence condition.*

Proof. It is obvious that it is enough to prove that $\Lambda^\times$ does not verify the Mackey convergence condition. In $\Lambda^\times$ we considere the sequence

(20) $(x_{ij}^{(k)}),\ k = 1,2,\ldots,$

defined in the following way:

$$x_{1j}^{(k)} = 0,\ j,k = 1,2,\ldots$$

$$x_{ij}^{(k)} = 0,\ j = 1,2,\ldots,k;\ i = 2,3,\ldots$$

$$x_{ij}^{(k)} = \frac{1}{i}\ b_{ij}^{(1)},\ j = k+1,\ k+2,\ldots,\ i = 2,3,\ldots$$

We shall see that (20) converges to the origin in $\Lambda^\times$. If B is a bounded set in Λ there is a sequence (h_m) of positive numbers such that, for every $(u_{ij}) \in B$,

$$\Sigma|u_{ij}|\ b_{ij}^{(m)} \leqslant h_m,\ m = 1,2,\ldots$$

Given a positive number ε we can find an integer p, $p \geqslant 2$, such that

$$h_1 < \frac{\varepsilon}{2}\, p.$$

Then, for every $(u_{ij}) \in B$ and $k \in N$, it follows that

$$|\Sigma u_{ij}x_{ij}^{(k)}| \leqslant |\sum_{i=2}^{p} \sum_{i} u_{ij}x_{ij}^{(k)}| + \sum_{i=p+1}^{\infty} \sum_{j} |u_{ij}|\ \frac{1}{i}\ b_{ij}^{(1)}$$

$$\leqslant |\sum_{i=2}^{p} \sum_{j} u_{ij}x_{ij}^{(k)}| + \frac{h_1}{p} \leqslant |\sum_{i=2}^{p} \sum_{j} u_{ij}x_{ij}^{(k)}| + \frac{\varepsilon}{2}.$$

We write (18) as

$$\lim_{j} (b_{ij}^{(1)};b_{ij}^{(i)}) = 0,\ i = 2,3,\ldots$$

and therefore there is a positive integer q such that

$$(b_{ij}^{(1)};b_{ij}^{(i)}) < \frac{\varepsilon}{2ph_i},\ j \geqslant q,\ i = 2,3,\ldots,p.$$

Then, if $k > q$ it follows, for every $(u_{ij}) \in B$, that

$$|\Sigma u_{ij}x_{ij}^{(k)}| \leqslant |\sum_{i=2}^{p} \sum_{j} u_{ij}x_{ij}^{(k)}| + \frac{\varepsilon}{2}$$

$$\leqslant \sum_{i=2}^{p} \sum_{j=q+1}^{\infty} |u_{ij}x_{ij}^{(k)}| + \frac{\varepsilon}{2} \leqslant \sum_{i=2}^{p} \sum_{j=q+1}^{\infty} |u_{ij}|\ \frac{b_{ij}^{(1)}}{i} + \frac{\varepsilon}{2}$$

$$\leq \sum_{i=2}^{p} \frac{\varepsilon}{2_p h_i} \sum_{j=q+1}^{\infty} |u_{ij}|\ b_{ij}^{(i)} + \frac{\varepsilon}{2} < \sum_{i=2}^{p} \frac{\varepsilon}{2_p h_i} h_i + \frac{\varepsilon}{2} < \varepsilon .$$

Thus (20) converges to the origin in $\Lambda^{\times}$.

Given any positive integer s the sequence (20) is in $\Lambda^{\times}{}_{\beta_s}n$. Let $||.||$ be the norm in $\Lambda^{\times}{}_{\beta_s}n$. By (17) there is a positive integer r such that

$$(a_{m(s+1)i}^{(p)}; a_{m(s+1)i}^{(q_{s+1})}) > \frac{c_{s+1}}{2}\ ,\ i \geq r.$$

Take any positive integer k. If h is an integer, $h > k$, $h > r$, we have that

$$||(x_{ij}^{(k)}|| = \sup \{|x_{ij}^{(k)}; b_{ij}^{(s)}| : i,j = 1,2,\ldots\}$$

$$\geq |x_{(s+1)h}^{(k)}; b_{(s+1)}^{(s)}| = \frac{1}{s+1} (b_{(s+1)h}^{(1)}; b_{(s+1)h}^{(s)})$$

$$= \frac{1}{s+1} (a_{m(s+1)h}^{(p)}; a_{m(s+1)h}^{(q_{s+1})}) > \frac{c_{s+1}}{2(s+1)}$$

from where it follows that (20) does not converge to the origin in $\Lambda^{\times}{}_{\beta_s}n$. The proof is complete.

The former results imply (21).

(21) *In an echelon space λ the following conditions are equivalent:*

a) λ *is quasinormable;*

b) ϕ^1 *is locally complete;*

c) ϕ^1 *is complete;*

d) $\lambda^{\times}$ *satisfies the Mackey convergence condition*

(22) *If the echelon space λ is not quasinormable, then there is a subspace G of $\lambda^{\times}$ verifying.*

a) G *is of countable codimension;*

b) G *is quasibarrelled;*

c) G *is not bornological*

Proof. We take a vector u in the closure of ϕ in $\lambda^{\times}$ which is not in ϕ^1. Let G be the linear hull of $\phi \cup \{u\}$ endowed with the topology induced by $\beta(\lambda^{\times},\lambda)$. G is of countable dimension and, since the subspace ϕ of $\lambda^{\times}$ is bornological, it follows that G is quasibarrelled. Suppose now that G is bornological. Let T be the linear form on G with $T(u) = 1$, $T^{-1}(0) = \phi$. We have that ϕ is a dense hyperplane of G and therefore T is not continuous. Thus there is in G a bounded closed absolutely convex subset A such that T is not continuous on G_A and therefore $\phi \cap G_A$ is dense G_A. We find a sequence v^r, $r = 1,2,\ldots$, in $\phi \cap G_A$ converging in G_A to an element $v \notin \phi$. Then $v \in \phi^1$.

We have that

$$v = hu+w,\ h \in k,\ h \neq 0,\ w \in \phi,$$

and consequently

$$u = h\bar{v}^{1} - h\bar{w}^{1} \in \phi^{1} + \phi = \phi^{1}$$

which is a contradiction. Thus G is not bornological.

(23) *Let* E *be a Fréchet space. We suppose that in* $E'[\sigma(E',E)]$, *if* A *is any bounded set, there is a bounded absolutely convex subset* $B \supset A$ *such that* $\sigma(E',E")$ *coincides in* A *with the weak topology of* E'_B. *Then, if* F *is a quasibarrelled subspace of* $E'[\beta(E',E)]$, F *is bornological.*

Proof. In $E'[\sigma(E',E)]$ let (B_m) be a fundamental increasing sequence of bounded closed absolutely convex sets such that $\sigma(E',E")$ coincides on B_m with the weak topology of $E'_{B_{m+1}}$, $m = 1,2,\ldots$

For every positive integer m, let V_m be the closure of $B_m \cap F$ in $E'_{B_{m+1}}$. Since B_m is absolutely convex, V_m coincides with the weak closure of $B_m \cap F$ in $E'_{B_{m+1}}$ and therefore V_m is $\sigma(E',E")$-closed; thus V_m is complete for the topology $\beta(E',E)$.

Setting

$$H = \cup\{V_m;\ m = 1,2,\ldots\},$$

H is a linear space. We suppose H endowed with the topology induced by $\beta(E'E)$.

Let V be a barrel in H. For every positive integer m, $V \cap H_{V_m}$ is a barrel in the Banach space H_{V_m} an therefore V absorbs V_m. Clearly $V \cap F$ absorbs $B_m \cap F$ and, since F is quasibarrelled, $V \cap F$ is a neighbourhood of the origin in F. Thus V is a neighbourhood of the origin in H. Then H is barrelled. Let T be a linear form on H which is continuous on H_{V_m}, $m = 1,2,\ldots.$ Fixing the positive integer m we have that $B_{m+1} \cap H$ is a bounded set of H and, according to Chapter One §3, 1.(3), there is a positive integer $p > m$ such that

$$B_{m+1} \cap H \subset p\,V_p.$$

The weak topology of H_{V_p} is finer than the topology induced on H_{V_p} by $\sigma(E",E')$. Since $H_{B_{m+1} \cap H}$ is a subspace of $E'_{B_{m+1}}$ it follows that $\sigma(E',E")$ coincides on V_m with the weak topology of H_{V_p}. We deduce that T is continuous on V_m for the topology of H. We apply HORVÁTH [2], 3. 10. Proposition to obtain that T is continuous on H. Thus H is the inductive limit of the sequence of Banach spaces (H_{V_m}).

Let f be a linear form on F whose restriction f_m to F_{B_m} F is continuous, m = 1,2,... Let g_m be the continuous extension of f_m to H_{V_m}. Let g be the linear form on H which coincides with g_{m+1} on H_{V_m}, m = 1,2, ... Since g_m coincides with g_{m+1} on H_{V_m}, g is well defined and obviously continuous on H. On the other hand, f is the restriction of g to F and therefore f is continuous. Finally F has the Mackey topology $\mu(F,F')$ and consequently F is bornological.

(24) *The space* λ *is quasinormable if and only if every quasibarrelled subspace of* $\lambda^\times$ *is bornological.*

Proof. It is an obvious conclusion of (22) and (23).

The results contained in this section except theorem (23), which appears here for the first time, can be found in VALDIVIA [13] and [14].

In VALDIVIA [15], it is proven that if λ is a Montel echelon space which is not a Schwartz space there is in $\lambda^\times$ a barrelled subspace which is not bornological. With slight modifications it can the proved that if λ is an echelon space which is not quasinormable there is in $\lambda^\times$ a barrelled subspace which is not bornological.

§ 3. ECHELON AND COECHELON SPACES OF ORDER p, $1 < p < \infty$

1. GENERAL PROPERTIES OF THE ECHELON AND COECHELON SPACES OF ORDER p, $1 < p < \infty$. In what follows p is a real number $1 < p < \infty$, and q its conjugate real number, i. e., $\frac{1}{p} + \frac{1}{q} = 1$. Let

(1) $\alpha_r = (a_m^{(r)})$, r = 1,2, ... be a sequence of elements ω satisfying

a) $\alpha_{r+1} \geq \alpha_r$, r = 1,2, ...;

b) $a_m^{(r)} > 0$, m, r = 1,2,...

We set

$$\lambda = \{(x_m) \in \omega : \Sigma |x_m a_m^{(r)}|^p < \infty, \ r = 1,2,\ldots\}$$

Let μ be the set of all those elements of ω (u_m) such that is a positive integer r with

$$\Sigma \left|\frac{u_m}{a_m^{(r)}}\right|^q < \infty.$$

(2) μ *is a normal sequence space*

Proof. By b) we have that e_m belongs to μ, $m = 1,2,\ldots$ We take $u = (u_m)$ and $v = (v_m)$ in μ and h in K. Let $w = (w_m)$ be an element of ω with

$$|w_m| \leqslant |u_m|, \ m = 1,2,\ldots$$

By the very definition of μ and remembering a) we can find a positive integer r such that

$$\Sigma \left|\frac{u_m}{a_m^{(r)}}\right|^q < \infty, \Sigma \left|\frac{v_m}{a_m^{(r)}}\right|^q < \infty$$

Then

$$\Sigma \left|\frac{w_m}{a_m^{(r)}}\right|^q \leqslant \Sigma \left|\frac{u_m}{a_m^{(r)}}\right|^q < \infty, \ \Sigma \left|\frac{hu_m}{a_m^{(r)}}\right|^q = |h|^q \ \Sigma \left|\frac{u_m}{a_m^{(r)}}\right|^q < \infty,$$

$$\Sigma \left|\frac{u_m + v_m}{a_m^{(r)}}\right|^q \leqslant \Sigma \left(\frac{2 \max (|u_m|, |v_m|)}{a_m^{(r)}}\right)^q$$

$$\leqslant 2^q \left(\Sigma \left|\frac{u_m}{a_m^{(r)}}\right|^q + \Sigma \left|\frac{v_m}{a_m^{(r)}}\right|^q\right) < \infty,$$

from where it follows that w, hu and u+v are in μ. Consequently μ is a normal sequence space.

(3) λ *is the* α-*dual of* μ.

Proof. Let (x_m) be an element of λ. Then

$$\Sigma |x_m a_m^{(r)}|^p < \infty, \ r = 1,2,\ldots$$

Let (u_m) be any alement of μ. There is a positive integer s such that

$$\Sigma \left|\frac{u_m}{a_m^{(s)}}\right|^q < \infty$$

Consequently. $(x_m a_m^{(s)})$ belongs to ℓ^p and $\left(\frac{u_m}{a_m^{(s)}}\right)$ belongs to ℓ^q, from where it follows

$$\Sigma|x_m u_m| \leq (\Sigma|x_m a_m^{(s)}|^p)^{\frac{1}{p}} (\Sigma|\frac{u_m}{a_m^{(s)}}|^q)^{\frac{1}{q}} < \infty,$$

and therefore (x_m) belongs to $\mu^\times$. Thus λ is contained in $\mu^\times$.

Let (y_m) be an element of $\mu^\times$. We take any vector (w_m) in ℓ^q. For every positive integer r we have that

$$\Sigma|\frac{w_m a_m^{(r)}}{a_m^{(r)}}|^q = \Sigma|w_m|^q < \infty,$$

and therefore $(w_m a_m^{(r)})$ belongs to μ and thus

$$\Sigma|y_m a_m^{(r)} w_m| < \infty$$

and so $(y_m a_m^{(r)})$ belongs to ℓ^p. Then

$$\Sigma|y_m a_m^{(r)}|^p < \infty$$

and therefore (y_m) belongs to λ. We have that $\lambda=\mu^\times$.

Result (4) is now obvious

(4) λ *is a perfect sequence space.*

For every positive integer r we set

$$A_r = \{(u_m) \in \omega : \Sigma|\frac{u_m}{a_m^{(r)}}|^q \leq 1\}.$$

Obviously A_r is a normal set. If (x_m) belongs to λ and if (u_m) is any element of A_r we have that

$$\Sigma|x_m u_m| \leq (\Sigma|x_m a_m^{(r)}|^p)^{\frac{1}{p}} (\Sigma|\frac{u_m}{a_m^{(r)}}|^q)^{\frac{1}{q}} \leq (\Sigma|x_m a_m^{(r)}|^p)^{\frac{1}{p}}$$

and therefore A_r is a bounded set of $\mu[\sigma(\mu,\lambda)]$. Let $T_r : \ell^q \longrightarrow \omega$ the mapping defined by

$$T_r((x_m)) = (x_m a_m^{(r)}),\ (x_m) \in \ell^q.$$

If B is the closed unit ball of ℓ^q and if (x_m) belongs to B then

$$\Sigma|\frac{x_m a_m^{(r)}}{a_m^{(r)}}|^q = \Sigma|x_m|^q \leq 1$$

and thus $T_r(B)$ is contained in A_r. On the other hand, if (u_m) belongs to A_r we have that $(\frac{u_m}{a_m^{(r)}})$ is obviously in B and

$$T_r\left(\left(\frac{u_m}{a_m^{(r)}}\right)\right) = (u_m)$$

Then $T_r(B) = A_r$. A_r is then a bounded normal absolutely convex subset of $\mu[\sigma(\mu,\lambda)]$. We represent by λ_r the linear hull of A_r provided with the topology derived from its gauge $|\cdot|_r$, i.e., if (u_m) belongs to λ_r then

$$|(u_m)|_r = \left(\Sigma\left|\frac{u_m}{a_m^{(r)}}\right|^q\right)^{\frac{1}{q}}$$

$T_r: \ell^q \longrightarrow \lambda_r$ is an isomorphism and therefore A_r is weakly compact in λ_r and consequently A_r is compact in $\mu[\sigma(\mu,\lambda)]$.

If U_r is the polar set of $r\,A_r$ in λ, $r = 1,2,\ldots$, we have that

$$\{U_r : r = 1,2,\ldots\}$$

is a fundamental system of neighbourhoods of the origin in λ $\sigma(\lambda,\mu)$-closed and absolutely convex for a topology metrizable T on λ. In this paragraph we suppose the space λ provided with the topology T.

If u is an element of μ we find a positive integer s such that u belongs to sA_s, Since sA_s is a normal set it follows that u^n is contained in sA_s. Consequently T is finer that $\nu(\lambda,\mu)$. By §1,4.(1), $\lambda[\nu(\lambda,\mu)]$ is a complete space and according to a result due to BOURBAKI [1] and W. ROBERTSON [1] λ is complete (cf. KÖTHE [1], Chapter Four, §18, Section 4).

Given a positive integer r we set, if (x_m) belongs to λ,

$$||(x_m)||_r = \left(\Sigma|x_m a_m^{(r)}|^p\right)^{\frac{1}{p}}.$$

Clearly $||.||_r$ is a norm on λ. The sequence (x_m) is in U_r if and only of

(5) $\qquad \Sigma|x_m u_m| \leq 1$

for every $(u_m) \in r\,A_r$ or equivalently if and only if (5) holds for every (u_m) with

$$\left(\frac{u_m}{a_m^{(r)}}\right) \in r\,B$$

and also if and only if $(r\,x_m a_m^{(r)})$ belongs to the closed unit ball of ℓ^p. Thus

$$U_r = \{x \in \lambda : ||x||_r \leq \frac{1}{r}\},\ r = 1,2,\ldots$$

Consequently the topology T can be defined by the system of norms $||\cdot||_r$, $r = 1,2,\ldots$

(6) *The sequence space* μ *is perfect.*

Proof. Let (u^s) be a Cauchy sequence in $\mu[\sigma(\mu,\lambda)]$. Since λ is a Fréchet space the sequence (u^s) is equicontinuous and therefore it converges to an element u in $\mu[\sigma(\mu,\lambda)]$. Then $\mu[\sigma(\mu,\lambda)]$ is sequentially complete and, applying §1, 4.(18), μ is a perfect space.

A consequence from (6) is that μ coincides with $\lambda^{\times}$. In this paragraph we shall use $\lambda^{\times}$ instead of μ. We suppose $\lambda^{\times}$ endowed with the strong topology $\beta(\lambda^{\times},\lambda)$.

λ is said to be an echelon space of order p defined by the system of steps (1). $\lambda^{\times}$ is the co-echelon space f order p defined by the system of steps (1).

Sometimes we shall give the system of steps (1) by means of double sequences

$$\alpha_r = (a_{ij}^{(r)}),\ r = 1,2,\dots$$

Then

$$\lambda = \{(x_{ij}) : \Sigma|x_{ij}a_{ij}^{(r)}|^p < \infty,\ r = 1,2,\dots\}$$

(7) *The echelon space is* λ *is totally reflexive.*

Proof. For every positive integer r the set A_r is weakly compact in λ_r. It is enough to apply §2,3.(1) to obtain the conclusion.

(8) *If* F *is a quasibarrelled subspace of* $\lambda^{\times}$ *then* F *is bornological.*

Proof. Applying §2, 4.(23) the conclusion follows since λ_r is reflexive, r = 1,2,...

(9) *The space* $\lambda^{\times}$ *is the inductive limit of the sequence* (λ_r) *of Banach spaces.*

Proof. Since λ is reflexive, $\lambda^{\times}$ is barrelled and, applying (8), $\lambda^{\times}$ is bornological. The conclusion follows having in mind that

$$\{r A_r : r = 1,2,\dots\}$$

is a fundamental family of bounded sets in $\lambda^{\times}$.

(10) *Let* A *be a bounded set of* λ. A *is relatively compact if and only if for every positive integer* r *the sequence*

$$(11) \qquad \sup \{ \sum_{m=s}^{\infty} |x_m a_m^{(r)}|^p : (x_m) \in A \}, \; s = 1,2,\ldots,$$

converges to zero.

Proof. Since A_m is a normal set of $\lambda^\times$, $m = 1,2,\ldots$, it follows that the closed normal hull of A is compact, §1,4.(29), if A is relatively compact. Therefore we suppose first that A is normal and compact. We suppose also that the sequence (11) does not converge to zero for a certain positive integer r. The sequence (11) is decreasing and therefore has a limit $2\delta>0$. We set $s_1 = 1$ and we find an element $x^1 = (x_m^{(1)})$ in A with

$$\sum_{m=s_1+1}^{\infty} |x_m^{(1)} a_m^{(r)}|^p > \delta.$$

Then there is an integer $s_2 > s_1$ such that

$$\sum_{m=s_1+1}^{s_2} |x_m^{(1)} a_m^{(r)}|^p > \delta.$$

Proceeding by recurrence se suppose the set of integers

$$s_1 < s_2 < \ldots < s_k < s_{k+1}$$

the elemnt of A: $x^j = (x_m^{(j)})$, $j = 1,2,\ldots, k$ with

$$\sum_{m=s_j+1}^{s_{j+1}} |x_m^{(j)} a_m^{(r)}|^p > \delta, \; j = 1,2,\ldots, k$$

already constructed.

We determine a vector $x^{h+1} = (x_m^{(k+1)})$ in A and an integer $s_{k+2}>s_{k+1}$ such that

$$\sum_{m=s_{k+1}+1}^{s_{k+2}} |x_m^{(k+1)} a_m^{(r)}|^p > \delta.$$

We set $z^j = (z_m^{(j)})$ with

$$z_m^{(j)} = 0, \; m \leq s_j, \; z_m^{(j)} = x_m^{(j)}, \; s_j < m \leq s_{j+1},$$

$$z_m^{(j)} = 0, \; s_{j+1} < m, \; j = 1,2,\ldots$$

The sequence (z^j) is in A. If we take two different positive integer j and k we have that

$$||z^j - z^k||_r = \Big(\sum_{m=s_j+1}^{s_{j+1}} |x_m^{(j)} a_m^{(r)}|^p + \sum_{m=s_k+1}^{s_{k+1}} |x_m^{(k)} a_m^{(r)}|^p\Big)^{\frac{1}{p}}$$

$$\geqslant \Big(\sum_{m=s_j+1}^{s_{j+1}} |x_m^{(j)} a_m^{(r)}|^p\Big)^{\frac{1}{p}} > \delta^{\frac{1}{p}}$$

which is in contradiction, since A is a compact set.

Reciprocally, we suppose that the sequence (11) converges to zero for every positive integer r. Let

(12) $\qquad y^h = (y_m^{(h)}), \quad h = 1,2,\ldots,$

be a sequence in A. Since A is bounded we apply a diagonal procedure to extract a subsequence

(13) $\qquad z^h = (z_m^{(h)}), \; h = 1,2,\ldots,$

from (12) such that

$$\lim_h z_m^{(h)} = z_m \in K, \; m = 1,2,\ldots$$

If $z = (z_m)$ we shall see that z belongs to λ and that (13) converges to z in λ which implies that A is relatively compact.

Given a positive integer r and $\varepsilon > 0$ we can find a positive integer k such that

$$\sup\Big\{\sum_{m=s}^{\infty} |x_m a_m^{(r)}|^p : (x_m) \in A\Big\} < \left(\frac{\varepsilon}{3}\right)^p, \; s = k, k+1,\ldots$$

Then, if s is an integer larger than k and if f is a positive integer, we have that

$$\sum_{m=s}^{s+f} |z_m^{(h)} a_m^{(r)}|^p < \left(\frac{\varepsilon}{3}\right)^p, \; h = 1,2,\ldots,$$

and consequently

$$\sum_{m=s}^{s+f} |z_m a_m^{(r)}|^p \leqslant \left(\frac{\varepsilon}{3}\right)^p$$

from where it follows

$$\Sigma |z_m a_m^{(r)})|^p < \infty.$$

Thus z belongs to λ. We find a positive integer h_0 such that

$$\sum_{m=1}^{k} |(z_m^{(h)} - z_m) a_m^{(r)}|^p < \left(\frac{\varepsilon}{3}\right)^p, \; h > h_0$$

Then, for $h > h_0$,

$$||z-z^h||_r \ (\Sigma|(z_m-z_m^{(h)})a_m^{(r)}|^p)^{\frac{1}{p}} < (\sum_{m=1}^{k}|(z_m-z_m^{(h)})a_m^{(r)}|^p)^{\frac{1}{p}}$$

$$+ (\sum_{m=k+1}^{\infty}|z_m a_m^{(r)}|^p)^{\frac{1}{p}} + (\sum_{m=k+1}^{\infty} z_m^{(h)} a_m^{(r)}|^p)^{\frac{1}{p}} < \frac{\varepsilon}{3} + \frac{\varepsilon}{3} + \frac{\varepsilon}{3} = \varepsilon$$

which completes the proof.

(14) *Let* A *be a bounded set of* ℓ^p. A *is relatively compact if and only if the sequence*

$$\sup\{\sum_{m=s}^{\infty}|x_m|^p : (x_m) \in A\},\ s = 1,2,\ldots,$$

converges to zero.

Proof. By taking $a_m^{(r)} = 1$, m, r = 1,2,..., we have $\lambda = \ell^p$. It is enough to apply the former result.

The echelon spaces of order p have been introduced by DIEUDONNÉ and GOMES [1] (cf. KÖTHE [1], Chapter Six, §31, Section 8). Given the steps (1) the author quoted above defines the echelon space of order p as

$$\{(x_m) \in \omega: \Sigma|x_m|^p a_m^{(r)} < \infty,\ r = 1,2,\ldots\}$$

The echelon space λ studied here coincides with the echelon space of order p of DIEUDONNÉ and GOMES for the system of steps

$$(|a_m^{(r)}|^p),\ r = 1,2,\ldots$$

2. MONTEL ECHELON SPACES OF ORDER p. In this section we suppose λ an echelon space of order p defined by the system 1.(1) of steps. The results of this section can also be found in KÖTHE [1] and DIEUDONNÉ and GOMES [1].

(1) *If* λ *is not a Montel space there is a positive integer* r *and a sequence of positive integers*

$$n_1 < n_2 < ,,, < n_m < \ldots$$

such that

$$\inf\{(a_{n_m}^{(r)};\ a_{r_m}^{(k)}) : m = 1,2,\ldots\} > 0,\ k = r, r+1,\ldots$$

Proof. If λ is not a Montel space we apply 1.(10) to obtain a bounded subset A of λ and a positive integer r such that the sequence 1.(11) does not converge to zero. Then there is $h > 0$ such that

$$\text{(2)} \qquad \sup \{ \sum_{m=s}^{\infty} |x_m a_m^{(r)}|^p : (x_m) \in A\} > h, \ s = 1,2,\dots$$

Since A is bounded in λ we have that

$$\sup \{\Sigma x_m a_m^{(k)}|^p : (x_m) \in A\} < M_k < \infty, \ k = 1,2,\dots$$

Given positive integers s and k, $k \geq r$, we set N_{sk} for the subset of all elements m of $\{s+1,s+2,\dots\}$ with

$$(a_m^{(r)}; a_m^{(k)})^p \leq \frac{h}{2^k M_k}.$$

Applying (2), we obtain an element $(y_m) \in A$ with

$$\sum_{m=s+1}^{\infty} |y_m a_m^{(r)}|^p > h.$$

$$\sum_{m \in N_{sk}} |y_m a_m^{(r)}|^p = \sum_{m \in N_{sk}} |y_m a_m^{(k)} (a_m^{(r)}; a_m^{(k)})|^p$$

$$\leq \frac{h}{2^k M_k} \sum_{m \in N_{sk}} |y_m a_m^{(k)}|^p \leq \frac{h}{2^k M_k} \Sigma |y_m a_m^{(k)}|^p < \frac{h}{2^k}$$

We see now that there is an integer $m(s) > s$ verifying

$$\text{(3)} \qquad (a_{m(s)}^{(r)}; a_{m(s)}^{(k)})^p > \frac{h}{2^k M_k}, \ k = r, r+1,\dots$$

Indeed, if (3) is not true we have that

$$\{s+1,s+2,\dots\} = \bigcup_{k=r}^{\infty} N_{sk}$$

and consequently

$$h < \sum_{m=s+1}^{\infty} |y_m a_m^{(r)}|^p \leq \sum_{k=r}^{\infty} \sum_{m \in N_{sk}} |y_m a_m^{(r)}|^p \leq \sum_{k=r}^{\infty} \frac{h}{2^k} \leq h$$

which is a contradiction.

We set $m(1) = n_1$, $m(n_s) = n_{s+1}$, $s = 1,2,\dots$ Then

$$\inf\{(a_{n_m}^{(r)};a_{n_m}^{(k)}) : m = 1,2,\ldots\} \geq \left(\frac{h}{2^k M_k}\right)^{\frac{1}{p}} > 0, \ k=r,\ r+1,\ldots$$

(4) *Given the space λ if we suppose the existence of a positive integer r and a sequence of positive integers*

$$n_1 < n_2 < \ldots < n_m < \ldots$$

with

$$\inf\{(a_{n_m}^{(r)};a_{n_m}^{(k)}) : m = 1,2,\ldots\} = m_k > 0, \ k=r,\ r+1,\ldots,$$

then λ is not a Montel space.

Proof. Let E be the sectional subspace of λ of all those elements (z_m) of λ with $z_m = 0$, $m \neq n_s$, $s = 1,2,\ldots$ For each $(a_m) \in \ell^p$ we set $\chi((a_m))$ for the sequence (x_s) such that

$$x_s = 0 \text{ if } s \neq n_m,\ m = 1,2,\ldots,$$

$$x_{n_m} = \frac{1}{a_{n_m}^{(r)}}\, a_m,\ m = 1,2,\ldots$$

For every positive integer $k \geq r$ we have that

$$||\chi((a_m))||_k = \left(\sum_{m=1}^{\infty} |x_{n_m} a_{n_m}^{(k)}|^p\right)^{\frac{1}{p}}$$

$$= \left(\sum_{m=1}^{\infty} \left|\frac{a_m}{(a_{n_m}^{(r)};a_{n_m}^{(k)})}\right|^p\right)^{\frac{1}{p}} \leq \frac{1}{m_k}\left(\Sigma|a_m|^p\right)^{\frac{1}{p}}$$

from where it follows that χ is a mapping from ℓ^p into E. Clearly χ is linear injective and continuous. We take any vector (y_m) of E. If

$$b_m = a_{n_m}^{(r)} y_{n_m},\ m = 1,2,\ldots,$$

it follows that

$$\left(\Sigma|b_m|^p\right)^{\frac{1}{p}} = \left(\sum_{m=1}^{\infty} |y_{n_m} a_{n_m}^{(r)}|^p\right)^{\frac{1}{p}}$$

$$= \left(\Sigma|y_m a_m^{(r)}|^p\right)^{\frac{1}{p}} = ||(y_m)||_r$$

and therefore $(b_m) \in \ell^p$, $X((b_m)) = (y_m)$ and the mapping $X^{-1}: E \longrightarrow \ell^p$ is continuous.

Thus λ has a sectional subspace isomorphic to ℓ^p and consequently λ is not a Montel space.

(5) *The space λ is not a Montel space if and only if it has a sectional subspace isomorphic to ℓ^p.*

Proof. If λ is not a Montel space the conditions of (1) are verified. Now we apply the proof of (4) to obtain that λ has a sectional subspace isomorphic to ℓ^p. Finally it is obvious that, if λ has a sectional subspace isomorphic to ℓ^p, then λ is not a Montel space.

3. TOTALLY MONTEL ECHELON SPACES OF ORDER p. A locally convex space E is totally Montel if every separated quotient of E is a Montel space.

In this section λ is an echelon space of order p defined by the system of steps 1.(1).

(1) *The space λ is a Schwartz space if and only if given any positive integer r there is an integer k>r such that*

$$\text{(2)} \qquad \lim\, (a_m^{(r)};a_m^{(k)}) = 0$$

Proof. Given any positive integer r we suppose the existence of an integer $k > r$ such that (2) is verified. Given $\varepsilon > 0$, $\varepsilon > 1$, we find a positive integer s with

$$(a_m^{(r)},a_m^{(k)}) < \varepsilon, \ m \geqslant s.$$

Let T_k be the mapping from ℓ^q into λ_k introduced in Section 1. If B is the closed unit ball of ℓ^q we have that $T_k(B) = A_k$ and since A_r is a $\sigma(\lambda^\times,\lambda)$-closed set contained in A_k we have that $T_k^{-1}(A_r)$ is a bounded closed subset of ℓ^q.

If (x_m) belongs to $T_k^{-1}(A_r)$ then $(x_m a_m^{(k)})$ belongs to A_r and therefore

$$T_r^{-1}((x_m a_m^{(k)})) = ((x_m a_m^{(k)};a_m^{(r)})) \in B.$$

Consequently

$$\frac{1}{\varepsilon}\sum_{m=s}^{\infty}|x_m|^q \leq \sum_{m=s}^{\infty}|x_m\frac{1}{\varepsilon}|^q \leq \sum_{m=s}^{\infty}|x_m|^q\,(a_m^{(k)};a_m^{(r)})^q$$

$$= \Sigma|x_m a_m^{(k)};a_m^{(r)}|^q \leq 1$$

and therefore

$$\sum_{m=s}^{\infty}|x_m|^q \leq \varepsilon$$

and, according to 1.(14) it follows that $T_k^{-1}(A_r)$ is a compact subset of ℓ^q. Thus A_r is a compact subset of λ_k and λ is a Schwartz space.

Now we suppose that λ is a Schwartz space. Given any positive integer r we find an integer $k > r$ such that A_r is a compact subset of λ_k. For each positive integer m, $a_m^{(r)}e_m$ clearly belongs to A_r. The sequence

(3) $\quad a_m^{(r)}e_m,\ m = 1,2,\ldots$

converges to zero in $\lambda^\times$ coordinatewise and since A_r is compact in λ_k it follows that (3) converges to the origin in λ_k. Then

$$\lim|a_m^{(r)}e_m|_k = \lim(a_m^{(r)};a_m^{(k)}) = 0$$

which completes the proof.

Since the conditions on the steps which characterize the Montel echelon spaces as well as the Schwartz echelon spaces are the same that the conditions on the echelon spaces of order p which are Montel or Schwartz respectively, then we can repeat the construction we did in §2, Section 3, to obtain the system of steps

(4) $\quad \beta_s = (a_{ij}^{(s)}),\ s = r,\ r+1,\ldots$

with

(5) $\quad (a_{hi}^{(r)};a_{hi}^{(r+h)}) > b_h > 0,\ i,h = 1,2,\ldots$

such that the echelon space Λ of order p defined by the system of steps (4) is isomorphic to a quotient of λ.

Let Δ be the linear space on K of the double sequences (x_{ij}) of K verifying, for $j = 1,2,\ldots,$

(6) $\quad \sum_{j=1}^{\infty}|x_{ij};a_{1j}^{(r)}|^q \leq b,\ x_{ij} = \dfrac{a_{ij}^{(r+1)}}{i^2 a_{ij}^{(r)}}x_{ij},\ i = 2,3,\ldots$

b being a positive number depending on the sequence (x_{ij}).

(7) *Δ is a linear subspace of* $\Lambda^{\times}$.

Proof. If the double sequence (x_{ij}) verifies (6) we have that

$$\Sigma |x_{ij};a_{ij}^{(r+1)}|^q \leq \sum_{j=1}^{\infty} |x_{1j},a_{1j}^{(r)}|^q + \sum_{i=2}^{\infty} \frac{1}{i^{2q}} \sum_{j=1}^{\infty} |x_{1j};a_{1j}^{(r)}|^q \leq b\Sigma \frac{1}{i^{2q}}$$

and therefore (x_{ij}) belongs to $\Lambda^{\times}$. On the other hand it is obvious that Δ is a vectorial space on K.

(8) Δ is closed in $\Lambda^{\times}[\sigma(\Lambda^{\times},\Lambda)]$.

Proof. Applying KREIN-SMULIAN'S theorem and remembering that Λ is separable it is enough to prove that Δ is sequentially closed in $\Lambda^{\times}[\sigma(\Lambda^{\times},\Lambda)]$. Let

(9) $\qquad (x_{ij}^{(k)}), \; k = 1,2,...,$

be a sequence in Δ converging to (y_{ij}) in $\Lambda^{\times}[\sigma(\Lambda^{\times},\Lambda)]$. Since the sequence (9) is equicontinuous we can find an integer $m > 1$ such that

(10) $\qquad \Sigma |x_{ij}^{(k)};a_{ij}^{(r+m)}|^q \leq m, \; k = 1,2,...$

Then, according to (5), (6) and (10), it follows that

(11) $$\sum_{j=1}^{\infty} |x_{1j}^{(k)};a_{1j}^{(r)}|^q = \sum_{j=1}^{\infty} |m^2 x_{mj}^{(k)};\, a_{mj}^{(r+1)}|^q$$

$$= m^{2q} \sum_{j=1}^{\infty} \left| \frac{(x_{mj}^{(k)};a_{mj}^{(r+m)})}{(a_{mj}^{(r+1)};a_{mj}^{(r+m)})} \right|^q \leq m^{2q} \sum_{j=1}^{\infty} \left| \frac{(x_{mj}^{(k)};a_{mj}^{(r+m)})}{(a_{mj}^{(r)};a_{mj}^{(r+m)})} \right|^q$$

$$\leq \frac{m^{2q+1}}{b_m}, \; k = 1,2,...$$

By (11) and by the coordinatewise convergence of (9) to (y_{ij}) it follows, for j = 1,2,..., that

$$\sum_{j=1}^{\infty} |y_{1j},\, a_{1j}^{(r)}|^q \leq \frac{m^{2q+1}}{b_m}, \quad y_{ij} = \frac{a_{ij}^{(r+1)}}{i^2 a_{1j}^{(r)}}\, y_{1j},$$

$i = 2,3,...$

and, thus, (y_{ij}) belongs to Δ.

(12) *Δ endowed with the topology induced by $\sigma(\Lambda^{\times},\Lambda)$ is isomorphic to ℓ^q $[\sigma(\ell^q,\ell^p)]$.*

Proof. Let $f: \ell^q \longrightarrow \omega$ be the mapping defined by

$$f((z_j))=(x_{ij}),(z_j) \in \ell^q,$$

suchthat, for $j = 1,2,\ldots,$

$$x_{1j} = a_{1j}^{(r)} z_j, x_{ij} = \frac{1}{i^2} a_{ij}^{(r+1)} z_j, \quad i = 2,3,\ldots$$

If $||.||$ denotes the norm in ℓ^q it follows that

$$\sum_{j=1}^{\infty} |x_{1j}, a_{1j}^{(r)}|^q \leq \sum_{j=1}^{\infty} |z_j|^q = ||(z_j)||^q,$$

$$x_{ij} = \frac{a_{ij}^{(r+1)}}{i^2 a_{1j}^{(r)}} x_{1j}, \quad i = 2,3,\ldots,$$

and therefore

$$f(\ell^q) \subset \Delta.$$

Given the element (x_{ij}) of Δ verifying (6) we set

$$z_j = \frac{x_{1j}}{a_{1j}^{(r)}}, \quad j = 1,2,\ldots,$$

Then (z_j) belongs to ℓ^q and $f((z_j)) = (x_{ij})$ and therefore $f(\ell^q) = \Delta$.

We have that

$$\Sigma |x_{ij}; a_{ij}^{(r+1)}|^q = \sum_{j+1}^{\infty} |x_{1j}; a_{1j}^{(r+1)}|^q$$

$$+ \sum_{i=2}^{\infty} \frac{1}{i^{2q}} \sum_{j=1}^{\infty} |x_{1j}; a_{1j}^{(r)}|^q \leq ||(z_j)||^q \sum_{i=1}^{\infty} \frac{1}{i^{2q}},$$

from where it follows that $f: \ell^q \longrightarrow \Lambda^{\times}$ is continuous and therefore $f: \ell^q [\sigma(\ell^q,\ell^p)] \longrightarrow \Lambda^{\times}[\sigma(\Lambda^{\times},\Lambda)]$ is continuous. Obviously f is injective. Finally, since $\Lambda^{\times}[\mu(\Lambda^{\times},\Lambda)]$ is B-complete and Δ is $\sigma(\Lambda^{\times},\Lambda)$-closed we have that $f: \ell^q[\sigma(\ell^q,\ell^p)] \longrightarrow \Lambda^{\times}[\sigma(\Lambda^{\times},\Lambda)]$ is an isomorphism.

(13) *If the space λ is not a Schwartz space then λ has a quotient isomorphic to* ℓ^p.

Proof. It λ is not a Montel space we apply 2.(5) to obtain that λ has a sectional subspace isomorphic to ℓ^p. Consequently λ has a quotient isomorphic to ℓ^p.

If λ is a Montel space we know that it has a subspace isomorphic to the space Λ described above and, then, it is enough to consider the space Λ. If L is the subspace of Λ orthogonal to Δ we have that the space Λ/L is isomorphic to ℓ^p, according to (12).

(14) *The space λ is totally Montel if and only if it is a Schwartz space.*

Proof. If λ is a Schwartz space it is obvious that every separated quotient of λ is a Schwartz space and therefore a Montel space.

If λ is not a Schwartz space we apply result (13) to obtain that λ is not totally Montel.

The results contained in this section, except (1), appear here for the first time. Result (1) can be found in FENSKE and SCHOCK [1].

4. NUCLEAR ECHELON SPACES OF ORDER p. In this section λ is an echelon spaces of order p defined by the system of steps 1.(1).

(1) *If the space λ is nuclear then, given a positive integer* r, *there is an integer* s > r *such that*

$$\Sigma(a_m^{(r)};a_m^{(s)}) < \infty.$$

Proof. Given $r \in N$ A_r is an equicontinuous closed absolutely convex subset of $\lambda^\times[\sigma(\lambda^\times,\lambda)]$. Since λ is nuclear we can find in $\lambda^\times[\sigma(\lambda^\times,\lambda)]$ an equicontinuous closed absolutely convex subset B such that $A_r \subset B$ and the canonical injection $J:\lambda_r \longrightarrow \lambda^\times_B$ is nuclear. Let $|.|$ be the gauge of the polar set of A_r in $(\lambda_r)'$ and let $||.||$ be the norm in $\lambda^\times_B$. We find an integer s > r with $B \subset s A_s$ and two sequences (z^h) and (y^h) in $(\lambda_r)'$ and $\lambda^\times_B$ respectively, being $y^h = (y_m^{(h)})$, h = 1,2,..., and such that

$$\Sigma|z^h|_x||y^h|| < \infty, \; J(x) = x = \Sigma <x,z^h> y^h, \; x \in \lambda_r.$$

If $T_r : \ell^q \longrightarrow \lambda_r$ is the mapping described in Section 1, let S_r:

$(\lambda_r)' \longrightarrow \ell^p$ be the transposed mapping of T_r. If $|||.|||$ is the norm of ℓ^p we have that

$$|z^h| = |||S_r(z^h)||| = |||(<e_m, S_r(z^h)>)|||$$

$$= |||(<T_r(e_m), z^h>)||| = |||(a_m^{(r)}<e_m, z^h>)||| = (\Sigma|a_m^{(r)}<e_m, z^h>|^p)^{\frac{1}{p}}.$$

Since B is contained in s A_s it follows that

$$||y^h|| \geqslant s|y^h|_s = s\ (\Sigma|y_m^{(h)};a_m^{(s)}|^q)^{\frac{1}{q}}$$

The vectors $a_m^{(r)}e_m$, m = 1,2,..., are in A_r and consequently

$$J(a_m^{(r)}e_m) = a_m^{(r)}e_m = \sum_h <a_m^{(r)}e_m, z^h>\ y^h$$

and therefore

$$(a_m^{(r)};a_m^{(s)}) = \sum_h <a_m^{(r)}e_m, z^h>\ (y_m^{(h)};a_m^{(s)})$$

from where it follows

$$\Sigma(a_m^{(r)};a_m^{(s)}) \leqslant \sum_{h=1}^{\infty}\sum_{m=1}^{\infty}|<a_m^{(r)}e_m, z^h>|.|\ y_m^{(h)};a_m^{(s)}|$$

$$\leqslant \sum_{h=1}^{\infty}\ (\sum_{m=1}^{\infty}|<a_m^{(r)}e_m, z^h>|^p)^{\frac{1}{p}}\ (\sum_{m=1}^{\infty}|y_m^{(h)};a_m^{(s)}|^q)^{\frac{1}{q}} \leqslant \frac{1}{s}\ \Sigma|z^h|.||y^h|| < \infty$$

and the proof is complete.

Let Λ be the echelon space defined by steps 1.(1), i.e.,

$$\Lambda = \{(x_m) \in \omega : \Sigma|x_m a_m^{(r)}| < \infty,\ r = 1,2,\ldots\}.$$

(2) *If given a positive integer* r *there is an integer* s > r *such that*

$$\Sigma(a_m^{(r)};a_m^{(s)}) < \infty$$

then λ *coincides with* Λ.

Proof. Let (x_m) be an element of Λ. Then, for every positive integer r, $(x_m a_m^{(r)})$ belongs to ℓ^1 and consequently $(x_m a_m^{(r)})$ belongs to ℓ^p and therefore

$$\Sigma|x_m a_m^{(r)}|^p < \infty;$$

thus (x_m) belongs to Λ.

Let (x_m) be an element of λ. Given any positive integer r we find an integer $s > r$ such that

$$\Sigma(a_m^{(r)} ; a_m^{(s)}) < \infty.$$

Then $((a_m^{(r)} ; a_m^{(s)}))$ belongs to ℓ^q and since $(x_m a_m^{(s)})$ belongs to ℓ^p it follows that

$$\Sigma|x_m a_m^{(r)}| = \Sigma|x_m a_m^{(s)}| . (a_m^{(r)}; a_m^{(s)})$$

$$\leq (\Sigma|x_m a_m^{(s)}|^p)^{\frac{1}{p}} (\Sigma(a_m^{(r)}; a_m^{(s)})^q)^{\frac{1}{q}} < \infty$$

and thus (x_m) belongs to Λ. We have obtained that λ coincides with Λ.

(3) *If given a positive integer* r *there is an integer* $s > r$ *such that*

$$\Sigma(a_m^{(r)}; a_m^{(s)}) < \infty.$$

then λ *is nuclear.*

Proof. According to § 2,3.(16), Λ is nuclear and, since λ coincides with Λ(see(2)), then λ is nuclear.

5. NOTE. Let λ be the echelon space of order p defined by DIEUDONNÉ and GOMES [1] defined by the system of steps 1.(1), i.e.,

$$\lambda = \{(x_m) \in \omega : \Sigma|x_m|^p a_m^{(r)} < \infty, \quad r = 1,2,..\}$$

We suppose the topology of λ defined by the system of norms

$$(\Sigma|x_m|^p a_m^{(r)})^{\frac{1}{p}}, \ (x_m) \in \lambda, \ r = 1,2,...$$

According to what has been said in the end of Section 1 in relation with the spaces of DIEUDONNÉ and GOMES and using results of former sections we have the following theorems:

(1) *If* λ *is not a Montel space there is an integer* r *and a sequence of positive integers*

$$n_1 < n_2 < \ldots < n_m < \ldots$$

such that

$$\inf \{(a_{n_m}^{(r)} ; a_{n_m}^{(k)}) : m = 1,2,\ldots\} > 0, \quad k = r, r+1, \ldots$$

(2) *If there is a positive integer* r *and a sequence of positive integers*

$$n_1 < n_2 < \ldots < n_m < \ldots$$

such that

$$\inf \{(a_m^{(r)}; a_m^{(s)} : m = 1,2,\ldots\} > 0, \quad k = r, r+1$$

then the space λ *is a not Montel space.*

(3) *The space* λ *is not a Montel space if and only if it has a sectional subspace isomorphic to* ℓ^p.

(4) *The space* λ *is a Schwartz space if and only if given any positive in teger* r *there is an integer* $s > r$ *such that*

$$\lim (a_m^{(r)} ; a_m^{(s)}) = 0$$

(5) *If* λ *is not a Schwartz space it as a quotient isomorphic to* ℓ^p.

(6) *The space* λ *is totally Montel if and only if it is a Schwartz space.*

(7) *The space* λ *is nuclear if and only if given a positive integer* r *there is an integer* $s > r$ *with*

$$\Sigma \ (a_m^{(r)} ; a_m^{(s)})^{\frac{1}{p}} < \infty$$

§ 4. ECHELON SPACES OF ORDER ZERO AND ECHELON SPACES OF INFINITE ORDER

1. GENERAL PROPERTIES OF ECHELON SPACES OF ORDER ZERO AND OF ECHELON SPACES OF FINITE ORDER. Let

(1) $$\alpha_r = (a_m^{(r)}), \qquad r = 1,2,\ldots$$

be a sequence of elements of ω satisfying

a) $\alpha_{r+1} \geqslant \alpha_r, \qquad r = 1,2, \ldots;$

b) $a_m^{(r)} > 0, \qquad m, r = 1,2, \ldots$

We set

$$\lambda_0 = \{(x_m \in \omega: \lim x_m a_m^{(r)} = 0, \quad r = 1,2,\ldots\}$$

$$\lambda_\infty = \{(x_m) \in \omega : \sup |x_m a_m^{(r)}| < \infty, \ r = 1,2,\dots\}$$

Let λ be the set of all elements of ω such that (u_m) belongs to λ if and only if there is a positive integer r such that

$$\Sigma |u_m ; a_m^{(r)}| < \infty$$

(2) *λ is a normal sequence space.*

Proof. By b) e_m belongs to λ, $m = 1,2,\dots$ We take $u = (u_m)$ and $v = (v_m)$ in λ and h in K. Let $w = (w_m)$ be an element of ω with

$$|w_m| \leqslant |u_m|, \ m = 1,2,\dots$$

By a), we can find a positive integer r such that

$$\Sigma |u_m ; a_m^{(r)}| < \infty, \ \Sigma |v_m ; a_m^{(r)}| < \infty.$$

Then

$$\Sigma |w_m ; a_m^{(r)}| \leqslant \Sigma |u_m ; a_m^{(r)}| < \infty, \ \Sigma |hu_m ; a_m^{(r)}| = |h| \ \Sigma |u_m ; a_m^{(r)}| < \infty,$$

$$\Sigma |u_m + v_m ; a_m^{(r)}| \leqslant \Sigma |u_m ; a_m^{(r)}| + \Sigma |v_m ; a_m^{(r)}| < \infty$$

from where it follows that w, hu and u+v are in λ. Consequently λ is a normal sequence space.

(3) *λ_∞ is the α-dual of λ.*

Proof. Let (x_m) be a vector of λ_∞. Then

$$\sup \{|x_m a_m^{(r)}| : m = 1,2,\dots\} < \infty, \ r = 1,2,\dots$$

Let (u_m) be any element of λ. Then there is a positive integer s such that

$$\Sigma |u_m ; a_m^{(s)}| < \infty.$$

Consequently

$$\Sigma |x_m u_m| = \Sigma |x_m a_m^{(s)} (u_m ; a_m^{(s)})|$$

$$\leqslant (\sup \{|x_m a_m^{(s)}| : m = 1,2,\dots\}) \ \Sigma |u_m ; a_m^{(s)}| < \infty$$

from where it follows that (x_m) belongs to $\lambda^\times$. Then λ_∞ is contained in $\lambda^\times$.

Let (y_m) be an element of $\lambda^\times$. We take any vector (w_m) of ℓ^1. For each positive integer r we have that $(w_m a_m^{(r)})$, belongs to λ and therefore

$$\Sigma |y_m a_m^{(r)} w_m| < \infty.$$

Thus, $(y_m a_m^{(r)})$ belongs to ℓ^∞, i.e.,

$$\sup \{|y_m a_m^{(r)}| : m = 1,2,\ldots\} < \infty$$

and therefore (y_m) is in λ_∞. Then $\lambda_\infty = \lambda^\times$.

Result (4) is now obvious.

(4) *λ_∞ is a perfect sequence space.*

For each positive integer r we set

$$A_r = \{(u_m) \in \omega : \Sigma |u_m; a_m^{(r)}| \leq 1\}$$

Clearly A_r is an absolutely convex normal subset of λ . If (x_m) belongs to λ_∞ and if (u_m) is any element of A_r we have that

$$\Sigma |x_m u_m| \leq (\sup\{|x_m a_m^{(r)}| : m = 1,2,\ldots\}) \, \Sigma |u_m; a_m^{(r)}|$$

$$\leq \sup \{|x_m a_m^{(r)}| : m = 1,2,\ldots\}$$

and therefore A_r is a bounded set of $\lambda[\sigma(\lambda,\lambda_\infty)]$. Obviously

$$\bigcup_{r=1}^{\infty} r A_r = \lambda.$$

If U_r is the polar set of $r A_r$ in λ_∞, $r = 1,2,\ldots$, we have that

$$\{U_r : r = 1,2,\ldots\}$$

is a fundamental system of $\sigma(\lambda_\infty,\lambda)$-closed absolutely convex neighbourhoods of the origin, in λ_∞ for a metrizable topology T on λ_∞. In what follows is provided with the topology T.

If u is an element of λ we find a positive integer s such that u belongs to sA_s. Since sA_s is a normal subset of ω we have that u^n is contained in sA_s. Consequently T is finer than $\nu(\lambda_\infty,\lambda)$.

We apply § 1,4.(1) to obtain that $\lambda_\infty[\nu(\lambda_\infty,\lambda)]$ is complete from where it follows that λ_∞ is a Fréchet space. There is no difficulty in showing that T can be described by the system of norms $||\cdot||_r$, r=1,2,..., such that

$$||(x_m)||_r = \sup\{|x_m a_m^{(r)}| : m = 1,2,\ldots\},\ (x_m) \in \lambda_\infty.$$

(5) *For every positive integer* r *the set* A_r *is* $\sigma(\lambda,\lambda_o)$*-compact.*

Proof. Let

(6) $$u^s = (u_m^{(s)}),\ s = 1,2,\ldots,$$

be a sequence in A_r. Since this set is $\sigma(\lambda,\phi)$-bounded a subsequence

(7) $$v^s = (v_m^{(s)}),\ s = 1,2,\ldots$$

can be extracted from (6) such that

$$\lim_s v_m^{(s)} = v_m \in K,\ m = 1,2,\ldots$$

Given a positive integer k we have that

$$\sum_{m=1}^{k} |v_m^{(s)};a_m^{(r)}| \leqslant 1,\ s = 1,2,\ldots,$$

and therefore

$$\sum_{m=1}^{k} |v_m;a_m^{(r)}| \leqslant 1$$

from where it follows that

$$\Sigma |v_m;a_m^{(r)}| \leqslant 1.$$

Then $v = (v_m)$ belongs to A_r and (7) converges to v in $\lambda[\sigma(\lambda,\phi)]$.

Let (x_m) be any element of λ_o. Given $\varepsilon > 0$ we find a positive integer h with

$$|x_m a_m^{(r)}| < \frac{\varepsilon}{3},\ m > h.$$

Now we determine a positive integer k such that

$$\sum_{m=1}^{h} |x_m(v_m - v_m^{(s)})| < \frac{\varepsilon}{3},\ s \geqslant k.$$

Then, for $s \geqslant k$, we have that

$$\Sigma |x_m(v_m - v_m^{(s)})| \leq \sum_{m=1}^{h} |x_m(v_m - v_m^{(s)})| + \sum_{m=h+1}^{\infty} |x_m a_m^{(r)}(v_m;a_m^{(r)})|$$

$$+\sum_{m=h+1}^{\infty}|x_m a_m^{(r)}(v_m^{(s)};a_m^{(r)})| \leq \frac{\varepsilon}{3}+\frac{\varepsilon}{3}\sum_{m=h+1}^{\infty}|v_m;a_m^{(r)}|+\frac{\varepsilon}{3}\sum_{m=h+1}^{\infty}|v_m,a_m^{(r)}|<\varepsilon$$

and therefore the sequence (7) converges to v in $\lambda[\sigma(\lambda,\lambda_o)]$. Then A_r is $\sigma(\lambda,\lambda_o)$-sequentially compact. Since ϕ is dense in $\lambda_o[\sigma(\lambda_o,\lambda)]$ it follows that A_r is $\sigma(\lambda,\lambda_o)$-compact.

Let

$$w^r = (w_m^{(r)}),\ r = 1,2,\ldots,$$

be a sequence in λ_o converging to $w = (w_m)$ in λ_∞. We fix a positive integer r. Given $\varepsilon > 0$ there is a positive integer s with

$$||w^s - w||_r < \frac{1}{2}\varepsilon.$$

We find a positive integer k such that

$$|w_m^{(s)}a_m^{(r)}| < \frac{1}{2}\varepsilon,\ m \geq k.$$

Then, for $m \geq k$, we have that

$$|w_m a_m^{(r)}| \leq |(w_m - w_m^{(s)})a_m^{(r)}| + |w_m^{(s)}a_m^{(s)}| \leq ||w^s - w||_r + |w_m^{(s)}a_m^{(r)}| < \varepsilon,$$

and thus w belongs to λ_o. Then λ_o is a closed subspace of λ_∞. In what follows we suppose that λ_o is endowed with the topology induced by T. Then λ_o is a Fréchet space which has a fundamental system neighbourhoods formed by the polar sets in λ_o of $r A_r$, $r = 1,2,\ldots$ According to result (5) $r A_r$ is $\sigma(\lambda,\lambda_o)$-compact and therefore λ can be identified with the topological dual of λ_o.

(8) *λ is the α-dual of λ_o.*

Proof. If $u = (u_m)$ belongs to $\lambda^\times_o$, we apply §1,3.(1) to obtain that u^n is $\sigma(\lambda_o^\times,\lambda_o)$-compact. Since $(u(r))$ is T-equicontinuous and $\sigma(\lambda^\times_o,\lambda_o)$-converges to u it follows that u belongs to λ.

Now result (9) is obvious

(9) *λ is a perfect space.*

Let $T_r : \ell^1 \longrightarrow \omega$ be the mapping defined by

$$T_r((x_m)) = (x_m a_m^{(r)}),\ (x_m) \in \ell^1.$$

Clearly T_r is linear and injective and maps the closed unit ball of ℓ^1 in A_r. Then, if λ_r denotes the linear hull of A_r provided with the topology induced by the gauge $|.|_r$ of A_r, i.e.,

$$|(u_m)|_r = \Sigma |u_m; a_m^{(r)}|, \quad (u_m) \in \lambda_r,$$

the space λ_r is isomorphic to ℓ^1.

A locally convex space E is distinguished if its strong dual $E'[\beta(E',E)]$ is a barrelled space.

(10) *The space λ_o is distinguished and its strong bidual can be identified with λ_∞.*

Proof. Given a positive integer r we have that

$$T_r (e_m) = a_m^{(r)} e_m, \quad m = 1,2,\ldots,$$

and linear hull of

$$\{e_m : m = 1,2,\ldots\}$$

is dense in ℓ^1 and therefore the subspace ϕ of λ is dense in λ_r. Then $\lambda[\beta(\lambda,\lambda_o)]$ is separable and according to a theorem of GROTHENDIECK (cf. KÖTHE [1], Chapter Six, §29, Section 3 and Section 4), $\lambda[\beta(\lambda,\lambda_o)]$ is barrelled and bornological and thus λ_o is distinguished.

The subspace ϕ^1 introduced in §1,9.(1) coincides with λ and therefore $\lambda[\mu(\lambda,\lambda_\infty)]$ is ultrabornological. Thus $\mu(\lambda,\lambda_\infty)$ coincides with $\beta(\lambda,\lambda_o)$ and consequently λ_∞ is the topological dual of $\lambda[\beta(\lambda,\lambda_o)]$. Then λ_∞ is the strong bidual of λ_o and the proof is finished.

We suppose λ endowed with the strong topology $\beta(\lambda,\lambda_o)$. We say that λ_o is an echelon space of order zero and λ_∞ is an echelon space of order ∞, both defined by the system of steps (1). We also say that λ is a co-echelon space of order zero or infinite defined by the system of steps (1).

Sometimes we given the system of steps (1) by means of double sequences

$$\alpha_r = (a_{ij}^{(r)}), \quad r = 1,2,\ldots$$

Then

$$\lambda_o = \{(x_{ij}) : \lim_{i+j\to\infty} x_{ij}a_{ij}^{(r)} = 0, \ r = 1,2,\ldots\}$$

$$\lambda_\infty = \{(x_{ij}) : \sup_{ij} |x_{ij}a_{ij}^{(r)}| < \infty, \ r = 1,2,\ldots\}$$

(11) *The space λ is the inductive limit of the sequence (λ_r) of Banach spaces.*

Proof. Since λ is ultrabornological it is enough to consider that

$$\{r\, A_r : r = 1,2,\ldots\}$$

is a fundamental system of bounded sets of λ.

(12) *Let A be a bounded set of λ_o. A is relatively compact if and only if for every positive integer r the sequence*

(13) $$\sup \{\sup_{m\geq s} |x_m a_m^{(r)}| : (x_m) \in A\},\ s = 1,2,\ldots$$

converges to zero.

Proof. Since A_m is a normal subset of λ, $m = 1,2,\ldots$, and since λ_o is a normal sequence space it follows that the closed normal hull of A is compact, §1, 4.(29), if A is relatively compact. Therefore we suppose first that A is a normal compact subset of λ_o. Suppose that the sequence (13) does not converge to zero for a certain positive integer r. Since the sequence (13) is decreasing it has a limit $2\delta>0$. Clearly there is a sequence of positive integers

$$s_1 < s_2 < \ldots < s_m < \ldots$$

and a sequence $x^s = (x_m^{(s)})$, $s = 1,2,\ldots$, of elements of A such that

(14) $$|x_{s_m}^{(m)} a_{s_m}^{(r)}| > \delta,\ m = 1,2,\ldots$$

If

$$z^j = x_{s_j}^{(j)} e_{s_j},\ j = 1,2,\ldots,$$

the sequence (z^j) belongs to A. Taking two different positive integers j and k we have that

$$||z^j - z^k||_r \geq |x_{s_j}^{(j)} a_{s_j}^{(r)}| > \delta$$

which is in contradiction with A being a compact subset of λ_o.

Reciprocally we suppose that the sequence (13) converges to zero for every positive integer r. Let

(15) $$y^h = (y_m^{(h)}),\ h = 1,2,\ldots,$$

be a sequence in A. Since A is a bounded subset of λ_o we use a diagonal procedure to extract a subsequence

$$(16)\qquad z^h = (z_m^{(h)}),\ h = 1,2,\dots,$$

from (15) such that

$$\lim_h z_m^{(h)} = z_m \in K,\ m = 1,2,\dots$$

Given a positive integer r and $\varepsilon > 0$ we can find a positive integer k such that

$$\sup\{\sup_{m\geq s}|x_m a_m^{(r)}| : (x_m) \in A\} < \frac{\varepsilon}{3}\ ,\ s = k,\ k+1,$$

Then

$$|(z_m^{(h)} - z_m)a_m^{(r)}| \leq \frac{\varepsilon}{3},\ m = k,\ k+1,\dots$$

from where it follows that $z = (z_m)$ belongs to λ_o. Finally we can find a positive integer t such that

$$|(z_m^{(h)} - z_m)a_m^{(r)}| < \frac{\varepsilon}{3},\ h \geq t,\ m = 1,2,\dots,\ k.$$

Then, if $h \geq t$,

$$||z - z^h||_r \leq \sup_{m\leq k}|(z_m^{(h)} - z_m)a_m^{(r)}| + \sup_{m>k}|z_m^{(h)}a_m^{(r)}| + \sup_{m>k}|z_m a_m^{(r)}| < \varepsilon$$

and thus (16) converges to z in λ_o. The proof is complete.

2. MONTEL ECHELON SPACES OF ORDER ZERO. λ_o and λ_∞ have the same meaning as is the former section

(1) *If λ is a reflexive space then λ_o is a Montel space.*

Proof. We suppose the proposition not true. We find in λ_o a bounded set A such that the sequence 1.(13) does not converge to zero for a certain positive integer r.

We find a positive number δ, a sequence of positive integers

$$s_1 < s_2 < \dots < s_m < \dots$$

and a sequence in A

$$x^j = (x_m^{(j)}),\ j = 1,2,\dots,$$

verifying 1.(14). We set

$$z_m = 0 \text{ if } m \neq s_j,\ z_{s_j} = x_{s_j}^{(j)},\ j = 1,2,\ldots$$

Since A is a bounded set in λ_o there is $h > 0$ with

$$||x^j||_r < h,\ j = 1,2,\ldots$$

If $z = (z_m)$ we have that

$$\sup\{|z_m a_m^{(r)}| : m = 1,2,\ldots\} = \sup\{|x_{s_j}^{(j)} a_{s_j}^{(r)}| : j = 1,2,\ldots\}$$

$$\leqslant \sup\{||x^j||_r : j = 1,2,\ldots\} \leqslant h$$

and thus z belongs to λ_∞. On the other hand, according to 1.(14), $(z_m a_m^{(r)})$ does not converge to zero and therefore $z \notin \lambda_o$. Since λ_o is distinct from λ_∞, λ_o is not reflexive and we arrive to a contradiction.

(2) *If λ_∞ is a reflexive space then λ_∞ is a Montel space.*

Proof. If λ_∞ is reflexive we have that $\lambda_\infty = \lambda_o$ and we apply the former result.

(3) *If λ_o is not a Montel space there is a positive integer* r *and a sequence of positive integers*

$$n_1 < n_2 < \ldots < n_m < \ldots$$

such that

(4) $$\inf\{(a_{n_m}^{(r)} ; a_{n_m}^{(s)}) : m = 1,2,\ldots\} > 0,\ s = r, r+1,\ldots$$

Proof. If λ_o is not a Montel space it follows from (1) that λ_o is not reflexive. Therefore λ_o is distinct from its strong bidual λ_∞. We take a vector (x_m) in λ_∞ which is not in λ_o. Then there is a positive integer r such that the sequence $(x_m a_m^{(r)})$ does not belong to c_o and consequently we can find a sequence of positive integers

$$n_1 < n_2 < \ldots < n_m < \ldots$$

and $h > 0$ such that

$$|x_{n_m} a_{n_m}^{(r)}| \geqslant h > 0,\ m = 1,2,\ldots$$

Given any positive integer $s \geqslant r$, $(x_m a_m^{(s)})$ is in ℓ^∞ and therefore there is

$k > 0$ with

$$|x_m a_m^{(s)}| \leq k, \ m = 1,2,\ldots$$

Then

$$(a_{n_m}^{(r)}; a_{n_m}^{(s)}) = \frac{|x_{n_m} a_{n_m}^{(r)}|}{|x_{n_m} a_{n_m}^{(s)}|} \geq \frac{h}{k}, \ m = 1,2,\ldots,$$

from where it follows (4).

(5) *If there is a positive integer* r *and a sequence of positive integers*

$$n_1 < n_2 < \ldots < n_m < \ldots$$

with

$$\inf \{(a_{n_m}^{(r)}; a_{n_m}^{(s)}) : m = 1,2,\ldots\} > M_s > 0, \ s = r, r+1,\ldots$$

then λ_o *is not a Montel space.*

Proof. Let (z_p) be an element of ω satisfying the following conditions

$$z_p = 0, \ p \neq n_m, \ z_{n_m} = \frac{1}{a_{n_m}^{(r)}}, \ m = 1,2,\ldots$$

Given an integer $s \geq r$ we have that

$$z_p a_p^{(s)} = 0, \ p \neq n_m, \ z_{n_m} a_{n_m}^{(s)} = \frac{a_{n_m}^{(s)}}{a_{n_m}^{(r)}} < \frac{1}{M_s}, \ m = 1,2,\ldots,$$

and thus (z_p) belongs to λ_∞. On the other hand,

$$z_{n_m} a_{n_m}^{(s)} = \frac{a_{n_m}^{(s)}}{a_{n_m}^{(r)}} \geq 1, \ m = 1,2,\ldots$$

and thus $(z_p) \notin \lambda_o$. Therefore $\lambda_o \neq \lambda_\infty$ and consequently λ_o is not a Montel space.

(6) *The space* λ_o *is not a Montel space if and only if it has a sectional subspace isomorphic to* c_o.

Proof. If λ_o has a sectional subspace isomorphic to c_o we have that λ_o is not a Montel space since c_o is not reflexive.

We suppose now that λ_o is not a Montel space. We apply (3) to obtain a positive integer r and a sequence of positive integers

$$n_1 < n_2 < \dots < n_m < \dots$$

such that

$$\inf \{(a_{n_m}^{(r)};a_{n_m}^{(s)}) : m = 1,2,\dots\} > M_s > 0,\ s = r,r+1,\dots$$

Let E be the sectional subspace of λ_o of all those elements (y_p) such that

$$y_p = 0 \text{ if } p \neq n_m,\ m = 1,2,\dots$$

Let $T: E \longrightarrow c_o$ bhe the injective linear mapping defined by

$$T((y_p)) = (y_{n_m} a_{n_m}^{(r)}), (y_p) \in E.$$

If $||.||$ denotes the norm of c_o we have that

$$||T((y_p))|| = \sup \{|y_{n_m} a_{n_m}^{(r)}| : m = 1,2,\dots\} = ||(y_p)||_r$$

and therefore T is continuous.

If (z_m) belongs to c_o and if s is any positive integer larger than r we have that

$$|z_m;a_{n_m}^{(r)}|\ a_{n_m}^{(s)} \leq |z_m| \frac{a_{n_m}^{(s)}}{a_{n_m}^{(r)}} \leq \frac{|z_m|}{M_s},\ m = 1,2,\dots,$$

and therefore the vector (u_p) of ω with

$$u_p = o \text{ if } p \neq n_m,\ u_{n_m} = (z_m;a_{n_m}^{(r)}),\ m = 1,2,\dots,$$

belongs to E and $T((u_p)) = (z_m)$. We have also that

$$||((u_p)||_s = \sup\{|z_m;a_{n_m}^{(r)}|\ a_{n_m}^{(s)} : m = 1,2,\dots\} \leq \frac{1}{M_s} ||T((u_p))||$$

and thus T is an isomorphism from E onto c_o.

(7) *The space λ_∞ is not a Montel space if and only if it has a sectional subspace isomorphic to ℓ^∞.*

Proof. If λ_∞ has a sectional subspace isomorphic to ℓ^∞ it is obvious that λ_∞ is not a Montel space.

We suppose now that λ_∞ is not a Montel space. Let (n_m) be the sequence of positive integers obtained in (6) and let E be the subspace of λ_∞ defined above. Let F be the sectional subspace of λ_o such that (y_p) belongs to F if and only if

$$y_p = 0 \text{ if } p \neq n_m,\ m = 1,2,\ldots$$

Since λ_∞ is the strong bidual of λ_0 it is easy to show that F is the strong bidual of E. Since E is isomorphic to c_0 it follows that F is isomorphic to the strong bidual of c_0, i.e., F is isomorphic to ℓ^∞.

3. TOTALLY REFLEXIVE ECHELON SPACES OF ORDER ZERO. In this section λ_0 and λ_∞ have the same meaning as in former sections.

(1) *The space λ_0 is a Schwartz space if and only if given a positive integer r there is an integer $k > r$ such that*

$$\lim (a_m^{(r)};a_m^{(k)}) = 0 \tag{2}$$

Proof. If λ_0 is a Schwartz space, given a positive integer r there is an integer $k > r$ such that A_r is compact in λ_k. The sequence

$$a_m^{(r)}e_m,\ m = 1,2,\ldots \tag{3}$$

is in A_r and converges coordinatewise to the origin, i.e., converges to the origin for $\sigma(\lambda,\phi)$. Then (3) converges to the origin in λ_k and therefore

$$\lim_m |a_m^{(r)}e_m|_k = \lim_m (a_m^{(r)};a_m^{(k)}) = 0$$

Reciprocally given a positive integer r we suppose the existence of a positive integer k such that (2) is verified. Given $\varepsilon > 0$, $\varepsilon < 1$, we find a positive integer s with

$$(a_m^{(r)};a_m^{(k)}) < \varepsilon,\ m \geqslant s.$$

Let $T_k : \ell^1 \longrightarrow \lambda_k$ be the mapping defined in Section 1. It is obvious that $T_k^{-1}(A_r)$ is a closed subset of ℓ^1 contained in the closed unit ball B of ℓ^1. If (x_m) belongs to $T_k^{-1}(A_r)$ then $(x_m a_m^{(k)})$ belongs to A_r and therefore

$$T_r^{-1}((x_m a_m^{(k)})) = ((x_m a_m^{(k)};a_m^{(r)})) \in B$$

Consequently

$$\frac{1}{\varepsilon}\sum_{m=s}^{\infty} |x_m| \leqslant \sum_{m=s}^{\infty} |x_m|\ (a_m^{(k)};a_m^{(r)}) = \sum_{m=s}^{\infty} |x_m a_m^{(k)};a_m^{(r)}| \leqslant 1$$

and therefore

$$\sum_{n=s}^{\infty}|x_m| < \varepsilon .$$

According to §1,4.(20) it follows that $T_k^{-1}(A_r)$ is a compact subset of ℓ^1. Thus A_r is a compact subset of λ_k and therefore λ_o is a Schwartz space.

(4) *The space* λ_∞ *is a Schwartz space if and only if given any positive integer* r *there is an integer* $k > r$ *such that* (2) *is verified.*

Proof. If λ_∞ is a Schwartz space then it is a Montel space (cf. HORVATH [1], Chapter 3,§15) and therefore $\lambda_o = \lambda_\infty$ and (2) is verified. Reciprocally, if (2) holds, then λ_∞ is a Schwartz space and therefore $\lambda_o = \lambda_\infty$ which completes the proof.

The conditions on the steps characterizing the Montel or Schwartz echelon spaces of order zero are the same as conditions holding for Montel or Schwartz echelon spaces. If λ_o is a Montel space which is not a Schwartz space we can use the construction of §2,3 to obtain a system of steps

(5) $$\beta_s = (a_{ij}^{(s)}),\ s = r, r+1,\ldots$$

with

(6) $$(a_{hi}^{(r)};a_{hi}^{(r+h)}) > b_h > 0,\ i,h = 1,2,\ldots$$

in such a way that the echelon space of order zero Λ defined by the system of steps (5) is isomorphic to a quotient of λ_o.

Let Δ be the linear space over K of the double sequences (x_{ij}) verifying, for $j = 1,2,\ldots,$

(7) $$\sum_{j=1}^{\infty}|x_{1j};a_{1j}^{(r)}| < b,\quad x_{ij} = \frac{a_{ij}^{(r+1)}}{i^2 a_{1j}^{(r)}}\,x_{1j},\ i = 2,3,\ldots,$$

b being a positive number depending on the sequence (x_{ij}).

Results (8) and (9) can be proven analogously to § 3, 3(7) and § 3, 3(8) respectively by setting 1 instead of q.

(8) Δ *is a linear subspace of* $\Lambda^\times$.

(9) *In* $\Lambda^\times[\sigma(\Lambda^\times,\Lambda)]$ *the subspace* Δ *is closed.*

(10) *The space* Δ *endowed with the topology induced by* $\sigma\,\Lambda^\times\,\Lambda$ *is isomorphic to* $\ell^1[\sigma(\ell^1,c_o)]$.

Proof. Let f by the mapping defined in §3.3(12) using 1 instead of q. Then $f: \ell^1 \longrightarrow \Delta$ is linear and bijective.

Let (u_{ij}) be an element of Λ. If $z = (z_j)$ belongs to ℓ^1 we have that

$$<(u_{ij}),f(z)> = \sum_{j=1}^{\infty} u_{1j}a_{1j}^{(r)}z_j + \sum_{i=2}^{\infty} \frac{1}{i^2} \sum_{j=1}^{\infty} u_{ij}a_{ij}^{(r+1)}z_j$$

$$= \sum_{j=1}^{\infty} z_j(a_{1j}^{(r)}u_{1j} + \sum_{i=2}^{\infty} \frac{1}{i^2} u_{ij}^{(r+1)}) = \sum_{j=1}^{\infty} z_j w_j$$

and

$$|w_j| \leqslant \sum_{i=1}^{\infty} \frac{1}{i^2} |u_{ij}| a_{ij}^{(r+1)}$$

Since

$$\lim_{i+j\to\infty} u_{ij} \; a_{ij}^{(r+1)} = 0$$

given $\varepsilon > 0$ there is a positive integer s such that

$$|u_{ij}| \; a_{ij}^{(r+1)} < \varepsilon \text{ for } i+j \geqslant s.$$

By taking $j \geqslant s$ it follows that

$$\sum_{i=1}^{\infty} \frac{1}{i^2} |u_{ij}| \; a_{ij}^{(r+1)} \leqslant \sum_{i=1}^{\infty} \frac{\varepsilon}{i^2}$$

and therefore (w_j) belongs to c_0 and thus $f: \ell^1[\sigma(\ell^1,c_0)] \longrightarrow \Delta$ is continuous. Finally, since Δ is $\sigma(\Lambda^\times,\Lambda)$-closed, we have that $\ell^1[\sigma(\ell^1,c_0)]$ is isomorphic to Δ for the topology induced by $\sigma(\Lambda^\times,\Lambda)$.

(11) *If λ_0 is not a Schwartz space then λ_0 has a quotient isomorphic to c_0.*

Proof. If λ_0 is not a Montel space we apply 2.(6) to obtain a sectional subspace of λ_0 isomorphic to c_0. Consequently λ_0 has a quotient isomorphic to c_0.

If λ_0 is a Montel space which is not Schwartz it has a quotient isomorphic to the space Λ described above. It is enough consider the space Λ. According to (10), if L denotes the subspace of Λ orthogonal to Δ we have that Λ/L is isomorphic to c_0.

(12) *The space λ_0 is totally reflexive if and only if it is a Schwartz space.*

Proof. If λ_0 is a Schwartz space it is obvious that every separated quotient of λ_o is a Schwartz space and therefore Montel.

If λ_o is not a Schwartz space we apply result (11) to obtain that λ_o is not totally reflexive.

4. NUCLEAR ECHELON SPACES OF ORDER ZERO. In this section λ_o denotes the echelon space of order zero defined by the system of steps 1.(1).

(1) *If the space* λ_o *is nuclear then given a positive integer* r *there is an integer* s > r *such that*

$$\Sigma(a_m^{(r)};a_m^{(s)}) < \infty.$$

Proof. The same proof in 3.(1) can be used by setting $q = 1$, $p = \infty$,

$$\left(\sum_{m=1}^{\infty} |\langle a_m^{(r)} e_m, z^h\rangle|^p\right)^{\frac{1}{p}} = \sup\{|\langle a_m^{(r)} e_m, z^h\rangle| : m = 1,2,\ldots\},$$

λ_o instead of λ and λ instead of $\lambda^\times$.

Let Λ be the echelon space defined by the system of steps 1.(1), i.e.,

$$\Lambda = \{(x_m) \in \omega : \Sigma|x_m a_m^{(r)}| < \infty,\ r = 1,2,\ldots\}$$

(2) *If given a positive integer* r *there is an integer* s > r *such that*

$$\Sigma(a_m^{(r)};a_m^{(s)}) < \infty$$

then λ_o *coincides with* Λ.

Proof. See the proof of §3,4.(2) setting $p = \infty$, λ_o instead of λ, $q=1$ and

$$(\Sigma|x_m a_m^{(s)}|^p)^{\frac{1}{p}} = \sup\{|x_m a_m^{(r)}| : m = 1,2,\ldots\}.$$

(3) *If given a positive integer* r *there is an integer* s > r *such that*

$$\Sigma(a_m^{(r)};a_m^{(s)}) < \infty$$

then λ_o *is nuclear.*

Proof. See the proof in §3,4.(3) setting λ_o instead of λ.

5. NOTE. The echelon spaces of infinite order λ_∞ have been introduced by DUBINSKY [1]. Results 2.(1), 3.(1) and 4.(1) are due to the quoted author although his methods of proof are different. The results about echelon spaces of order zero have been taken from our work "Espacios de sucesiones" supported by FUNDACION AGUILAR.

DUBINSKY [2] studies the perfect Fréchet spaces and those perfect Fréchet spaces which are Montel. CROFTS [1] completes the work of DUBINSKY and characterizes the perfect Fréchet spaces which are Schwartz spaces.

Other directions in the development of the theory of sequences spaces can be seen in the bibliography collected in KÖTHE [1].

§ 5. EXAMPLES

1. NON-COMPLETE NORMED (LF)-SPACES. In this section we shall consider double sequences of elements of K. Let p be a real number, $1 \leq p < \infty$. If $p > 1$ we set q for the conjugate real number of p, i.e.,

$\frac{1}{p} + \frac{1}{q}$ and if $p = 1$, $q = \infty$. Given a double sequence $x = (x_{ij})$ we set

$$||x||_p = (\Sigma|x_{ij}|^p)^{\frac{1}{p}}, \quad ||x||_q = (\Sigma|x_{ij}|^q)^{\frac{1}{q}}, \; q \neq \infty,$$

$$||x||_q = \sup |x_{ij}|, \; q = \infty.$$

If $q = \infty$ and if M is a non-void subset of $\{(i, j) : i, j = 1,2,..\}$ we write

$$(\sum_{(i, j) \in N} |x_{ij}|^q)^{\frac{1}{q}} = \sup \{|x_{ij}| : (i, j) \in M\}.$$

We set

$$\Lambda = \{(x_{ij}) : ||(x_{ij})||_p < \infty\}$$

Obviously Λ endowed with the norm $||.||_p$ is isomorphic to ℓ^p and Λ^x endowed with the norm $||.||_q$ is isomorphic to ℓ^q.

Given a positive integer m let λ_m be the linear space of all double sequences (x_{ij}) such that

(1) $\|(x_{ij})\|_p < \infty$, $\sup \{j|x_{ij}| : j = 1,2,\ldots\} < \infty$, $i \geq m$.

According to (1), if (u_{ij}) verifies $\|(u_{ij})\|_q < \infty$ then (u_{ij}) belongs to the α-dual $\lambda_m^\times$ of λ_m. Consequently $\lambda_m^{\times\times}$ is contained in Λ.

If (z_{ij}) is in Λ and $(z_{ij}) \in \lambda_m$, there is an integer $r \geq m$ such that

$$\sup \{j|z_{rj}| : j = 1,2,\ldots\} = \infty$$

and therefore we can take a sequence of positive integers

$$j_1 < j_2 < \ldots < j_s < \ldots$$

such that

(2) $j_s\, |z_{rj_s}| > s^2$, $s = 1,2,\ldots$

Let (v_{ij}) be a double sequence such that

$$v_{ij} = 0,\ i \neq r,\ v_{rj} = 0,\ j \neq j_s,\ v_{rj_s} = \frac{j_s}{s^2},\ s = 1,2,\ldots$$

If (x_{ij}) belongs to λ_m we have that

$$\Sigma|x_{ij}v_{ij}| = \sum_{s=1}^{\infty} |x_{rj_s}v_{rj_s}| = \sum_{s=1}^{\infty} \frac{1}{s^2} |j_s x_{rj_s}| < \infty$$

and thus (v_{ij}) belongs to λ_m. On the other hand, according to (2),

$$\Sigma|z_{ij}v_{ij}| = \sum_{s=1}^{\infty} |z_{rj_s} \frac{j_s}{s^2}| = < \infty$$

and therefore (z_{ij}) is not in λ_m. The following result is now obvious.

(3) λ_m *is a perfect sequence space.*

We set $\lambda = \bigcup_{m=1}^{\infty} \lambda_m$ and e_{ij} to denote the double sequence of all those elements which are zero except the element in position (i,j) which is one, $i,j = 1,2,\ldots$ The linear hull of $\{e_{ij} : i,j = 1,2,\ldots\}$ is ϕ. We write

$$B = \{(u_{ij}) : \|(u_{ij})\|_q < 1\}$$

and for every positive integer r

$$A_{mr} = B \cup \{je_{ij} : i = m, m+1,\ldots, m+r;\ j = 1,2,\ldots\}$$

Clearly A_{mr} is in $\Lambda^\times$. Let V_{mr} be the polar set of rA_{mr} in Λ. We set $U_{mr} = V_{mr} \cap \lambda_m$.

(4) A_{mr} *is a bounded subset of* $\lambda^{\times}_{m}[\sigma(\lambda^{\times}_{m},\lambda_m)]$.

Proof. It is obvious that B is $\sigma(\lambda^{\times}_{m},\lambda_m)$-bounded. On the other hand, if $x = (x_{ij})$ belongs to λ_m we have that

$$\sup\{|<x,je_{ij}>| : i = m,m+1,\dots,m+r;\ j = 1,2,\dots\}$$

$$\leq \sum_{i=m}^{m+r} \sup\{|jx_{ij}| : j = 1,2,\dots\} < \infty$$

which completes the proof.

Since A_{mr} is absorbing in $\Lambda^{\times}$ and since $<\lambda_m,\Lambda^{\times}>$ is a dual pair with the canonical bilinear form, it follows that

$$\{U_{mr} : r = 1,2,\dots\}$$

is a fundamental system of neighbourhoods of the origin for a locally convex topology T_m.

(5) *If* $x=(x_{ij})$ *is an element of* Λ *which is not in* λ_m *there is a positive integer* r *such that*

$$\sup\{|<x,u>\} : u \in A_{mr}\} = \infty .$$

Proof. Since x is not in λ_m there is an integer $s \geq m$ such that

$$\sup\{|jx_{sj}| : j = 1,2,\dots\} = \infty.$$

By taking $m+r \geq s$ we have that $je_{sj} \in A_{mr}$, $j = 1,2,\dots$, and

$$\sup \Sigma|<x,je_{sj}>|: j = 1,2,\dots\} = \sup\{|jx_{sj}| : j = 1,2,\dots\} = \infty$$

and thus, the conclusion follows.

(6) *The space* $\lambda_m[T_m]$ *is a Fréchet space.*

Proof. Let (x^s) be a Cauchy sequence in $\lambda_m[T_m]$. Then (x^s) is a Cauchy sequence in $\Lambda[\mu(\Lambda,\Lambda^{\times})]$ and, since this last space is isomorphic to ℓ^p, we have that (x^s) converges to x in $\Lambda[\mu(\Lambda,\Lambda^{\times})]$.

Since $\{x^s : s = 1,2,\dots\}$ is a bounded set in $\lambda_m[T_m]$, given a positive integer r there is $h > 0$ such that

$$hx^s \in U_{mr},\ s = 1,2,\dots,$$

and therefore hx belongs to V_{mr}; thus x is a linear form on $\Lambda^{\times}$ bounded on A_{mr}. We apply (5) to obtain that **x** is in λ_m. Finally (x^s) is a Cauchy sequence in $\lambda_m\ [T_m]$ converging to x in $\lambda_m[\mu\ (\Lambda,\Lambda^\times)\ ]$ and since U_{mr} is closed in this space, r = 1,2,..., it follows that (x^s) converges to x in λ_m $[T_m]$. The proof is complete.

Let B_{mr} be the closed absolutely convex hull of A_{mr} in $\lambda_m^\times[\sigma(\lambda_m^\times,\lambda_m)]$, r = 1,2,... Let $u = (u_{ij})$ be an element of $\lambda_m^\times$. If is any positive integer, we set u(s) to denote the double sequence (v_{ij}) with

$$v_{ij} = u_{ij},\ i+j \leq s,\ v_{ij} = 0,\ i+j > s.$$

According to §1,3.(1), the sequence (u(s)) converges to u in $\lambda_m^\times[\sigma(\lambda_m^\times,\lambda_m)]$. The set {u (s) : s = 1,2,...} is $\sigma(\lambda_m^\times,\lambda_m)$-bounded and therefore T_m-equicontinuous and thus there is a positive integer r such that u(s) belongs to r B_{mr}, s = 1,2,...; consequently u belongs to r B_{mr}. It follows that

$$\lambda_m^\times = \bigcup_{r=1}^{\infty} r\,B_{mr}$$

(7) *T_m coincides with $\beta(\lambda_m,\ \lambda_m^\times)$.*

Proof. Obviously T_m is coarser than $\beta(\lambda_m\lambda_m^\times)$. On the other hand, let A be a closed bounded absolutely convex set of $\lambda_m[\sigma\ (\lambda_m^\times,\lambda_m)\ ]$. Let E be the linear hull of A endowed with the topology deduced from the Minkowski functional of A. Since $\lambda_m^\times[\nu\ (\lambda_m^\times,\lambda_m)]$ is complete, E is a Banach space. We have that

$$\bigcup_{r=1}^{\infty} E \cap (r\,B_{mr}) = E$$

and, since r $B_{mr} \cap E$ is an absolutely convex closed subset of E, we apply BAIRE's theorem to obtain a positive integer s such that $sB_{ms} \cap E$ is a neighbourhood of the origin in E. Then there is a number h > 0 with $A \subset h\ B_{ms}$ and thus T_m is finer than $\beta(\lambda_m,\lambda_m^\times)$.

(8) *The space $\lambda_m[\beta\ (\lambda_m,\lambda_m^\times)\ ]$ is not separable.*

Proof. Let P be the collection of all non-void subset of

$$\{1,2^2,3^2,\ldots,r^2,\ldots\}$$

If P belongs to P let x(P) be the double sequence (x_{ij}) with

$$x_{ij} = 0,\ i \neq m,\ x_{mj} = 0,\ j \notin P,\ x_{mj} = \frac{1}{j^2},\ j \in P$$

Then x (P) belongs to λ_m. We suppose that $\lambda_m[\beta(\lambda_m,\lambda_m^\times)]$ is separable. Then there is a sequence (x^s) of elements of λ_m such that

$$\bigcup_{s=1}^{\infty} (x^s + \frac{1}{4}U_{m1}) = \lambda_m.$$

Since P is non-countable there are $P_1, P_2 \in P$, $P_1 \neq P_2$, and a positive integer r such that

$$x(P_1), (P_2) \in x^r + \frac{1}{4}\ U_{m1}.$$

Then

$$2(x(P_1) - x(P_2)) \in U_{m1}.$$

Let n be a positive integer such that $n^2 \in P_1$, $n^2 \notin P_2$. We have that $n^2 e_{mn^2} \in A_{m1}$ and therefore

$$1 \geq |\langle 2(x(P_1) - x(P_2)), n^2 e_{mn^2}\rangle| = 2,$$

which is a contradiction.

(9) *The topological dual of* $\lambda_m[\beta(\lambda_m,\lambda_m^\times)]$ *is distinct from* $\lambda_m^\times$.

Proof. Suppose that the topological dual of $\lambda_m[\beta(\lambda_m,\lambda^\times_m)]$ is $\lambda_m^\times$. Then $\beta(\lambda_m,\lambda_m^\times)$ coincides with $\mu(\lambda_m,\lambda_m^\times)$ and thus ϕ is dense in $\lambda_m[\beta(\lambda_m,\lambda_m^\times)]$ which is in contradiction with (8).

(10) *The space* λ_{m+1} *is distinct from* λ_m.

Proof. The double sequence $x = (x_{ij})$ with

$$x_{ij} = 0,\ i \neq m,\ x_{mj} = 0,\ j \neq r^3,\ x_{mr^3} = \frac{1}{r^2},\ r = 1,2,\ldots,$$

verifies

$$||x||_p = \left(\Sigma \frac{1}{r^{2p}}\right)^{\frac{1}{p}} < \infty,$$

$$jx_{ij} = 0,\ i > m,$$

and therefore x belongs to λ_{m+1}. On the other hand,

$$r^3 x_{mr^3} = r,\ r = 1,2,\ldots,$$

and thus $x \notin \lambda_m$. The proof is complete.

We denote by $\mathcal{U}$ the topology on λ such that $\lambda[\mathcal{U}]$ is the inductive limit of the sequence $(\lambda_m[\beta(\lambda_m, \lambda_m^\times)])$.

(11) *ϕ is dense in $\lambda[\mathcal{U}]$.*

Proof. Let $x = (x_{ij})$ be an element of λ. Given an absolutely convex neighbourhood of the origin W in $\lambda[\mathcal{U}]$ we shall see that $(x+W)\cap\phi \neq \emptyset$ We find a positive integer m such that x belongs to λ_m. Let r be a positive integer with $U_{mr} \subset \frac{1}{2} W$. We find an integer $s > m+r$ such that

$$\left(\sum_{i=s}^{\infty} \sum_{j=1}^{\infty} |x_{ij}|^p\right)^{\frac{1}{p}} < \frac{1}{r}$$

We set

$$z_{ij} = x_{ij},\ i = 1,2,\ldots, s-1,\ z_{ij} = 0,\ i \geq s,\ j = 1,2,\ldots$$

If $z = (z_{ij})$ we have that z belongs to λ_m. If $u = (u_{ij}) \in r\,A_{mr}$ and $u \notin r\,B$ then $u_{ij} = 0$ for $i > m+r$ and, accordingly, $\langle x-z,u\rangle = 0$; if u belongs to r B we have that

$$|\langle x-z,u\rangle| = \left|\sum_{i=s}^{\infty} \sum_{j=1}^{\infty} x_{ij}\, u_{ij}\right|$$

$$\leq \left(\sum_{i=s}^{\infty} \sum_{j=1}^{\infty} |x_{ij}|^p\right)^{\frac{1}{p}} \left(\sum_{i=s}^{\infty} \sum_{j=1}^{\infty} |u_{ij}|^q\right)^{\frac{1}{q}} \leq \frac{1}{r}\, r = 1$$

and thus $x-z$ belongs to U_{mr}.

In the space λ_s we find a positive integer h such that $U_{sh} \subset \frac{1}{2} W$. We find a positive integer k such that

$$\left(\sum_{i=1}^{\infty} \sum_{j=k}^{\infty} |x_{ij}|^p\right)^{\frac{1}{p}} < \frac{1}{h}$$

We set

$$t_{ij} = z_{ij},\ j = 1,2,\ldots, h-1;\ t_{ij} = 0,\ j \geq h,\ i = 1,2,\ldots$$

If $t = (t_{ij})$, then $t \in \phi \subset \lambda_s$. If $u = (u_{ij}) \in h\,A_{sh}$ and $u \in h\,B$, then $u_{ij}=0$ for $i < s$ and therefore $\langle z-t,u\rangle = 0$; if u belongs to h B we have that

$$|\langle z-t,u\rangle| = \left|\sum_{i=1}^{\infty} \sum_{j=k}^{\infty} z_{ij}u_{ij}\right| \leq \sum_{i=1}^{\infty} \sum_{j=k}^{\infty} |x_{ij}u_{ij}|$$

$$\leq \left(\sum_{i=1}^{\infty} \sum_{j=k}^{\infty} |x_{ij}|^p \right)^{\frac{1}{p}} \left(\sum_{i=1}^{\infty} \sum_{j=k}^{\infty} |u_{ij}|^q \right)^{\frac{1}{q}} < \frac{1}{h} h = 1$$

and therefore t belongs to $W \cap \phi$ which completes the proof.

(12) $\lambda^{\times}$ *coincides with* $\Lambda^{\times}$.

Proof. Since λ is contained in Λ we have that $\Lambda^{\times}$ is contained in $\lambda^{\times}$. We suppose the existence of an element $u = (u_{ij})$ in $\lambda^{\times}$ with $||u||_q = \infty$. If there is a positive integer r such that

$$(\Sigma |u_{rj}|^q)^{\frac{1}{q}} = \infty$$

we can find a vector (x_j) in ℓ^p such that

$$\Sigma |x_j u_{rj}| < \infty.$$

We set

$$x_{ij} = 0,\ i \neq r,\ x_{rj} = x_j,\ j = 1,2,\ldots$$

Then (x_{ij}) is in λ and

$$\Sigma |x_{ij} u_{ij}| = \Sigma |x_j u_{rj}| = \infty$$

which is a contradiction.

If there is a positive integer r such that

$$(\Sigma |u_{ir}|^q)^{\frac{1}{q}} = \infty ,$$

we can find a vector (x_i) in ℓ^p such that

$$\Sigma\ |x_i u_{ir}| = \infty.$$

We set

$$x_{ij} = 0,\ j \neq r,\ x_{ir} = x_i,\ i = 1,2,\ldots$$

Then (x_{ij}) is in λ and

$$\Sigma |x_{ij} u_{ij}| = \Sigma |x_i u_{ir}| = \infty$$

which is a contradiction.

We suppose now that

$$\left(\sum_{i=1}^{\infty} |u_{ij}|^q \right)^{\frac{1}{q}} < \infty,\ j = 1,2,\ldots$$

$$\left(\sum_{j=1}^{\infty} |u_{ij}|^{q}\right)^{\frac{1}{q}} < \infty, \ i = 1,2,\ldots$$

given a positive integer s we find an integer $m(s) > s$ such that

$$\left(\sum_{i,j=s}^{m(s)-1} |u_j|^q\right)^{\frac{1}{q}} > s.$$

We set $s_1=1$, $m(s_j) = s_{j+1}$, $j = 1,2,\ldots$ We set

$$M_m = \{(i,j) : i,j = s_m, s_m+1, \ldots, s_{m+1}-1\} , \ m = 1,2,\ldots$$

Let (v_{ij}) be the double sequence with

$$v_{ij} = u_{ij}, \ (i,j) \in M_m, \ m = 1,2,\ldots$$

$$v_{ij} = 0, \ (i,j) \in \bigcup_{m=1}^{\infty} M_m.$$

Then (v_{ij}) belongs to $\lambda^\times$ and

$$\left(\Sigma |v_{ij}|^q\right)^{\frac{1}{q}} = \left(\sum_{m=1}^{\infty} \sum_{i,j=s_m}^{s_{m+1}-1} |u_{ij}|^{q}\right)^{\frac{1}{q}} = \infty.$$

We can find a vector (x_{ij}) in Λ such that

$$x_{ij} = 0, \ (i,j) \notin \bigcup_{m=1}^{\infty} M_m, \ \Sigma |x_{ij} v_{ij}| = \infty.$$

Then (x_{ij}) belongs to λ and we arrive to a contradiction.

(13) *λ provided with the topology of the norm $||.||_p$ is the inductive limit of the sequence $(\lambda_m[\beta(\lambda_m,\lambda_m^\times)])$ of Fréchet spaces.*

Proof. Let f be a linear form on λ, continuous on each $\lambda_m[\beta(\lambda_m,\lambda_m^\times)]$. Let g be the restriction of f to ϕ. If x belongs to λ there is a positive integer m such that x is in λ_m and therefore the normal hull x^n of x is in λ_m. Since x^n is $\sigma(\lambda_m,\lambda_m^\times)$-compact and absolutely convex we have that x^n is a bounded set in $\lambda_m[\beta(\lambda_m,\lambda_m^\times)]$. Consequently f is bounded on x^n. According to §1, 8.(1), $\phi[\mu(\phi,\lambda^\times)] = \phi[\beta(\Lambda,\Lambda^\times)]$ is the inductive limit of the family of normed spaces

$$\{\phi_x : x \in \lambda\}$$

and since g is continuous on each ϕ_x, g is continuous on $\phi[\beta(\Lambda,\Lambda^\times)]$. Then g can be extended to a linear continuous form h on λ for the topology of the

the norm $||\cdot||_p$ which coincides with $\beta(\Lambda,\Lambda^\times)$.

Obviously, the topology of the norm $||\cdot||_p$ in λ is coarser than $\mathcal{U}$ and therefore h is continuous on $\lambda[\mathcal{U}]$. Since ϕ is dense in $\lambda[\mathcal{U}]$ and since h coincides with f on ϕ we have that h = f. The conclusion follows easily.

(14) *For every positive integer* m, $\lambda_m[\beta(\Lambda,\Lambda^\times)]$ *is not barrelled.*

Proof. Given a positive integer m, the injection mapping $T:\lambda_m[\mathcal{T}_m] \longrightarrow \lambda_m[\beta(\Lambda,\Lambda^\times)]$ is continuous. Therefore if $\lambda_m[\beta(\Lambda,\Lambda^\times)]$ is barrelled we apply PTAK's open mapping theorem to conclude that T is an isomorphism and thus $\lambda_m[\beta(\Lambda,\Lambda^\times)]$ is a Fréchet space, which is in contradiction with (10).

(15) λ *endowed with the topology of the norm* $||\cdot||_p$ *is not complete.*

Proof. If $\lambda[\beta(\Lambda,\Lambda^\times)]$ is complete then λ coincides with Λ and according to BAIRE's theorem there is a positive integer r such that $\lambda_r[\beta(\Lambda,\Lambda^\times)]$ is a Baire space and therefore barrelled. This is a contradiction (see(14)).

The construction of the space λ is due to W. ROELCKE. The proof given here of λ, endowed with the topology of the norm $||\cdot||_p$, being inductive limit of the sequence $(\lambda_m[\mathcal{T}_m])$ of Fréchet spaces is slightly different from the one Prof. ROELCKE gave to us. Since this section is included in the Chapter devoted to sequence spaces we have included here results (3), (7), (8) and (9).

The space λ, endowed with the topology of the norm $||\cdot||_p$ is ultrabornological and therefore ordered-convex-Baire. On the other hand, λ is not suprabarrelled since it is union of the increasing sequence $(\lambda_m[\beta(\Lambda,\Lambda^\times)])$ of non-barrelled spaces.

2. A NORMED SUPRABARRELLED SPACE WHICH IS NOT CONVEX-BAIRE. Let P be a subset of positive integers. For every positive integer m we denote by m(P) the number of elements of the set $\{r \in P : r \leq m\}$. P is said to have zero density if

$$\lim_m \frac{m(P)}{m} = 0.$$

For every sequence $x = (x_m)$ of K we set

$$||x|| = \Sigma|x_m|.$$

Let λ be the set of all sequence (x_m) of K with $||(x_m)|| <\infty$ and such that the set of indices m for which $x_m \neq 0$ has zero density. It is obvious that λ is a normal sequence space. We suppose λ endowed with the topology of the norm $||.||$.

Let B be the closed unit ball of ℓ^1. It is obvious that λ is a dense subspace of ℓ^1 and that $\lambda \cap B$ is a neighbourhood of the origin in λ. Let P be the family of all subsets of N having zero density. If P belongs to P we set $\lambda(P)$ to denote the sectional subspace of λ of all those elements of λ (z_m) with $z_m = 0$ for $m \notin P$. Clearly $\lambda(P)$ is a Banach space.

In the sequence space λ let U be an absolutely convex subset such that $U \cap \lambda(P)$ is a neighbourhood of the origin in $\lambda(P)$ for every $P \in P$. We suppose that U does not absorb the set $\{e_1, e_2, \ldots, e_m, \ldots\}$. We set $n_1=1$. Proceeding by recurrence, suppose we have obtained the integers

$$n_1 < n_2 < \ldots < n_r$$

We find an integer $n_{r+1} > (r+1)^2 + n_r$ such that

(1) $$e_{n_{r+1}} \notin (r+1)U .$$

The set

$$P = \{n_1, n_2, \ldots, n_m, \ldots\}$$

belongs to P. Indeed, if m is an integer larger than n_2 there is a positive integer r such that

$$n_{r+1} \leq m < n_{r+2}$$

from where it follows

$$\lim_m \frac{m(P)}{m} \leq \lim_r \frac{r+1}{(r+1)^2} = 0$$

Since $B \cap \lambda(P)$ is a bounded set in $\lambda(P)$ we have that $U \cap \lambda(P)$ absorbs $B \cap \lambda(P)$. Since e_j belongs to B, $j = 1,2,\ldots$, (1) provides a contradiction. Thus U absorbs the set $\{e_1, e_2, \ldots, e_m, \ldots\}$.

(2) *The sequence space λ is the inductive limit of the family of Banach spaces $\{\lambda(P) : P \in P\}$*

Proof. It is obvious that the family $\{\lambda(P) : P \in P\}$ is ordered by inclusion. In the sequence space λ, let U be an absolutely convex subset such that $U \cap \lambda(P)$ is a neighbourhood of the origin in $\lambda(P)$ for every $P \in \mathcal{P}$. According to what has been said before, there is a number $h > 0$ such that

$$he_j \in \frac{1}{2}U, \; j = 1,2,\ldots$$

Given any element (x_m) od $hB \cap \lambda$ let S be the subset of all those positive integers m such that $x_m \neq 0$. Clearly S belongs to $\mathcal{P}$. Since $U \cap \lambda(S)$ is a neighbourhood of the origin in $\lambda(S)$ we have that the closure V of $\frac{1}{2}U \cap \lambda(S)$ in $\lambda(S)$ is contained in U. Since

$$\Sigma\left|\frac{x_m}{h}\right| \leq 1,$$

it follows that

$$\sum_{j=1}^{s} \frac{x_j}{h} he_j = \sum_{j=1}^{s} x_j e_j \in \frac{1}{2}U, \; s = 1,2,\ldots$$

On the other hand,

$$\sum_{j=1}^{s} x_j e_j \in \lambda(S)$$

and, according to

$$\lim_s ||(x_m) - \sum_{j=1}^{s} x_j e_j|| = \lim_s \sum_{m=s+1}^{\infty} |x_j| = 0,$$

it follows that $(x_m) \in V \subset U$. Thus $hB \cap \lambda \subset U$ and consequently U is a neighbourhood of the origin in λ. Now the proof is complete.

(3) *$\lambda^\times$ coincides wit ℓ^∞.*

Proof. Since λ is dense in ℓ^1 the topological dual of λ coincides with ℓ^∞. Since λ is containedin ℓ^1 we have that ℓ^∞ is contained in $\lambda^\times$. We suppose now the existence of an element $u = (u_m)$ of $\lambda^\times$ which is not in ℓ^∞. Then u is not continuous on λ and therefore there is $P \in \mathcal{P}$ such that u is not continuous on the Banach space $\lambda(P)$. Since $\lambda(P)$ is a sectional subspace of ℓ^1 its topological dual can be identified with the sequences (v_m) of ℓ^∞ with $v_m = 0$ for each $m \notin P$. Thus there is a sequence $(x_m) \in \lambda(P)$ with

$$\Sigma|x_m u_m| = \infty$$

which is a contradiction.

(4) $\lambda[\mu(\lambda,\lambda^{\times})]$ *is a normed suprabarrelled space.*

Proof. According to (3), $\lambda^{\times}$ can be identified with the topological dual of λ and therefore $\mu(\lambda,\lambda^{\times})$ coincides with the topology of λ. Thus $\lambda[\mu(\lambda,\lambda^{\times})]$ is a normed space. Now we consider an increasing sequence of subspaces of λ.

(5) $$E_m,\ m = 1,2,\ldots,\ \text{with}\ \bigcup_{m=1}^{\infty} E_m = \lambda.$$

If F_m denotes the closure of E_m in λ we have that there is a positive integer r such that F_r is a neighbourhood of the origin in λ, i.e., $\lambda=F_r$ since λ is ultrabornological and therefore ordered-convex-Baire. Thus we can select a subsequence (G_m) from (5) with G_m dense in λ, $m = 1,2$.

We set $H_m = G_m + \phi$ and we consider it as a subspace of λ, $m = 1,2,\ldots$ For every positive integer m, we suppose the existence if a barrel T_m in H_m which does not absorb the set $\{e_1,e_2,\ldots,e_r,\ldots\}$. We set $n_1=1$. Proceeding by recurrence suppose we have obtained the integers

$$n_1 < n_2 < \ldots < n_r.$$

We find a positive integer s with $s^2 \leq r < (s+1)^2$ and we set $k = r - s^2$. Select an integer $n_{r+1} > (r+1)^2 + n_r$ with

(6) $$e_{n_{r+1}} \notin (r+1)\,T_{k+1}$$

We saw before that the set $P = \{n_1,n_2,\ldots,n_r,\ldots\}$ has zero density. Let W_r the closure of T_r in λ, $r = 1,2,\ldots$ In the Banach space $\lambda(P)$, the set $W_r \cap \lambda(P)$ is closed and absolutely convex and $\bigcup_{r=1}^{\infty} W_r \cap \lambda(P)$ is absorbing, therefore, there is positive integer m such that $W_m \cap \lambda(P)$ is a neighbourhood of the origin. Since

$$\{e_{n_1},e_{n_2},\ldots,e_{n_r},\ldots\}$$

is a bounded set of $\lambda(P)$ there is $h > 0$ with

$$e_{n_j} \in h\,W_m,\ j = 1,2,\ldots$$

and since $W_m \cap H_m = T_m$ and $\phi \subset H_m$ we have that

(7) $$e_{n_j} \in h\,T_m,\ j = 1,2,\ldots$$

We find a positive integer r such that $2_r > m$ and $r^2+m > h$. Then $r^2 \leq r^2 + m - 1 < (r+1)^2$ and according to (6),

$$e_{n_{r^2+n}} \notin (r^2+m)\,T_m$$

which is in contradiction with (7).

Accordingly, there is a positive integer s such that every barrel in H_s absorbs the set $\{e_1,e_2,\ldots,e_m,\ldots\}$. Since $B\cap\phi$ is obviously the absolutely convex hull of this set, given a barrel T in H_s there is $h > 0$ such that

$$(8) \qquad h B \cap \phi \subset T$$

If W is the closure of T in λ it follows from (8) that $h B\cap\lambda$ is contained in W and thus $T = W\cap H_s$ is a neighbourhood of the origin in H_s. Therefore H_s is barrelled. Finally G_s is barrelled since it is a countable codimensional subspace of H_s. We conclude that λ is suprabarrelled.

(9) *The normed space* $\lambda[\mu(\lambda,\lambda^\times)]$ *is not convex-Baire.*

Proof. Set

$$L_m = \{(x_r) \in \lambda: x_m = 0\},\ m = 1,2,\ldots$$

L_m is a closed hyperplane of λ. Given any element (z_r) of λ there is a $P\in\mathcal{P}$ such that (z_r) belongs to $\lambda(P)$. Since P is distinct from N we take a positive integer $s \notin P$. Then $z_s = 0$ and, accordingly, (z_r) belongs to L_s. Thus

$$\bigcup_{m=1}^{\infty} L_m = \lambda$$

which completes the proof since L_m is closed and convex and has no interior point in λ, $m = 1,2,\ldots$

3. A NORMED CONVEX-BAIRE SPACE WHICH IS NOT BAIRE. Take a real number p, $0 < p < 1$. Given a sequence $x = (x_m)$ of K we set

$$|x| = \Sigma|x_m|^p.$$

Set

$$\lambda = \{(x_m) \in \omega : |(x_m)| < \infty\}.$$

If (u_m) and (v_m) are elements of the Banach space $\ell^{\frac{1}{p}}$ we apply MINKOWSKI's inequality to obtain

$$(1) \qquad (\Sigma|u_m+v_m|^{\frac{1}{p}})^p \leq (\Sigma|u_m|^{\frac{1}{p}})^p + (\Sigma|v_m|^{\frac{1}{p}})^p.$$

Let $x = (x_m)$ and $y = (y_m)$ be elements of λ. If we set in (1)

$$u_1 = |x_r|^p,\ u_m = 0,\ m = 2,3,\ldots$$

$$v_1 = 0,\ v_2 = |y_r|^p,\ v_m = 0,\ m = 3,4,\ldots,$$

we have that

$$(|x_r| + |y_r|)^p \leq |x_r|^p + |y_r|^p \quad r = 1,2,\ldots$$

from where it follows

(2) $$|(x_m+y_m)| = \Sigma|x_r+y_r|^p \leq \Sigma(|x_r|+|y_r|)^p$$

$$\leq\Sigma|x_r|^p+\Sigma|y_r|^p = |(x_m)| + |(y_m)|.$$

(3) λ *is a normal sequence space.*

Proof. We take $h \in K$, $x = (x_m)$ and $y = (y_m)$ in λ and the sequence $z = (z_m)$ with $|z_m| \leq |x_m|$, $m = 1,2,\ldots$ Then

$$|hx| = \Sigma|hx_m|^p = |h|^p\ |x| < \infty$$

$$|x+y| \leq |x| + |y| < \infty$$

$$|z| = \Sigma|z_m|^p \leq \Sigma\ |x_m|^p = |x| < \infty$$

and, accordingly, hx, $(x+y)$ and z belongs to λ. Obviously e_m belongs to λ, $m = 1,2,\ldots$ Thus λ is a normal sequence space.

For each $x, y \in \lambda$ we set $d(x,y) = |x-y|$.

(4) d *is a metric on* λ.

Proof. If x,y,z are elements of λ we have that

$$d(x,y)=|x-y| = |x-z+z-y| \leq |x-z| + |z-y| = d(x,z)+d(z,y)$$

$$d(x,y) = |x-y| = |y-x| = d(y,x),$$

and, finally, it is obvious that $d(x,y)$ is zero if and only if x coincides with y. The proof is complete.

ℓ^p denotes the linear space λ endowed with the metric d. If we set

$$U_r = \{x \in \lambda:\ |x| \leq \frac{1}{r}\},$$

a fundamental system of neighbourhoods of the origin in ℓ^p is

$$\{U_r : r = 1,2,\ldots\}.$$

Since d is invariant by translation, given a vector x in λ a fundamental system of neighbourhoods of the point x in ℓ^p is

$$\{x+U_r : r = 1,2,\dots\}$$

(5) ℓ^p *is complete.*

Proof. Let $x^r = (x_m^{(r)})$, r = 1,2,..., be a Cauchy sequence in ℓ^p. For every three positive integers m, r and s we have that

$$|x_m^{(r)}-x_m^{(s)}|^p \leqslant |x^r-x^s|$$

from where it follows

$$\lim_r x_m^{(r)} = x_m \in K, \; m = 1,2,\dots$$

Given $\varepsilon > 0$ we find a positive integer s such that

$$|x^r-x^k| < \varepsilon, \; r,k \geqslant s.$$

Let q be a positive integer such that

$$\sum_{m=q}^{\infty} |x_m^{(s)}|^p < \varepsilon.$$

If $k \geqslant s$, $h \geqslant r \geqslant q$, then

$$\sum_{m=r}^{h} |x_m^{(k)}|^p \leqslant \sum_{m=r}^{h} |x_m^{(k)}-x_m^{(s)}|^p + \sum_{m=r}^{h} |x_m^{(s)}|^p \leqslant |x^k-x^s|+\varepsilon < 2\varepsilon$$

and consequently

$$\sum_{n=q}^{h} |x_m|^p \leqslant 2\varepsilon$$

and therefore $x = (x_m)$ belongs to ℓ^p and also

$$\sum_{m=q}^{\infty} |x_m|^p \leqslant \varepsilon$$

Finally, we determine a positive integer t such that

$$\sum_{m-1}^{q-1} |x_m-x_m^{(k)}|^p < \varepsilon \text{ for } k \geqslant t.$$

Then, if $k \geqslant t$

$$|x-x^k| = \sum_{m=1}^{q-1} |x_m-x_m^{(k)}|^p + \sum_{m=q}^{\infty} |x_m-x_m^{(k)}|^p$$

$$\leqslant \varepsilon + \sum_{m=q}^{\infty} |x_m|^p + \sum_{m=q}^{\infty} |x_m^{(k)}|^p < 4\varepsilon$$

and consequently (x^r) converges to x in ℓ^p.

(6) *$\lambda^\times$ coincides with ℓ^∞,*

Proof. If $x = (x_m)$ belongs to λ we have that $|x|<\infty$ and therefore there is a positive integer r such that $|x_m|^p < 1$ for $m > r$. Then

$$\Sigma|x_m| \leq \sum_{m=1}^{r} |x_m| + \sum_{m=r+1}^{\infty} |x_m|^p < \infty$$

and therefore x is in ℓ^1. Thus $\lambda \subset \ell^1$ and consequently $\lambda^\times \supset \ell^\infty$. Suppose the existence of a vector (u_m) in $\lambda^\times$ which is not in ℓ^∞. We find a sequence of positive integers

$$m_1 < m_2 < \ldots < m_r < \ldots$$

with

$$|u_{m_r}| > r^{\frac{2}{p}}, \quad r = 1,2,\ldots$$

If we set

$$x_m = 0, \; m \neq m_r, x_{m_r} = r^{-\frac{2}{p}}, \quad r = 1,2,\ldots,$$

we have that

$$\Sigma\, |x_m|^p = \Sigma \frac{1}{r^2} < \infty$$

and therefore (x_m) belongs to λ. On the other hand,

$$\Sigma|x_m u_m| = \Sigma|x_{m_r} u_{m_r}| = \infty$$

which is a contradiction. Thus $\lambda^\times = \ell^\infty$.

If $x = (x_m)$ belongs to ℓ^1 we write

$$||x|| = \Sigma|x_m|$$

and we suppose λ endowed with the topology derived from $||.||$.

(7) *$\mu(\lambda,\lambda^\times)$ coincides with the topology of λ.*

Proof. Since λ is a dense subspace of ℓ^1 we apply (6) to obtain that $\lambda^\times$ is the topological dual of λ and the conclusion follows.

(8) *The topology of λ is coarser than the topology of ℓ^p.*

Proof. Let $x^r = (x_m^{(r)})$, $r = 1,2,\ldots$, be a sequence in λ converging to the origin in ℓ^p. There is a positive integer s such that $|x^r| < 1$ for $r > s$. If $r > s$ we have that

$$||x^r|| = \Sigma|x_m^{(r)}| \leq \Sigma|x_m^{(r)}|^p \leq |x^r|$$

from where it follows that (x^r) converges to the origin in λ. Finally the conclusion follows if we consider the invariance by translation of the topology of ℓ^p.

(9) *The space λ is convex-Baire.*

Proof. Let (A_m) be a sequence of closed convex subsets of λ covering λ. Since ℓ^p is a complete metric space and since A_m is closed in ℓ^p, $m = 1,2,\ldots$, there is a positive integer r such that A_r has an interior point x in ℓ^p. Then $A_r - x$ is a neighbourhood of the origin in ℓ^p. Find a positive integer s such that U_s is contained in $A_r - x$. If B denotes the unit ball of ℓ^1 we take

$$(z_m) \in \left(\frac{1}{s}\right)^{\frac{1}{p}} B \cap \lambda.$$

Obviously it follows that

$$\left(\frac{1}{s}\right)^{\frac{1}{p}} (z_m;|z_m|)\, e_m \in U_s \subset A_r - x,\ m = 1,2,\ldots$$

Since $A_r - x$ is a convex set containing the origin and since

$$\sum_{m=1}^{k} s^{\frac{1}{p}} |z_m| \leqslant 1,\ k = 1,2,\ldots,$$

We have that

$$\sum_{m=1}^{k} s^{\frac{1}{p}} |z_m| \left(\frac{1}{s}\right)^{\frac{1}{p}} (z_m;|z_m|) e_m = \sum_{m=1}^{k} z_m e_m \in A_r - x.$$

Finally

$$\lim_k ||(z_m) - \sum_{h=1}^{k} z_h e_h|| = \lim_k \sum_{m=k+1}^{\infty} |z_m| = 0$$

and, since $A_r - x$ is closed in λ, it follows that (z_m) belongs to $A_r - x$. Thus

$$\left(\frac{1}{s}\right)^{\frac{1}{p}} B \cap \lambda \subset A_r - x$$

and consequently x is an interior point of A_r in λ. The proof is complete.

(10) *λ is not a Baire space.*

Proof. Given a positive integer s let $x^r = (x_m^{(r)})$, $r = 1,2,\ldots$, be a sequence in U_s converging to $x = (x_m)$ in λ. Then

$$\lim_r x_m^{(r)} = x_m \in K,\ m = 1,2,$$

Consequently, given a positive integer k we have that

$$\lim \sum_{m=1}^{k} |x_m^{(r)}|^p = \sum_{m=1}^{k} |x_m|^p \leq \frac{1}{s}$$

and therefore

$$\Sigma |x_m|^p \leq \frac{1}{s}$$

Thus x belongs to U_s and so U_s is closed in λ.

Suppose now that λ is a Baire space. Since

$$\bigcup_{m=1}^{\infty} U_s = \lambda$$

there is a positive integer q such that q U_s has an interior point and thus

$$\left(\frac{1}{2}\right)^{\frac{1}{p}} U_s$$

has an interior point. Therefore

$$\left(\frac{1}{2}\right)^{\frac{1}{p}} (U_s + U_s)$$

is a neighbourhood of the origin. If

$$x,y \in \left(\frac{1}{2}\right)^{\frac{1}{p}} U_s$$

we have that

$$2|x+y| = |2^{\frac{1}{p}} x + 2^{\frac{1}{p}} y| \leq |2^{\frac{1}{p}} x| + |2^{\frac{1}{p}} y| \leq \frac{2}{s}$$

and therefore x+y belongs to U_s from where it follows

$$\left(\frac{1}{2}\right)^{\frac{1}{p}} (U_s + U_s) \subset U_s$$

and thus U_s is a neighbourhood of the origin. Then, according to (8), the topology of ℓ^p coincides with $\mu(\lambda,\lambda^\times)$. On the other hand, we can find a sequence of positive integers $n_1 < n_2 < \ldots < n_r < \ldots$ such that

$$\sum_{m=n_r}^{n_{r+1}} \frac{1}{m} > r, \; r = 1,2,\ldots$$

and setting

$$x_m^{(r)}=0,\ m=1,2,\ldots,\ n_{r-1}, x_m^{(r)} = \left(\frac{1}{m}\right)^{\frac{1}{p}},\ m = n_r, n_r+1,\ldots,n_{r+1},$$

$$x_m^{(r)} = 0,\ m = n_{r+1}+1,\ n_{r+1}+2,\ldots$$

we have that $(x_m^{(r)})$ belongs to λ, $r = 1,2,\ldots$, and

$$\lim_r |(x_m^{(r)})| = \lim \sum_{m=n_r}^{n_{r+1}} \frac{1}{m} = \infty$$

$$\lim_r ||(x_m^{(r)})|| = \lim \sum_{m=n_r}^{n_{r+1}} \left(\frac{1}{m}\right)^{\frac{1}{p}} = 0$$

and consequently the sequence $(x_m^{(r)})$, $r = 1,2,\ldots$, converges to the origin in λ and does not converge to the origin in ℓ^p, which is a contradiction. The proof is complete.

The space λ endowed with the topology derived from the norm $||.||$ is used by P. DIEROLF, S. DIEROLF and L.DREWNOWSKI [1] to give an example of a normal unordered Baire-like space which is not ultrabarrelled in the sense of W. ROBERTSON [2] and therefore a space which is not Baire.

4. A FRECHET SPACE WHOSE STRONG DUAL PROVIDED WITH ITS MACKEY TOPOLOGY IS NOT BORNOLOGICAL. Given the sequence

(1) $\alpha_r = (a_{ij}^{(r)}),\ r = 1,2,\ldots,$

such that for every pair of positive integers j and r

$$a_{ij}^{(r)} = 2^{rj},\ i = 1,2,\ldots,r;\ a_{ij}^{(r)} = 1,\ i = r+1,\ r+2,\ldots$$

we represent by λ the echelon space defined by the system (1) of steps.

Given the double sequence (z_{ij}) with $z_{rs} = 1$, $z_{ij} = 0$, $(i,j)\neq(r,s)$, we set $f_{rs} = (z_{ij})$ when (z_{ij}) is considered as an element of λ and $e_{rs}=(z_{ij})$ when is supposed to be in $\lambda^{\times}$.

We set

$$B_r=\{(u_{ij})\in\lambda^{\times}: |u_{ij}| \leq a_{ij}^{(r)},\ i,j = 1,2,\ldots\}\ ,\ r = 1,2,\ldots$$

If r is a positive integer and if (m_i) is a strictly increasing sequence of positive integers we set

$$M(r,(m_i)) = \{f_{ij} : i > r,\ j \geqslant m_i\}.$$

The sets $M(r,(m_i))$ are a basis of a filter M in λ. Let $\mathcal{U}$ be an ultrafilter in λ finer than M. Let E be the algebraic dual of $\lambda^\times$. We identify λ with a subspace of E in the usual way. Given an element $u=(u_{ij}) \in \lambda^\times$ we set

$$M(r,(m_i),u) = \{<f_{ij},u> : f_{ij} \in M(r,(m_i))\}.$$

The sets $M(r,(m_i),u)$ with $M(r,(m_i)) \in \mathcal{U}$ are a basis of an ultrafilter $\mathcal{V}(u)$ in K. On the other hand, there is a positive integer s and h > 0 such that u belongs to $h B_s$. Consequently, if r > s it follows that $|u_{rj}| \leqslant h$ and therefore $M(r,(m_i),u)$ is a bounded set of K. Then $\mathcal{V}(u)$ converges in K to an element f(u) from where it follows that $\mathcal{U}$ converges to f in $E[\sigma(E,\lambda^\times)]$. Given a positive integer k, if $v = (v_{ij})$ belongs to B_k and r > k, we have that

$$|<f_{rj},v>| = |v_{rj}| \leqslant 1$$

and therefore

$$\text{(2)} \qquad |f(u)| \leqslant 1 \text{ for all } u \in \bigcup_{r=1}^{\infty} B_r.$$

Thus f is a linear form on $\lambda^\times$ bounded on every bounded subset of $\lambda^\times[\beta(\lambda^\times,\lambda)]$.

Now suppose that f is continuous on $\lambda^\times[\beta(\lambda^\times,\lambda)]$. Then there is a bounded set A in the echelon space λ such that f is in the closure of A in $E[\sigma(E,\lambda^\times)]$. If $A°$ is the polar set of A in $\lambda^\times$ we have that

$$|f(w)| \leqslant 1,\ w \in A°$$

For every positive integer i we find a positive integer h(i) such that

$$B_i \subset 2^{h(i)} A°.$$

Let $v = (v_{ij})$ be the element of B_1 defined by

$$v_{ij} = 1,\ j > h(i)+2,\ v_{ij} = 0,\ j \leqslant h(i)+2,\ i = 1,2,\ldots$$

The sequence

$$\text{(3)} \qquad \sum_{i,j=1}^{r} 2v_{ij}e_{ij},\ r = 1,2,\ldots,$$

is contained in $2B_1$ and converges coordinatewise to $2v$. Since B_1 is $\sigma(\lambda^\times,\lambda)$--compact, (3) converges to $2v$ in $\lambda^\times[\sigma(\lambda^\times,\lambda)]$. On the other hand,

$$2^{ij}e_{ij} \in B_i \subset 2^{h(i)}A^\circ$$

and therefore

$$2^{ij-h(i)}e_{ij} \in A^\circ$$

If we take $j > h(i)+2$ it follows that

$$ij-h(i) \geq (i-1)j+3$$

and consequently

$$2^{(i-1)j+3}e_{ij} \in A^\circ$$

Then

$$2^{(i-1)j+3}v_{ij}e_{ij} \in A^\circ, \quad i,j = 1,2,\ldots$$

Since A° is absolutey convex and since

$$\Sigma\frac{1}{2^{(i-1)j+2}} = \sum_{i=1}^{\infty} \frac{1}{2^2} \sum_{j=1}^{\infty} \left(\frac{1}{2^{i-1}}\right)^j = \sum_{i=1}^{\infty} \frac{1}{2^i} = 1$$

We have that

$$\sum_{i,j=1}^{r} \frac{1}{2^{(i-1)j+2}} \cdot 2^{(i-1)j+3} v_{ij}e_{ij} = \sum_{i,j=1}^{r} 2v_{ij}e_{ij} \in A^\circ, \quad r=1,2,\ldots$$

Since A° is $\sigma(\lambda^\times,\lambda)$-closed it follows that $2v$ belongs to A°.

If f_{ij} is any element of $M(1,(h(i)+3))$ we have that $j > h(i)+2$ and therefore

$$\langle f_{ij},2v\rangle = 2v_{ij} = 2$$

Consequently, $\langle f,2v\rangle = 2$ which is in contradiction with (2). Thus f is not continuous on $\lambda^\times[\beta(\lambda^\times,\lambda)]$. Accordingly, $\lambda^\times[\beta(\lambda^\times,\lambda)]$ endowed with its Mackey topology is not a bornological space. Then $\lambda[\beta(\lambda^\times,\lambda)]$ is not barrelled (cf. KÖTHE [1], Chapter Six, §29, Section 4), i.e., λ is not distinguished.

(4) *There exists a complete separable* (LB)-*space whose strong bidual endowed with its Mackey topology is not bornological.*

Proof. Let ϕ be the linear hull of

$$\{e_{ij}: i,j = 1,2,\ldots\}$$

and let ψ be the closure of ϕ in $\lambda^{\times}[\beta(\lambda^{\times},\lambda)]$. We suppose ψ endowed with the topology induced by $\beta(\lambda^{\times},\lambda)$. According to §2,1.(12), ψ is a complete (LB)-space, obviously separable, and by §2,1.(13) the strong dual of ψ is λ. Thus the strong bidual of ψ coincides with $\lambda^{\times}[\beta(\lambda^{\times},\lambda)]$. The conclusion follows from what has been said above.

(5) *If* G *is the strong bidual of* λ *and if* F *is the subspace of* G *orthogonal to* ϕ, *then* F *is a Banach space topological complement of* λ *in* G.

Proof. By §2,1.(7), F is a subspace of G which is a topological complement of λ. Let us see that F is a Banach space. We set

$$B = \{(u_{ij}) \in \lambda^{\times}: |u_{ij}| \leq 1, \ i,j = 1,2,\ldots\}.$$

Let B° be the polar set of B in G. Since B is $\sigma(\lambda^{\times},\lambda)$-bounded we have that B° is a neighbourhood of the origin in G and therefore B° $\cap$ F is a neighbourhood of the origin in F.

We take any element $w = (w_{ij})$ in $\lambda^{\times}$. We find $h > 0$ and a positive integer r such that w belongs to h B_r. Let $z = (z_{ij})$ the element of $\lambda^{\times}$ such that, for every positive integer j,

$$z_{ij} = 0, \ i = 1,2,\ldots, r; \ z_{ij} = w_{ij}, \ i = r+1, r+2,\ldots$$

Then z is in hB. For every positive integer j we set

$$y_{ij} = 2^j w_{ij}, \ i = 1,2,\ldots, r; \ y_{ij} = 0, \ i = r+1,r+2,\ldots,$$

then

$$|y_{ij}| \leq 2^j \, |w_{ij}| \leq h2^{(r+1)j}, \ i = 1,2,\ldots$$

from where it follows that (y_{ij}) belongs to h B_{r+1}. Let M be a bounded set of λ. There is a $k > 0$ such that

$$\Sigma|x_{ij}y_{ij}| < k \ \text{for all} \ (x_{ij}) \in M$$

Given $\varepsilon > 0$ we find a positive integer s such that $\frac{k}{2^s} < \varepsilon$. Then, if $x=(x_{ij})$ is any element of M, we have that

$$|\langle x,w-z-\sum_{i=1}^{r}\sum_{j=1}^{s} w_{ij}e_{ij}\rangle| = |\sum_{i=1}^{r}\sum_{j=s+1}^{\infty} x_{ij}w_{ij}|$$

$$= \left| \sum_{j=s+1}^{\infty} \frac{1}{2^j} \sum_{i=1}^{r} x_{ij} y_{ij} \right| \leq \sum_{j=s+1}^{\infty} \frac{k}{2^j} = \frac{k}{2^s} < \varepsilon,$$

and thus w-z is in the closure ψ of ϕ in $\lambda^\times[\beta(\lambda^\times,\lambda)]$ and therefore each element of F takes the value zero in w-z. If we take any element $g \in B^\circ \cap F$ it results

$$|\langle g,w\rangle| = |\langle g,z\rangle| \leq h$$

from where we deduce that $B^\circ \cap F$ is $\sigma(G,\lambda^\times)$-bounded and, since F is a Fréchet space, it follows that F is a Banach space. The proof is now complete.

Every Fréchet space is isomorphic to a closed subspace of a countable product of Banach spaces (cf. KÖTHE [1], Chapter Four, §19, Section 9). Thus there is a sequence (E_n) of Banach spaces such that λ is isomorphic to a closed subspace H of

$$E = \Pi \{E_n : n = 1,2,\ldots\}$$

Let L be the linear subspace of E' orthogonal to H. The space $E'[\beta(E',E)]$ is isomorphic to

$$\oplus \{E'_n[\beta(E'_n,E_n)] : n = 1,2,\ldots\}$$

and therefore bornological. Consequently $E'[\beta(E',E)]/L$ is bornological. On the other hand, $(E'/L)[\beta(E'/L,H)]$ is isomorphic to $\lambda^\times[\beta(\lambda^\times,\lambda)]$ which is not bornological and thus

$$E'[\beta(E',E)]/L \neq (E'/L)[\beta(E'/L,H)]$$

Having in mind that $\lambda^\times[\beta(\lambda^\times,\lambda)]$ endowed with its Mackey topology is not bornological we can even conclude that the topological dual of $E'\beta[(E,E)]/L$ is strictly larger than the topological dual of $(E'/L)[\beta(E'/L,H)]$.

We consider now the double sequences

(6) $\beta_r = (b_{ij}^{(r)}),\ r = 1,2,\ldots$

with $b_{ij}^{(r)} = j,\ i = 1,2,\ldots,r;\ b_{ij}^{(r)} = 1,\ i = r+1, r+2,\ldots.$

Let Λ be the echelon space defined by the system of steps (6). KÖTHE uses Λ to give an example of non-distinguished Fréchet space (cf. KÖTHE [1], Chapter Six, §31, Section 9). GROTHENDIECK [2] proves that on $\Lambda^\times[\beta(\Lambda^\times,\Lambda)]$ there is a linear form bounded on the bounded sets which is not continuous.

The argument used here to prove that f is bounded on every bounded subset of $\lambda^{\times}[\beta(\lambda^{\times},\lambda)]$ and non-continuous on that space is the same used by GROTHENDIECK [2] for $\Lambda^{\times}[\beta(\Lambda^{\times},\Lambda)]$.

5. A FRECHET SPACE WHOSE STRONG DUAL HAS A TOPOLOGY DISTINCT FROM ITS MACKEY TOPOLOGY. Consider the steps 4.(1). We set

$$\lambda = \{(x_{ij}) : \sum_i (\sum_j |a_{ij}^{(r)} x_{ij}|^2)^{\frac{1}{2}} < \infty, \ r = 1,2,\dots\}$$

Let Λ be the set of all double sequences (u_{ij}) such that there is a positive integer r, depending on (u_{ij}), such that

$$\sup_i \sum_j |u_{ij};a_{ij}^{(r)}|^2 < \infty.$$

Let e_{ij} be the double sequence taking value one in position (i,j) and zero in positions $(h,k) \neq (i,j)$ and let ϕ be the linear hull of

$$\{e_{ij} : i,j = 1,2,\dots\}.$$

(1) *Λ is a normal sequence space.*

Proof. Obviously, e_{ij} belongs to Λ, i,j = 1,2,... We take $b \in K$, $u = (u_{ij})$, $v = (v_{ij}) \in \Lambda$, $w = (w_{ij})$ with $|w_{ij}| \leq |u_{ij}|$, i,j = 1,2,... We can find a positive integer r such that

$$\sup_i \sum_j |u_{ij};a_{ij}^{(r)}|^2 < \infty, \quad \sup_i \sum_j |v_{ij};a_{ij}^{(r)}|^2 < \infty.$$

We have that

$$\sup_i(\sum_j |u_{ij}+v_{ij};a_{ij}^{(r)}|^2)^{\frac{1}{2}} \leq \sup_i |(\sum_j [u_{ij};a_{ij}^{(r)}|^2)^{\frac{1}{2}}$$

$$+ (\sum_j |v_{ij};a_{ij}^{(r)}|^2)^{\frac{1}{2}}| \leq \sup_i (\sum_j |u_{ij};a_{ij}^{(r)}|^2)^{\frac{1}{2}}$$

$$+ \sup_i (\sum_j |v_{ij};a_{ij}^{(r)}|^2)^{\frac{1}{2}} < \infty, \quad \sup_i \sum_j |bu_{ij};a_{ij}^{(r)}|^2$$

$$\leq |b|^2 \sup_i \sum_j |u_{ij};a_{ij}^{(r)}|^2 < \infty, \quad \sup_i \sum_j |w_{ij};a_{ij}^{(r)}|^2$$

$$\leq \sup_i \sum_j |u_{ij};a_{ij}^{(r)}|^2 < \infty,$$

from where it follows that

$$u+v,\ bu,\ w \in \Lambda$$

and therefore Λ is a normal sequence space.

(2) $\Lambda^\times$ *coindices with* λ.

Proof. If (x_{ij}) belongs to λ and (u_{ij}) belongs to Λ there is a positive integer r and $h > 0$ such that

$$\sum_j |u_{ij};a_{ij}^{(r)}|^2 < h^2,\ i = 1,2,\ldots$$

Therefore

$$\Sigma|x_{ij}u_{ij}| = \Sigma|a_{ij}^{(r)}x_{ij}|.|u_{ij};a_{ij}^{(r)}|$$

$$\leq \Sigma(\sum_{i\ j}|a_{ij}^{(r)}x_{ij}|^2)^{\frac{1}{2}}(\sum_j|u_{ij};a_{ij}^{(r)}|^2)^{\frac{1}{2}} \leq h\ \sum_i(\sum_j|a_{ij}^{(r)}x_{ij}|^2)^{\frac{1}{2}} < \infty$$

and thus (x_{ij}) belongs to $\Lambda^\times$ and consequently $\lambda \subset \Lambda^\times$.

Let (z_{ij}) be an element of $\Lambda^\times$. We suppose the existence of a positive integer r such that

$$\sum_i (\sum_j |a_{ij}^{(r)}z_{ij}|^2)^{\frac{1}{2}} = \infty.$$

If for a positive integer s

$$\sum_j |a_{sj}^{(r)}z_{sj}|^2 = \infty$$

we find an element $(b_j) \in \ell^2$ with

$$\sum_j |b_j a_{sj}^{(r)}z_{sj}| = \infty.$$

We set

$$v_{ij} = 0,\ i \neq s,\ v_{sj} = b_j a_{sj},\ j = 1,2,\ldots$$

Then (v_{ij}) belongs to Λ and

$$\Sigma|z_{ij}v_{ij}| = \sum_j |z_{sj}v_{sj}| = \sum_j |b_j a_{sj}^{(r)}z_{sj}| = \infty$$

which is a contradiction.

Now we suppose thar for every index i

$$(\sum_j |a_{ij}^{(r)} z_{ij}|^2)^{\frac{1}{2}} = m(i) < \infty.$$

We set

$$w_{ij} = ((a_{ij}^{(r)})^2 z_{ij};\ m(i)), i,j = 1,2,\ldots,$$

Then

$$\sum_j |w_{ij};a_{ij}^{(r)}|^2 = \sum_j |a_{ij}^{(r)} z_{ij};m(i)|^2 \leq 1,\ i = 1,2,\ldots,$$

and thus (w_{ij}) belongs to Λ. On the other hand,

$$\Sigma|z_{ij}w_{ij}| = \sum_{m(i)\neq 0} \frac{1}{m(i)} \sum_j |a_{ij}^{(r)} z_{ij}|^2 = \sum_{m(i)\neq 0} (\sum_j |a_{ij}^{(r)} z_{ij}|^2)^{\frac{1}{2}} = \infty$$

and we arrive to a contradiction. Thus $\Lambda^{\times} \subset \lambda$ and the proof is complete.

Next result is now obvious

(3) *λ is a perfect sequence space.*

For every positive integer we set

$$B_r = \{(u_{ij}) : \sum_j |u_{ij};a_{ij}^{(r)}|^2 \leq 1,\ i = 1,2,\ldots\}$$

(4) *B_r is a normal bounded absolutely convex subset of $\Lambda[\sigma(\Lambda,\lambda)]$.*

Proof. If $u = (u_{ij})$, $v = (v_{ij}) \in B_r$, $h,k \in K$ with $|h|+|k| \leq 1$, (w_{ij}) with $|w_{ij}| \leq |u_{ij}|$, $i,j = 1,2,\ldots$, and $(x_{ij}) \in \lambda$ we have that

$$(\sum_j |hu_{ij}+kv_{ij};a_{ij}^{(r)}|^2)^{\frac{1}{2}} \leq (\sum_j |hu_{ij};a_{ij}^{(r)}|^2)^{\frac{1}{2}}$$

$$+(\sum_j |hv_{ij};a_{ij}^{(r)}|^2)^{\frac{1}{2}} = |h|\ (\sum_j |u_{ij};a_{ij}^{(r)}|^2)^{\frac{1}{2}}$$

$$+ |k|\ (\sum_j |v_{ij};a_{ij}^{(r)}|^2)^{\frac{1}{2}} \leq |h| + |k| \leq 1$$

and therefore hu+kv belongs to B_r and consequently B_r is absolutely convex.

On the other hand,

$$\Sigma|x_{ij}u_{ij}| = \Sigma|a_{ij}^{(r)}x_{ij}|.|u_{ij};a_{ij}^{(r)}|$$

$$\leqslant \sum_i (\sum_j |a_{ij}^{(r)}x_{ij}|^2)^{\frac{1}{2}} (\sum_j |u_{ij};a_{ij}^{(r)}|^2)^{\frac{1}{2}} \leqslant \sum_i (\sum_j |a_{ij}^{(r)}x_{ij}^{(r)}|)^{\frac{1}{2}}$$

and therefore B_r is a bounded subset of $\Lambda[\sigma(\Lambda,\lambda)]$. Finally,

$$\sum_j |w_{ij};a_{ij}^{(r)}|^2 \leqslant \sum_j |u_{ij};a_{ij}^{(r)}|^2 \leqslant 1$$

and therefore B_r is a normal set.

(5) B_r *is compact in* $\Lambda[\sigma(\Lambda,\lambda)]$.

Proof. Since B_r is $\sigma(\Lambda,\lambda)$-bounded it is enough to show that B_r is $\sigma(\Lambda,\lambda)$-complete. Let

(6) $\{u^s = (u_{ij}^{(s)}) : s \in S, \geqslant\}$

be a Cauchy net in $\Lambda[\sigma(\Lambda,\lambda)]$ contained in B_r. Then

$$\lim_s u_{ij}^{(s)} = u_{ij} \in K, \; i,j = 1,2,\ldots$$

For every positive integer k we have that

$$\lim_s \sum_{i,j=1}^{k} |u_{ij}^{(s)};a_{ij}^{(r)}|^2 = \sum_{i,j=1}^{k} |u_{ij};a_{ij}^{(r)}|^2 \leqslant 1$$

and therefore

$$\Sigma|u_{ij};a_{ij}^{(r)}|^2 \leqslant 1$$

from where it follows that $u = (u_{ij})$ belongs to B_r.

Given $\varepsilon > 0$ and $x = (x_{ij}) \in \lambda$ we find $t \in S$ such that

$$\Sigma|x_{ij}(u_{ij}^{(s)} - u_{ij}^{(t)})| < \frac{1}{4}\varepsilon, \; s \geqslant t.$$

Select a positive integer h with

$$\sum_{i+j>h} |x_{ij}u_{ij}| < \frac{1}{4}\varepsilon, \quad \sum_{i+j>h} |x_{ij}u_{ij}^{(t)}| < \frac{1}{4}\varepsilon$$

We find an element $s_0 \in S$, $s_0 \geqslant t$, such that

$$\sum_{i+j\leq h} |x_{ij}(u_{ij}-u_{ij}^{(s)})| < \varepsilon, \ s \geq s_0.$$

Then, for $s \geq s_0$, it follows that

$$|\langle x,u-u^s\rangle| \leq \Sigma|x_{ij}(u_{ij}-u_{ij}^{(s)})| \leq \sum_{i+j\leq h} |x_{ij}(u_{ij}-u_{ij}^{(s)})|$$

$$+ \sum_{i+j>h} |x_{ij}\, u_{ij}| + \sum_{i+j>h} |x_{ij}u_{ij}^{(t)}| + \sum_{i+j>h} |x_{ij}(u_{ij}^{(s)}-u_{ij}^{(t)})| < \varepsilon$$

and therefore the net (6) converges to u in $\Lambda[\sigma(\Lambda,\lambda)]$. The proof is complete.

Now we denote by U_r the polar set of $r\,B_r$ in λ, $r = 1,2,\ldots$ If u belongs to Λ there is a positive integer s such that $u \in s\,B_s$ and, since this set is normal, the normal hull u^n of u is contained in $s\,B_s$. Then

$$\{U_r : r = 1,2,\ldots\}$$

is a fundamental system of neighbourhoods of the origin for a metrizable topology T finer than $\nu(\lambda,\Lambda)$. We suppose λ endowed with the topology T. By §1, 4.(1), $\lambda[\nu(\lambda,\Lambda)]$ is complete and thus λ is a Fréchet space.

(7) *Λ is the topological dual of λ.*

Proof. It follows easily from (5).

(8) *Λ coincides with $\lambda^\times$.*

Proof. Let (u^r) be a Cauchy sequence in $\Lambda[\sigma(\Lambda,\lambda)]$. Since λ is a Fréchet space, the sequence is equicontinuous in λ. Therefore there is a positive integer s with $u^r \in s\,B_s$, $r = 1,2,\ldots$, and consequently (u^r) converges to a vector u of $s\,B_s$. Then $\Lambda[\sigma(\Lambda,\lambda)]$ is sequentially complete and, according to §1,4.(18), Λ is perfect. Thus $\Lambda=\lambda^\times$.

We represent by ψ the closure of ϕ in $\lambda^\times[\beta(\lambda^\times,\lambda)]$. We set

$$B = \{(u_{ij}) : \sum_j |u_{ij}|^2 \leq 1, \ i = 1,2,\ldots\}.$$

In $\lambda^\times[\sigma(\lambda^\times,\lambda)]$, B is compact since it is a closed subset of B_1. Proceeding as in (4), we show that B is normal and absolutely convex. Let H be the linear hull of B with the topology derived from its gauge $\|.\|$. Then H is a Banach space and if $u = (u_{ij})$ belongs to H we have that

$$||u|| = \inf\{h > 0 : u \in h B\}$$

$$= \{h > 0 : \sum_j |u_{ij}|^2 \leq h^2,\ i = 1,2,\ldots\} = \sup_i(\sum_j |u_{ij}|^2)^{\frac{1}{2}}.$$

For each sequence of positive integers (m_i), we set

$$A((m_i)) = \{(u_{ij}) \in H : u_{ij} = 0,\ j > m_i,\ i = 1,2,\ldots\}$$

$$B((m_i)) = \{(u_{ij}) \in H : u_{ij} = 0,\ j \leqslant m_i,\ i = 1,2,\ldots\}$$

Given a positive integer m, we set

$$L(m) = \{(u_{ij}) \in H : u_{ij} = 0,\ i \geqslant m\}$$

and

$$L = \cup\{L(m) : m = 1,2,\ldots\}.$$

Let f be a continuous linear form on H such that $f(u) = 0$ for every $u \in L$.

(9) *There is a sequence of positive integers* (m_i) *such that* $f(u) = 0$ *for every* $u \in B((m_i))$.

Proof. We set

$$||f|| = \sup\{|f(v)| : v \in B\}.$$

Without loss of generality we assume $||f|| = 1$. A sequence $u^r = (u_{ij}^{(r)})$, $r = 1,2,\ldots$, can be determined in B such that $f(u^r)$ is real and

$$f(u^r) > 1 - \frac{1}{r},\ r = 1,2,\ldots$$

For every index i, let m_i be a positive integer such that

$$\sum_{j=m_i+1}^{\infty} |u_{ij}^{(r)}|^2 < \frac{1}{r^2},\ r = 1,2,\ldots,i.$$

Suppose the existence of an element $v = (v_{ij})$ in $B((m_i))$ with $||v|| = 1$ and $f(v)$ stritly positive and equal to a. Then $a \leq 1$ and therefore there is a real number b, $0 < b < 1$, such that

$$a^2 > 1-b.$$

Then

$$a\sqrt{1-b} > 1-b$$

and consequently $b + a\sqrt{1-b} > 1$. For every pair of positive integers i and r

we set

$$w_{ij}^{(r)} = u_{ij}^{(r)},\ j \leq m_i,\ w_{ij}^{(r)} = 0,\ j = m_i+1,\ m_i+2,\dots$$

Set $w^r = (w_{ij}^{(r)})$. For every pair of positive integers j and r we set

$$y_{ij}^{(r)} = u_{ij}^{(r)},\ z_{ij}^{(r)} = w_{ij}^{(r)},\ i > r$$

$$y_{ij}^{(r)} = z_{ij}^{(r)} = 0,\ i \leq r$$

Set $y^r = (y_{ij}^{(r)})$, $z^r = (z_{ij}^{(r)})$. Obviously, w^r, y^r and z^r are in B. We have that

$$|f(u^r-w^r)| = |f(y^r-z^r)| \leq ||y^r-z^r|| = \sup_{i>r} \Big(\sum_{j=m_i+1}^{\infty} |u_{ij}^{(r)}|^2\Big)^{\frac{1}{2}} < \frac{1}{r}$$

and therefore

$$|f(w^r)| \geq |f(u^r)|-|f(w^r-u^r)| \geq 1 - \frac{2}{r}$$

and then

$$\lim |f(bw^r+\sqrt{1-b}\ v)| = b \lim f(w^r) + \sqrt{1-b}\ f(v) = b+a\sqrt{1-b} > 1$$

and thus there is a positive integer s such that

$$|f(b\ w^s+\sqrt{1-b}\ \ v)| > 1.$$

On the other hand,

$$||bw^s+\sqrt{1-b}\ v||^2 = \sup_i\Big(\sum_{j=1}^{m_i} |bw_{ij}^{(r)}|^2 + \sum_{j=m_i+1} (1-b)|v_{ij}|^2\Big)$$

$$\leq b^2+1-b < b^2+1-b^2 = 1$$

which is a contradiction. Thus f vanishes in $B((m_i))$.

(10) *Given* $\varepsilon > 0$ *and the sequence of positive integers* (m_i) *there are* $g \in H'$ *and* $u \in B\ ((m_i))$ *verifying:*

a) g *vanishes on* $L \cup \phi \cup A\ ((m_i))$;

b) $||g|| = ||u|| = 1$, $|\langle u,g\rangle| > 1-\varepsilon$.

Proof. The subspace of H of all double sequences (v_{ij}) with $v_{ij} = 0$, $j > 1$, $i = 1,2,\dots$, is isomorphic to ℓ^∞ and therefore H is not separable. Since $L \cup \phi$ is a separable subset of H there is $f \in H'$, $||f||=1$, vanishing in $L \cup \phi$. If (y_{ij}) belongs to H we set $T((y_{ij})) = (z_{ij})$ with

$$z_{ij} = y_{i(m_i+j)}, \quad i,j = 1,2,\ldots$$

It is obvious that T is a linear mapping from H into H. On the other hand,

$$(11) \qquad \|T((y_{ij}))\| = \sup_i (\sum_j |y_{i(m_i+j)}|^2)^{\frac{1}{2}} \leq \|(y_{ij})\|$$

and therefore T is continuous. We have that $T^{-1}(0)=A((m_i))$, $T(L)=L$ and $T(\phi)=\phi$. We set $g=f\ T$. Then g is a linear form on H vanishing on $L \cup \phi \cup A((m_i))$ and according to (11), $\|g\| \leq \|f\| = 1$.

Given a real number $\varepsilon > 0$, we find an element $w = (w_{ij})$ in H with $\|w\| = 1$ and $|\langle w,f\rangle| > 1-\varepsilon$. For every positive integer i we set

$$u_{ij} = 0,\ j \leq m_i,\ u_{i(m_i+j)} = w_{ij},\ j = 1,2,\ldots$$

Then $u = (u_{ij})$ belongs to $B((m_i))$, $T(u) = w$, $\|u\| = \|w\| = 1$ and

$$|\langle w,f\rangle| = |f(T(u))| = |\langle u,g\rangle| > 1-\varepsilon.$$

Since ε is arbitrary we have that $\|g\| = 1$. The proof is complete.

Let G be the linear space of all linear form on $\lambda^{\times}[\beta(\lambda^{\times},\lambda)]$ which are bounded on every bounded set of this space. We suppose G endowed with its strong topology and λ identified with a subspace of G in the usual way. Let F be the subspace of G orthogonal to ϕ.

(12) *If* f *belongs to* F *then* f *vanishes on* L.

Proof. For every positive integer r let H_r be the subspace of H of all those vectors (x_{ij}) with $x_{ij} = 0$, $i \neq r$, $j = 1,2,\ldots$ It is obvious that H_r is isomorphic to ℓ^2 and therefore $\phi \cap H_r$ is dense in H_r. Since f is continuous in H and vanishes on ϕ it also vanishes on H_r. Thus f vanishes on L.

(13) F *is a Banach space which is a topological complement of* λ *in* G.

Proof. Since ϕ is dense in $\lambda^{\times}[\sigma(\lambda^{\times},\lambda)]$ it follows that $\lambda \cap F = \{0\}$. If S is a vector of G, we have that $(S(e_{ij}))$ is an element of the α-dual λ of $\lambda^{\times}$ according to the proof of §2,8.(1). Then

$$S = S-(S(e_{ij}))+(S(e_{ij})),\ S-(S(e_{ij})) \in F,\ (S(e_{ij})) \in \lambda.$$

It follows easily that F is a topological complement of λ in G since G is a Fréchet space.

If B° is the polar set of B in G, we have that $B^\circ \cap F$ is a neighbourhood of the origin in F. Taking $w = (w_{ij})$ in $\lambda^\times$ we find $h > 0$ and a positive integer r with $hw \in B_r$. Let $z = (z_{ij})$ be the element of $\lambda^\times$ such that, for every positive integer j,

$$z_{ij} = 0,\ i = 1,2,\dots,r;\ z_{ij} = w_{ij},\ i = r+1,r+2,\dots$$

Then z belongs to h B. If f is any element of $B^\circ \cap F$ we have that

$$|\langle f,w\rangle| = |\langle f,z\rangle| \leqslant h$$

since f vanishes on L and w-z belongs to L. It follows that $B^\circ \cap F$ is $\sigma(G, \lambda^\times)$-bounded and, consequently, F is a Banach space.

(14) G *is the topological dual of* $\lambda^\times[\beta(\lambda^\times,\lambda)]$.

Proof. If g belongs to G, we have that $g = k+f, k \in \lambda$, $f \in F$. It is enough to prove that f is $\beta(\lambda^\times,\lambda)$-continuous. We apply (9) to obtain a sequence (m_i) of positive integers such that

$$f(u) = 0, \text{ for each } u \in B((m_i)).$$

We set

$$M = \{(v_{ij}) \in \lambda^\times : \sum_{j=1}^{m_i} |v_{ij}|^2 \leqslant 1,\ i = 1,2,\dots\}.$$

Obviously M is absorbing, absolutely convex and $\sigma(\lambda^\times,\lambda)$-closed and therefore M is the polar set in $\lambda^\times$ of a bounded set of λ and thus M is a neighbourhood of the origin in $\lambda^\times[\beta(\lambda^\times,\lambda)]$. Given any element $v = (v_{ij})$ of M we set, for every positive integer i,

$$w_{ij} = v_{ij},\ j \leqslant m_i,\ w_{ij} = 0,\ j > m_i.$$

Then $w = (w_{ij})$ is in B and v-w belongs to $B((m_i))$ and therefore

$$|f(v)| = |f(v-w) + f(w)| = |f(w)| \leqslant \|f\|.$$

Consequently, f is bounded on M and therefore $\beta(\lambda^\times,\lambda)$-continuous. We conclude that G is the topological dual of $\lambda^\times[\beta(\lambda^\times,\lambda)]$.

(15) *The Mackey topology of* $\lambda^\times[\beta(\lambda^\times,\lambda)]$ *is distinct from* $\beta(\lambda^\times,\lambda)$.

Proof. Since the topological dual of $\lambda^\times[\beta(\lambda^\times,\lambda)]$ is G it follows that

$\lambda^\times[\mu(\lambda^\times,G)]$ is ultrabornological and therefore $B^\circ \cap F$ is $\sigma(G,\lambda^\times)$-compact. Suppose the existence of a bounded set A in λ with closure D in $G[\sigma(G,\lambda^\times)]$ containing $B^\circ \cap F$. Given a positive integer r there is $h > 0$ such that

$$\sum_i \left(\sum_j |a_{ij}^{(r+1)} x_{ij}|^2\right)^{\frac{1}{2}} < h \text{ for each } (x_{ij}) \in A.$$

Then

$$|x_{rj}| \leq h2^{-j(r+1)}, j = 1,2,\ldots$$

We find a positive integer m_r such that $h\,2^{-j} \leq 1$, for $j \geq m_r$, and

$$\sum_r 2^{-rm_r} < 2^{-1}$$

Consequently, $|x_{rj}| \leq 2^{-jr}$ and therefore

$$\sum_i \left(\sum_{j=m_i+1}^{\infty} |x_{ij}|^2\right)^{\frac{1}{2}} \leq \sum_i\left(\sum_{j=m_i+1}^{\infty} 2^{-2ji}\right)^{\frac{1}{2}} \leq \sum_i 2^{-im_i} \leq 2^{-1}$$

According to (10) there are $f \in H'$ and $u = (u_{ij}) \in B((m_i))$ such that f vanishes on $L \cup \phi \cup A((m_i))$ and $||f|| = 1$, $||u|| = 1$, $|\langle f,u\rangle| > \frac{1}{2}$.

If $w = (w_{ij})$ belongs to $\lambda^\times$ we find a positive integer s and $k > 0$ such that w belongs to $k\,B_s$. If

$$v_{ij} = 0,\ i \leq s,\ v_{ij} = w_{ij},\ i = s+1, s+2,\ldots,\ j = 1,2,\ldots$$

it results that $v = (v_{ij})$ belongs to H. We set $\langle g,w\rangle = \langle f,w\rangle$. Since f vanishes on L, g is well defined on $\lambda^\times$ and bounded on every bounded of $\lambda^\times[\beta(\lambda^\times,\lambda)]$. Obviously, g is linear on $\lambda^\times$. On the other hand, since f vanishes on $\phi \cup A((m_i))$ and $||f|| = 1$, we have that g belongs to $B^\circ \cap F$ and $\langle g,z\rangle = 0$ for every $z \in A((m_i))$; also we have

$$|\langle g,u\rangle| = |\langle f,u\rangle| > \frac{1}{2}. \tag{16}$$

Finally, if $x = (x_{ij})$ belongs to A, we have that

$$|\langle x,u\rangle| \leq \Sigma|x_{ij}u_{ij}| = \sum_i \sum_{j=m_i+1}^{\infty} |x_{ij}u_{ij}|$$

$$\leq \sum_i \left(\sum_{j=m_i+1}^{\infty} |x_{ij}|^2\right)^{\frac{1}{2}} \left(\sum_{j=m_i+1}^{\infty} |u_{ij}|^2\right)^{\frac{1}{2}} \leq \sum_i \left(\sum_{j=m_i+1}^{\infty} |x_{ij}|^2\right)^{\frac{1}{2}} \leq \frac{1}{2}$$

and since g belongs to D,

$$|\langle g,u\rangle| < \frac{1}{2},$$

which is in contradiction with (16). We conclude that $B^\circ \cap F$ is not equicontinuous in $\lambda^\times[\beta(\lambda^\times,\lambda)]$ and, thus, $\beta(\lambda^\times,\lambda) \neq \mu(\lambda^\times,G)$.

(17) *There is a complete separable*(LB)-*space whose strong bidual has not its Mackey topology.*

Proof. Let ψ be the closure of ϕ in $\lambda^\times[\beta(\lambda^\times,\lambda)]$ endowed with the topology induced by $\beta(\lambda^\times,\lambda)$. We have that ψ coincides with the closure of ϕ in the bornological space $\lambda^\times[\mu(\lambda^\times,G)]$.

We apply §1,8.(4), to obtain that ψ is bornological from where it follows easily that ψ is an (LB)-space which is obviously complete and separable. Finally $\lambda^\times[\beta(\lambda^\times,\lambda)]'$ is the strong bidual of ψ and $\beta(\lambda^\times,\lambda)$ is distinct from $\mu(\lambda^\times,G)$.

The space λ is a slight modification of an example of a Fréchet space given by Y. KOMURA [1] whose strong dual has a topology which does not coincide with its Mackey topology. In this section we followed KOMURA's ideas and we added some new results.

6. NON-COMPLETE (LB)-SPACES. We set

(1) $$\alpha_r = (a_{ij}^{(r)}),\ r = 1,2,\ldots,$$

$$a_{ij}^{(r)} = j^r,\ i = 1,2,\ldots,r,\ a_{ij}^{(r)} = i^r,\ i = r+1, r+2,\ldots, j=1,2,\ldots$$

Let λ be the echelon space defined by the system of steps (1). Given positive integers r and k, $k \geq r$, let s be an integer larger than k. Then

$$\lim_j (a_{sj}^{(r)};a_{sj}^{(k)}) = s^{r-k} \neq 0$$

and, by §2,3.(2), λ is not a Schwartz space.

Given a positive integer r and a sequence

(2) $$((i_m,j_m))$$

in $N\times N$ such that $(i_p,j_p) \neq (i_q,j_q)$ if $p,q \in N$, $p \neq q$, we have the following cases: 1) there is a subsequence of (2) $((h,h_m))$; 2) there is a subsequence of (2) $((h_m, k_m))$ with $h_1 < h_2 < \ldots$ If 1), for $r \geq h$, it follows that

$$\lim_m (a_{hh_m}^{(r)}; a_{hh_m}^{(r+1)}) = \lim_m \frac{1}{h_m} = 0$$

If 2),

$$\lim_m (a_{h_m k_m}^{(r)}; a_{h_m k_m}^{(r+1)}) = \lim_m \frac{1}{h_m} = 0$$

According to §2,2.(2), λ is a Montel space.

For every positive integer r, we set

$$M_r = \{(u_{ij}) : \lim_{i+j\to\infty} (u_{ij}; a_{ij}^{(r)}) = 0\}$$

M_r is a bounded absolutey convex subset of $\lambda^\times$ whose linear hull Λ_r, endowed with the topology derived from the gauge of M_r, is a Banach space isomorphic to c_0. The inductive limit of the sequence (Λ_r) of Banach space coincides obviously with the ultrabornological space ϕ^1 used in §1, 9. Since λ is a Montel space which is not Schwartz it is not totally reflexive (see §2,3.(15)). Since §2,3.(1) does not hold, λ is not quasinormable. We apply §2,3.(21)to obtain that ϕ^1 is not complete. Therefore we have obtained an (LB)-space which is not complete and whose completion is a Montel space.

We set

$$\beta_r = (b_{ij}^{(r)}),\ r = 1,2,\ldots,$$

with

$$b_{ij}^{(r)} = j,\ i = 1,2,\ldots, r,\ b_{ij}^{(r)} = 1,\ i = r+1, r+2,\ldots, j=1,2,\ldots$$

Let λ be the echelon space defined by this system of steps. This space is not distinguished (cf. KÖTHE [1], Chapter Six, §31, Section 7). Considering the subspace ϕ^1 of $\Lambda^\times$ we obtain the non-complete (LB)-space of KÖTHE (cf. KÖTHE [1], Chapter Six, §31, Section 6). Then the non-complete (LB)-space of KÖTHE has a non-barrelled strong bidual.

7. CLOSED SUBSPACES OF STRICT (LB)-SPACES. Let λ be the echelon space which has been described in the former section. Since λ is a Montel space which is not Schwartz, we apply §2,3.(14), to obtain a closed subspace F of λ such that λ/F is isomorphic to ℓ^1. Let G be the subspace of $\lambda^\times$ orthogonal

to F. If

$$A_r = \{(u_{ij}) : |u_{ij};a_{ij}^{(r)}| \leq 1,\ i,\ j = 1,2,\ldots\},\ r = 1,2,\ldots,$$

and if λ_r is the linear hull of A_r endowed with the topology derived from the gauge of A_r, we set $E = \overset{\infty}{\underset{r=1}{\oplus}} \lambda_r$. Every element of E can be written as $u = (u^1,u^2,\ldots,u^r,\ldots)$ with $u_r \in \lambda_r$, $r = 1,2,\ldots$, where there is a positive integer m, depending on u, such that $o = u^{m+1} = u^{m+2} \ldots$ We set $Tu = \sum_r u_r$. $T : E \longrightarrow \lambda^\times$ is linear and onto and its restriction to every λ_r is continuous. Thus, T is continuous on E. On the other hand, E and $\lambda^\times$ are (LB)-spaces and thus T is an homomorphism.

Let H be $T^{-1}(G)$ with the topology induced by E. Since λ/F is not reflexive there is a continuous linear form f on $G[\beta(G,\lambda/F)]$ which is not $\sigma(G,\lambda/F)$-continuous, i.e., f is not continuous on G with topology induced by $\sigma(\lambda^\times, \lambda)$. For every $x \in H$ we set $g(x) = f(T(x))$. It is obvious that g is a linear form on H. If A denotes a bounded set on H then T(A) is a bounded set of $\lambda^\times[\sigma(\lambda^\times,\lambda)]$ and also of $G[\sigma(G,\lambda/F)]$. Then f is bounded on T(A) and therefore g is bounded on A. It follows that g is continuous on every Banach space of the form $(\overset{s}{\underset{r=1}{\oplus}} \lambda_r) \cap H$, $s = 1,2,\ldots$, Thus, g is sequentially continuous on H.

Now we suppose that g is continuous on H. We find a continuous linear form k on E whose restriction to H coincides with g. Since g vanishes on $T^{-1}(0)$ there is a linear form h on $\lambda^\times$ with $k = h \circ T$ and, since T is an homomorphism, h is $\sigma(\lambda^\times,\lambda)$-continuous and its restriction to G, coinciding with f, is not $\sigma(G,\lambda/F)$-continuous. That is a contradiction.

Since λ_r is isomorphic to ℓ^∞, $r = 1,2,\ldots$, for the topological direct sum $(\ell^\infty)^{(N)}$ of a countable infinity of spaces equaling ℓ^∞ we obtain the following result:

(1) *There is in* $(\ell^\infty)^{(N)}$ *a closed subspace* L *and a linear form on* L *which is sequentially continuous but not continuous.*

Given the system of steps 6.(1) and a real number p, $1 < p$, we set

$$\Lambda = \{(x_{ij}) : \Sigma|a_{ij}^{(r)}x_{ij}|^p < \infty,\ r = 1,2,\ldots\}.$$

Then Λ is an echelon space of orden p. We suppose Λ and $\Lambda^\times$ provided with their strong topologies. According to §3,2.(1) and §3,2.(4), Λ is a Montel space and, by §3,3.(1), Λ is not a Schwartz space. Now we apply §3,3.(13), to obtain a closed subspace F of Λ such that Λ/F is isomorphic to ℓ^p. Let G be the subspace of $\Lambda^\times$ orthogonal to F. If $\frac{1}{p}+\frac{1}{q}=1$, we write

$$B_r = \{(u_{ij}) : \Sigma|u_{ij};a_{ij}^{(r)}|^q \leq 1\},\ r = 1,2,\ldots$$

Let Λ_r be the linear hull of B_r endowed with the topology derived from the gauge of B_r. Set $E = \bigoplus_{r=1}^{\infty} \Lambda_r$. Proceeding as we did before we obtain an homomorphis T from E onto $\Lambda^\times$.

The closed unit disk B of ℓ^p is weakly compact and non-compact. If S: $\Lambda \longrightarrow \Lambda/F$ denotes the canonical mapping, since Λ is a Montel space, it does not exist a bounded subset of Λ with image by S containing B. Therefore the topology of G is not its Mackey topology.

If $H = T^{-1}(G)$ is provided with the topology induced by E it is obvious that $H/T^{-1}(0)$ is isomorphic to G and, consequently, H has not the Mackey topology. It follows that H is not the inductive limit of the sequence of Banach space $(\bigoplus_{r=1}^{s} \Lambda_r) \cap H$, s = 1,2,...

Since Λ_r is isomorphic to ℓ^p, r = 1,2,..., the topological direct sum $(\ell^p)^{(N)}$ of a countable infinity of spaces equal to ℓ^p has the following properties:

(2) *There is a closed subspace in* $(\ell^p)^{(N)}$ *whose topology is not the Mackey topology.*

(3) *There is a closed subspace* L *in* $(\ell^p)^{(N)}$ *which is not an* (LB)*-space.*

$(\ell^p)^{(N)}$ can be identified with the strong dual of the Fréchet space $(\ell^q)^N$. If Z is a closed subspace of $(\ell^p)^{(N)}$ and if χ is the subspace of $(\ell^q)^N$ orthogonal to Z then $(\ell^q)^N/X$ is a separable Fréchet space and therefor every sequentially continuous linear form on Z is continuous and thus the subspace L of (3), endowed with its Mackey topology, is an (LB)-space.

(4) *There is a closed subspace* L *in* $(\ell^\infty)^{(N)}$ *whose topology is not the Mackey topology and* L *with its Mackey topology is an* (LB)*-space.*

Proof. Since every separable Banach space is isomorphic to a closed subspace of ℓ^∞ (cf. KÖTHE, Chapter Four, §22, Section 4) we have that $(\ell^p)^{(N)}$ is isomorphic to a closed subspace of $(\ell^\infty)^{(N)}$. The result is now immediate.

Given the system of steps 6.(1) we set

$$\lambda_0 = \{(x_{ij}) : \lim_{i+j\to\infty} a_{ij}^{(r)} x_{ij} = 0, \ r = 1,2,\ldots\}$$

Then λ_0 is an echelon space of order zero. We suppose λ_0 and its α-dual $\lambda_0^\times$ endowed with their strong topologies. According to §4,2,(3) and §4,2.(5) λ_0 is a Montel space and, by §4,3.(1), λ_0 is not a Schwartz space. Now we apply §4,3.(11) to obtain a closed subspace F of λ_0 such that λ_0/F is isomorphic to c_0. Let G be the subspace of $\lambda_0^\times$ orthogonal to F. We set

$$C_r = \{(u_{ij}) : \Sigma |u_{ij};a_{ij}^{(r)}| \leqslant 1\}, \ r = 1,2,\ldots$$

If μ_r denotes the linear hull of C_r endowed with the topology deduced from the gauge of C_r, we set $E = \bigoplus_{r=1}^{\infty} \mu_r$. We proceed as we did before to obtain an homomorphism T from E onto $\lambda_0^\times$. Since λ_0/F is not reflexive we repeat our arguments to conclude the existence of a sequentially continuous linear form g on $H = T^{-1}(G)$, endowed with the topology induced by E, which is not continuous.

On the other hand, the subset of c_0

$$\{(x_n) : \Sigma|x_n| \leqslant 1\}$$

is weakly compact and non-compact. Therefore, if $Y : \lambda_0 \longrightarrow \lambda_0/F$ is the canonical mapping there is a weakly compact subset of λ_0/F which is not contained in the image by Y of any bounded subset of λ_0. Consequently, the topology of G is not the Mackey topology. Since μ_r is isomorphic to ℓ^1, r=1, 2,..., the following result is now obvious.

(5) *There is a closed subspace* L *in* $(\ell^1)^{(N)}$ *satisfying:*

a) *There is a linear form on* L *which is sequentially continuous but not continuous.*
b) *The topology of* L *is not the Mackey topology.*

Results (1), (2) and (3) can be found in GROTHENDIECK [2]. Result (5) appears here for the first time.

§. 6. A CLASS OF SPACES OF VECTOR SEQUENCES

1. THE SPACES $J\{E\}$ AND $J[E]$. The linear spaces to be considered are defined on the field K. Let Y be a linear space. We denote by $\omega(Y)$ the family of all sequences of Y. If $x = (x_m)$, $y = (y_m) \in \omega(Y)$ and $h \in K$ we set $x + y = (x_m + y_m)$, $hx = (h x_m)$. We suppose $\omega(Y)$ endowed with the linear structure derived from the former algebraic operations. If r is a positive integer, we set

$$x_m\{r\} = x_m,\quad m = r,\ r+1,\ \ldots,\quad x_m\{r\} = 0,\ m = 1,2,\ldots,r-1$$

$$x_m[r] = x_m,\ m = 1,2,\ldots,r-1,\quad x_m[r] = 0,\quad m = r,r+1,\ \ldots$$

$$x\{r\} = (x_m\{r\}),\quad x[r] = (x_m[r]).$$

If z belongs to Y we write ze_r to denote the element (u_m) of $\omega(Y)$ such that $u_m = 0$, $m \neq r$, $u_r = z$. We represent by H the set of all finite sequences of positive integers which are strictly increasing and with an odd number of elements, i.e., an element of H is $(r_1,r_2,\ldots,r_{2m+1})$ with r_j a positive integer $j = 1,2,\ldots,2m + 1$, and $r_1 < r_2 < \ldots < r_{2m+1}$.

In this paragraph E denotes a separated locally convex space. We define the topology on E by the fundamental system of seminorms $\{p_i : i \in I\}$ If $x = (x_m) \in \omega(E)$ and $i \in I$ we write

$$q_i(x) = \sup\{(\sum_{j=1}^{m} p_i(x_{r_{2j-1}} - x_{r_{2j}})^2 + p_i(x_{r_{2m+1}})^2)^{\frac{1}{2}} : (r_1,r_2,\ldots,r_{2m+1}) \in H\}$$

and we set $J\{E\} = \{x \in \omega(E) : q_i(x) < \infty,\ i \in I\}$.

(1) $J\{E\}$ *is a is a linear subspace of* $\omega(E)$.

Proof. We take $x = (x_m)$, $y = (y_m)$, $h \in K$, $i \in I$. Then

$$q_i(x+y) = \sup\{(\sum_{j=1}^{m} p_i(x_{r_{2j-1}} + y_{r_{2j-1}} - y_{r_{2j}})^2 + p_i(x_{r_{2m+1}}$$

$$+ y_{r_{2m+1}})^2)^{\frac{1}{2}} : (r_1,r_2,\ldots,r_{2m+1}) \in H\} \leq \sup\{(\sum_{j=1}^{m} p_i(x_{r_{2j-1}} - x_{r_{2j}})^2$$

$$+ p_i(x_{r_{2m+1}})^2)^{\frac{1}{2}} : (r_1,r_2,\ldots,r_{2m+1}) \in H\}$$

$$+ \sup \{(\sum_{j=1}^{m} p_i(y_{r_{2j-1}}-y_{r_{2j}})^2+p_i(y_{r_{2m+1}})^2)^{\frac{1}{2}}:(r_1,r_2,\ldots,r_{2m+1})\in H\}$$

$$= q_i(x)+q_i(y) < \infty,$$

$$q_i(hx)=\sup \{(\sum_{j=1}^{m} p_i(hx_{r_{2j-1}}-hx_{r_{2j}})^2+p_i(hx_{r_{2m+1}})^2)^{\frac{1}{2}}$$

$$: (r_1,r_2,\ldots,r_{2n+1}) \in H\} = |h|\ q_i(x) < \infty$$

and thus x+y and hx belong to J {E} and the proof is complete.

According to the proof given in (1) we have

(2) $\{q_i : i \in I\}$ *is a family of seminorms on* J {E}.

In what follows we shall suppose J {E} endowed with the locally convex topology defined by the system of seminorms $\{q_i : i \in I\}$.

(3) J {E} *is a separated topological space.*

Proof. If $x = (x_m)$ is a non-zero vector of J {E} there is a positive integer r such that $x_r \neq 0$. Let i be an element of I with $p_i(x_r) \neq 0$. The finite sequence r belongs to H and therefore

$$q_i(x) \geq p_i(x) > 0.$$

The conclusion follows.

(4) *If* $x=(x_m)$ *belongs to* J {E} *the sequence* (x_m) *is a Cauchy sequence in* E.

Proof. Suppose $x = (x_m)$ is not a Cauchy sequence in E. We find $i \in I$, $\varepsilon > 0$ and a sequence

$$r_1 < r_2 < \ldots < r_m < \ldots$$

of positive integers with

$$p_i(x_{r_{2j-1}}-x_{r_{2j}}) > \varepsilon,\ j = 1,2,\ldots$$

Given any positive integer m we have that

$$q_i(x) \geq (\sum_{j=1}^{m} p_i(x_{r_{2j-1}}-x_{r_{2j}})^2+p_i(x_{r_{2m+1}})^2)^{\frac{1}{2}}$$

$$\geq (\sum_{j=1}^{m} p_i(x_{r_{2j}}-x_{r_{2j}})^2)^{\frac{1}{2}} > \sqrt{m}\ \varepsilon$$

and thus $q_i(x) = \infty$ which is a contradiction.

Let

(5) $\{x^s = (x_m^s) : s \in S, \geqslant\}$

be a Cauchy net in J {E} such that, for every positive integer m, the net

$\{x_m^s : s \in S, \geqslant\}$

converges to x_m in E.

(6) $x = (x_m)$ *is an element of* J {E} *and the net* (5) *converges to* x *in this space.*

Proof. Given $i \in I$ and $\varepsilon > 0$ we find an element $t \in S$ such that

$$q_i(x^s - x^r) < \varepsilon, \quad s, r \in S; \; s, r \geqslant t.$$

Given $(r_1, r_2, \ldots, r_{2m+1})$ in H we have, for $r, s \geqslant t$,

$$\sum_{j=1}^{m} p_i(x^s_{r_{2j-1}} - x^s_{r_{2j}} - x^r_{r_{2j-1}} + x^r_{r_{2j}})^2 + p_i(x^s_{r_{2m+1}} - x^r_{r_{2m+1}})^2 < \varepsilon^2,$$

from where it follows

$$(7) \qquad \sum_{j=1}^{m} p_i(x^s_{r_{2j-1}} - x^s_{r_{2j}} - x_{r_{2j-1}} + x_{r_{2j}})^2 + p_i(x^s_{r_{2m+1}} - x_{r_{2m+1}})^2 < \varepsilon^2.$$

Then

$$(8) \qquad \left(\sum_{j=1}^{m} p_i(x_{r_{2j-1}} - x_{r_{2j}})^2 + p_i(x_{r_{2m+1}})^2\right)^{\frac{1}{2}}$$

$$\leqslant \left(\sum_{j=1}^{m} p_i(x^t_{r_{2j-1}} - x^t_{r_{2j}} - x_{r_{2j-1}} + x_{r_{2j}})^2 + p_i(x^t_{r_{2m+1}} - x_{r_{2m+1}})^2\right)^{\frac{1}{2}}$$

$$+ \left(\sum_{j=1}^{m} p_i(x^t_{r_{2j-1}} - x^t_{r_{2j}})^2 + p_i(x^t_{r_{2m+1}})^2\right)^{\frac{1}{2}} < \varepsilon + q_i(x^t).$$

From (8) we have that

$$q_i(x) \leqslant \varepsilon + q_i(x^t) < \infty$$

and therefore x belongs to J {E}. From a consequence of (7) we deduce that

$$q_i(x^s - x) \leqslant \varepsilon, \quad s \geqslant t,$$

and thus the net (5) converges to x in J {E}.

(9) *If* E *is quasicomplete then* J{E} *is quasicomplete.*

Proof. Let

(10) $$\{(x_m^s) : s \in S, \geq\}$$

be a bounded Cauchy net in J {E}. For every positive integer m the net

(11) $$\{x_m^s : s \in S, \geq\}$$

is a bounded Cauchy net in E. Since E is quasicomplete, the net (11) converges to x_m in E. We apply (6) to obtain that $x = (x_m)$ belongs to J {E} and (10) converges in J {E} to x.

(12) *If* E *is complete then* J {E} *is complete.*

Proof. We proceed as we did in (9). No boundedness condition on the nets (10) and (11) are required

Analogously it follows that

(13) *If* E *is sequentially complete* J{E} *is sequentially complete.*

We denote by J [E] the subspace of J {E} of all those elements $x=(x_m)$ converging to the origin. Let L be the subspace of J {E} of all those sequences (x_m) of E with $x_1 = x_2 = \ldots = x_m = \ldots$

(14) *If* $x=(x_m)$ *belongs to* J[E] *we have that*

$$\lim_r x\{r\} = 0, \lim_r x[r] = x.$$

Proof. Since $x\{r\} + x[r] = x$ it is enough to show that $\lim_r x\{r\} = 0.$

Take $i \in I$. Given $\varepsilon > 0$ we find $(r_1, r_2, \ldots, r_{2m+1})$ in H with

$$\sum_{j=1}^{m} p_i(x_{r_{2j-1}} - x_{r_{2j}})^2 + p_i(x_{r_{2m+1}})^2 > q_i(x)^2 - \frac{\varepsilon^2}{2}.$$

Since (x_r) converges to zero in E, we have that

$$\lim_r \sum_{j=1}^{m} p_i(x_{r_{2j-1}} - x_{r_{2j}})^2 + p_i(x_{r_{2m+1}} - x_r)^2$$

$$= \sum_{j=1}^{m} p_i(x_{r_{2j-1}}-x_{r_{2j}})^2+p_i(x_{r_{2m+1}})^2$$

and therefore there is $s > r_{2m+1}$ such that, if $r \geqslant s$,

$$\sum_{j=1}^{m} p_i\ (x_{r_{2j-1}}-x_{r_{2j}})^2+p_i(x_{r_{2m+1}}-x_r)^2$$

$$> q_i(x)^2 - \frac{\varepsilon^2}{2},\ p_i(x_r)^2 < \frac{\varepsilon^2}{2}.$$

We shall see that $q_i(x\ \{r\}) < \varepsilon$ for $r > s$. If it is not true we take $r > s$ and $(s_1,s_2,\ldots,s_{2p+1})$ in H with

$$\sum_{j=1}^{p} p_i(x_{s_{2j-1}}\{r\} - x_{s_{2j}}\{r\})^2+p_i(x_{s_{2p+1}}\{r\}\ )^2 \geqslant \varepsilon^2.$$

Since $x_m\ \{r\} = 0$ for $m < r$, there is a first element q in $\{1,2,\ldots,2p+1\}$ for which $s_q \geqslant r$. If q is odd and equal to 2h-1 then, since

$$(r_1,r_2,\ldots,r_{2m+1},s,s_{2h-1},s_{2h},\ldots,s_{2p+1})$$

belongs to H, we have that

$$q_i(x)^2 \geqslant \sum_{j=1}^{m} p_i(x_{r_{2j-1}}-x_{r_{2j}})^2+p_i(x_{r_{2m+1}}-x_s)^2$$

$$+ \sum_{j=h}^{p} p_i(x_{s_{2j-1}}-x_{s_{2j}})^2+p_i(x_{s_{2p+1}})^2$$

$$> q_i(x)^2 - \frac{\varepsilon^2}{2} + \sum_{j=1}^{p} p_i(x_{s_{2j-1}}\ \{r\} - x_{s_{2j}}\ \{r\}\ )^2+p_i(x_{s_{2p+1}}\ \{r\}\ )^2$$

$$\geqslant q_i(x)^2 - \frac{\varepsilon^2}{2} + \varepsilon^2 > q_i(x)^2.$$

This is a contradiction.

If q is even and equal to 2h then, since

$$(r_1,r_2,\ldots,r_{2m+1},s_{2h},s_{2h+1},\ldots,s_{2p+1})$$

belongs to H, we have that

$$q_i(x)^2 > \sum_{j=1}^{m} p_i(x_{r_{2j-1}}-x_{r_{2j}})^2+p_i(x_{r_{2m+1}}-x_{s_{2h}})^2$$

$$+ \sum_{j=h+1}^{p} p_i(x_{s_{2j-1}}-x_{s_{2j}})^2 + p_i(x_{s_{2p+1}})^2$$

$$> q_i(x)^2 - \frac{\varepsilon^2}{2} - p_i(x_{s_{2h}})^2 + \sum_{j=1}^{p} p_i(x_{s_{2j-1}}\{r\}-x_{s_{2j}}\{r\})^2$$

$$+ p_i(x_{s_{2p+1}}\{r\})^2 \geq q_i(x)^2 - \frac{\varepsilon^2}{2} - \frac{\varepsilon^2}{2} + \varepsilon^2 = q_i(x)^2$$

which is a contradiction. The proof is complete.

(15) *The space* L *is isomorphic to* E.

Proof. It F is the mapping from E into L such that

$$Tx = (x,x,\ldots,x,\ldots),\ x \in E,$$

it is obvious that T is linear and bijective. On the other hand, if i belongs I, $x_m = x$, $m = 1,2,\ldots$, $x \in E$ and $(r_1,r_2,\ldots,r_{2m+1})$ belongs to H, we have that

$$\sum_{j=1}^{m} p_i(x_{r_{2j-1}}-x_{r_{2j}})^2+p_i(x_{r_{2m+1}})^2 = p_i(x)^2$$

and therefore $p_i(x) = q_i\ (Tx)$ and we are done.

(16) *If* E *is sequentially complete then* J{E} *is the topological direct sum of* J{E} *and* L.

Proof. If (x_m) belongs to J [E] ∩ L we have that $x_1=x_2=\ldots=x_m=\ldots$ and $\lim_m x_m = 0$ in E and thus $x_m = 0$, $m = 1,2,\ldots$ On the other hand, let $z = (z_m)$ be an element of J {E}. By (4), (z_m) converges to z_0 in E. The sequence (z_m-z_0) is an element of J [E] and $(z_0,z_0,\ldots)$ belongs to L and therefore J [E]+ L = J{E}. Finally, let T be the projection from J{E} onto L along J[E]. Take any i in I. From

$$p_i(z_m) \leq q_i(z),\ m = 1,2,\ldots,$$

we obtain that

$$\lim_m p_i(z_m) = p_i(z_0) = p_i(T(z)) \leq q_i(z),$$

and thus T is continuous. The conclusion follows.

(17) *If* E *is sequentially complete then* J{E} *is isomorphic to* J [E].

Proof. If $x = (x_m)$ is in $J\{E\}$ let x_o be the limit in E of the sequence (x_m). Let $U: J\{E\} \longrightarrow J[E]$ be linear injective mapping defined by

$$(x_m) \longrightarrow (x_o, x_1 - x_o, x_2 - x_o, \ldots, x_m - x_o, \ldots).$$

If $z = (z_m) \in J[E]$ we have that $v = (z_2 + z_1, \ldots, z_{m+1} + z_1, \ldots)$ belongs to $J\{E\}$ and $U(v) = z$. Thus U is onto. We set $U(x) = u = (u_m)$. Let i be an element of I. If $r_1 > 1$ we have that

$$\left(\sum_{j=1}^{m} p_i(u_{r_{2j-1}} - u_{r_{2j}})^2 + p_i(u_{r_{2m+1}})^2\right)^{\frac{1}{2}}$$

$$< \left(\sum_{j=1}^{m} p_i(u_{r_{2j-1}} - u_{r_{2j}})^2\right)^{\frac{1}{2}} + p_i(u_{r_{2m+1}})$$

$$\leq q_i(x) + p_i(u_{r_{2m+1}} + x_o) + p_i(x_o) \leq 2q_i(x) + p_i(x_o).$$

If $r_1 = 1$ it follows that

$$\left(\sum_{j=1}^{m} p_i(u_{r_{2j-1}} - u_{r_{2j}})^2 + p_i(u_{r_{2m+1}})^2\right)^{\frac{1}{2}}$$

$$\leq p_i(u_1 - u_{r_2}) + \left(\sum_{j=2}^{m} p_i(u_{r_{2j-1}} - u_{r_{2j}})^2\right)^{\frac{1}{2}} + p_i(u_{r_{2m+1}})$$

$$\leq p_i(x_o) + p_i(u_{r_2}) + q_i(x) + p_i(u_{r_{2m+1}}) \leq 3\, q_i(x) + 3p_i(x_o).$$

In any case,

$$q_i(u) \leq 3q_i(x) + 3p_i(x_o)$$

and, according to proof of (16), it follows $p_i(x_o) \leq q_i(x)$ and therefore

$$q_i(U(x)) \leq 6\, q_i(x).$$

Thus U is continuous.

On the other hand,

$$\left(\sum_{j=1}^{m} p_i(x_{r_{2j}} - x_{r_{2j}})^2 + p_i(x_{r_{2m+1}})^2\right)^{\frac{1}{2}}$$

$$\leq \Big(\sum_{j=1}^{m} p_i(x_{r_{2j-1}}-x_{r_{2j}})^2\Big)^{\frac{1}{2}} + p_i(x_{r_{2m+1}}-x_o)+p_i(x_o)$$

$$\leq 2q_i(u)+p_i(u_1) \leq 3q_i(u).$$

Therefore $q_i(x) \leq 3q_i(u)$ from where it follows the continuity of U^{-1}. Now the proof is complete.

We shall give in (18) and (20) some properties of certain subsets of $J\{E\}$ which we shall use later.

(18) *If* $x=(x_m)$ *belongs to* $J\{E\}$ *then*

$$\{x\,[r] : r = 1,2,\ldots\}$$

is a bounded subset of $J\,[E]$.

Proof. Given a positive integer r, $i \in I$ and $(r_1,r_2,\ldots,r_{2m+1}) \in H$ we shall show that $q_i(x)$ is an upper bound of

$$(19) \qquad \Big(\sum_{j=1}^{m} p_i(x_{r_{2j-1}}[r] - x_{r_{2j}}[r])^2+p_i(x_{r_{2m+1}}[r])^2\Big)^{\frac{1}{2}}$$

If $r_{2m+1}< r$, (19) coincides with

$$\Big(\sum_{j=1}^{m} p_i(x_{r_{2j-1}}-x_{r_{2j}})^2+p_i(x_{r_{2m+1}})^2\Big)^{\frac{1}{2}}$$

and therefore (19) is less or equal than $q_i(x)$. If $r \leq r_1$, (19) is zero. Now suppose $r_1 < r \leq r_{2m+1}$. If $r-1$ is odd and equals $2q+1$ then (19) coincides with

$$\Big(\sum_{j=1}^{q} p_i(x_{r_{2j-1}}-x_{r_{2j}})^2+p_i(x_{r_{2q+1}})^2\Big)^{\frac{1}{2}}$$

which is less or equal than $q_i(x)$. If $r-1$ is even and equals $2q$ we have that (19) coincides with

$$\Big(\sum_{j=1}^{q} p_i(x_{r_{2j-1}}-x_{r_{2j}})^2\Big)^{\frac{1}{2}}$$

which is lessor equal than $q_i(x)$. The proof is complete.

For our next proposition we take a bounded subset of $J\{E\}$:

$$A = \{x^s = (x_m^s) : s \in S\}.$$

Given a sequence of integers

$$1 = n_1 < n_2 < \dots < n_r < \dots$$

let $y = (y_m)$ be an element of $\omega(E)$ such that

$$y_m = \frac{1}{r} x_m^{s(r)}, \quad m = n_r, n_r + 1, \dots, n_{r+1}-1,$$

$$s(r) \in S, \; r = 1,2,\dots$$

(20) *The sequence* (y_m) *belongs to* $J[E]$.

Proof. Given $i \in I$ there is h positive with

$$q_i(x^s) \leq h, \; s \in S$$

Obviously

(21) $$p_i(y_m) = \frac{1}{r} p_i(x_m^{s(r)}) \leq \frac{1}{r} q_i(x^{s(r)}) \leq \frac{h}{r},$$

$$m = n_r, n_{r+1}, \dots, n_{r+1}-1.$$

If $n_r \leq i_1 \leq i_2 < \dots < i_{2q-1} < i_{2q} < n_{r+1}$ we have that

$$\sum_{j=1}^{q} p_i(y_{i_{2j-1}} - y_{i_{2j}})^2 = \frac{1}{r^2} \sum_{j=1}^{q} p_i(x_{i_{2j-1}}^{s(r)} - x_{i_{2j}}^{s(r)})^2$$

$$\leq \frac{1}{r^2} q_i(x^{s(r)})^2 \leq \frac{h^2}{r^2}.$$

If $n_r \leq i_1 < n_{r+1} \leq n_{r+p} \leq i_2 < n_{r+p+1}$ it follows that

$$p_i(y_{i_1} - y_{i_2}) \leq p_i(y_{i_1}) + p_i(y_{i_2})$$

$$= \frac{1}{r} p_i(x_{i_1}^{s(r)}) + \frac{1}{r+p} p_i(x_{i_2}^{s(r+p)})$$

$$\leq \frac{1}{r} q_i(x^{s(r)}) + \frac{1}{r+p} q_i(x^{s(r+p)}) \leq \left(\frac{1}{r} + \frac{1}{r+p}\right)h \leq \frac{2h}{r}.$$

Then

$$q_i(y) = \sup\left\{ \left(\sum_{j=1}^{m} p_i(y_{r_{2j-1}} - y_{r_{2j}})^2 + p_i(y_{r_{2m+1}})^2 \right)^{\frac{1}{2}} \right.$$

$$\left. :(r_1, r_2, \dots, r_{2m+1}) \in H \right\} < \left(\sum_{r=1}^{\infty} \frac{5h^2}{r^2} + \sum_{r=1}^{\infty} \frac{h^2}{r^2} \right)^{\frac{1}{2}} < \infty$$

Finally, from (21) we deduce that (y_m) converges to the origin in E.

2. TOPOLOGICAL DUAL AND BIDUAL OF J[E]. We denote by J'[E] the topological dual of J[E] endowed with the strong topology. J"[E] is the strong bidual of J[E]. J[E] is identified with a vectorial subspace of J"[E] in the usual way. If f is a linear form on J[E] we write

$$f_m(x) = f(xe_m),\ x \in E,\ m = 1,2,\ldots$$

From every positive integer r, we set

$$f_m\{r\} = f_m,\ m = r, r+1,\ldots,\ f_m\{r\} = 0,\ m = 1,2,\ldots, r-1,$$

$$f_m[r] = f_m,\ m = 1,2,\ldots, r-1,\ f_m[r] = 0,\ m = r, r+1,\ldots$$

$$f\{r\} = (f_m\{r\}) \text{ and } f[r] = (f_m[r]).$$

Let $A = \{x^s = (x_m^s) : s \in S\}$ be a bounded absolutely convex subset of J[E].

(1) *If* f *belongs to* J'[E], *given* $\varepsilon > 0$ *there is a positive integer* q *with*

$$|\langle x^s\{r\}, f\rangle| < \varepsilon,\ s \in S,\ r \geq q.$$

Proof. Suppose the conclusion is not true. There is $\delta > 0$ such that, given any positive integer p, an integer $r > p$ and $s \in S$ can be determined with

$$|\langle x^s\{r\}, f\rangle|\ \delta..$$

Take $n_1 = 1$ and s_1 any element of S. Select an integer $n_2 > 1$ and $r_2 \in S$ such that

$$|\langle x^{r_2}\{n_2\}, f\rangle| > \delta.$$

Proceeding by recurrence we suppose

$$1 = n_1 < n_2 < \ldots < n_p < n_{p+1},$$

and elements of S, $s_1, s_2, \ldots, s_p, r_{p+1}$ obtained such that

$$|\langle x^{r_{p+1}}\{n_{p+1}\}, f\rangle| > \delta.$$

According to 1.(14), the sequence $(x^{r_{p+1}}\{r\})$ converges to the origin in J[E] and therefore there is an integer $m_{p+2} > n_{p+1}$ with

$$|<x^{r_{p+1}}\{m\}, f>| < \frac{\delta}{4}, \; m \geq m_{p+2}.$$

For $m \geq m_{p+2}$ $|<x^{r_{p+1}}\{n_{p+1}\} - x^{r_{p+1}}\{m\}, f>| \geq \delta - \frac{\delta}{4} > \frac{\delta}{2}$.

Determine an integer $n_{p+2} > m_{p+2}$ and $r_{p+2} \in S$ with

$$|<x^{r_{p+2}}\{n_{p+2}\}, f>| > \delta$$

In order to find s_{p+1} let $k \in K$ be a number of modulus one with

$$k <x^{r_{p+1}}\{n_{p+1}\} - x^{r_{p+1}}\{n_{p+2}\}, f>$$

real and positive. Since A is absolutely convex there is $s_{p+1} \in S$ with $x^{s_{p+1}} = k\, x^{r_{p+1}}$. Then

$$<x^{s_{p+1}}\{n_{p+1}\} - x^{s_{p+1}}\{n_{p+2}\}, f>$$

is real and larger than $\frac{\delta}{2}$. Once we have determined (n_p) and (s_p) we set

$$y_m = \frac{1}{r}\, x_m^{s_r}, \; m = n_r, n_r+1, \ldots, n_{r+1}-1, \; r = 1,2,\ldots$$

By 1.(20), $y = (y_m)$ belongs to $J[E]$ and, by 1.(14),

(2) $\{y[r] : r = 1,2,\ldots\}$

is a bounded subset of $J[E]$. On the other hand

$$|<y[n_{r+1}], f>| = |\sum_{j=1}^{r} <\frac{1}{j}\, x^{s_j}\{n_j\} - \frac{1}{j}\, x^{s_j}\{n_{j+1}\}, f>|$$

$$\geq |\sum_{j=2}^{r} <\frac{1}{j}\, x^{s_j}\{n_j\} - \frac{1}{j}\, x^{s_j}\{n_{j+1}\}, f>| - |<x^{s_1}\{n_1\} - x^{s_1}\{n_2\} f>|$$

$$\geq \sum_{j=2}^{r} \frac{\delta}{2j} - |<x^{s_1}\{n_1\} - x^{s_1}\{n_2\}, f>|$$

which is in contradiction with the boundedness condition on (2).

(3) *If* f *belongs to* $J'[E]$ *and* $x = (x_m)$ *belongs to* $J[E]$ *we have that*

$$<x, f> = \sum <x_m, f_m>$$

Proof. By 1,(14), the sequence $(x\{r\})$ converges to the origin in $J[E]$. Therefore

$$\lim_{r} |\langle x,f\rangle - \sum_{m=1}^{r-1} \langle x_m, f_m\rangle |$$

$$= \lim_{r} |\langle x,f\rangle - \sum_{m=1}^{r-1} \langle e_m x_m, f_m\rangle |$$

$$= \lim_{r} |\langle x - \sum_{m=1}^{r-1} e_m x_m, f\rangle| = \lim_{r} |\langle x\{r\}, f\rangle| = 0$$

and thus

$$\langle x,f\rangle = \Sigma\langle x_m, f_m\rangle.$$

The former result allow us to state that every continuous linear form f on J[E] is determined by the sequence (f_m).

(4) *If* f *belongs to* J'[E] *and* $x = (x_m)$ *belongs to* J[E] *the series*

(5) $\Sigma\langle x_m, f_m\rangle$

is convergent.

Proof. By 1.(18),

$$\{x[r] : r = 1,2,\ldots\}$$

is a bounded subset of J[E]. Given $\varepsilon > 0$ we apply (1) to obtain a positive integer q such that, for every integer $r \geq q$ and every positive integer p,

$$|\langle x[r+p+1] - x[r+1], f\rangle| = |\sum_{m=r+1}^{r+p} \langle x_m, f_m\rangle| < \varepsilon.$$

and the series (5) is convergent.

Let M be the algebraic dual of J'[E]. Let $\chi : J[E] \longrightarrow M$ be the mapping defined by

$$\langle \chi(x), f\rangle = \Sigma\langle x_m, f\rangle, \quad x = (x_m) \in J\{E\},\ f \in J'[E].$$

(6) χ *is an injective linear mapping.*

Proof. If $x = (x_m)$, $y = (y_m) \in J\{E\}$, $f \in J'[E]$ and $h \in K$, we have that

$$\langle\chi(x+y), f\rangle = \Sigma\langle x_m + y_m, f_m\rangle = \Sigma\langle x_m, f_m\rangle + \Sigma\langle y_m, f_m\rangle = \langle\chi(x), f\rangle + \langle\chi(y), f\rangle$$

$$= \langle\chi(x) + \Upsilon(y), f\rangle.$$

$$<\chi(hx),f> = \Sigma<hx_m,f_m> = h\Sigma<x_m,f_m> = h<\chi(x),f> = <h\,\chi(x),f>$$

and therefore χ is linear.

If $x = (x_m) \neq 0$ is an element of J{E} we find a positive integer r with $x_r \neq 0$ and therefore there is an element $g_r \in E'$ with $<x_r,g_r> \neq 0$. For every $z = (z_m)$ belongs to J[E] we set

$$<z,k> = <z_r,g_r>.$$

Clearly k is an element of J'[E] with $k_m = 0$, $m \neq r$, $k_r = g_r$. Also

$$<\chi(x),k> = \Sigma<x_m,k_m> = <x_r,g_r> \neq 0$$

and therefore χ es injective.

According to (3), the restriction of χ to J[E] is the canonical injection of J[E] into M. In what follows, we shall identify J{E} with the subspace χ(J{E}) of M.

(7) J{E} *is contained in* J"[E].

Proof. If $x = (x_m)$ belongs to J{E} and f to J'[E] it follows

$$<x,f> = \Sigma<x_m,f_m> = \lim_r \sum_{m=1}^{r-1} <x_m,f_m> = \lim <x[r], f>$$

and then x is the limit in $M[\sigma(M, J'[E])]$ of the sequence (x[r]) of J[E]; thus x belongs to J"[E].

If f is an element of J'[E] and if r is a positive integer, f[r] and f{r} are the linear forms on J[E] satisfying

$$<x,f[r]> = \sum_{m=1}^{r-1} <x_m,f_m>, \quad <x,f\{r\}> = \sum_{m=r}^{\infty} <x_m,f_m>$$

if $x = (x_m)$ belongs to J[E]. It is obvious that f[r] is in J'[E] and, since f coincides with f[r] + f{r}, f {r} also belongs to J'[E].

(8) *If* f *is an element of* J'[E] *then* (f[r]) *is a sequence in* J'[E] *converging to* f.

Proof. If B is a bounded set of J[E], given $\varepsilon > 0$ we apply (1) to obtain a positive integer q with

$$|<x\{r\}, f>| < \varepsilon, \quad x \in B, \ r \geq q.$$

For every $x = (x_m) \in B$ we have that, if $r \geq q$,

$$|\langle x, f - f[r]\rangle| = |\langle x, f\{r\}\rangle|$$

$$= |\sum_{m=r}^{\infty} \langle x_m, f_m\rangle| = |\langle x\{r\}, f\rangle| < \varepsilon.$$

Therefore $(f[r])$ converges to f in $J'[E]$.

Let g be any element of E'. Given a positive integer r we set g^r to denote the linear form on $J[E]$ with

$$\langle x, g^r\rangle = \langle x_r, g\rangle, \; x = (x_m) \in J[E].$$

Clearly g^r belongs to $J'[E]$ and

$$g_r^{\,r} = g, \; g_m^{\,r} = 0, \; m \neq r.$$

If w belongs to $J''[E]$, we write w_r to denote the linear form on E' defined by

$$\langle w_r, g\rangle = \langle w, g^r\rangle \;, \; g \in E'.$$

(9) w_r *belongs to the bidual* E'' *of* E.

Proof. Since w is continuous on $J'[E]$ there is a bounded set B in $J[E]$ such that, if $B°$ denotes the polar set of B in $J'[E]$,

$$|\langle w, f\rangle| \leq 1, \text{ for every } f \in B°.$$

The set

$$B_r = \{x_r \in E: (x_m) \in B\}$$

is a bounded set of E. Let $B°_r$ be the polar set in E' of B_r. If g belongs to $B°_r$ and $x = (x_m)$ belongs to B we have that x_r belongs to B_r and therefore

$$|\langle x, g^r\rangle| = |\langle x_r, g\rangle| \leq 1$$

and thus g^r belongs to $B°$. Accordingly, for every $g \in B°_r$,

$$|\langle w_r, g\rangle| = |\langle w, g^r\rangle| \leq 1$$

and consequently $w_r \in E''$.

(10) *If* f *belongs to* J'[E] then

(11) $$\langle w, f\rangle = \Sigma\langle w_m, f_m\rangle.$$

Proof. According to (8), (f[r]) is a sequence converging in J'[E] to f. Thus,

$$\langle w,f\rangle = \lim_r \langle w, f[r]\rangle = \lim_r \sum_{m=1}^{r-1} \langle w_m, f_m\rangle = \Sigma\langle w_m, f_m\rangle.$$

(12) *If* E *is semireflexive, the sequence* $u = (w_r)$ *is an element of* J{E}.

Proof. If E is semireflexive, then w_r belongs to E, r = 1,2,... and therefore (u [r]) is a sequence contained in J[E]. On the other hand, if f belongs to J'[E] we have that

$$\langle w,f\rangle = \lim_r \sum_{m=1}^{r-1} \langle w_m, f_m\rangle = \lim_r \langle u[r], f\rangle,$$

according to (11); then (u[r]) converges to w in J"[E][σ(J"[E], J'[E]) . Consequently, (u[r]) is a bounded sequence in J[E].

Given i ∈ I, we find h > 0 with

$$q_i(u[r]) < h, \ r = 1,2,\ldots$$

Then, if $(r_1, r_2, \ldots, r_{2m+1})$ belongs to H we take an integer $s > r_{2m+1}$ and we have

$$\left(\sum_{j=1}^{m} p_i(w_{r_{2j-1}} - w_{r_{2j}})^2 + p_i(w_{r_{2m+1}})^2\right)^{\frac{1}{2}} \leq q_i(x[s]) < h$$

and thus (w_r) belongs to J {E} .

(13) *If* E *is semireflexive then* J{E} *coincides with the linear space* J"[E].

Proof. According to (7) it is enough to show that J"[E] is contained in J{E}. Take w ∈ J"[E] and determine the sequence $u = (w_r)$ which, according to (12), is an element of J{E}. On the other hand, if f belongs to J'[E]

$$\langle u,f\rangle = \Sigma\langle w_r, f_r\rangle = \langle w,f\rangle$$

and therefore (w_r) can be identified with w, i.e., w belongs to J{E}.

(14) *If* f *belongs to* J'[E] *and if* E *is barrelled then the series* Σf_m *is convergent in* E'[σ (E',E)].

Proof. If x belongs E, the sequence $(x,x,\ldots,x,\ldots)$ is an element of J{E} and therefore $\Sigma\langle x,f_m\rangle$ is convergent; then the sequence $(\sum_{m=1}^{r} f_m)$ is Cauchy in $E'[\sigma(E',E)]$. Since E is barrelled, this sequence converges in $E'[\sigma(E',E)]$ to an element Σf_m. The proof is complete.

If E is a Fréchet space, we can take I as a countable set and, therefore, J{E} is a metrizable space. We apply 1.(12) to obtain that J{E} is a Fréchet space. If E is a Banach space, J{E} is a Banach space.

(15) *If* E *a reflexive Fréchet space the space* J"[E] *coincides with* J{E}.

Proof. By (13), the linear spaces J"[E] and J{E} coincide. Let $x^r = (x_m^r)$, $r = 1,2,\ldots$, be a sequence in J{E} converging to the origin. Let y^r be the projection of x^r on J[E] along L and let $z^r = (z_m^r)$, $z_1^r = z_2^r = \ldots = z_m^r, \ldots$, be the projection of x^r on L along J[E], $r = 1,2,\ldots$ According to 1.(16) the sequence (y^r) converges in J[E] to the origin and (z^r) converges in L to the origin, thus (z_1^r) converges in E to the origin.

If f belongs to J'[E] we apply (14) to obtain

$$\lim \langle z_1^r, \Sigma f_m\rangle = 0$$

On the other hand,

$$\langle x^r,f\rangle = \langle y^r,f\rangle + \langle z^r,f\rangle = \langle y^r,f\rangle + \Sigma\langle z_m,f_m\rangle$$

$$= \langle y^r,f\rangle + \langle z_1^r,\Sigma f_m\rangle$$

and therefore

$$\lim \langle x^r,f\rangle = \lim\langle y^r,f\rangle + \lim\langle z_1^r,\Sigma f_m\rangle = 0$$

which implies that the topology of J{E} is finer than the topology σ(J{E}, J'[E]).

If A is a bounded subset of J'[E] and if A° denotes the polar set of A in J{E} we have that A° is absorbing and σ(J{E}, J'[E])-closed in J{E}. Consequently, the topology of J{E} is finer than the topology of J"[E]. Finally, J"[E] is the strong bidual of a Fréchet space and therefore J"[E] is also a Fréchet space. The closed graph theorem implies the coincidence of J{E} and J"[E].

(16) *If* E *is a reflexive Fréchet space there is a Fréchet space* F *satisfying*

1) F *is isomorphic to its strong bidual* F",

2) *if we identify* F *with a subspace of* F" *through the canonical injection, then* F *has a topological complement in* F" *isomorphic to* E

Proof. We take F = J[E] and apply (15) to obtain that J{E} can be identified with the strong bidual F" of F. Accordingly 1.(17), F is isomorphic to F" and by 1.(15) and 1.(16), F has a topological complement in F" isomorphic to E.

If E coincides with the field of the real numbers, then F is the quasi-reflexive space due to R.C. JAMES [1]. The original paper [1] suggested us the introduction of the spaces of vector sequences studied in this paragraph. These results have been taken from our work "Espacios de sucesiones" supported by the Fundación Aguilar.

Banach spaces related with R.C. JAMES's example can be found in R.C. JAMES [2], W.J. DAVIS, T. FIGIEL, W.B. JOHNSON and A. PELCZYNSKI [1] and J. LINDENSTRAUSS [1].

In PIETSCH [3] other classes of vector-valued sequences spaces can be found as well as results about scalar sequence spaces which appear in paragraph 1.

CHAPTER THREE
SPACES OF CONTINUOUS FUNCTIONS

In the first paragraph we represent diverse spaces of infinitely differentiable functions. The second paragraph is devoted to the study of spaces of functions of class C^m, $1 \leq m < \infty$. *In the last paragraph we represent spaces of* K-*valued continuous functions defined on metric spaces including the theorem of Milutin.*

§ 1. SPACES OF INFINITELY DIFFERENTIABLE FUNCTIONS AND SPACES OF DISTRIBUTIONS

1. THE SPACE s. Let s be the echelon space defined by system of steps

$$(m^r), \; r = 1, 2, \ldots$$

We denote by s' the topological dual s endowed with the strong topology.

(1). s *is a nuclear Fréchet space.*

Proof. Since S is an echelon space it is a Fréchet space. On the other hand, given a positive integer r we have that

$$\Sigma \frac{m^r}{m^{r+2}} = \Sigma \frac{1}{m^2} < \infty$$

and according to Chapter Two, § 2, 3. (16), s is nuclear.

A sequence (a_m) of elements of K is rapidly decreasing if, for every positive integer r, $\lim m^r a_m = 0$.

(2) *The sequence* (a_m) *of* K *is rapidly decreasing if and only if, for every positive integer* r,

(3) $\Sigma \, m^r \, |a_m| < \infty$.

Proof. Let (a_m) be a sequence in K verifying (3).For every positive integer r it is obvious that $\lim m^r a_m = 0$ and thus (a_m) is rapidly decreasing. On the other hand, if (a_m) is rapidly decreasing, given a positive integer r the sequence $(m^{r+2} a_m)$ converges to zero and accordingly there is $h > 0$ with

$$m^{r+2} |a_m| \leq h, \ m = 1,2,\ldots$$

from where it follows

$$\Sigma \, m^r \, |a_m| \leq \Sigma \, \frac{h}{m^2} < \infty$$

which completes the proof.

By (2), s is formed by all rapidly decreasing sequences of K. Given any sequence (a_m) of K we set

$$|(a_m)|_r = \Sigma \, m^r \, |a_m|$$

for every positive integer r. Then $|\cdot|_r$, $r = 1,2,\ldots$ is a system of norms describing the topology of s.

A sequence (b_m) of K is slowly increasing if there is a positive integer r, depending on the sequence, such that

$$\sup \left\{ \frac{|b_m|}{m^r} : m = 1,2,\ldots \right\} < \infty.$$

It is obvious that a sequence of K is slowly increasing if and only if it belongs to s'.

(4) *The topological product* K x s *is isomorphic to* s.

Proof. If b belongs to K and (a_m) belongs to s we set

$$T\,(b,(a_m)) = (b_m)$$

with $b_1 = b$, $b_{m+1} = a_m$, $m = 1,2,\ldots$ We write $c_1 = 0$, $c_{m+1} = a_m$, $m = 1,2,\ldots$ Given a positive integer r, we have that

$$|((c_m)|_r = \Sigma \, m^r \, |c_m| = \Sigma \, (m+1)^r \, |a_m|$$

$$\leq \Sigma \, (2m)^r \, |a_m| = 2^r \, |(a_m)|_r$$

and so (c_m) belongs to s. If e_m denotes the sequence whose m-th therm is one and vanishes elsewhere, $m = 1,2,\ldots$, we have that e_1 belongs to s and therefore $(b_m) = be_1 + (c_m)$ is in s then $T : K\times s \to s$ is a linear injective mapping. On the other hand,

$$|(b_m)|_r \leq |b| \cdot |e_1|_r + |(c_m)|_r \leq |b| + 2^r |(a_m)|_r$$

and T is continuous

If (d_m) is any element of s we set

$$Z((d_m)) = (d_1, (p_m))$$

with $p_m = d_{m+1}$, m = 1,2,... Then $Z: s \to K \times$ is a linear injective mapping. On the other hand,

$$|d_1| + |(p_m)|_r \leq |d_1| + |(d_m)|_r \leq 2|(d_m)|_r$$

and Z is continuous. Finally,

$$Z\circ T(b,(a_m)) = Z((b_m)) = (b, (a_m))$$

and thus Z is the inverse mapping of T and consequently $T : K\times s \to s$ is an isomorphism.

(5) *The topological product* $s\times s$ *is isomorphic to* s.

Proof. If (a_m) and (b_m) are in s we set

$$T((a_m),(b_m)) = ((c_m))$$

with $c_{2m-1} = a_m$, $c_{2m} = b_m$, m = 1,2,... Writing

$$p_{2m-1} = a_m, p_{2m} = 0, \; q_{2m-1} = 0, \; q_{2m} = b_m, \; m = 1,2,..$$

we have that

$$|(p_m)|_r = \Sigma m^r |p_m| = \Sigma(2m-1)^r |a_m| \leq 2^r |(a_m)|_r,$$

$$|(q_m)|_r = \Sigma m^r |q_m| = \Sigma (2m)^r |b_m| = 2^r |(b_m)|_r$$

for every positive integer r; then (p_m) and (q_m) are in s and therefore $(c_m) = (p_m) + (q_m)$ is in s. On the other hand,

$$|(c_m)|_r \leq |(p_m)|_r + |(q_m)|_r \leq 2^r |(a_m)|_r + 2^r |(b_m)|_r$$

and T is continuous.

If (d_m) denotes any element of s we set

$$Z((d_m)) = ((d_{2m-1}), (d_{2m})).$$

Then

$$|(d_{2m-1})|_r = \Sigma m^r |d_{2m-1}| \leq |(d_m)|_r,$$

$$|(d_{2m})|_r = \Sigma m^r |d_{2m}| \leq |(d_m)|_r,$$

and therefore $((d_{2m-1}), (d_{2m}))$ belongs to $s \times s$. Thus $Z : s \to s \times s$ is a linear injective mapping. On the other hand,

$$|(d_{2m-1})|_r + |(d_{2m})|_r \leq 2|(d_m)|_r$$

from where we get the continuity of Z. Finally

$$Z \circ T ((a_m),(b_m)) = Z((c_m)) = ((a_m),(b_m))$$

and we have that Z is the inverse mapping of T and consequently T is an isomorphism from $s \times s$ onto s.

Let Δ be the linear space over K of all double sequences of elements of K. For every $x = (x_{ij})$ of Δ we set

(6) $X(x) = (x_{11}, x_{12}, x_{21}, \ldots, x_{1n}, x_{2(n-1)}, \ldots, x_{n1}, \ldots)$

Then $X : \Delta \to \omega$ is a linear injective mapping. Given positive integers i and j the set of all pairs of positive integers (h,k) with $h+k = i+j$ equals to

$$\{(1,i+1-1), (2,i+j-2), \ldots, (i+j-1,1)\}$$

and there are $i+j-1$ of them. Thus, given the term x_{hk} of the sequence $X(x)$ which lies in position m, we have that m is larger than the number of terms x_{ij} with $i+j < h+k$ and m is less or equal than the number of terms x_{ij} with $i + j \leq h+k$. Thus

$$1+2+\ldots+(h+k-2) = \frac{1}{2}(h+k-2)(h+k-1) < m,$$

$$m \leq 1+2+\ldots+(h+k-1) = \frac{1}{2}(h+k-1)(h+k)$$

and therefore

(7) $4m \geq 2m+2 > (h+k-2)(h+k-1)+2 \geq h+k,$

(8) $m < (h+k)^2.$

For every (z_m) of ω we set

(9) $Y((z_m)) = (d_{nm})$

with $d_{11} = z_1$, $d_{12} = z_2$, $d_{21} = z_3, \ldots$, such that every term of the double sequence (d_{nm}) is a term of the sequence (z_m) and every term of this sequence is a term of the double sequence (d_{nm}) verifying also that d_{hk} is posterior to d_{ij} in (z_m) if $i + j < h + k$ or $i + j = h + k$ and $i < h$. Then $Y: \omega \longrightarrow \Delta$ is a linear injective mapping and Y is the inverse of X. Thus X is an algebraic isomorphism of Δ onto ω.

Let t be the echelon space of double sequences of K defined by the system of steps

$$(m^r n^r), \ r = 1,2,\ldots$$

If (a_{mn}) belongs to Δ we set

$$|(a_{mn})|_r = \Sigma\, m^r n^r \ |a_{mn}| \ , \quad r = 1,2,\ldots$$

Then $|\cdot|_r$, $r = 1,2,\ldots$, is a system of norms on t defining its topology.

(10) *The echelon space* s *is isomorphic to the echelon space* t.

Proof. Let T be the restriction to t of the mapping X defined by (6) If $x = (x_{mn})$ is an element of t and if $T(x) = (b_m)$, given a positive integer r, we apply (8) to obtain that

$$|T(x)|_r = \Sigma\, m^r \ | b_m | \leq \Sigma\ (h+k)^{2r} \ | x_{hk}|$$

$$\leq \Sigma\ (h+h)^{2r} \ (k + k)^{2r} \ |x_{hk}| = 4^{2r} \ \Sigma\ h^{2r} \ k^{2r} \ |x_{hk}|,$$

thus

(11) $$|T(x)|_r \leq 4^{2r} \ |x|_{2r}$$

and therefore T(x) is in s.

Let Z be the restriction to s of the mapping Y defined by (9). If $z = (z_m)$ is an element of s and if $Z(z) = (d_{mn})$, given a positive integer r, we apply (7) to obtain that

$$|(d_{mn})|_r = \Sigma\, m^r n^r \ |d_{mn}| \leq \Sigma \ (m+n)^{2r} \ |d_{mn}| \leq \Sigma (4h)^{2r} \ |z_h|,$$

thus Z(z) belongs to the and

(12) $$|Z(z)|_r < 4^{2r} \ | z |_{2r}$$

Since $X:\Delta \longrightarrow \omega$ is an algebraic isomorphism it follows that $T:t \longrightarrow s$ is an algebraic isomorphism. According to (11) and (12), T is an isomor - phism from t onto s.

Given two locally convex spaces E and F we suppose $E \otimes F$ provided with the projective topology. $E \otimes_{\varepsilon} F$ is the tensor product endowed with the ε-topology. $E \hat{\otimes} F$ and $E \hat{\otimes}_{\varepsilon} F$ are the respective completions.

Since s is a nuclear Fréchet space the ε-topology coincides with the projective topology on $s \otimes s$ (cf. PIETSCH [1], 7.3.2) but we shall not use this property in what follows.

If z belongs to $s \otimes s$ we set

$$| z |_r = \inf \{ \sum_{j=1}^{m} |x^j|_r |y^j|_r \}$$

where the infimum is taken over all possible representations of the element z of the form

$$z = \sum_{j=1}^{m} x^j \otimes y^j, \quad x^j, y^j \in s, \quad j = 1,2,\ldots,m.$$

Then $|\cdot|_r$, $r = 1,2,\ldots$, is a system of seminorms on $s \otimes s$ defining its topology (cf. SCHAEFER [1], Chapter III, 6.3).

If (x_m) and (y_m) are elements of s we set

$$B((x_m), (y_m)) = (x_m y_m).$$

Then, given a positive integer r,

$$|B((x_m),(y_m))|_r = \sum m^r n^r |x_m y_n| = (\sum m^r |x_m|)(\sum n^r |y_n|)$$

and therefore B is a bilinear continuous function from $s \times s$ into t.

Let Ψ be the canonical bilinear mapping from $s \times s$ in $s \otimes s$. Then there is a continuous linear mapping $W: s \otimes s \to t$ with $W \circ \Psi = B$ (cf. SCHAEFER [1], Chapter III, 6.1).

(13) W *is an isomorphism from* $s \otimes s$ *into* t.

Proof. Let z be an element of $s \otimes s$. Then

$$z = \sum_{j=1}^{p} x^j \otimes y^j, \ x^j = (x_m^{(j)}), \ y^j = (y_m^{(j)}) \in s, \ j = 1,2,\ldots,p.$$

Given a positive integer we have that

$$|z|_r = | \sum_{j=1}^{p} x^j \otimes y^j |_r = | \sum_{j=1}^{p} (\sum_m x_m^{(j)} e_m) \otimes y^j |_r$$

$$\leq \sum_m \Big| \sum_{j=1}^{p} x_m^{(j)} e_m \otimes y^j \Big|_r = \sum_m \Big| \sum_{j=1}^{p} x_m^{(j)} e_m \otimes \big(\sum_n y_n^{(j)} e_n\big) \Big|_r$$

$$= \sum_m \Big| \sum_n \sum_{j=1}^{p} x_m^{(j)} y_n^{(j)} e_m \otimes e_n \Big|_r \leq \sum \Big| \sum_{j=1}^{p} x_m^{(j)} y_n^{(j)} e_m \otimes e_n \Big|_r$$

$$= \sum \Big| \sum_{j=1}^{p} x_m^{(j)} y_n^{(j)} \Big| \cdot |e_m|_r |e_n|_r = \sum m^r n^r \Big| \sum_{j=1}^{p} x_m^{(j)} y_n^{(j)} \Big|$$

$$= \Big| \Big(\sum_{j=1}^{p} x_m^{(j)} y_n^{(j)} \Big) \Big|_r = \Big| \sum_{j=1}^{p} B(x^j, y^j) \Big|_r = \sum_{j=1}^{p} W(x^j \otimes y^j) |_r = |W(z)|_r$$

from where it follows that W is injective and open from $s \otimes s$ onto $W(s \otimes s)$ The conclusion follows observing that W is continuous.

(14) $s \hat{\otimes} s$ *is isomorphic to* s.

Proof. $W(e_m \otimes e_n)$ is the element of t having zero terms except for the term localted in position (m,n) in which it takes the value one. Then $W(s \otimes s)$ is dense in t. Since $s \hat{\otimes} s$ and t are Frechet spaces (and therefore complete) it follows that W can be extended to an isomorphism from $s \hat{\otimes} s$ onto t. We apply (10) to obtain that $s \hat{\otimes} s$ is isomorphic to s.

2. SOME PROPERTIES OF COMPLEMENTATION IN LOCALLY CONVEX SPACES. Let E and F be a locally convex spaces. Let J be an infinite set of cardinal α. E^J and $E^{(J)}$ are the topological product and topological direct sum of α spaces equal to E respectively. In particular, for the set N of the natural numbers, E^N and $E^{(N)}$ are the topological product and topological direct sum of an infinite countable family of spaces equal to E respectively. We set $E \simeq F$ if E is isomorphic to F.

(1). *Let* H *be a complemented subspace of* E^J *and let* L *be a complemented subspace of* H. *If* L *is isomorphic to* E^J *then* H *is isomorphic to* E^J.

Proof. Let H_1 be a topological complement of H in E^J and let L_1 be a topological complement of L in H. Then

$$H \simeq L_1 \times L \simeq L_1 \times E^J \simeq L_1 \times E^J \times E^J \simeq H \times E^J.$$

On the other hand,

$$E^J \simeq (E^J)^J \simeq (H\times H_1)^J \simeq H\times H^J\times H_1{}^J \simeq H\times E^J.$$

Thus H is isomorphic to E^J.

(2). *Let* H *be a complemented subspace of* $E^{(J)}$ *and let* L *be a complemented subspace of* H. *If* L *is isomorphic to* $E^{(J)}$ *then* H *is isomorphic to* $E^{(J)}$

Proof. The same proof given in (1) is valid changing J by (J).

(3). *Let* H *be a complemented subspace of* E. *Let* L *be a complemented subspace of* H. *If* L *is isomorphic to* E, *then* H^J *is isomorphic to* E^J *and* $H^{(J)}$ *is isomorphic to* $E^{(J)}$.

Proof. It is obvious that H^J and $H^{(J)}$ are complemented subspaces of E^J and $E^{(J)}$ respectively. On the other hand, L^J and $L^{(J)}$ are complemented subspaces of H^J and $H^{(J)}$ respectively. The proof is complete applying (1) and (2).

(4). *If the dimension of* F *equals one, then*

$$E \hat{\otimes} F \simeq E \hat{\otimes}_\varepsilon F \simeq \hat{E},$$

$\hat{E}$ *being the completion of* E.

Proof. Let $y \neq 0$ be a vector of F. We take $x_j \in E$, $y_j \in F$, $j = 1, 2, \ldots, m$, and find scalars h_j with $h_j y = y_j$, $j = 1,2,\ldots,m$. Then

$$\sum_{j=1}^{m} x_j \otimes y_j = \sum_{j=1}^{m} h_j x_j \otimes y = \left(\sum_{j=1}^{m} h_j x_j\right) \otimes y$$

and thus every element of $E \otimes F$ can be written as $x \otimes y$ with $x \in E$ and this representation is obviously unique. Let $T: E \to E \otimes F$ be the mapping defined by

$$T(x) = x \otimes y, \quad x \in E.$$

Let q be a norm on F with $q(y)=1$. If p denotes any continuous seminorm of E we have

$$p(x) = p(x)q(y) = (p \otimes q)(x \otimes y)$$

for every x of E and thus T is an isomorphism. As a consequence $\hat{E} \otimes F$ is isomorphic to $\hat{E}$ and therefore complete coinciding with $E \hat{\otimes} F$. Then

$$E \hat{\otimes} F \simeq \hat{E} \otimes F \simeq \hat{E}.$$

Finally if

$$U = \{x \in E : p(x) \leq 1 \}$$
$$V = \{z \in F : q(z) \leq 1 \}$$

and if U^o and V^o denote the polar sets of U in V in E´ respectively, then (cf. PIETSCH [1], 7.1.2).

$$\mathscr{C}_{(U,V)}(x \otimes y) = \sup \{ |\langle x,u\rangle \langle y,v\rangle| : u \in U^o, v \in V^o \}$$
$$= p(x)\, q(y) = p(x)$$

and the ε-topology coincides with the Π-topology on $E \otimes F$. The proof is complete.

(5). *If* F_1 *and* F_2 *are locally convex spaces such that* $F_1 \times F_2$ *is isomorphic to* F, *then*

$$E \hat{\otimes} F \simeq (E \hat{\otimes} F_1) \times (E \hat{\otimes} F_2).$$

Proof. (cf. KÖTHE [1], Chapter Eight, §41, Section 6).

(6). *If* F_1 *and* F_2 *are locally convex spaces such that* $F_1 \times F_2$ *is isomorphic to* F, *then*

$$E \hat{\otimes}_\varepsilon F \simeq (E \hat{\otimes}_\varepsilon F_1) \times (E \hat{\otimes}_\varepsilon F_2).$$

Proof. (cf. KÖTHE [1], Chapter Eight, §44, Section 5).

(7). *Let* E *be a locally convex space satisfying:*

a) $E \times K$ *is isomorphic to* E;

b) $E \times E$ *is isomorphic to* E;

c) $E \hat{\otimes} E$ *is isomorphic to* E.

Let G *be a complemented subspace of* $E \hat{\otimes} F$. *Let* H *be a complemented subspace of* G. *If* H *is isomorphic to* $E \hat{\otimes} F$, *then* G *is isomorphic to* $E \hat{\otimes} F$.

Proof. Let G_1 be a topological complement of G in $E \hat{\otimes} F$. Let H_1 be a topological complement of H in G. Since G is complete we have that

$$E\hat{\otimes}F \simeq (E\hat{\otimes}E) \hat{\otimes} F \simeq E \hat{\otimes} (E\hat{\otimes}F) \simeq E \hat{\otimes} (G \times G_1)$$

$$\simeq (E\hat{\otimes}G) \times (E\hat{\otimes}G_1) \simeq ((E\times K) \hat{\otimes} G) \times (E\hat{\otimes}G_1)$$

$$\simeq (E \hat{\otimes} G) \times (K \hat{\otimes} G) \times (E \hat{\otimes} G_1) \simeq (E \hat{\otimes} G) \times (G \hat{\otimes} K) \times (E \hat{\otimes} G_1)$$

$$\simeq G \times (E \hat{\otimes}(G \times G_1)) \simeq G \times (E \hat{\otimes}(E \hat{\otimes} F)) \simeq G \times (E \hat{\otimes} F).$$

On the other hand,

$$G \simeq H_1 \times H \simeq H_1 \times (E \hat{\otimes} F) \simeq H_1 \times ((E \times E) \hat{\otimes} F)$$

$$\simeq H_1 \times (E \hat{\otimes} F) \times (E \hat{\otimes} F) \simeq G \times (E \hat{\otimes} F).$$

Accordingly, G is isomorphic to $E \hat{\otimes} F$.

(8). *Let* G *be a complemented subspace of* s. *Let* H *be a complemented subspace of* G. *If* H *is isomorphic to* s, *then* G *is isomorphic to* s.

Proof. It is enough to apply (7) taking $E=F=s$, since s satisfies a), b), c) in (7), as we have shown before.

(9). *Suppose* E *satisfies* a),b) in (7). *Suppose that* $E \hat{\otimes}_\varepsilon E$ *is isomorphic to* E. *Let* G *be a complemented subspace of* $E \hat{\otimes} F$. *Let* H *be a complemented subspace of* G. *If* H *is isomorphic to* $E \hat{\otimes}_\varepsilon F$ *then* G, *is isomorphic to* $E \hat{\otimes}_\varepsilon F$.

Proof. See the proof of (7).

Result (8) can be found in VALDIVIA [16] where two different proofs are given;one of them is analogous to the one given in (7). Results (1) and (2) can be seen in VALDIVIA [27].

3. THE SPACE $\lambda_o(a)$. Let $a = (a_m)$ be a sequence of real numbers with $a_m \geq 1$, $m = 1,2,..$ We denote by $\lambda_o(a)$ the echelon space of order zero defined by the system of steps

$$(a_m^r),\ r = 1,2,...$$

If $x = (x_m)$ is any sequence in K we set

$$||x||_r = \sup \{a_m^r |x_m| : m = 1,2,...\}.$$

Then $||\cdot||_r$ $r = 1,2,...$, is a system of norms on $\lambda_o(a)$ defined the topology in this space.

(1). *The space* $\lambda_o(a)$ *is a Schwartz space if and only if is a Montel space.*

Proof. If $\lambda_o(a)$ is a Schwartz space then it is a Montel space. If $\lambda_o(a)$ is not a Schwartz space, we apply Chapter Two, §4, 3.(1),to obtain a a positive integer r such that $\left(\frac{a_m^r}{a_m^{r+p}}\right)$ does not converges to zero for eve ry positive integer p. Then $(\frac{1}{a_m})$ does not converge to zero and therefore there exists a sequence of positive integers

$$m_1 < m_2 < \ldots < m_j < \ldots$$

with

$$\inf\{\frac{1}{a_{m_j}} : j = 1,2,\ldots\} = h>0$$

Then, for each positive integer q, we have that

$$\inf\{\frac{a_{mj}}{a_{mj}^q} : j = 1,2,\ldots\} = h^{q-1} > 0$$

and by virtue of Chapter Two, §3, 2.(5), $\lambda_o(a)$ is not a Montel space.

(2) λ_o (a) *is a Schwartz space if and only if* $\lim a_m = \infty$

Proof. If $\lim a_m = \infty$, given a positive integer r, we have that

$$\lim \frac{a_m^{\ r}}{a_m^{\ r+1}} = \lim \frac{1}{a_m} = 0$$

and therefore $\lambda_o(a)$ is a Schwartz space. Reciprocally if $\lambda_o(a)$ is a Schwartz space, given the step (a_m) there is a positive integer p such that

$$\lim \frac{a_m}{a_m^{\ p+1}} = \lim \frac{1}{a_m^p} = 0$$

and accordingly $\lim a_m = \infty$ (cf. Chapter Two, §4, 3.1).

(3) λ_o (a) *is nuclear if and only if there is a positive integer* q *with*

(4) $\Sigma \frac{1}{a_m^{\ q}} < \infty$.

Proof. If λ_o (a) is nuclear, by virtue of Chapter Two, §4. 4.(1), given the step (a_m) there is a step (a_m^{q+1}) such that

$$\Sigma \frac{a_m}{a_m^{\ q+1}} = \Sigma \frac{1}{a_m^q} < \infty$$

and reciprocally, if (4) is verified, then given a positive integer r we have that

$$\Sigma \frac{a_m^{\ r}}{a_m^{\ r+q}} = \Sigma \frac{1}{a_m^{\ q}} < \infty$$

and therefore, according to Chapter Two, §4,4.(3), $\lambda_o(a)$ is nuclear.

(5) *s is isomorphic to a complemented subspace of* $s \hat{\otimes} \lambda_o(a)$

Proof. Let H be a closed hyperplane of $\lambda_o(a)$. Then $\lambda_o(a)$ is isomorphic to HxK and therefore

$$s \hat{\otimes} \lambda_o(a) \simeq s \hat{\otimes}(H \times K) \simeq (s\hat{\otimes}H) \times (s\hat{\otimes}K) \simeq (s\hat{\otimes}H) \times s$$

and the conclusion follows.

If (x_{ij}) is a double sequence in K we set

$$p_r(x) = \sup \{ \sum_i i^r a_j^{\ r} |x_{ij}| : j = 1,2,\ldots\}$$

for every positive integer r.

Let $t_o(a)$ be the linear space over K of the double sequences (x_{ij}) of K with

$$(6) \qquad \lim_j \sum_i i^r a_j^{\ r} |x_{ij}| = 0, \ r = 1,2,\ldots$$

Then p_r, r = 1,2,..., is a family of norms on $t_o(a)$ defining a topology on $t_o(a)$ which is locally convex and metrizable. We suppose $t_o(a)$ endowed with this topology.

(7) $t_o(a)$ *is a Fréchet space.*

Proof. Let $x^m = (x_{ij}^{(m)})$, m = 1,2..., be a Cauchy sequence in $t_o(a)$. Fixing positive integers i and j we deduce from

$$|x_{ij}^{(m)} - x_{ij}^{(p)}| \leq p_1(x^m - x^p)$$

that $x_{ij}^{(m)}$, m = 1,2,..., is a Cauchy sequence in K. If

$$x_{ij} = \lim_m x_{ij}^{(m)}$$

we set $x = (x_{ij})$. Giving a positive integer r and $\varepsilon > 0$ we find a positive integer q with

$$p_r(x^p - x^m) < \varepsilon, \ p, m \geq q.$$

For those values of m and p we have that

$$\sum_{i=1}^{n} i^r a_j{}^r \, |x_{ij}^{(p)} - x_{ij}^{(m)}| \leq p_r\,(x^p - x^m) \leq \varepsilon,$$

making p tend to infinity we have that

$$(8) \qquad \sum_{i=1}^{n} i^r a_j{}^r \, |x_{ij} - x_{ij}^{(m)}| \leq \varepsilon$$

from where it follows

$$\sum_{i=1}^{n} i^r a_j^{(r)} \, |x_{ij}^{(m)}| \leq \sum_{i=1}^{n} i^r a_j^r | \, x_{ij} - x_{ij}^{(m)}|$$

$$+ \sum_{i=1}^{n} i^r a_j{}^r \, |x_{ij}^{(m)}| \leq \sum_{i} i^r a_j{}^r \, |x_{ij}^{(m)}| + \varepsilon$$

and therefore

$$\lim_{j} \sum_{i} i^r a_j{}^r \, |x_{ij}| = 0.$$

Thus x belongs to $t_o(a)$. Finally, from (8) we have that

$$p_r\,(x - x^m) \leq \varepsilon$$

and, accordingly, the sequence (x^m) converges to x in $t_o(a)$.

For every $x = (x_m) \in s$ and $y = (y_m) \in \lambda_o(a)$ we set

$$M(x,y) = (x_i y_j).$$

Then, given the positive integer r,

$$\lim_{j} \sum_{i} i^r a_j{}^r \, |x_i y_j| = \sum_{i} i^r \, |x_i| \, \lim_{j} a_j{}^r \, |y_j| = 0$$

which implies that $M(x,y)$ is in $t_o(a)$. On the other hand,

$$p_r\,(M(x,y)) = \sup\,\{\sum_{i} i^r a_j \, |x_i y_j| : j = 1,2,\ldots\}$$

$$= \sum i^r \, |x_i| \, \sup\,\{a_j{}^r \, |y_j| \, : \, j = 1,2,\ldots\} = |x|_r \, |y|_r$$

and accordingly $M : s \times \lambda_o\,(a) \longrightarrow t_o(a)$ is a bilinear continuous mapping.

If z denotes an arbitrary element of $s \otimes \lambda_o\,(a)$ we set

$$||z||_r = \inf\,\{\sum_{j=1}^{m} |x^j|_r \, ||y^j||_r\}$$

where the infimum is taken over all the representations of the form

$$z = \sum_{j=1}^{m} x^j \otimes y^j,\ x^j \in s,\ y^j \in \lambda_o(a),\ j = 1,2,\dots, m,$$

of the element z.

Let ϕ: $s \times \lambda_o(a) \longrightarrow s \otimes \lambda_o(a)$ be the canonical bilinear mapping. Then there is a continuous linear mapping $V : s \otimes \lambda_o(a) \longrightarrow t_o(a)$ with $V_o\phi = M$.

(9) *V is an isomorphism from* $s \otimes \lambda_o(a)$ *into* $t_o(a)$.

Proof. Let z be an element of $s \otimes \lambda_o(a)$. Then

$$z = \sum_{j=1}^{p} x^j \otimes y^j, x^j = (x_m^{(j)}) \in s,\ y^j = (y_m^{(j)}) \in \lambda_o(a),\ j=1,2,\dots,p.$$

Given a positive integer r we have that

$$||z||_r = ||\sum_{j=1}^{p} x^j \otimes y^j||_r = ||\sum_{j=1}^{p} (\sum_m x_m^{(j)} e_m) \otimes y^j||_r \leq \sum_m ||\sum_{j=1}^{p} x_m^{(j)} e_m \otimes y^j||_r$$

$$\leq \sum_m |e_m|_r ||\sum_{j=1}^{p} x_m^{(j)} y^j||_r = \sum_m m^r \sup\{a_q^r \mid \sum_{j=1}^{p} x_m^{(j)} y_q^{(j)}| : q = 1,2,\dots\}$$

$$= \sup \{\sum_m m^r a_q^r \mid \sum_{j=1}^{p} x_m^{(j)} y_q^{(j)}| : q = 1,2,\dots\} = p_r((\sum_{j=1}^{p} x_m^{(j)} y_q^{(j)}))$$

$$= p_r(\sum_{j=1}^{p} M(x^j,y^j)) = p_r(\sum_{j=1}^{p} V(x^j \otimes y^j)) = p_r(V(z))$$

from where it follows that V is injective and open from $s \otimes \lambda_o(a)$ onto V $(s \otimes \lambda_o(a))$. The conclusion follows since V is continuous.

(10) $s \hat{\otimes} \lambda_o(a)$ *is isomorphic to* $t_o(a)$.

Proof. $V(e_m \otimes e_p)$ is the element of $t_o(a)$ having zero terms except for the term located in position (m,p) which it takes the value one, m,p= 1,2,... Let (x_{ij}) be any element of $t_o(a)$. Since (6) holds, given $\varepsilon>0$ and a positive integer r, we find a positive integer p such that

$$\sum_i i^r a_j^r |x_{ij}| < \frac{\varepsilon}{2},\ j > p.$$

We can find a positive integer q such that

$$\sum_{i=q+1}^{\infty} i^r a_j^{\ r} |x_{ij}| < \frac{\varepsilon}{2}, \ j = 1,2,\ldots,p.$$

Then, if

$$(y_{ij}) = x - \sum_{i=1}^{q} \sum_{j=1}^{p} x_{ij} V(e_i \otimes e_j)$$

We have that

$$p_r((y_{ij})) = \sup\{\sum_i i^r a_j^{\ r} |y_{ij}| : j = 1,2,\ldots\}$$

$$\leq \sup \sum_i i^r a_j^{\ r} |x_{ij}| : j = p+1, p+2,\ldots\}$$

$$+ \sup \sum_{i=q+1} i^r a_j^{\ r} |x_{ij}| : j = 1,2,\ldots, p\} < \frac{\varepsilon}{2} + \frac{\varepsilon}{2} = \varepsilon$$

and therefore $V(s \otimes \lambda_0(a))$ is dense in $t_0(a)$. Since $s \hat{\otimes} \lambda_0(a)$ and $t_0(a)$ are complete, V can be extended to an isomorphism from $s \hat{\otimes} \lambda_0(a)$ onto $t_0(a)$ and the conclusion follows.

If we take the sequence $a = (a_m)$ to be

$$1 = a_1 = a_2 = \ldots = a_m = \ldots$$

the space $\lambda_0(a)$ coincides with the Banach space c_0. We write t_0 instead of $t_0(a)$ in this case. If $x = (x_{ij})$ is an arbitrary sequence of K we write

$$|||x|||_r = \sup\{\sum_i i^r |x_{ij}| : j = 1,2,\ldots\}.$$

Then $|||\cdot|||_r$, $r = 1,2,\ldots$, is a family of norms on t_0 describing its topology.

Result (11) is a particular case of (10).

(11) $s \hat{\otimes} c_0$ *is isomorphic to* t_0.

(12) $s \hat{\otimes} \lambda_0(a)$ *is isomorphic to a complemented subspace of* $s \hat{\otimes} c_0$.

Proof. According to (10) and (11) it is enough to see that $t_0(a)$ is isomorphic to a complemented subspace of t_0.

We set h_j to denote the largest integer which is less or equal than a_j, $j = 1,2,\ldots$

Let F be the subspace of t_0 of all double sequences (y_{ij}) with

$$y_{kj} = 0, \ k \neq h_j + i, \ i, j = 1,2,\ldots$$

Obviously, F is a complemented subspace of t_0. If $x = (x_{ij})$ is an element of $t_0(a)$ we set

$$T(X) = (z_{ij})$$

such that

$$x_{ij} = z_{(hj+i)j},\ z_{kj} = 0,\ k \neq h_k+i,\ i,j = 1,2,\ldots$$

Then,given a positive integer r,

$$\sum_i i^r |z_{ij}| \leqslant \sum_i (h_j+i)^r |x_{ij}| \leqslant \sum_i (a_j+i)^r |x_{ij}|$$

$$\leqslant \sum_i (a_j+a_j)^r (i+i)^r |x_{ij}| = 4^r \sum_i i^r a_j^r |x_{ij}|$$

and therefore

$$\lim_j \sum_i i^r |z_{ij}| \leqslant 4^r \lim_j \sum_i i^r a_j^r |x_{ij}| = 0$$

and thus $T : \lambda_0(a) \longrightarrow F$ is a linear injective mapping. We have also

$$|||T(x)|||_r = \sup\{\sum_i i^r |z_{ij}| : j = 1,2,\ldots\} \leqslant 4^r p_r(x)$$

and consequently T is continuous.

Given an element $y = (y_{ij})$ of F, let $u = (u_{ij})$ be the double sequence with

$$u_{ij} = y_{(h_j+i)j},\ i,j = 1,2,\ldots$$

Then

$$\sum_i i^r a_j^r |u_{ij}| = \sum_i i^r a_j^r |y_{(h_j+i)j}| \leqslant \sum_i (a_j+i)^{2r} |y_{(h_j+i)j}|$$

$$\leqslant \sum_i (h_j+i+1)^{2r} |y_{(h_j+i)j}| \leqslant 4^r \sum_i (h_j+i)^{2r} |y_{(h_j+i)j}| = 4^r \sum_i i^{2r} |y_{ij}|$$

and therefore

$$\lim_j \sum_i i^r a_j^r |u_{ij}| \leqslant 4^r \lim_j \sum_i i^{2r} |y_{ij}| = 0$$

from where it follows that u belongs to $t_0(a)$. Obviously $T(u) = y$. Thus, $T : t_0(a) \longrightarrow F$ is onto. Finally,

$$p_r(u) = \sup\{\sum_i i^r a_j^r |u_{ij}| : j = 1,2,\ldots\}$$

$$\leq 4^r \sup \{\sum_i i^{2r} |y_{ij}| : j = 1,2,\ldots\} = 4^r |||y|||_{2r}$$

and accordingly T is open. Thus, T is an isomorphism from $t_0(a)$ onto F. Now the proof is complete.

We represent by $\lambda_0(a,s)$ the set of sequences (x_m) of s such that, for every continuous seminorm q on s, we have that $(q(x_m))$ belongs to $\lambda_0(a)$. If

$$(x_m),(y_m) \in \lambda_0(a,s),\ k \in K,$$

then (x_m+y_m) and (kx_m) are, obviously, in $\lambda_0(a,s)$ and, thus, $\lambda_0(a,s)$ can be endowed with a structure of linear space over K.

If (x_p) is a sequence in $\lambda_0(a,s)$ and if x_p is the following sequence of K:

$$x_{1p}, x_{2p}, \ldots, x_{mp}, \ldots,$$

then, given a positive integer r, we set

$$q_r((x_p)) = \sup \{a_p^{\ r} |x_p|_r : p = 1,2,\ldots\}$$

$$= \sup\{ \sum_m m^r a_p^{\ r} |x_{mp}| : p = 1,2,\ldots\}.$$

It is obvious that q_r, r = 1,2,..., is a family of norms on $\lambda_0(a,s)$ defining a metrizable locally convex topology on this space. We suppose $\lambda_0(a,s)$ endowed with this topology.

If $|.|_r^*$, r = 1,2,..., is a system of seminorms on s defining its topology, given a positive integer h there are integers n and k, n>k>h, and positive constants P and Q such that, for every x ∈ s,

$$|x|_h^* \leq P|x|_k \leq Q \sum_{r=1}^{n} |x|_r^*$$

and therefore

$$q_h^*((x_m)) = \sup \{a_p^{\ h}|x_p|_h^* : p = 1,2,\ldots\}$$

$$\leq P \sup \{a_p^{\ k} |x_p| : p = 1,2,\ldots\} = P_{q_k}((x_m))$$

$$\leq Q \sum_{r=1}^{n} \sup \{a_p^{\ r} |x_p|_r^* : p = 1,2,\ldots\}$$

from where it follows that q_r^*, r = 1,2,..., is a system of seminorms on

$\lambda_o\ (a,s)$ defining its topology.

(13) $\lambda_o(a,s)$ *is isomorphic to* $s\ \hat{\otimes}\ \lambda_o(a)$.

Proof. By (10), it is enough to show that $\lambda_o(a,s)$ is isomorphic to $t_o(a)$. If (x_p) is a sequence in s and if x_p is the sequence in K:

$$x_{1p}, x_{2p}, \ldots, x_{mp}, \ldots,$$

we set

$$S\ ((x_p)) = (x_{mp}).$$

For every positive integer r we have that

$$(14) \qquad \sum_m m^r a_p^{\ r}\ |x_{mp}| = a_p^{\ r} \sum_m m^r\ |x_{mp}| = a_p^{\ r}\ |x_p|_r$$

from where it follows

$$\lim_p \sum_m m^r a_p^{\ r}\ |x_{mp}| = \lim_p a_p^{\ r}\ |x_p|_r = 0$$

and therefore $S : \lambda_o(a,s) \longrightarrow t_o(a)$ is well defined. Obviously S is linear and injective. From (14) it follows that

$$p_r((x_{mp})) = \sup\{|\sum_m m^r a_p^{\ r}\ |x_{mp}| : p = 1,2,\ldots\}$$

$$= \sup\ \{a_p^{\ r}\ |x_p|_r\ :\ p = 1,2,\ldots\} = q_r\ ((x_p))$$

and thus S is an isomorphism from $\lambda_o(a,s)$ into $t_o(a)$. On the other hand, if $y = (y_{mp})$ is in $t_o(a)$ and if we write y_p to denote the sequence y_{1p}, $y_{2p}, \ldots, y_{mp}, \ldots$ then

$$\sum_m m^r\ |y_{mp}| \leq \sum_m m^r a_p^{\ r}\ |y_{mp}| \leq p_r\ ((y_{mp}))$$

and therefore y_p belongs to s, p = 1,2,... Moreover

$$\lim_p a_p^{\ r}\ |y_p|_r = \lim_p \sum_m m^r a_p^{\ r}\ |y_{mp}| = 0$$

and thus (y_p) belongs to $\lambda_o(a,s)$. Obviously $S((y_p)) = (y_{mp})$. Thus, S into and the conclusion follows.

4. A REPRESENTATION OF THE SPACE $s\ \hat{\otimes}\ \lambda_o(a)$ WHEN IS NUCLEAR. Analogously to 3.(5) we have that $s\ \hat{\otimes}\ \lambda_o(a)$ has a complemented subspace isomorphic to

$\lambda_0(a)$. Since every subspace of a nuclear space is nuclear, we have that, if $s \hat{\otimes} \lambda_0(a)$ is nuclear, $\lambda_0(a)$ is nuclear (cf. SCHAEFER [1], Chapter III, 7.4). Consequently we apply 3.(3) to obtain a positive integer q such that

$$\Sigma \frac{1}{a_m{}^q} < \infty.$$

Since the echelon space of order zero defined by the system of steps

$$(a_m{}^{rq}), r = 1,2,\ldots$$

coincides with $\lambda_0(a)$ we can suppose in the rest of this section that

$$\Sigma \frac{1}{a_m} < \infty.$$

(1) $s \hat{\otimes} \lambda_0(a)$ *is isomorphic to a complemented subspace of* s.

Proof. If σ is a permutation of the positive integers, the space $\lambda_0(a)$ is isomorphic the echelon space of order zero defined by the system of steps

$$(a_{\sigma(m)}^r), \ r = 1, 2, \ldots$$

by means of the following isomorphism

$$(x_m) \longrightarrow (x_{\sigma(m)}), \ (x_m) \in \lambda_0(a).$$

Then we can suppose

$$a_1 \leqslant a_2 \leqslant \ldots \leqslant a_m \leqslant \ldots$$

Since $\Sigma \frac{1}{a_m} < \infty$ we have that

$$\lim \frac{m}{a_m} = 0$$

and consequently we can find a positive integer h with

$$m \leqslant h\, a_m, \ m = 1,2,\ldots$$

If h_j denotes the largest integer which is less or equal than $a_j + j$ the sequence (h_j) is strictly increasing and $h_j > a_j$, $j = 1,2,\ldots$

Let G be the subspace of t of all those elements (x_{ij}) verifying

$$x_{ik} = 0, \ k \neq h_j, \ i,j = 1,2,\ldots$$

Let H be the subspace of t of those elements (x_{ij}) verifying

$$x_{ih_j} = 0,\ i,j = 1,2,\ldots$$

It is obvious that t is the topological direct sum of G and H. It is enough to prove that $t_0(a)$ and G are isomorphic since t is isomorphic to s and $s \hat{\otimes} \lambda_0(a)$ is isomorphic to $t_0(a)$.

If $x = (x_{ij})$ is in $t_0(a)$ we set

$$U(x) = (u_{ij})$$

with $x_{ij} = u_{ih_j}$, $u_{ik} = 0$, $k \neq h_j$, $i,j = 1,2,\ldots$

Then

$$|U(x)|_r = \Sigma i^r j^r\ |u_{ij}| = \Sigma i^r h_j^{\ r}\ |u_{ih_j}| = \Sigma i^r h_j^{\ r} |x_{ij}|$$

$$\leq \Sigma i^r (a_j + j)^r |x_{ij}| \leq \Sigma i^r (a_j + ha_j)^r\ |x_{ij}| = (1+h)^r\ \ \Sigma i^r a_j^{\ r}\ |x_{ij}|$$

$$= (1+h)^r \sum_i i^r \sum_j a_j^{\ r}\ |x_{ij}| = (1+h)^r\ \sum_i i^r \sum_j \frac{1}{a_j} a_j^{\ r+1} |x_{ij}|$$

$$= (1+h)^r \sum_i i^r\ \sum_j \frac{1}{a_j}\ \ \sup\ \{a_j^{\ r+1} |x_{ij}|\ :\ j = 1,2,\ldots\}$$

$$= (1+h)^r (\sum_m \frac{1}{a_m})\ \sup\ \{\sum_i i^r a_j^{\ r+1} |x_{ij}|\ :\ j = 1,2,\ldots\}$$

$$= (1+h)^r\ (\sum_m \frac{1}{a_m})\ P_{r+1}(x).$$

Thus $U : t_0(a) \longrightarrow G$ is a continuos linear injective mapping.

Take an element $v = (v_{ij})$ of G and set

$$y_{ij} = v_{ihj},\ i,j = 1,2,\ldots$$

Then

$$\sum_i i^r a_j^{\ r} |y_{ij}| \leq \sum_i i^r h_j^{\ r}\ |y_{ij}| = \sum_i i^r h_j^{\ r} |v_i h_j|$$

and since

$$\sum_j \sum_i i^r h_j^{\ r}\ |v_{ihj}| = \Sigma i^r j^r\ |v_{ij}| < \infty$$

it follows that

$$\lim_j \sum_i i^r a_j^{\ r}\ |y_{ij}| \leq \lim_j \sum_i i^r h_j^{\ r}\ |v_{ih_j}| = 0$$

and therefore $y = (y_{ij})$ is in $t_o(a)$. Obviously $U(y) = v$ and therefore $U : t_o(a) \longrightarrow G$ is onto. Finally

$$P_r(y) = \sup \{\sum_i i^r a_j^r |y_{ij}| : j = 1,2,\dots\}$$
$$\leq \sum i^r j^r |v_{ij}| = |v|_r$$

and consequently U is an isomorphism from $t_o(a)$ onto G. Now the proof is complete.

(3) $s \hat{\otimes} \lambda_o(a)$ *is isomorphic to* s.

Proof. It is an obvious consequence from (2), 3.(5) and 2.(8).

5. A REPRESENTATION OF THE SPACE $s \hat{\otimes} \lambda_o(a)$ WHEN IT IS NOT REFLEXIVE. If $s \hat{\otimes} \lambda_o(a)$ is not reflexive then $\lambda_o(a)$ is not Schwartz (cf. SCHAEFER [1], Chapter IV, 9.9) and, by virtue of 3.(2), (a_m) does not diverge to infinite. Since $a_m \geq 1$, $m = 1,2,\dots$, we can find a sequence of positive integers

$$m_1 < m_2 < \dots < m_j < \dots$$

and a positive integer h such that

$$a_{mj} \leq h, \; j = 1,2,\dots$$

(1) $s \hat{\otimes} c_o$ *is isomorphic to a complemented subspace of* $s \hat{\otimes} \lambda_o(a)$.

Proof. Since $s \hat{\otimes} c_o$ is isomorphic to t_o and $s \hat{\otimes} \lambda_o(a)$ is isomorphic to $t_o(a)$ it is enough to prove that t_o is isomorphic to complemented subspace of $t_o(a)$.

Let L be the subspace of $t_o(a)$ whose elements verify

$$x_{ij} = 0, \; j \neq m_p, \; i,p = 1,2,\dots$$

If $u = (u_{ij})$ belongs to t_o we set

$$W(u) = (x_{ij})$$

with

$$x_{ij} = 0, \; j \neq m_p, \; x_{imp} = u_{ip}, \; i,p = 1,2,\dots$$

Then given a positive integer r we have that

$$\sum_i i^r a_j^r |x_{ij}| = 0, \; j \neq m_p, \; p = 1,2,\dots$$

$$\sum_i i^r a_{m_j}^r |x_{im_j}| \leq h^r \sum_i i^r |u_{ij}|$$

and therefore

$$\lim_j \sum_i i^r a_j^r |x_{ij}| = 0$$

implying that (x_{ij}) is in $t_o(a)$ and consequently $W : t_o \longrightarrow L$ is well defined and, obviously, linear and injective. Moreover

$$P_r(W(u)) = \sup \{\sum_i i^r a_j^r |x_{ij}| : j = 1,2,\ldots\} = \sup \{\sum_i i^r a_{m_j}^r |x_{im_j}|$$

$$: j = 1,2,\ldots\} \leq h^r \sup \{\sum_i i^r |u_{ij}| : j = 1,2,\ldots\} = h^r |||u|||_r$$

and thus W is continuous.

Take an element $v = (v_{ij})$ of L and set

$$y_{ij} = v_{im_j}, \; i,j = 1,2,\ldots$$

Then

$$\sum_i i^r |y_{ij}| = \sum_i i^r |v_{im_j}| \leq \sum_i i^r a_{m_j}^r |v_{im_j}|$$

and consequently

$$\lim_j \sum_i i^r |y_{ij}| \leq \lim_j \sum_i i^r a_{m_j}^r |v_{im_j}| = 0$$

from where it follows that $y = (y_{ij})$ is in t_o. Obviously $W(y) = v$ and therefore W is an isomorphism from t_o onto L. Finally,

$$|||y|||_r = \sup \{\sum_i i^r |y_{ij}| : j = 1,2,\ldots\}$$

$$\leq \sup \{\sum_i i^r a_{m_j}^r |v_{im_j}| : j = 1,2, \ldots$$

$$\leq \sup \{\sum_i i^r a_j^r |v_{ij}| : j = 1,2,\ldots\} = P_r (W(y))$$

and therefore W is an isomorphism from t_o onto L. The proof is complete.

(2) $s \hat{\otimes} \lambda_o(a)$ *is isomorphic to* $s \hat{\otimes} c_o$.

Proof. Since s satisfies a), b), c) of 2.(7) it is enough 2.(7),(1) and 3.(12) to obtain the conclusion.

6. THE SPACE $\lambda_\infty(a)$. Let $a = (a_m)$ be a sequence of real numbers with $a_m \geq 1$, $m = 1,2,\ldots$ We denote by $\lambda_\infty(a)$ the echelon space of order infinite defined by the system of steps

$$(a_m^{\ r}),\ r = 1,2,\ldots$$

If $\lambda_\infty(a)$ is reflexive, then coincides with the space $\lambda_0(a)$ described in Section 4.

Let $t_\infty(a)$ be the linear space over K of the double sequences $x = (x_{ij})$ of K such that, for every positive integer r,

$$p_r(x) = \sup \{\sum_i i^r a_j^{\ r}\ |x_{ij}|: j = 1,2,\ldots\} < \infty.$$

Then p_r, $r = 1,2,\ldots$, is a family of norms on $t_\infty(a)$ defining on this space a metrizable locally convex topology. We suppose $t_\infty(a)$ endowed with this topology.

With slight modifications in 3.(7) and 3.(10), results (1) and (2) follow.

(1) $t_\infty(a)$ *is a Fréchet space.*

(2) $s\hat{\otimes}\lambda_\infty(a)$ *is isomorphic to* $t_\infty(a)$.

If we take the sequence $a = (a_m)$ to be

$$1 = a_1 = a_2 = \ldots a_m = \ldots$$

$\lambda_\infty(a)$ coincides with ℓ^∞. We write t_∞ instead $t_\infty(a)$ in this case. We have

(3) $s \hat{\otimes} \ell^\infty$ *is isomorphic to* t_∞.

Next result can be proven analogously as 3.(12).

(4) $s \hat{\otimes} \lambda_\infty(a)$ *is isomorphic to a complemented subspace of* $s \hat{\otimes} \ell^\infty$.

We represent by $\lambda_\infty(a,s)$ the set of those sequences (x_m) of s such that, for every continuous seminorm q on s, we have that $(q(x_m))$ belongs to $\lambda_\infty(a)$. If

$$(x_m),(y_m) \in \lambda_\infty(a,s),\ k \in K,$$

we have that (x_m+y_m) and $(k\ x_m)$ belong to $\lambda_\infty(a,s)$ and therefore $\lambda_\infty(a,s,)$ can be endowed with a structure of linear space over k.

If (x_p) is a sequence in $\lambda_\infty(a,s)$ and if x_p in the sequence in K:

$$x_{1p}, x_{2p},\ldots, x_{mp},\ldots,$$

for every positive integer r we set

$$q_r((x_p)) = \sup \{a_p{}^r |x_p|_r : p = 1,2,\ldots\}$$

$$= \sup\{\sum_m m^r a_p{}^r |x_{mp}| : p = 1,2,..\} .$$

It is immediate that q_r, $r = 1,2,\ldots$, is a family of norms on $\lambda_\infty(a,s)$ defining a metrizable locally convex topology on this space. We suppose $\lambda_\infty(a, s)$ endowed with this topology.

If $|.|_r^*$, $r = 1,2,\ldots$, is a system of seminorms on s defining its topology and if we set

$$q_r^* ((x_p)) = \sup \{a_p{}^r |x_p|_r^* : p = 1,2,\ldots\}$$

we have that q_r^*, $r = 1,2,\ldots$, is a system of seminorms on $\lambda_\infty(a,s)$ defining the topology of this space.

Analogously, to 3.(13) and 5.(1) the next two results follow.

(5) *$\lambda_\infty(a,s)$ is isomorphic to* $s \hat{\otimes} \lambda_\infty(a)$.

(6) *If $\lambda_\infty(a,s)$ is not reflexive, then* $s \otimes \ell^\infty$ *is isomorphic to a complemented subspace of* $s \hat{\otimes} \lambda_\infty(a)$.

(7) *If $\lambda_\infty(a)$ is not reflexive, then* $s \hat{\otimes} \lambda_\infty(a)$ *is isomorphic to* $s \hat{\otimes} \ell^\infty$.

Proof. Since s satisfies a),b)and c) in 2.(7) it is enough to apply 2.(7), (4) and (6) to obtain the desired conclusion.

Result (7) can be obtained by observing that $s \hat{\otimes} \lambda_\infty(a)$ is the strong bidual of $s \hat{\otimes} \lambda_o(a)$ and that $s \hat{\otimes} \ell^\infty$is the strong bidual of $s \hat{\otimes} c_o$ and apply 5.(2).

7. THE SPACE M_n. In the n-dimensional euclidean space R^n we represent its norm by

$$||x|| = (x_1{}^2 + x_2{}^2 + \ldots + x_m{}^2)^{\frac{1}{2}}$$

if $x = (x_1, x_2,\ldots,x_n)$.

Given the multi-index $\alpha = (q_1,q_2,\ldots,q_n)$ where q_j is a non-negative integer, $j = 1,2,\ldots, n$, we denote by $|\alpha|$ the sum $q_1+q_2+\ldots+q_n$. If f is a function defined on an open set Ω of R^n with values in K and if it exists

$$(1) \qquad \frac{\partial^{|\alpha|} f(x)}{\partial x_1{}^{q_1} \partial x_2{}^{q_2} \ldots \partial x_n{}^{q_n}}$$

when $x = (x_1,x_2,\ldots,x_n)$ varies in Ω, we set $D^{\alpha}f(x)$ instead of (1). Given the multi-index $\beta = (r_1,r_2,\ldots,r_n)$ we set

$$\alpha+\beta = (q_1+r_1,q_2+r_2,\ldots,q_n+r_n), \alpha-\beta = (q_1-r_1,q_2-r_2,\ldots,q_n-r_n)$$

If $r_j \leq q_j$, $j = 1,2,\ldots$, we write $\beta \leq \alpha$.

We denote by M_n the linear space in K of all the functions f of n real variables and valued in K which are infinitely differentiable in R^n, even and 2Π-periodic with respect to each variable. For every positive integer r we set

(2) $$|f|_r = \sum_{|\alpha|\leq r} \sup\{|D^{\alpha}f(x_1,x_2,\ldots,x_n)| : 0 \leq x_j \leq \Pi, \ j = 1,2,\ldots,n\}.$$

Obviously, $|.|_r$, $r = 1,2,\ldots$, is a family of norms on M_n defining a metrizable locally convex topology on M_n. In what follows we suppose M_n endowed with the topology.

(3) *M_n is a Fréchet space.*

Proof. Let (f_p) be a Cauchy sequence in M_n. Given any multi-index α it follows from (2) that $(D^{\alpha}f_p)$ is a sequence of K-valued functions defined on R^n which is a Cauchy sequence for the topology of the uniform convergence in R^n and therefore there is an infinetely differentiable function $f: R^n \longrightarrow K$ such that $(D^{\alpha}f_p)$ converges to $D^{\alpha}f$ uniformly on R^n. Obviously f belongs to M_n. On the other hand, given a positive integer r and $\varepsilon > 0$ there is a positive integer m such that, if p and q are integers larger than m,

$$|f_p - f_q|_r \leq \varepsilon$$

from where it follows that

$$|f_p - f|_r \leq \varepsilon, \ p \geq m,$$

and therefore (f_p) converges to f in M_n. Now the conclusion follows.

Given a non-negative integer p, let f be an element of M_{p+1} and let

(4) $$\frac{1}{2} f_o(x_1,x_2,\ldots,x_p) + \sum_{m=1}^{\infty} f_m(x_1,x_2,\ldots,x_p) \cos mx_{p+1}$$

be the Fourier series of $f(x_1,x_2,\ldots,x_{p+1})$ with respect to the variable x_{p+1}. Then

(5) $$f_m(x_1,x_2,\ldots,x_p) = \frac{2}{\pi}\int_0^{\pi} f(x_1,x_2,\ldots,x_{p+1}) \cos mx_{p+1}dx_{p+1},$$

$$m = 0,1,2,\ldots,$$

and therefore f_m is even and 2π-periodic with respect to each of its variables, $m = 0,1,2,\ldots$ If $p = 0$ we set $\alpha = 0$ and if $p > 0$, α is a multi-index $(q_1,q_2,\ldots,q_p)$. Set $\beta = (q_1,q_2,\ldots,q_p,0)$. Then, for $m = 0,1,2,\ldots,$

(6) $$D^{\alpha}f_m(x_1,x_2,\ldots,x_p) = \frac{2}{\pi}\int_0^{\pi} D^{\beta}f(x_1,x_2,\ldots,x_{p+1}) \cos mx_{p+1}dx_{p+1}$$

and therefore, if $p > 0$,

$$f_m \in M_p.$$

If $m > 0$, given a non negative integer r we integrate by parts (6) $r+2$ times to obtain that, if $\gamma = r + 2$ when $p = 0$ and $\gamma = (q_1,q_2,\ldots,q_p,r+2)$ when $p > 0$,

$$m^{r+2}D^{\alpha}f_m(x_1,x_2,\ldots,x_p) = \frac{2}{\pi}\int_0^{\pi} D^{\gamma}f(x_1,x_2,\ldots,x_{p+1}) \cos (mx_{p+1} + (r + 2)\frac{\pi}{2})\, dx_{p+1}$$

from where it follows

$$|m^rD^{\alpha}f_m(x_1,x_2,\ldots,x_p)| \leq \frac{2}{m^2} \sup \{|D^{\gamma}f(x_1,x_2,\ldots,x_{p+1})|: 0 \leq x_j \leq \pi, j = 1,2,\ldots,n\}$$

and consequently, if $|\alpha| \leq r$,

(7) $$| m^rD^{\alpha}f_m(x_1,x_2,\ldots,x_p)| \leq m^r |f_m|_r \leq \frac{2}{m^2} |f|_{2r+2}$$

and therefore the series

$$\sum_{m=1}^{\infty} m^rD^{\alpha}f_m(x_1,x_2,\ldots,x_p) \cos (mx_{p+1}+r \frac{\pi}{2})$$

converges absolutely and uniformly in R^n to $D^{\delta}f$ being $\delta = r$ when $p = 0$ and $\delta = (q_1,q_2,\ldots,q_p,r)$ if $p > 0$. From where it follows that the series (4) converges absolutely to f in M_{p+1}.

(8) M_1 *is isomorphic to* s.

Proof. If f belongs to M_1, let

$$\frac{b_o}{2} + \sum_{m=1}^{\infty} b_m \cos mx \tag{9}$$

be the Fourier series of f(x). Given a positive integer r, it follows from (5) and (7) that

$$|b_o| \leqslant 2|f|_1, \; m^r \; |b_m| \leqslant \frac{2}{m^2} \; |f|_{2r+2}, \; m = 1,2,\ldots, \tag{10}$$

and therefore the sequence (c_m) such that

$$c_1 = \frac{b_o}{2}, \; c_{m+1} = b_m, \; m = 1,2,\ldots,$$

is rapidly decreasing. Thus (c_m) is in s. We set

$$Tf = (c_m)$$

It is obvious that $T : M_1 \longrightarrow s$ is a linear mapping. On the other hand, if f is non-zero and since (9) converges uniformly to f in R, it follows that the Fourier series of f is non-zero and therefore T is injective. It follows from (10) that

$$|(c_m)|_r = \frac{|b_o|}{2} + \sum_{m=1}^{\infty} (m+1)^r |b_m| \leqslant |f|_1 + 2^r \sum_{m=1}^{\infty} m^r |b_m|$$

$$\leqslant |f|_1 + 2^{r+1} |f|_{2r+2} \sum_{m=1}^{\infty} \frac{1}{m^2}$$

and thus T is continuous.

Let (d_m) be a rapidly decreasing sequence. For every non-negative integer r the trigonometric series

$$\Sigma m^r d_{m+1} \cos\left(mx + r\frac{\pi}{2}\right)$$

converges absolutely and uniformly in R and therefore the function g defined by

$$g(x) = d_1 + \sum_{m=1}^{\infty} d_{m+1} \cos mx, \; x \in R,$$

belongs to M_1 and $Tg = (d_m)$. Consequently T is onto. Finally we apply the open mapping theorem to conclude that T is an isomorphism from M_1 onto s, since M_1 and s are Fréchet spaces. The proof is complete.

If f is an element of M_n and if g belongs to M_1 we set B(f,g) to denote the function defined on R^{n+1} with values in K such that

$$B(f,g)(x_1,x_2,\ldots,x_{n+1}) = f(x_1,x_2,\ldots,x_n)g(x_{n+1})(x_1,x_2,\ldots,x_{n+1}) \in R^{n+1}$$

(11) $B : M_n \times M_1 \longrightarrow M_{n+1}$ *is a continuous bilinear mapping.*

Proof. It is obvious that B is a bilinear mapping. On the other hand, given a positive integer r, if f belongs to M_n and g belongs to M_1 and if we set $\alpha = (q_1,q_2,\ldots,q_{n+1})$, q_j being a non-negative integer, j = 1,2,..., n+1, $\beta = (q_1,q_2,\ldots,q_n)$ and $x = (x_1,x_2,\ldots,x_n)$, we have that

$$|B(f,g)|_r = \sum_{|\alpha|\leq r} \sup\{|D^\beta f(x)g(x_{n+1})| : 0 \leq x_j \leq \Pi, j = 1,2,\ldots,n+1\}$$

$$\leq \sum_{|\alpha|\leq r} \sup\{|D^\beta f(x)| : 0 \leq x_j \leq, j=1,2,\ldots,n\} \sup\{|D^{q_{n+1}} g(x_{n+1})| : 0 \leq x_{n+1} \leq \pi\} \leq |f|_r |g|_r$$

from where it follows the continuity of B.

If z is an element of $M_n \otimes M_1$ we set, for every positive integer r,

$$|z|_r = \inf \{ \sum_{j=1}^{q} |f_j|_r |g_j|_r \}$$

where the infimum is taken over all the representations of z of the form

$$z = \sum_{j=1}^{q} f_j \otimes g_j, \ f_j \in M_n, \ g_j \in M_1, \ j = 1,2,\ldots,q.$$

Then $|.|_r$, r = 1,2,...,is a system of seminorms on $M_n \otimes M_1$ defining the topology of this space.

Let ψ be the canonical bilinear mapping from $M_n \times M_1$ into $M_n \otimes M_1$. Then there is a continuous linear mapping $\chi : M_n \otimes M_1 \longrightarrow M_{n+1}$ such that $\chi \circ \psi = B$.

(12) $\chi : M_n \otimes M_1 \longrightarrow M_{n+1}$ *is an isomorphism into.*

Proof. Let r be any positive integer. Let z be an element of $M_n \otimes M_1$.

Then

$$z = \sum_{j=1}^{q} f_j \otimes g_j, f_j \in M_n, g_j \in M_1,\ j = 1,2,\ldots,q.$$

Let

$$\frac{a_o^{(j)}}{2} + \sum_{m=1}^{\infty} a_m^{(j)} \cos m x$$

be the Fourier series of $g_j(x)$, $j = 1,2,\ldots, q$. If h is the element of M_1 such that $h(x) = 1$, $x \in R$, we have that

$$z = \sum_{j=1}^{q} f_j \otimes g_j = \sum_{j=1}^{q} f_j \otimes \left(\frac{a_o^{(j)}}{2} h + \sum_{m=1}^{\infty} a_m^{(j)} \cos(m.)\right)$$

$$= \left(\frac{1}{2} \sum_{j=1}^{q} a_o^{(j)} f_j\right) \otimes h + \sum_{m=1}^{\infty} \left(\sum_{j=1}^{q} a_m^{(j)} f_j\right) \otimes \cos (m.)$$

Therefore

$$\chi(z)(x_1,x_2,\ldots,x_{n+1}) = \frac{1}{2} \sum_{j=1}^{q} a_o^{(j)} f_j(x_1,x_2,\ldots,x_n)$$

$$+ \sum_{m=1}^{\infty} \sum_{j=1}^{q} a_m^{(j)} f_j(x_1,x_2,\ldots,x_n) \cos mx_{n+1}$$

and according to (6) an (7)

$$\left|\sum_{j=1}^{q} a_o^{(j)} f_j\right|_r \leq 2|\chi(z)|_r,\quad m^r \left|\sum_{j=1}^{q} a_m^{(j)} f_j\right|_r \leq \frac{2}{m^2} |\chi(z)|_{2r+2}$$

and therefore

$$|z|_r \leq \left|\frac{1}{2} \sum_{j=1}^{q} a_o^{(j)} f_j\right|_r |h|_r + \sum_{m=1}^{\infty} \left|\sum_{j=1}^{q} a_m^{(j)} f_j\right|_r |\cos(m.)|_r$$

$$\leq |\chi(z)|_r + \sum_{m=1}^{\infty} \left|\sum_{j=1}^{q} a_m^{(j)} f_j\right|_r \sum_{\alpha=0}^{r} \sup\{|D^{\alpha}\cos mx| : 0 \leq x \leq \pi\}$$

$$\leq |\chi(z)|_r + (r+1) \sum_{m=1}^{\infty} m^r \left|\sum_{j=1}^{q} a_m^{(j)} f_j\right|_r$$

$$\leq |\chi(z)|_r + (r+1)\ |\chi(z)|_{2r+2} \sum_{m=1}^{\infty} \frac{1}{m^2}$$

from where it follows the χ is open and injective from $M_n \otimes M_1$ into M_{n+1}.

The conclusion follows since χ is continuous.

(13) $M_n \hat{\otimes} M_1$ *is isomorphic to* M_{n+1}.

Proof. Given the element f of M_{n+1}, let (4) be the Fourier series of $f(x_1,x_2,\ldots x_{n+1})$ with respect to x_{n+1}. If h is the element of M_1 defined in the proof of (12) we have that, for every positive integer k,

$$\frac{1}{2} f_0 \otimes h + \sum_{m=1}^{k} f_m \otimes \cos(m.)$$

belongs to $M_n \otimes M_1$ and its image by X converges to f in M_{n+1} when k tends to infinity. Thus $X(M_n \otimes M_1)$ is dense in M_{n+1}. Since $M_n \hat{\otimes} M_1$ and M_{n+1} are complete it follows that X can be extended to an isomorphism from $M_n \hat{\otimes} M_1$ onto M_{n+1}.

(14) M_n *is isomorphic to* s.

Proof. By (8), M_1 is isomorphic to s. Proceeding by recurrence we suppose that M_p is isomorphic to s for a positive integer p. We apply (13) to obtain that M_{p+1} is isomorphic to $M_p \hat{\otimes} M_1$ which in turn is isomorphic to $s \hat{\otimes} s$. We apply now 1.(14) to obtain that M_{p+1} is isomorphic to s.

8. THE SPACES $\mathcal{E}(H)$ AND $\mathcal{D}(H)$. In what follows, let H be a compact set of the space R^n. If H coincides with the closure of its interior $\overset{\circ}{H}$, $\mathcal{E}(H)$ is the linear space over K of all the continuous functions f defined on H and valued in K which are infinitely differentiable in $\overset{\circ}{H}$ such that, for every multi-index $\alpha = (q_1,q_2,\ldots,q_n)$, $D^\alpha f$ can be continuously extended from $\overset{\circ}{H}$ to H.

Given the real numbers $a_j < b_j$, $j = 1,2,\ldots, n$, we set

$$L = \{(x_1,x_2,\ldots,x_n) : a_j \leq x_j \leq b_j,\ j = 1,2,\ldots, n\}.$$

Let E be the linear space over K of all the continuous functions defined on L and valued in K which have continuous partial derivatives of all orders. In the boundary of L those derivatives are defined laterally.

(1) *The linear space* E *coincides with* $\mathcal{E}(L)$.

Proof. Let f be an element of E. Since f is continuous and it has continuous partial derivatives in L of all orders, we have that f is infinitely differentiable in $\overset{\circ}{L}$. Given the multi-index α, $D^\alpha f$ is a continuous extension to L of the partial derivative of order α of f in $\overset{\circ}{L}$. Therefore f is in $\mathcal{E}(L)$.

If g is an element of $E(L)$ it is continuous and each partial derivative of any order of g in $\overset{\circ}{L}$ continuously extended to L belongs ovbiously to $E(L)$. Therefore in order to show that g belongs to E it is enough to prove that g has partial derivatives of first order in L which are continuous. Let h be a continuous function in L coinciding with $\frac{\partial g}{\partial x_j}$ in $\overset{\circ}{L}$. We take two distinct points in L:

$$x = (x_1,x_2,\ldots,x_{j-1},a,x_{j+1},\ldots,x_n),$$

$$y = (x_1,x_2,\ldots,x_{j-1},b,x_{j+1},\ldots,x_n).$$

Let

$$x^m = (x_{1m},x_{2m},\ldots,x_{(j-1)m},p_m,x_{(j+1)m},\ldots,x_{nm})$$

$$y^m = (x_{1m},x_{2m},\ldots,x_{(j-1)m},q_m,x_{(j+1)m},\ldots,x_{nm}) \qquad m= 1,2,\ldots$$

be sequences of positive in $\overset{\circ}{L}$ such that (x^m) converges to x and (y^m) converges to y. For every positive integer m we have that

(2) $$g(y^m)-g(x^m) = (b_m-a_m)h(u^m)$$

being

$$u^m = (x_{1m},x_{2m},\ldots,x_{(j-1)m},c_m,x_{(j+1)m},\ldots,x_{nm})$$

with $c_m \in [p_m,q_m]$. Since L is compact we extract from (u^m) a subsequence, which we denote by (u^m), which converges to

$$u = (x_1,x_2,\ldots,x_{j-1},c,x_{j+1},\ldots,x_n)$$

It is immediate that $c \in [a,b]$. If m tends to infinity in (2) we have that

$$g(y)-g(x) = (b-a)h(u).$$

When b tends to a we have that c tends to a and therefore

$$\lim \frac{g(y)-g(x)}{b-a} = h(x)$$

when b tends to a, $b \neq a$. Consequently g has derivative respect to x_j in L and coincides with h. Now the conclusion follows.

We consider that H coincides with the closure of its interior. If f is an element of $E(H)$ and α is a multi-index, we represent by $D^{\alpha}f$ the extension to H of the partial derivative of order α of f in $\overset{\circ}{H}$. For every positive integer r we set

$$|f|_r = \sum_{|\alpha|\leq r} \sup\{|D^{\alpha}f(x)| : x \in H\}.$$

Then $|\cdot|_r$, r = 1,2,...,is a family of norms on $E(H)$ defining on this space a metrizable locally convex topology. In what follows we suppose $E(H)$ endowed with this topology.

(3) $E(H)$ *is a Fréchet space.*

Proof. Let (f_p) be a Cauchy sequence in $E(H)$. For every multi-index α, $(D^{\alpha}f_p)$ is a sequence of K-valued functions defined on H which is a Cauchy sequence for the topology of the uniform convergence on H. Consequently there is a continuous function f defined on H and infinitely differentiable on $\overset{\circ}{H}$ such that $D^{\alpha}f$ can be continuously extended to H and $(D^{\alpha}f_p)$ converges uniformly on H to this extension. Now it is easy to check that (f_p) converges to f in $E(H)$. The proof is complete.

Let g be a continuous function from R^n into K. The support of g is the closure of the set

$$\{x \in R^n : g(x) \neq 0\}.$$

We represent the support of g by supp g. Given any compact H of R^n we denote by $\mathcal{D}(H)$ the linear space over K of all the K-valued functions defined on R^n which are infinitely differentiable and with support contained in H. If f belongs to $\mathcal{D}(H)$ and if r any positive integer we set

$$|f|_r = \sum_{|\alpha|\leq r} \sup\{|D^{\alpha}f(x)| : x \in R^n\}$$

$$= \sum_{|\alpha|\leq r} \{|D^{\alpha}f(x)| : x \in H\}.$$

Then $|.|_r$, r = 1,2,..., is a system of norms on $\mathcal{D}(H)$ defining on this space a metrizable locally convex topology. In what follows we suppose $\mathcal{D}(H)$ endowed with this topology. If M denotes the closure of $\overset{\circ}{H}$ then $\mathcal{D}(H)$ coincides obviously with $\mathcal{D}(M)$. If $H \neq \emptyset$ belongs to $\mathcal{D}(H)$ we set Tf to denote its restriction to M. T: $\mathcal{D}(H) \longrightarrow E(M)$ is a linear mapping such that

$$|Tf|_r = |f|_r,\ r = 1,2,...$$

Obviously T ($\mathcal{D}(H)$) is a closed subspace of $E(M)$ and therefore we have that

(4) $\mathcal{D}(H)$ *is a Fréchet space.*

Sometimes we shall identify $\mathcal{D}(H)$ with $T(\mathcal{D}(H))$ when we deal with isomorphism problems.

9. FUNCTIONS WITH COMPACT SUPPORT. In this section we denote by f the real valued function defined on R with

$$f(x) = 0,\ x \leq 0,\ f(x) = e^{-\frac{1}{x}},\ x > 0.$$

We have that f is infinitely differentiable in R.

(1) *Given real numbers* $a_1 < b_1 < b_2 < a_2$ *there is an infinitely differentiable real function* g *defined on* R *whose support coincides with the closed internal* $[a_1,a_2]$ *such that*

$$g(x) > 0,\ a_1 < x < a_2,\ g(x) = 1,\ b_1 \leq x \leq b_2.$$

Proof. Let g_1 and g_2 be the functions defined by

$$g_1(x) = \frac{f(x-a_1)}{f(b_1-x)+f(x-a_1)},\ g_2(x) = \frac{f(a_2-x)}{f(x-b_2)+f(a_2-x)}$$

for every x in R.

Since

$$f(b_1-x)+f(x-a_1)\neq 0, f(x-b_2)+f(a_2-x)\neq 0, x \in R,$$

it follows that g_1 and g_2 are infinitely differentiable in R. On the other hand,

$$g_1(x) = 0,\ x \leq a_1, g_1(x) > 0,\ x > a_1, g_1(x) = 1,\ x \geq b_1,$$

$$g_2(x) = 1,\ x \leq b_2, g_2(x) > 0,\ x < a_2, g_2(x) = 0,\ x \geq a_2.$$

Therefore the function g verifying

$$g(x) = g_1(x),\ x< b_1, g(x) = g_2(x),\ x \geq b_2,\ g(x) = 1\ b_1 \leq x < b_2$$

satisfies the conditions above.

Given real numbers $a_j < b_j < c_j < d_j$, $j = 1,2,\ldots,n$, let A and B be the n-dimensional closed intervals

$$\{(x_1,x_2,\ldots,x_n) : a_j \leq x_j \leq d_j,\ j = 1,2,\ldots,n\}$$

$$\{(x_1,x_2,\ldots,x_n) : b_j \leqslant x_j \leqslant c_j,\ j = 1,2,\ldots,n\}$$

respectively

(2) *There is an infinitely differentiable real function* h *defined on* R^n *whose support coincides with* A *and such that*

$$h(x) > 0,\ x \in \mathring{A},\ h(x) = 1,\ x \in B$$

Proof. For every positive integer j, $1 \leqslant j \leqslant n$, we apply (2) to obtain an infinitely differentiable real function h_j defined on R whose support coincides with $[a_j,d_j]$, $h_j(x) > 0$ for $a_j < x < d_j$ and $h_j(x) =1$, for $b_j \leqslant x \leqslant c_j$. Let h be the function defined on R^n such that

$$h(x_1,x_2,\ldots,x_m) = h_1(x_1)\ h_2(x_2)\ldots h_n(x_n), (x_1,x_2,\ldots,x_n) \in R^n.$$

Then h satisfies the wanted conditions.

Let Ω be a non-void open set of R^n. Let

(3) $\{O_i : i \in I\}$

be a family of non void open sets contained in Ω and covering Ω. A sequence (g_m) of real functions defined on R^n are a partition of the unity of class C^∞ associated to the covering (3) if the following conditions are satisfied:

a) For every positive integer m, g_m is infinitely differentiable in R^n.

b) Given any x of Ω there is a neighbourhood U_x of x and a positive integer q such that

$$U_x \cap \text{supp } g_m = \emptyset,\ m > q,$$

i.e., the family

$$\{\text{supp } g_m : m = 1,2,\ldots\}$$

is locally finite.

c) For every positive integer m, g_m has compact support and $g_m(x) \geqslant 0$, $x \in R^n$.

d) Given a positive integer m there is an $i \in I$ such that supp $g_m \subset O_i$.

e) For every $x \in \Omega$, $\Sigma g_m(x) = 1$.

(4) *If* (g_m) *is a partition of the unity of class* C^∞ *associated to the covering* (3) *of* Ω *and if* H *is any compact set of* Ω *there is a positive integer* q *such that*

$$\text{supp } g_m \cap H = \emptyset, \quad m > q$$

Proof. For every x in H we find a neighbourhood U_x of x and a positive integer q_x such that supp $g_m \cap U_x = \emptyset$, $m > q_x$. From the family $\{U_x : x \in H\}$, we extract a finite subfamily $\{U_{x_1}, U_{x_2}, \ldots, U_{x_p}\}$ such that $\bigcup_{j=1}^{p} U_{x_j} \supset H$. If $q = \max \{q_{x_1}, q_{x_2}, \ldots, q_{x_p}\}$ then, given a positive integer m larger than q, it follows that

$$\text{supp } g_m \cap H \quad \text{sup } g_m \cap \left(\bigcup_{j=1}^{p} U_{x_p}\right) = \bigcup_{j=1}^{p} (\text{supp } g_m \cap U_{x_j}) = \emptyset$$

(5) *Given the covering* (3) *of the open set* Ω *there is a partition of the unity* (g_m) *of class* C^∞ *associated to the covering such that the support of* g_m *is an* n - *dimensional cube,* m = 1,2.

Proof. Let (H_m) be a sequence of non-void compacts sets of Ω such that H_m is contained in $\mathring{H}_{m+1}$, m = 1,2,..., and $\bigcup_{m=1}^{\infty} H_m = \Omega$. Given any x of H_1 we find a compact cube P_x contained in $O_i \cap \mathring{H}_2$ for some $i \in I$ such that $x \in \mathring{P}_x$. From the family $\{P_x : x \in H_1\}$ we extract a finite number of cubes $Q_1, Q_2, \ldots, Q_{m_1}$ such that $\bigcup_{j=1}^{m_1} \mathring{Q}_j \supset H_1$.

Proceeding by recurrence, given a positive integer p, we suppose the compact cubes

$$Q_1, Q_2, \ldots, Q_{m_p}$$

obtained such that Q_j is contained in $O_i \cap \mathring{H}_{p+1}$ for some index i depending of j, j = 1, 2, ..., m_p, and

$$\bigcup_{j=1}^{m_p} \mathring{Q}_j \supset H_p$$

Given $x \in H_{p+1} \sim H_p$ we find a compact cube R_x contained in $O_i \cap \mathring{H}_{p+2}$ for some index i, $R_x \cap H_p = \emptyset$ and $x \in \mathring{R}_x$. The family of open sets

$$\mathring{Q}_1, \mathring{Q}_2, \ldots, \mathring{Q}_{m_p}, \mathring{R}_x, \quad x \in H_{p+1} \sim H_p,$$

covers obviously H_{p+1} and therefore there is an integer $m_{p+1} > m_p$ such that

$$Q_1, Q_2, \ldots, Q_{m_p}, Q_{m_p+1}, \ldots, Q_{m_{p+1}}$$

is a subfamily of

$$\{Q_1, Q_2, \ldots, Q_{mp}, R_x : x \in H_{p+1} \sim H_p\}$$

such that

$$\bigcup_{j=1}^{m_{p+1}} \overset{\circ}{Q}_j \supset H_{p+1}.$$

Then Q_j is contained in $O_i \cap \overset{\circ}{H}_{p+2}$ for some index i depending on j, j=1,2, ,...,m_{p+1}.

Now we show that the sequence (Q_m) is locally finite. Given any x of Ω we find a positive integer p such that $x \in \overset{\circ}{H}_p$. Then

$$Q_m \cap \overset{\circ}{H}_p \subset Q_m \cap H_p = \emptyset \text{ for } m > m_p$$

Obviously

$$\bigcup_{m=1}^{\infty} \overset{\circ}{Q}_m = \Omega \tag{6}$$

We apply now (2) to obtain a function h_m, m = 1,2,..., defined on R^n which is infinitely differentiable in R^n whose support is Q_m and such that for every x of $\overset{\circ}{Q}_m$, $h_m(x) > 0$. Since the sequence of cubes (Q_m) is locally finite, $h = \Sigma h_m$ is a well defined function on R^n which is infinitely differentiable in Ω. By (6), we have that $h(x) > 0$ for every x of Ω. Then, if g_m is the function defined on R^n with

$$g_m(x) = \frac{h_m(x)}{h(x)}, \text{ for } x \in \Omega,$$

$$g_m(x) = 0, \text{ for } x \in R^n \sim \Omega,$$

we have that g_m is infinitely differentiable in R^n, its support is Q_m, it is strictly positive on $\overset{\circ}{Q}_m$ and, if $x \in \Omega$,

$$\Sigma g_m(x) = \Sigma \frac{h_m(x)}{h(x)} = 1$$

(g_m) is the desired sequence.

Let g be a continuous function from Ω into K. The support of g is the closure in Ω of the set

$$\{x \in \Omega : g(x) \neq 0\}.$$

We represent the support of g by supp g.

Given the covering (3) of Ω, a family $\{g_i : i \in I\}$ of real functions defined on Ω is a partition of the unity of class C^∞ subordinated to the covering (3) if the following conditions are satisfied:

1) For every $i \in I$, g_i is infinitely differentiable in Ω.
2) The family

   $\operatorname{supp}\{g_i : i \in I\}$
   is locally finite
3) For every $i \in I$ and $x \in \Omega$, $g_i(x) \geq 0$.
4) For every $x \in \Omega$, $\Sigma g_i(x) = 1$.
5) $\operatorname{supp} g_i \subset O_i$, $i \in I$

(7) *Given the covering* (3) *of* Ω *there is a partition of the unity* $\{g_i : i \in I\}$ *of class* C^∞ *subordinated to this covering.*

Proof. Let (g_m) be the sequence obtained in (5). Let T be a mapping from N into I such that supp g_m is contained in $O_{T(m)}$, $m \in N$. If $i \in I$ and $T(m) \neq i$, for every positive integer m, let g_i be the function defined on Ω such that $g_i(x) = 0$, $x \in \Omega$. If $i \in I$ and $i \in T(N)$, let g_i be the function defined on Ω such that

$$g_i(x) = \sum_{m \in T^{-1}(i)} g_m(x), \quad x \in \Omega.$$

Then, for every $i \in I$, g_i is infinitely differentiable in Ω, $\operatorname{supp} g_i = \cup \{\operatorname{supp} g_m : m \in T^{-1}(i)\} \subset O_i$ if $i \in T(N)$, $\operatorname{supp} g_i = \emptyset \subset O_i$ if $i \notin T(N)$ and $g_i(x) \geq 0$, $x \in \Omega$. On the other hand, the family $\{\operatorname{supp} g_i : i \in I\}$ is locally finite and, for every $x \in \Omega$, $\Sigma g_i(x) = \Sigma g_m(x) = 1$. The proof is complete.

Now we shall construct a family of n-dimensional cubes covering Ω having properties that we shall use later.

If $\Omega = R^n$ we consider all the cubes

(8) $\{(x_1, x_2, \dots, x_n) : a_j \leq x_j \leq a_j + 1, a_j$ *integer*, $j = 1, 2, \dots, n\}$ *ordered in a sequence* (B_m) *of pairwise different elements.*

If $\Omega \neq R^n$, let β_1 be the family of all the cubes of the form (8) such that, if $Q \in \beta_1$, then the distance from Q to $R^n \sim \Omega$ is larger or equal than $\sqrt{n}$. Proceeding by recurrence, suppose we have obtained the families of cubes β_h, $1 \leq h \leq k$. We denote by β_{k+1} the collection of all cubes of the form

$$\{(x_1,x_2,\dots,x_n) : \frac{a_j}{2^k} \leq x_j \leq \frac{a_j+1}{2^k}, \ a_j \text{ integer}, j=1,2,\dots,n\}$$

such that if $Q \in \beta_{k+1}$ then the distance from Q to $R^n \sim \Omega$ is larger or equal than $\frac{\sqrt{n}}{2^k}$ and Q is not contained in any element of β_h, $1 \leq h \leq k$; therefore Q does not intersect the interior of any element of β_h, $1 \leq h \leq k$. We order the cubes of $\bigcup_{k=1}^{\infty} \beta_k$ in a sequence (B_m) of pairwise different elements.

In any of the cases we have consider for Ω we suppose that, for every positive integer m,

$$B_m = \{(x_1,x_2,\dots,x_n): \frac{a_j(m)}{2^{k(m)}} \leq x_j \leq \frac{a_j(m)+1}{2^{k(m)}}, \ j = 1,2,\dots,n\},$$

$a_j(m)$ being an integer and $k(m)$ a non-negative integer. Then the diameter of B_m is $\frac{\sqrt{n}}{2^{k(m)}}$.

(9) *Let* z *be a point in* R^n *whose distance to the cube* B_m *is less than* $\frac{1}{2^{k(m)+1}}$. *Then there is a cube* B_q *such that*

$$z \in B_q, \ B_q \cap B_m \neq \emptyset \text{ and } k(q) \leq h(m) + 1$$

Proof. Ig z belons to any cube B_q of the family

$$(10) \qquad \bigcup_{p=1}^{k(m)+1} \beta_p$$

then $k(q) \leq k(m)$ and thus the diameter of B_q is larger or equal than the diameter of B_m. Since the distance between z and B_m is less than $\frac{1}{2^{k(m)+1}}$ we have that $B_q \cap B_m \neq \emptyset$.

If z does not belongs to any element of the family (10), we find a cube

$$M = \{(x_1,x_2,\dots,x_n) : \frac{b_j}{2^{k(m)+1}} \leq x_j \leq \frac{b_j+1}{2^{k(m)+1}}, \ b_j \text{ integer}, \ j = 1,2,\dots, n\}$$

such that z is in M. Then the distance from M to B_m is less than $\frac{1}{2^{k(m)+1}}$

and accordingly M intersects B_m. Let δ and d be the distances from M and B_m to $R^n \sim \Omega$, respectively. Then

$$\delta \geq d - \frac{\sqrt{n}}{2^{k(m)+1}} \geq \frac{\sqrt{n}}{2^{k(m)}} - \frac{\sqrt{n}}{2^{k(m)+1}} = \frac{\sqrt{n}}{2^{k(m)+1}}$$

from where it follows that M is cube B_q of $\beta_{k(m)+2}$ and thus $k(q) = k(m)+1$, which completes the proof.

(11) *Given positive integers* m *and* p, *if* $B_m \cap B_p \neq \emptyset$, *then* $k(p) \leq k(m)+1$.

Proof. If $B_m \cap B_p \neq \emptyset$ we take a point z in $\overset{\circ}{B}_p$ whose distance to B_m is less than $\frac{1}{2^{k(m)+1}}$. We apply (9) to obtain an element B_q such that

$$z \in B_q,\ B_q \cap B_m \neq \emptyset \text{ and } k(q) \leq k(m)+1$$

Then B_q intersects $\overset{\circ}{B}_p$ and accordingly $B_q = B_p$, from where the conclusion follows.

(12) $\bigcup_{m=1}^{\infty} B_m = \Omega$.

Proof. If $\Omega = R^n$ the result is obvious. Suppose that $\Omega \neq R^n$ and that the property is not true in this case. We take a point z of Ω which is not in $\bigcup_{m=1}^{\infty} B_m$. Let d be the distance from z to $R^n \sim \Omega$. Let q be a positive integer such that

$$\frac{\sqrt{n}}{2^q} < d.$$

We can find integers $p_1, p_2, \ldots, p_n$ such that

$$z \in M = \{(x_1, x_2, \ldots, x_m) : \frac{p_j}{2^{q+1}} \leq x_j \leq \frac{p_j+1}{2^{q+1}}, j=1,2,\ldots,n\}.$$

The cube M is not contained in any cube of β_m, m = 1,2,..., q+1. On the other hand, if d_1 is the distance from M to $R^n \sim \Omega$ we have that

$$d_1 \geq d - \frac{\sqrt{n}}{2^{q+1}} > \frac{\sqrt{n}}{2^q} - \frac{\sqrt{n}}{2^{q+1}} = \frac{\sqrt{n}}{2^{q+1}}$$

from where it follows that M belongs to β_{q+2} which is a contradiction. Thus $\bigcup_{m=1}^{\infty} B_m \supset \Omega$. It is inmediate that $\bigcup_{m=1}^{\infty} B_m \subset \Omega$. The proof is complete.

For every positive integer m we set

$$A_m = \{(x_1,x_2,\dots,x_n) : \frac{a_j(m)}{2^{k(m)}} - \frac{1}{\sqrt{n}2^{k(m)+3}} \leqslant x_j$$

$$\leqslant \frac{a_j(m)+1}{2^{k(m)}} + \frac{1}{\sqrt{n}2^{k(m)+3}}, \ j = 1,2,\dots,n\}.$$

Since $B_m \subset \mathring{A}_m$ it follows that

$$\bigcup_{m=1}^{\infty} \mathring{A}_m \supset \Omega.$$

On the other hand, if z is a point of $\mathring{A}_m$ the distance from z to B_m is less $\frac{1}{2^{k(m)+2}}$ and, according to (9), we have that z belongs to Ω. Thus

(13) $$\bigcup_{m=1}^{\infty} \mathring{A}_m = \Omega.$$

(14) *Let* m *and* p *be positive integers. If* $B_m \cap B_p = \emptyset$ *then* $\mathring{A}_m \cap \mathring{A}_p = \emptyset$

Proof. Suppose $k(m) \geqslant k(p)$. If the property is not true we take a point z of irrational coordinates in $\mathring{A}_m \cap \mathring{A}_p$. Let d_1 and d_2 be the distances from z to B_m and to B_p respectively. From $B_m \cap B_p = \emptyset$ we deduce that the distance d between B_m and B_p is larger or equal than $\frac{1}{2^{k(m)}}$. We have that

$$d_1 \leqslant \frac{1}{2^{k(m)+3}}, \ d_2 \leqslant \frac{1}{2^{k(p)+3}}$$

Then we apply (8) to obtain two cubes B_q and B_{q2} such that

$$z \in B_q, \ B_q \cap B_m \neq \emptyset, \ k(q) \leqslant k(m) + 1,$$

(15) $$z \in B_{q_1}, \ B_{q_1} \cap B_p \neq \emptyset, \ k(q_1) \leqslant k(p) + 1$$

The coordinates of z are irrational and therefore z belongs to $\mathring{B}_q$; thus

$$\mathring{B}_q \cap B_{q_1} \neq \emptyset$$

from where it follows that $q = q_1$.

Now we apply (11) to obtain that

$$k(m) \leq k(q)+1$$

and according to (15).

$$k(m) \leq k(q)+1 \leq k(p)+2.$$

Then

$$\frac{1}{2^{k(m)}} \leq d \leq d_1+d_2 \leq \frac{1}{2^{k(m)+3}} + \frac{1}{2^{k(p)+3}} \leq \frac{1}{2^{k(m)+3}} + \frac{1}{2^{k(m)+1}} < \frac{1}{2^{k(m)}} \text{ --}$$

which is a contradiction.

(16) *Given the integers*

$$1 \leq r_1 < r_2 < \ldots < r_p$$

such that $p > 4^n$ *then*

(17) $$\bigcap_{j=1}^{p} \overset{\circ}{A}_{r_j} = \emptyset.$$

Proof. There are 4^n cubes of the form

$$\{(x_1,x_2,\ldots,x_n): \frac{b_j}{2^{k(r_1)+1}} \leq x_j \leq \frac{b_j+1}{2^{k(r_1)+1}},\ b_j \text{ integer}, j=1,2,\ldots,n\}$$

intersecting B_{r_1}. Suppose these cubes are

(18) $$M_1, M_2, \ldots, M_{4^n}.$$

If for every positive integer j, $1 \leq j \leq p$, $\overset{\circ}{A}r_1 \cap \overset{\circ}{A}r_j$ is non-void, then we apply (14) and (11) to obtain that

$$Br_j \cap Br_1 \neq \emptyset,\ k(rj) \leq k(r_1)+1.$$

Consequently B_{r_j} contains some element of (18) and therefore the number of cubes of

$$\overset{\circ}{A}_{rj},\ j = 1,2,\ldots,p$$

intersecting $\overset{\circ}{A}_{r1}$ is less or equal than 4^n. Accordingly (17) is verified.

Now we set

$$I = \{(x_1,x_2,\ldots,x_n) : -1 \leq x_j \leq 1,\ j = 1,2,\ldots, n\}.$$

For every positive integer m, let $g_m : R^n \longrightarrow R^n$ be the mapping defined by

$$g_m(x) = \left(\frac{1}{2^{k(m)+1}}\left(2a_1(m)+1+x_1+\frac{x_1}{4\sqrt{n}}\right),\ldots,\frac{1}{2^{k(m)+1}}\left(2a_n(m)+1+x_n+\frac{x_n}{4\sqrt{n}}\right)\right)$$

for every $x = (x_1,x_2,\ldots,x_n)$. The functions g_m maps I in A_m.

We denote by G the linear space over K of the sequences (f_m) of $\mathcal{D}(I)$ such that, for every multi-index α and for every positive integer q

$$\lim_m 2^{k(m)q}\sup\{|D^\alpha f_m(x)| : x \in I\} = 0$$

We set

$$T((f_m)) = \Sigma f_m \circ g_m^{-1}$$

Since

$$\{\mathring{A}_m : m = 1,2,\ldots\}$$

is a locally finite covering of Ω, $T((f_m))$ is a function defined on R^n which is infinitely differentiable in Ω. For every multe-index, α we have that

(19)
$$D^\alpha T((f_m)) = D^\alpha\, \Sigma f_m \circ g_m^{-1} = \Sigma D^\alpha(f_m \circ g_m^{-1})$$
$$= \Sigma\left(\frac{4\sqrt{n}\,2^{k(m)+1}}{4\sqrt{n}+1}\right)^{|\alpha|}(D^\alpha f_m)\circ g_m^{-1}$$
$$= \left(\frac{8\sqrt{n}}{4\sqrt{n}+1}\right)^{|\alpha|}\Sigma 2^{k(m)|\alpha|}(D^\alpha f_m)\circ g_m^{-1}$$

(20) *If* (f_m) *belongs to* G *then given* $\varepsilon > 0$ *and a multi-index* α *there is a compact set* H *in* Ω *such that*

$$|D^\alpha T((f_m))(x)| < \varepsilon\ ,\ x \in \Omega \sim H.$$

Proof. Given $\varepsilon > 0$ and a multi-index α we find a positive integer q such that

$$2^{k(m)|\alpha|}\sup\{|D^\alpha f_m(x)| : x \in I\} < \frac{1}{4^n}\left(\frac{4\sqrt{n}+1}{8\sqrt{n}}\right)^{|\alpha|}\varepsilon,\ m \geq q.$$

If H is a compact subset of Ω verifying

$$A_m \subset H,\ m < q,$$

and if z is any point of $\Omega \sim H$ there are p positive integers $r_1,r_2,\ldots,r_p$,

$p \leqslant 4^n$, such that

$$z \notin A_m, \; m \neq r_j, r_j \geqslant q, \; j = 1,2,\ldots,p,$$

according to (16). Then by (19) it follows that

$$|D^\alpha(\Sigma f_m \circ g_m^{-1})|(z) = \left(\frac{8\sqrt{n}}{4\sqrt{n}+1}\right)^{|\alpha|} \sum_{j=1}^{p} 2^{k(r_j)|\alpha|} |(D^\alpha f_{rj}) \circ g_{rj}^{-1}(z)|$$

$$\leqslant \left(\frac{8\, n}{4\sqrt{n}+1}\right)^{|\alpha|} 4^n \frac{1}{4^n} \left(\frac{4\sqrt{n}+1}{8\sqrt{n}}\right)^{|\alpha|} \varepsilon = \varepsilon.$$

The proof is complete.

(21) *If* (f_m) *belongs to* G, *the function*

$$f = \Sigma f_m \circ g_m^{-1}$$

is infinitely differentiable in R^n *and vanishes in* $R^n \sim \Omega$ *as well as all its partial derivatives of all orders.*

Proof. According to (20), f is obviously a continuous function in R^n and it vanishes in $R^n \sim \Omega$. Proceeding by recurrence suppose that, for a multi-index $\alpha = (p_1, p_2, \ldots, p_n)$, there exists $D^\alpha f$ in R^n, it is continuous in this space and vanishes in $R^n \sim \Omega$. Given $\varepsilon > 0$ we apply (20) to obtain a compact set H in Ω such that

$$|D^\beta f(x)| < \varepsilon, \; x \in \Omega \sim H,$$

being $\beta = (p_1, \ldots, p_{j-1}, p_j+1, p_{j+1}, \ldots, p_n)$. If $z = (z_1, z_2, \ldots, z_n)$ is any point of $R^n \sim \Omega$ let h be a positive number such that

$$y = (z_1, \ldots, z_{j-1}, z_j+k, z_{j+1}, \ldots, z_n) \notin H, |k| < h.$$

We have that

$$\frac{D^\alpha f(y) - D^\alpha f(z)}{k} = 0, \; y \notin \Omega, \; 0 < |k| < h.$$

If y belongs to Ω let z_1 be the point of the closed segment $[y,z]$ which is closed to y and which is on the boundary of $[y,z] \cap \Omega$ in $[y,z]$. Then $D^\alpha f$ is continuous in $[y, z_1]$ and there exists $D^\beta f$ in the open segment $]y, z_1[$. Therefore we apply the mean value theorem to obtain.

$$|D^\alpha f(y) - D^\alpha f(z)| = |D^\alpha f(y) - D^\alpha f(z_1)| \leqslant |k|\varepsilon, \; 0 < |k| < h$$

and therefore

$$\frac{|D^{\alpha}f(y)-D^{\alpha}f(z)|}{|k|} < \varepsilon, \ 0 < |k| < h,$$

and thus there exists $D^{\beta}f$ in R^n and it vanishes in $R^n \sim \Omega$. The function $D^{\beta}f$ is continuous in R^n according to (20).

(22) *If* F *is a closed subset of* R^n *there is a real function* f *defined on* R^n *and infinitely differentiable such that* $f(x) > 0$, *for every* $x \in R^n \sim F$, *and it vanishes in* F *as well as all its partial derivatives of all orders.*

Proof. If $F = R^n$ we take f as the identically nulle function. If $F \neq R^n$ let Ω be $R^n \sim F$. Let f be an infinitely differentiable function defined on R^n whose support is I and $f(x) > 0$ for every x in $\mathring{I}$. We set

$$f_m = 2^{-m(k(m)+1)}f, \ m = 1,2,...$$

Given a multi-index α and a positive integer q we have that

$$\lim_m 2^{k(m)q}\sup\{|D^{\alpha}f_m(x)| : x \in I\}$$

$$= \lim_m 2^{-m}2^{-(m-q)k(m)}\sup\{|D^{\alpha}f\ (x)| : x \in I\} = 0$$

and therefore (f_m) is in G and thus $f = \Sigma f_m \circ g_m^{-1}$ is infinitely differentiable in R^n and vanishes in F as well as all its partial derivatives of all orders. On the other hand, given a positive integer m, $f_m \circ g_m^{-1}$ is strictly positive in $\mathring{A}_m$ and therefore $f(x) > 0$ for every $x \in \Omega$. The proof is complete.

(23) *Let* F *be a non-void closed subset of* R^n. *Let* Ω *be an open subset of* R^n *distinct from* F *and containing* F. *Then there is an infinitely differentiable real funtion* f *defined on* R^n *such that* $f(x) = 1$, $x \in F$, $f(x) > 0$, $x \in \Omega$, *and that it vanishes in* $R^n \sim \Omega$ *as well as all its partial derivatives of all orders.*

Proof. We apply (22) to obtain two infinitely differentiable real functions f_1 and f_2 defined on R^n such that

$$f_1(x) > 0, \ x \in \Omega, \ f_2(x) > 0, \ x \in R^n \sim F$$

and for every multi-index α

$$D^{\alpha}f_1(x) = 0, \ x \in R^n \sim \Omega, \ D^{\alpha}f_2(x) = 0, \ x \in F.$$

Then

$$f = \frac{f_1}{f_1+f_2}$$

is the desired function.

10. EXTENSION LINEAR OPERATOR. Given the real numbers

$$b_1 < d_1, a_j < b_j, j = 1,2,\dots, n,$$

let A, B and D be the n-dimensional intervals

$$\{(x_1,x_2,\dots,x_n) : a_j \leq x_j \leq b_j, j = 1,2,\dots,n\}$$

$$\{(x_1,x_2,\dots,x_n) : a_1 \leq x_1 \leq d_1, a_j \leq x_j \leq b_j,\ j = 2,3,\dots,n\}$$

$$\{(x_1,x_2,\dots,x_n) : b_1 < x_1 \leq d_1, a_j \leq x_j \leq b_j,\ j = 2,3,\dots, n\},$$

respectively. Let E be the subspace of $E(A)$ of all those functions which vanish in the face of A

$$\{(x_1,x_2,\dots,x_n) : x_1 = a_1,\ a_j \leq x_j \leq b_j,\ j = 2,3,\dots,n\}$$

as well as their partial derivatives of all orders. If the function f belongs to E we suppose f extended to the set

$$M = \{(x_1,x_2,\dots,x_n) : -\infty < x_1 \leq b_1, a_j \leq x_j \leq b_j,\ j = 2,3,\dots, n\}$$

setting $f(x) = 0$, when $x = (x_1,x_2,\dots,x_n)$ is a point of M with $x_1 \leq a_1$. Obviously the extended function is infinitely differentiable in M. If f belongs to E we set X f to denote the function defined in B by

$$(1)\qquad Xf(x) = \sum_{m=1}^{\infty} (-1)^{m+1} \frac{f(b_1+3^m(b_1-x_1),x_2,\dots,x_n)}{(2m-1)!} \left(\frac{\pi}{2}\right)^m,$$

$$x = (x_1,x_2,\dots,x_m) \in D,$$

$$X f(x) = f(x),\ x \in A.$$

(2) *If* f *belongs to* E, *then* Xf *belongs to* $E(B)$.

Proof. When $(x_1,x_2,\dots,x_n)$ varies in D, given the positive integer m, the point $(b_1+3^m(b_1-x_1),x_2,\dots,x_m)$ varies in M. Therefore, given a multi-index $\alpha = (p_1,p_2,\dots,p_n)$, the partial derivative of order α of the m-th term of the series (1) exists and it is continuous in D. This term is bounded by the m-th term of the convergent series

$$\sum_{m=1}^{\infty} \frac{(3^{p_1})^m}{(2m-1)!} \left(\frac{\pi}{2}\right)^m \sup\{|D^{\alpha}f(y)| : y \in A\} \tag{3}$$

and therefore (1) can be derived term by term α times in D. Consequently Xf is infinitely differentiable in D. Obviously, Xf is infinitely differentiable in A.

On the other hand, given the point $z = (b_1, z_2, \ldots, z_n)$ of A we have that

$$\lim_{\substack{x \to z \\ x \in D}} D^{\alpha}Xf(x) = \sum_{m=1}^{\infty} (-1)^{m+1} \frac{(-3^m)^{p_1} D^{\alpha}f(z)}{(2m-1)!} \left(\frac{\pi}{2}\right)^m$$

$$= (-1)^{p_1} D^{\alpha}f(z) \sum_{m=1}^{\infty} (-1)^{m+1} \frac{\left(\frac{\pi}{2} 3^{p_1}\right)^m}{(2m-1)!}$$

$$= (-1)^{p_1} D^{\alpha}f(z) \operatorname{sen}\frac{\pi}{2} 3^{p_1} = D^{\alpha}f(z)$$

from where it follows, according to 8.(1), that Xf belongs to $E(B)$.

(4) $X : E \longrightarrow E(B)$ *is a continuous linear operator.*

Proof. Obviously, $X : E \longrightarrow E(B)$ is a linear operator. Given a positive integer r, if f is in E, it follows that

$$|Xf|_r \leq e^{\frac{\pi}{2} 3^r} |f|_r$$

since (3) is a majorizing series of the partial derivative of order α of (1). Thus T is continuous.

Now we take a real number a such that

$$a_1 < a < b_1 < d_1.$$

We apply 9.(2) to obtain an infinitely differentiable real function g defined in R whose support is $[a_1, d_1]$ which takes the value one in $[a, b_1]$. If f is any element of $E(A)$ we set Yf to denote the element of E such that

$$Yf(x_1, x_2, \ldots, x_n) = f(x_1, x_2, \ldots, x_n) g(x_1), (x_1, x_2, \ldots, x_n) \in A.$$

(5) $Y : E(A) \longrightarrow E$ *is a continuous linear operator.*

Proof. Obviously $Y : E(A) \longrightarrow E$ is a linear operator. Given the multi-index $\alpha = (p_1,p_2,\ldots,p_n)$, we write $\alpha^j = (j, p_2,\ldots,p_n)$, $j = 0,1,2,\ldots,p_1$. Then

$$D^{\alpha}(Yf)(x_1,x_2,\ldots,x_n) = \sum_{j=0}^{p_1} \binom{p_1}{j} D^{\alpha^j} f(x_1,x_2,\ldots,x_n) D^{p_1-j} g(x_1)$$

from where it follows that, for every positive integer r,

$$|Yf|_r \leq |g|_r \sum_{j=1}^{p_1} \binom{p_1}{j} |f|_r = 2^{p_1} |g|_r |f|_r$$

and therefore Y is continuous.

For every f of $E(A)$ we set

$$Tf(x_1,x_2,\ldots,x_n) = (XoY)f(x_1,x_2,\ldots,x_n), (x_1,x_2,\ldots,x_n) \in D,$$

$$Tf(x_1,x_2,\ldots,x_n) = f(x_1,x_2,\ldots,x_n), (x_1,x_2,\ldots,x_n) \in A.$$

Since f and Yf coincide on

$$\{(x_1,x_2,\ldots,x_n) : a \leq x_1 \leq b_1, a_j \leq x_j \leq b_j, j = 2,3,\ldots,n\}$$

it follows that they have the same partial derivatives of all orders in the points of A of the form $(b_1,x_2,\ldots,x_n)$ and consequently Tf is an element of $E(B)$.

(6) $T : E(A) \longrightarrow E(B)$ *is a continuous linear operator.*

Proof. According to (4) and (5), given a multi-index α and a positive integer r with $|\alpha| \leq r$ there is a constant k such that

$$|D^{\alpha}(XoY)f(x_1,x_2,\ldots,x_n)| \leq k|f|_r, (x_1,x_2,\ldots,x_n) \in D$$

On the other hand,

$$|D^{\alpha}f(x_1,x_2,\ldots,x_n)| \leq |f|_r, (x_1,x_2,\ldots,x_n) \in A.$$

Then

$$|D^{\alpha}(Tf)(x_1,x_2,\ldots,x_n)| \leq (k+1)|f|_r, (x_1,x_2,\ldots,x_n) \in B,$$

and therefore there is $h > 0$ such that

$$|Tf|_r \leq h|f|_r$$

and the conclusion follows.

We say that T is an extension operator since given a function f belonging to the space $E(A)$, Tf is a function defined on B whose restriction to A coincides with f.

Set the real numbers

$$a_j < h_j < b_j, \ j = 2,3,\dots, n,$$

and

$$H_1 = \{(x_1,h_2,\dots,h_n) : a_1 < x_1 < b_1\}$$

$$H_2 = \{(x_1,h_2,\dots,h_n) : a_1 < x_1 < d_1\}.$$

We shall use result (7) in the following section.

(7) *If* f *is an element of* $E(A)$ *vanishing in* H_1 *as well as all its partial derivatives of all orders, then* Tf *vanishes in* H_2 *as well as all its derivatives of all orders.*

Proof. If f verifies the stated condition then Yf vanishes in the set

$$\{(x_1,h_2,\dots,h_n) : -\infty < x_1 < b_1\}$$

as well as all its derivatives of all orders. Consequently (XoY)f vanishes in H_2 as well as all its derivatives of all orders. Now the conclusion is immediate.

We take a real number $c_1 < a_1$ and we set

$$C = \{(x_1,x_2,\dots,x_n) : c_1 < x_1 < d_1, \ a_j < x_j < b_j, \ j = 2,3,\dots,n\}.$$

Let $\psi: R^n \longrightarrow R^n$ be the mapping defined by

$$\psi(x_1,x_2,\dots,x_n) = -(x_1,x_2,\dots,x_n), (x_1,x_2,\dots,x_n) \in R^n$$

Then $-a_1 < -c_1$ and

$$\psi(B) = \{(x_1,x_2,\dots,x_n) : -d_1 < x_1 < -a_1, \ -b_j < x_j < -a_j, j=2,3,..n\}$$

$$\psi(C) = \{(x_1,x_2,\dots,x_n) : -d_1 < x_1 < -c_1, \ -b_j < x_j < -a_j, j=2,3,..n\}$$

and therefore an extension operator $U : E(\psi(B)) \longrightarrow E(\psi(C))$ which is linear and continuous can be constructed by the method explained above. If f belongs to $E(B)$, let Vf be the element of $E(C)$ defined by $U(f \circ \psi^{-1}) \circ \psi$. It is not difficult to prove the following result

(8) $V: E(B) \longrightarrow E(C)$ *is a continuous linear extension operator.*

(9) *There is a continuous linear extension operator* $W : E(A) \longrightarrow E(C)$.

Proof. Take for W the linear operator $V \circ T$.

We take real numbers

$$c_1 < a_j < b_j < d_j,\ j = 1,2,\dots, n.$$

We set

$$L = \{(x_1,x_2,\dots,x_n) : c_j \leq x_j \leq d_j,\ j = 1,2,\dots, n\}.$$

We set, for every positive integer p with $1 \leq p \leq n$,

$$A_p = \{(x_1,x_2,\dots,x_n): c_j \leq x_j \leq d_j, j=1,2,\dots,p,$$

$$a_j \leq x_j \leq b_j,\ j=p+1,\dots,n\}.$$

Then $L = A_n$, $C = A_1$. Using the former method continuous, a linear extension operator $W_p : E(A_p) \longrightarrow E(A_{p+1})$ can be constructed.

(10) *There is a continuous linear extension operator* $Z : E(A) \longrightarrow E(L)$.

Proof. Take for Z the operator $W_{n-1} \circ W_{n-2} \circ \dots \circ W_1 \circ W$.

Extension linear operators can be found in SEELEY [1], MITIAGIN [1], OGRODZKA [1] and VALDIVIA [16]. As far as we know this is the first time that the function sen $\frac{\pi}{2}$ x has been used to construc continuous linear extension operators.

11. SOME COMPLEMENTED SUBSPACE OF $E(H)$. Let P_n be the linear space over K of all infinitely differentiable K-valued functions defined on R^n which are 2π-periodic with respect to each of its variables. Given a positive integer r we set

$$|f|_r = \sum_{|\alpha|\leq r} \sup\{|D^\alpha f(x_1,x_2,\dots,x_n)| : -\pi \leq x_j \leq \pi,\ j=1,2,\dots,n\}$$

for every f in P_n. Then $|.|_r$, $r = 1,2,\dots$, is a system of norms in P_n for a metrizable locally convex topology. In what follows we suppose P_n endowed with this topology. It can be checked that P_n is a Fréchet space (see the proof of the completeness of $\mathcal{M}_n$).

We set

$$H = \{(x_1,x_2,\ldots,x_n) : -2\pi < x_j < 2\pi,\ j = 1,2,\ldots,n\}.$$

Let F be the subspace of $\mathcal{E}(H)$ of all those functions f which are restriction to H of some function of $\mathcal{P}_n$. Obviously F is isomorphic to $\mathcal{P}_n$ which is a Fréchet space and therefore F is a closed subspace of $\mathcal{E}(H)$.

We set

$$P = \{(p_1,p_2,\ldots,p_n) : p_j \text{ integer},\ j = 1,2,\ldots,n\}.$$

If $p = (p_1,p_2,\ldots,p_n)$ is any element of P we set

$$H_p = \{(x_1,x_2,\ldots,x_n) : 2p_j\pi < x_j < (2p_j+4)\pi, j=1,2,\ldots,n\}.$$

Let $g_p : R^n \longrightarrow R^n$ be the mapping defined by

$$g_p(x) = (x_1+(2p_1+2)\pi,\ldots,x_n+(2p_n+2)\pi),$$

$$x = (x_1,x_2,\ldots,x_n) \in R^n$$

Then g_p maps H in H_p. Obviously, if f belongs to $\mathcal{D}(H)$ we have that $f \circ g_p^{-1}$ belongs to $\mathcal{D}(H_p)$. If $q = (q_1,q_2,\ldots,q_n)$ is an element of P and if $x = (x_1, x_2,\ldots,x_n)$ belongs to R^n we have that

$$g_{p+q}(x) = (x_1+(2p_1+2q_1+2)\pi,\ldots,x_n+(2p_n+2q_n+2)\pi)$$

$$= g_p(x_1+2q_1\pi,\ldots,x_n+2q_n\pi).$$

(1) *If* f *belongs to* $\mathcal{D}(H)$ *we have that*

$$Xf = \sum_{p \in P} f \circ g_p^{-1}$$

belongs to $\mathcal{P}_n$.

Proof. Obviously Xf is defined on R^n and it is infinitely differentiable Given a point $x = (x_1,x_2,\ldots,x_n)$ of R^n an element $q = (q_1,q_2,\ldots, q_n)$ of P we have that

$$Xf(x) = \sum_{p \in P} (f \circ g_p^{-1})(x) = \sum_{p \in P} (f \circ g_{p-q}^{-1})(x)$$

$$= \sum_{p \in P} (f \circ g_p^{-1})(x_1+2q_1\pi,\ldots,x_n+2q_n\pi)$$

$$= Xf(x_1+2q_1\pi,\ldots,x_n+2q_n\pi).$$

The proof is complete.

(2) $X : \mathcal{D}(H) \longrightarrow P_n$ *is a continuous linear operator.*

Proof. It is not difficult to check the existence of a positive integer m such that if x belongs to R^n then x belongs to a number of elements of the family

$$\{H_p : p \in P\}$$

which is not larger than m. Then, if f belongs to $\mathcal{D}(H)$ it follows that

$$|D^\alpha(Xf)(x)| = |D^\alpha(\sum_{p \in P} f \circ g_p^{-1})(x)|$$

$$\leq \sum_{p \in P} |D^\alpha(f \circ g_p^{-1})(x)| \leq m \sup\{|D^\alpha f(y)| : y \in H\}$$

for every multi-index α.

Consequently, given a positive integer r we have that

$$|Xf|_r \leq m\,|f|_r$$

and the conclusion follows since X is obviously linear.

Let h be an element of $\mathcal{D}(H)$ such that $h(x) > 0$ for every $x \in \overset{\circ}{H}$. We set

$$k = \sum_{p \in P} h \circ g_p^{-1}$$

The function K belongs to P_n and it is strictly positive for every point of R^n. We set $\phi = \frac{h}{k}$. For every f of $E(H)$ we set $Yf = f\phi$. $Y : E(H) \longrightarrow \mathcal{D}(H)$ is a continuous linear operator. We write Tf to denote the restriction of the function $(X \circ Y)f$ to H. It follows that $T : E(H) \longrightarrow E(H)$ is a continuous linear operator.

(3) T *is a continuous projection from* $E(H)$ *into itself such that* $T(E(H)) = F$.

Proof, Obviously $T(E(H))$ is contained in F. Let g be an element of P_n. We write g^* to denote the restriction of g to H. Then

$$(X \circ Y)g^* = \sum_{p \in P} (g^* \phi) \circ g_p^{-1}$$

$$= \sum_{p \in P} (g\phi) \circ g_p^{-1} = \sum_{p \in P} (g \circ g_p^{-1}).(\phi \circ g_p^{-1})$$

and since $g \circ g_p^{-1} = g$ it follows that

$$(X \circ Y)g^* = g \sum_{p \in P} \phi \circ g_p^{-1} = g \sum_{p \in P} \frac{h \circ g_p^{-1}}{k \circ g_p^{-1}}$$

$$= g \sum_{p \in P} \frac{h \circ g_p^{-1}}{k} = g$$

and therefore $Tg^* = g^*$. Since T is continuous and linear the conclusion follows.

(4) *P_n is isomorphic to a complemented subspace of $E(H)$.*

Proof. We have that

$$T^{-1}(o) + F = E(H), \quad T^{-1}(o) \cap F = \{0\}.$$

Since F is isomorphic to P_n the conclusion follows.

(5) *M_n is a complemented subspace of P_n.*

Proof. Given a positive integer j, $1 \leq j \leq n$, let E_j be the subspace of P_n of all those functions $f(x_1,x_2,\ldots,x_n)$ which are even with respect to every variable $x_1,x_2,\ldots,x_j$. Obviously E_j is a closed subspace of P_n and M_n coincides with E_n.

Let F_1 be the subspace of P_n of all those functions which are odd with respect to the variable x_1. Then F_1 is a closed subspace of P_n. If f belongs to $E_1 \cap F_1$ we have that

$$f(x) = f(-x_1,x_2,\ldots,x_n), f(x) - f(-x_1,x_2,\ldots,x_n)$$

for every $x = (x_1,x_2,\ldots,x_n)$ of R^n and adding both equalities it follows that $2f(x) = 0$ and therefore $E_1 \cap F_1 = 0$. On the other hand, if f belongs to P_n, we set g_1 and g_2 to denote the functions defined by

$$g_1(x) = \frac{1}{2}(f(x_1,x_2,\ldots,x_n)+f(-x_1,x_2,\ldots,x_n))$$

$$g_2(x) = \frac{1}{2}(f(x_1,x_2,\ldots,x_n)-f(-x_1,x_2,\ldots,x_n))$$

for every $x = (x_1,x_2,\ldots,x_n)$ of R^n. Then $f = g_1+g_2, g_1 \in E_1, g_2 \in F_1$, and therefore P_n is the topological direct sum of E_1 and F_1; thus E_1 is a complemented subspace of P_n. Proceeding by recurrence suppose that, for a positive integer p, $1 \leqslant p < n$, E_p is a complemented subspace of P_n. Let F_{p+1} be the subspace of E_p of all those functions which are odd with respect to the variable x_{p+1}. Then E_p is the topological sum of E_{p+1} and F_{p+1}. Thus E_{p+1} is a complemented subspace of P_n. Consequently, M_n is a complemented subspace of P_n.

(6) $E(H)$ *has a complemented subspace isomorphic to* s.

Proof. It is an inmediate consequence of (4), (5) and 7.(14).

We set

$$M = \{(x_1,x_2,\ldots,x_n) : 0 \leqslant x_j \leqslant \pi,\ j = 1,2,\ldots, n\}.$$

Given the real numbers

$$0 < a_j < b_j < \pi,\ j = 1,2,\ldots,n,$$

we write

$$J = \{(x_1,x_2,\ldots,x_n) : a_j \leqslant x_j \leqslant b_j,\ j = 1,2,\ldots,n\}.$$

Let $V : E(J) \longrightarrow E(M)$ be a continuous linear extension operator. Let ψ be a function of $D(M)$ such that $\psi(x) = 1$ when x varies in J. For every f of $E(J)$ we set

$$Tf = \psi .\ Vf.$$

It is immediate that $T : E(J) \longrightarrow D(M)$ is a continuous linear operator (see 12.(2)).

(7) $D(M)$ *has a complemented subspace isomorphic to* $E(J)$.

Proof. Since $T : E(J) \longrightarrow D(M)$ is an extension operator we have that

$$|f|_r \leqslant |Tf|_r$$

for every positive integer r and every f in $E(J)$. Therefore T is an isomorphic from $E(J)$ into $D(M)$.

Let $Z: D(M) \longrightarrow E(J)$ be the mapping which associates to every function f of $D(M)$ its restriction Zf to J. Z is a continuous linear operator from $D(M)$ into $E(J)$.

Now it is immediate to check that T o Z is a projection from $\mathcal{D}(M)$ into itself such that (T o Z) $(\mathcal{D}(M))$ = $T(E(J))$. Consequently the subspace $T(E(J))$ of $\mathcal{D}(M)$ is complemented in $\mathcal{D}(M)$ and it is isomorphic to $E(J)$.

(8) *$E(J)$ is isomorphic to a complemented subspace of* s.

Proof. Since s is isomorphic to M_n it is enough to show that $E(J)$ is isomorphic to a complemented subspace of M_n.

Let G be the subspace of $E(M)$ of all the restrictions of the elements of M_n to M. Then G is isomorphic to M_n and consequently G is isomorphic to s, according to 7.(14).

It is immediate that $\mathcal{D}(M)$ is contained in G and therefore the extension operator T defined above can be considered as an operator from $E(J)$ into G. We represent by $Z: G \longrightarrow E(J)$ the operator which associates to every f of G its restriction Zf to J. Then T o Z is a projection from G into itself such that (T o Z) (G) = $T(E(J))$. Consequently the subspace $T(E(J))$ of G is complemented and isomorphic to $E(J)$.

(9) *If Q is an n-dimensional compact interval of R^n, then $E(Q)$ is isomorphic to* s.

Proof. There is an affine mapping $\phi:R^n \longrightarrow R^n$ such that $\phi(Q) = J$. We associate to every f of $E(Q)$ the function Λf of $E(J)$ defined by f o ϕ. Λ is an isomorphism from $E(Q)$ onto $E(J)$ and thus we apply (8) to obtain that $E(Q)$ is isomorphic to a complemented subspace of s.

By (6), $E(Q)$ has a complemented subspace isomorphic to s. Now we apply 2.(8) to obtain that $E(Q)$ is isomorphic to s.

(10) *P_n is isomorphic to* s.

Proof. According to (4) and (9), P_n is isomorphic to a complemented subspace of s. By (5), P_n has a complemented subspace isomorphic to s. Now we apply 2.(8) to obtain that P_n is isomorphic to s.

If we fix r (n-1)-dimensional faces in H, $0 \leq r < n$, $C_1, C_2, \ldots, C_r$, let G_1 be the subspace of $E(H)$ whose elements f vanish in C_j, $j = 1,2,\ldots,r$, as well as all their derivatives of all orders.

Suppose the face of H

$$\{x_1,x_2,\ldots,x_n) : x_1 = -2\pi, -2\pi \leq x_j \leq 2\pi, j=2,3,\ldots,n\}$$

distinct from C_j, $j = 1,2,\dots,r$. We set

$$Q_1=\{(x_1,x_2,\dots,x_n) : -2\pi < x_1 < 0, -2\pi < x_j < 2\Pi, j=2,3,\dots,n\}$$

$$Q = (x_1,x_2,\dots,x_n) : 0 < x_1 < 2\pi, - 2\pi < x_j < 2\pi, j=2,3,\dots,n\}$$

Let $W: E(Q_1) \longrightarrow E(H)$ be an extension operator of the type constructed before. We take an infinitely differentiable real function ψ defined on R^n with value one in Q_1 and whose support is disjoint with the face

(11) $\{(x_1,x_2,\dots,x_n) : x_1 = 2\pi, - 2\pi < x_j < 2\pi, j=2,3,\dots,n\}$.

Let G_2 be the subspace of $E(Q_1)$ of all the restriction to Q_1 of the elements of G_1, For every f of G_2 we set $Tf = \psi . Wf$.

(12) $T : G_2 \longrightarrow G_1$ *is a continuous linear operator.*

Proof. Let f be an element of G_2. If some face C_j is the face (11), Tf vanishes obvioysly in (11) as well as all its derivatives of all orders. If the face C_j is distinct from (11) there is an integer m, $1 < m < n$, such that C_j is defined by one of the following expresions:

$\{(x_1,x_2,\dots,x_n) : x_m = -2\pi, -2\pi < x_j < 2\pi, j = 1,2,\dots,$

$m-1,m+1,\dots,n\}$

$\{(x_1,x_2,\dots,x_n) : x_m = 2\pi, -2\pi < x_j < 2\pi, j=1,2,\dots,$

$m-1,m+1,\dots,n\}$.

Then f vanishes in $C_j \cap Q_1$ as well as all its derivatives of all orders and applying 10.(7), Tf vanishes in C_j as well as all its derivatives of all orders. Thus Tf belongs to G_1. Finally T is obviously linear and continuous.

Let G_3 be the subspace of G_1 of all the functions vanishing in Q_1. Then G_3 is isomorphic to subspace of $E(Q)$ whose elements f vanish in the r+1 faces

$\{(x_1,x_2,\dots,x_n) : x_1 = 0, -2\pi < x_j < 2\pi, j = 1,\dots,n\}$

$C_j \cap Q, j = 1,2,\dots,r,$

as well as all their derivatives of all orders.

On the other hand, G_1 is the topological direct sum of $T(G_2)$ and G_3

since T is an extension operator. Therefore we have proved the following result:

(13) *There is an n-dimensional compact interval Q of R^n and there are r+1(n-1) -dimensional faces of Q, $F_1, F_2, \ldots, F_{r+1}$ such that the subspace G_3 of $E(Q)$ whose elements f vanish in F_j, $1 \leq j \leq r+1$, as well as all their partial derivatives of all orders is isomorphic to a complemented subspace of G_1.*

By repeated application of result (13) we find n-dimensional compact intervals in R^n

$$P_o, P_1, \ldots, P_{2n}$$

and subspaces L_j of $E(P_j)$, $j = 0, 1, \ldots, 2n$, such that the elements of L_j vanish in j (n-1)-dimensional fases of P_j as well as all their derivatives of all orders and such that, for $1 \leq j \leq 2n$, L_j is isomorphic to a complemented subspace of L_{j-1}. Consequently L_{2n} is isomorphic to a complemented subspace of $E(P_o)$. On the other hand, it is obvious that L_{2n} coincides with $D(P_{2n})$. Result (14) follows easily

(14) *If Q is an n-dimensional compact interval of R^n then $D(Q)$ is isomorphic to a complemented subspace of $E(Q)$.*

(15) *If Q is an n-dimensional compact interval of R^n, then $D(Q)$ is isomorphic to* s.

Proof. It is an obvious consequence of (7), (8), (14), and 2.(8).

Result (10) can be found in GROTHENDIECK [1]. Theorems (9) and (15) are due to MITIAGIN [1]. The proofs given here are original.

12. REPRESENTING THE SPACES $E(\Omega)$ AND $E'(\Omega)$. Let Ω be a non-void open set in the space R^n. We denote by $E(\Omega)$ the linear space over K of all K-valued infinitely differentiable functions defined on Ω. Given a positive integer r and a compact set H contained in Ω we set

$$|f|_{r,H} = \sum_{|\alpha| \leq r} \sup\{|D^\alpha f(x)| : x \in H\}$$

for every f of $E(\Omega)$.

When H varies over all the compact sets of Ω and r runs through the positive integers we have that $|.|_{r,H}$ is a system of seminorms on $E(\Omega)$ defining a locally convex topology. In what follows we suppose $E(\Omega)$ endowed with this topology. It is not difficult to check that $E(\Omega)$ is complete.

Now we take a sequence of compact sets of Ω

(1) (L_m), *with* $L_m \subset \overset{\circ}{L}_{m+1}, m=1,2,\ldots, \bigcup_{m=1}^{\infty} L_m = \Omega$

The family $|.|_r$, L_m, $r,m = 1,2,\ldots$, is a fundamental system of seminorms for the topology of $E(\Omega)$. Consequently $E(\Omega)$ is a Fréchet space. $E'(\Omega)$ denotes the strong dual of $E(\Omega)$.

Let (Q_m) be a sequence of n-dimensional compact cubes in Ω such that

$$\{\overset{\circ}{Q}_m : m = 1,2,\ldots\}$$

is a locally finite covering of Ω. Let $\{g_m : m = 1,2,\ldots\}$ be a partition of the unity of class C^∞ subordinated to this covering (see Section 9).

Let E be the topological product of $\mathcal{D}(Q_m)$, m=1,2,... If f_m belongs to $\mathcal{D}(Q_m)$, m = 1,2,... If f_m belongs to $\mathcal{D}(Q_m)$, m= 1,2,.., we set

$$X((f_m)) = \Sigma f_m$$

It is immediate to see that $X : E \longrightarrow E(\Omega)$ is a linear operator. On the other hand, given a compact set H of Ω there is a positive integer p such that

$$H \cap Q_m = \emptyset, \; m > p.$$

Then, given a positive integer r,

$$|X((f_m))|_{r,H} = \sum_{|\alpha|\leq r} \sup\{|D^\alpha \Sigma f_m(x)| : x \in H\}$$

$$\leq \sum_{m=1}^{p} \sum_{|\alpha|\leq r} \sup\{|D^\alpha f_m(x)| : x \in Q_m\}$$

and therefore X is continuous.

Given a positive integer m let $X_m : E(\Omega) \longrightarrow \mathcal{D}(Q_m)$ be the operator which associates to every g of $E(\Omega)$ the element gg_m.

(2) $X_m : E(\Omega)$ —— $\mathcal{D}(Q_m)$ *is a continuous linear operator.*

Proof. Given the multi-indices $\alpha = (p_1,p_2,\ldots,p_n)$ and $\beta = (q_1,q_2,\ldots,q_n)$ we set $\alpha-\beta$ to denote $(p_1-q_1,p_2-q_2,\ldots,p_n-q_n)$. If g is any element of

$E(\Omega)$ we have that

$$D^\alpha(gg_m) = \sum_{\beta<\alpha} A_\beta D^\beta g D^{\alpha-\beta} g_m$$

where A_β does not depend on g. Therefore, given a positive integer r, there is h >o such that

$$|gg_m|_r = \sum_{|\alpha|\leq r} \sup\{|D^\alpha g(x) g_m(x)| : x \in Q_m\}$$

$$\leq h \sum_{|\alpha|\leq r} \sup\{|D^\alpha g(x)| : x \in Q_m\}$$

from where follows that X_m is continuous (X_m is obviously linear).

For every g of $E(\Omega)$ we set

$$Yg = (gg_m).$$

We obtain from (2) that $Y: E(\Omega) \longrightarrow E$ is a continuous linear operator. On the other hand,

$$X((gg_m)) = \Sigma\, gg_m = g\Sigma g_m = g$$

and therefore Y is an isomorphism from $E(\Omega)$ into E.

If we set $T = Y \circ X$ it is immediate to check that T is a continuous projection from E into itself such that

$$T(E) = Y(E(\Omega)).$$

(3) $E(\Omega)$ *is isomorphic to a complemented subspace of* E.

Proof. Obviously $Y(E(\Omega))$ is a complemented subspace of E. The conclusion follows having in mind that $E(\Omega)$ is isomorphic to $Y(E(\Omega))$.

We select a subsequence (P_m) of pairwise disjoint cubes from the sequence (Q_m). For every positive integer m, let H_m be an n-dimensional closed cube concentric with P_m and contained in $\overset{\circ}{P}_m$. Let $T_m : E(H_m) \longrightarrow \mathcal{D}(P_m)$ be a continuous linear extension operator, m = 1,2,...

The operator from $E(\Omega)$ into $E(H_m)$ which associates to every f of $E(\Omega)$ its restriction f_m to H_m is obviously continuous and linear. Therefore if we set

$$Z f = (f_m),$$

then $Z : E(\Omega) \longrightarrow F = \prod_{m=1}^{\infty} E(H_m)$ is a continuous linear operator.

Let $U : F \longrightarrow \mathcal{E}(\Omega)$ be the operator defined by

$$U((f_m)) = \Sigma T_m f_m$$

for every (f_m) of F. Then U is linear and continuous. Since

$$Z(\Sigma T_m f_m) = (f_m)$$

we have that U is an isomorphism from F into $\mathcal{E}(\Omega)$. We set $W = U \circ Z$. It is immediate that W is a continuous projection from $\mathcal{E}(\Omega)$ into itself such that $W(\mathcal{E}(\Omega)) = U(F)$.

(4) F *is isomorphic to a complemented subspace of* $\mathcal{E}(\Omega)$.

Proof. Obviously $W(\mathcal{E}(\Omega))$ is a complemented subspace of $\mathcal{E}(\Omega)$. The conclusion follows having in mind that F is isomorphic to U(F).

According to 11.(9) and 11.(15), $\mathcal{E}(H_m)$ and $\mathcal{D}(H_m)$ are isomorphic to s, m = 1,2,... Therefore, results (5) and (6) are straightforward consequences of (3) and (4),respectively.

(5) $\mathcal{E}(\Omega)$ *is isomorphic to a complemented subspace of* s^N.

(6) s^N *is isomorphic to a complemented subspace of* $\mathcal{E}(\Omega)$.

(7) $\mathcal{E}(\Omega)$ *is isomorphic to* s^N.

Proof. It is enough to apply (5), (6) and 2.(1).

(8) $\mathcal{E}'(\Omega)$ *is isomorphic to* $(s')^{(N)}$.

Proof. It is an obvious consequence of (7).

Results (7) and (8) can be found in VALDIVIA [16].

13. REPRESENTING THE SPACES $\mathcal{D}(\Omega)$ AND $\mathcal{D}'(\Omega)$. Let Ω be a non-void open set of R^n. We denote by $\mathcal{D}(\Omega)$ the linear space over K of all the K-valued infinitely differentiable functions defined on R^n and whose support is a compact set contained in Ω. We provide $\mathcal{D}(\Omega)$ with the locally convex topology such that $\mathcal{D}(\Omega)$ is the inductive limit of the family of Fréchet spaces

$$\{\mathcal{D}(H) : H \text{ compact contained in } \Omega\}.$$

Since the family of compact sets 12.(1) is fundamental in Ω we have that $\mathcal{D}(\Omega)$ is the inductive limit of the sequence $(\mathcal{D}(L_m))$ and consequently $\mathcal{D}(\Omega)$ is an (LF)-space.

We take the sequence of cubes (Q_m) and the family of functions $\{g_m : m = 1,2,\ldots\}$ defined in the former section. Let G be the topological direct sum of $\mathcal{D}(Q_m)$, m = 1,2,... If (f_m) belongs to G we set

$$X((f_m)) = \Sigma f_m$$

$X : G \longrightarrow \mathcal{D}(\Omega)$ is a linear operator. For every positive integer m, the restriction of X to the subspace $\mathcal{D}(Q_m)$ of G is an isomorphism into. Therefore X is a continuous linear operator.

Given a positive integer m, let $X_m : \mathcal{D}(\Omega) \longrightarrow \mathcal{D}(Q_m)$ be the operator which associates to every g of $\mathcal{D}(\Omega)$ the element $g\, g_m$. Proceeding as we did in 12.(2) we have that X_m is a continuous linear operator. Consequently, if we set

$$Yg = (gg_m)$$

for every g of $\mathcal{D}(\Omega)$ and if we observe that each compact set H of Ω intersect a finite number of elements of (Q_m) it follows that $Y : \mathcal{D}(H) \longrightarrow G$ is a continuous linear operator. Therefore $Y : \mathcal{D}(\Omega) \longrightarrow G$ is a continuous linear operator. On the other hand,

$$X((gg_m)) = \Sigma gg_m = g\, \Sigma g_m = g$$

and therefore Y is an isomorphism from $\mathcal{D}(\Omega)$ into G.

We set $T = Y \circ X$. It is immediate that T is a continuous projection from G into itself such that $T(G) = Y(\mathcal{D}(\Omega))$.

(1) *$\mathcal{D}(\Omega)$ is isomorphic to a complemented subspace of* G.

Proof. We proceed as we did in 12.(3).

Let P_m and H_m be the cubes defined in the former section and let $T_m: E(H_m) \longrightarrow \mathcal{D}(P_m)$ be a extension operator, $m = 1,2,...$

The mapping from $\mathcal{D}(\Omega)$ into $E(H_m)$ which associates to every f of $\mathcal{D}(\Omega)$ its restriction f_m to H_m is obviously linear and continuos. We set

$$Zf = (f_m)$$

If we take a compact set H of Ω it is easy to check that $Z : \mathcal{D}(H) \longrightarrow L = \bigoplus_{m=1}^{\infty} E(H_m)$ is a continuous linear operator and therefore $Z : \mathcal{D}(\Omega) \longrightarrow L$ is also a continuos linear operator.

If $U : L \longrightarrow \mathcal{D}(\Omega)$ is the mapping defined by

$$U((f_m)) = \Sigma\, T_m f_m$$

for every (f_m) of L, then U is linear and continuous and, since

$$Z\ (\Sigma T_m f_m) = (f_m),$$

U is an isomorphism from L into $\mathcal{D}(\Omega)$.

If W is U o Z, it is inmediate that W is a continuous projection from $\mathcal{D}(\Omega)$ into itself such that $W(\mathcal{D}(\Omega)) = U(L)$.

(2) L *is isomorphic to a complemented subspace of* $\mathcal{D}(\Omega)$.

Proof. We proceed as we did in 12.(4).

Results (3) and (4) are straightforward consequences of (1) and (2), respectively.

(3) $\mathcal{D}(\Omega)$ *is isomorphic to a complementedsubspace of* $s^{(N)}$.

(4) $s^{(N)}$ *is isomorphic to a complemented subspace of* $\mathcal{D}(\Omega)$.

(5) $\mathcal{D}(\Omega)$ *is isomorphic to* $s^{(N)}$

Proof. It is enough to apply (3), (4) and 2.(8) to obtain the conclusion.

(6) $\mathcal{D}'(\Omega)$ *is isomorphic to* $(s')^N$

Proof. It is an obvious consequence of (5).

Results (5) and (6) can be found in VALDIVIA [16].

14. REPRESENTING THE SPACES $\mathcal{D}_+, \mathcal{D}_-, \mathcal{D}'_+$ AND $\mathcal{D}'_-$. Given the real number a, we set G_a to denote the subspace of $E(R)$ of all those functions with support contained in the closed interval $[a,+\infty[$. Obviously, G_a is closed in $E(R)$. We set $\mathcal{D}_+$ to denote the inductive limit of the family of Fréchet spaces

$$\{G_a : a \in R\}.$$

If we take a sequence of real numbers (a_m) diverging to $-\infty$ it is obvious that $\mathcal{D}_+$ is the inductive limit of the family of Fréchet spaces

$$\{G_{a_m} : m = 1,2,...\}$$

and therefore $\mathcal{D}_+$ is an (LF)-space. We write $\mathcal{D}'_+$ for the strong dual of $\mathcal{D}_+$.

Given a positive integer m, we set I_m and J_m to denote the compact intervals $[m,m+1]$ and $[m,m+2]$ respectively. Let S_m: $E(I_m) \longrightarrow \mathcal{D}([m-1,m+2])$ be a continuous linear extension operator. If f belongs to $E(I_m)$ we set $T_m f$ to denote the restriction of $S_m f$ to J_m. Then T_m: $E(I_m) \longrightarrow E(J_m)$ is a continuous linear extension operator such that $T_m f$ vanishes in the point m+2 as well as all it derivatives of all orders.

Let E_m be the subspace of $\mathcal{E}(I_m)$ of all those functions vanishing in the point m as well as all their derivatives of all orders.

(1) E_m *is isomorphic to* s.

Proof. E_m is isomorphic to $T_m(E_m)$ and s is isomorphic to $\mathcal{D}(J_m)$.Let F_m be the subspace of $\mathcal{D}(J_m)$ of all those functions vanishing in I_m. Since $\mathcal{D}(J_m)$ is the topological direct sum of $T_m(E_m)$ and F_m it follows that E_m is isomorphic to a complemented subspace of s. On the other hand, let $Z_m : \mathcal{E}([m+\frac{1}{4}, m+\frac{3}{4}]) \longrightarrow \mathcal{D}(I_m)$ be an extension operator. Then the functions of E_m vanishing in $[m+\frac{1}{4}, m+\frac{3}{4}]$ form a subspace of E_m which is a topological complement of $Z_m(\mathcal{E}([m+\frac{1}{4}, m+\frac{3}{4}]))$. Since this last space is isomorphic to s we have that E_m is isomorphic to a complemented subspace of E_m Finally we apply 2.(8) to obtain that E_m is isomorphic to s.

If f is in E_m we suppose $T_m f$ extended to R by setting

$$T_m f(x) = 0, \quad x \in R \sim J_m$$

Then $T_m f$ is an element of $\mathcal{E}(R)$ with support contained in J_m.

We set G for the locally convex space

$$\left(\bigoplus_{m=-1}^{-\infty} E_m\right) \times \prod_{m=o}^{\infty} E_m$$

Since E_m is isomorphic to s we have that G is isomorphic to $s^{(N)} \times s^N$.

Let $Z : G \longrightarrow \mathcal{D}_+$ be the mapping such that if

$$(2) \qquad u = ((f_m)_{m=-1}^{-\infty}, (f_m)_{m=o}^{\infty})$$

belongs to G then $Z\,u = \sum_{m=-\infty}^{\infty} T_m f_m$.

(3) $Z : G \longrightarrow \mathcal{D}_+$ *is an injective continuous linear operator.*

Proof. Obviously Z is a linear mapping. If the element (2) of G is distinct from zero let r be the least integer such that f_r is not zero. We take a point x in I_r with $f_r(x) \neq 0$. Then $Zu(x) = f_r(x) \neq 0$ and therefore Z is injective.

Given a positive integer p, we suppose that the element (2) belongs to the subspace of G

$$(4) \qquad \left(\bigoplus_{m=-1}^{-p} E_m\right) \times \prod_{m=o}^{\infty} E_m.$$

Let H be a compact set in $|-p,\infty|$. We take an integer q such that H is contained in $|-p,q|$. If r is any positive integer we have that

$$\sum_{|\alpha|\leq r} \sup\{|D^{\alpha}Zu(x)| : x \in H\} \leq \sum_{m=-p}^{q} \sum_{|\alpha|\leq r} \sup\{|D\, T_m f_m(x)| : x \in J_m\} = \sum_{m=-p}^{q} |D^{\alpha}T_m f_m|_r$$

from where it follows that Z is continuous from the space (4) into G_{-p}. Thus Z is continuous.

Let g be an element of $\mathcal{D}_+$ distinct from zero. We take an integer j such that $g(x) = 0$ for $x < j$. We set $g_m = 0$ for $m < j$. Let g_j be the restriction of g to $[j, j+1]$ and let g_{j+p+1} be the restriction of

$$(5) \qquad g-T_j g_j - T_{j+1}g_{j+1} - \ldots - T_{j+p}g_{j+p}$$

to $[j+p+1, j+p+2]$ for $p = 0,1,2,\ldots$ We show that (5) vanishes in the interval $]-\infty, j+p+1]$. For $p = 0$, g and $T_j g_j$ coincide in $]-\infty, j+1]$ and therefore $g-T_j g_j$ vanishes in this interval. Consequently the property is true for $p = 0$. Now we suppose the property true for a non-negative integer p. Since (5) and $T_{j+p+1}g_{j+p+1}$ coincide in $]-\infty, j+p+2]$ we have that

$$g-T_j g_j - \ldots - T_{j+p+1}g_{j+p+1}$$

vanishes in $]-\infty, j+p+2]$ and consequently the property is true for p+1.

Given any point x of R we find a positive integer q such that x belongs to $]-\infty, j+q+1]$. Then (5) vanishes in x for $p > q$ and therefore

$$g(x) = \sum_{p=0}^{\infty} T_{j+p}g_{j+p}(x).$$

(6) Z *is onto*.

Proof. Given the element g of $\mathcal{D}_+$ considered above we have that

$$v = ((g_m)_{m=-1}^{-\infty}, (g_m)_{m=0}^{\infty})$$

belongs obviously to G. On the other hand,

$$Zv = \sum_{m=-\infty}^{\infty} T_m g_m = \sum_{p=0}^{\infty} T_{j+p}g_{j+p} = g$$

and the proof is complete.

(7) *The space* $\mathcal{D}_+$ *is isomorphic to* $s^{(N)} \times s^N$.

Proof. $Z : G \longrightarrow \mathcal{D}_+$ is an bijective continuous linear operator. Since G is ultrabornological and $\mathcal{D}_+$ is an (LF)-space we apply the closed graph theorem to obtain that Z is an isomorphism. Since G is isomorphic to $s^{(N)} \times s^N$ the conclusion follows.

(8) *The space* $\mathcal{D}'_+$ *is isomorphic to* $s'^N \times s'^{(N)}$.

Proof. It is a straightfoward consequence of (7).

Given the real number a let H_a be the subspace of $\mathcal{E}(R)$ of all those functions with support contained in the closed interval $]-\infty,a]$. We set $\mathcal{D}_-$ to denote the inductive limit of the family of Fréchet spaces

$$\{H_a : a \in R\}$$

We denote by $\mathcal{D}'_-$ the strong dual of $\mathcal{D}_-$.

If f belongs to $\mathcal{D}_+$ we set Tf for the element of $\mathcal{D}_-$ with

$$Tf(x) = f(-x)$$

It is immediate that T is an isomorphism from $\mathcal{D}_+$ onto $\mathcal{D}_-$. Consequently, results (9) and (10) are obtained from (7) and (8) respectively.

(9) *The space* $\mathcal{D}_-$ *is isomorphic to* $s^{(N)} \times s^N$.

(10) *The space* $\mathcal{D}'_-$ *is isomorphic to* $s'^N \times s'^{(N)}$.

Results (7), (8), (9) and (10) can be found in VALDIVIA [17].

15. A REPRESENTATION OF THE SPACE $\mathcal{D}(H)$. Let H be a compact subset of R^n. We suppose that the interior of H, denoted by Ω, is non-void. In this section we shall prove that $\mathcal{D}(H)$ is isomorphic to s.

(1) $\mathcal{D}(H)$ *has a complemented subspace isomorphic to* s.

Proof. Let A and B be n-dimensional compact cubes contained in Ω such that $A \subset \overset{\circ}{B}$. Let $T: \mathcal{E}(A) \longrightarrow \mathcal{D}(B)$ be a continuous linear extension operator. Then T is a continuous linear extension operator from $\mathcal{E}(A)$ into $\mathcal{D}(H)$ $T(\mathcal{E}(A))$ is a subspace of $\mathcal{D}(H)$ having the subspace of $\mathcal{D}(H)$ whose elements vanish in A as topological complement. Since s is isomorphic to $\mathcal{E}(A)$

and since this space is isomorphic to T(E(A)) the conclusion follows.

Let (B_m) and (A_m) be the sequence of cubes contained in Ω constructed in Section 9. Then $2^{\frac{1}{k(m)}}$ is the length of the edge of B_m.

(2) *If for a positive integer r, k(r) is larger than zero then the distance of B_r to $R^n \sim \Omega$ is less or equal than* $\frac{\sqrt{n}}{2^{k(r)-2}}$.

Proof. We suppose the distance δ from B_r to $R^n \sim \Omega$ larger than $\frac{\sqrt{n}}{2^{k(r)-2}}$. We set

$$M = \{(x_1,x_2,\ldots,x_n) : \frac{b_j}{2^{k(r)-1}} \leqslant x_j \leqslant \frac{b_j+1}{2^{k(r)-1}},\ j = 1,2,\ldots,n\}$$

where b_j coincides with $\frac{a_j(r)}{2}$ when $a_j(r)$ is even and b_j equals $\frac{a_j(r)-1}{2}$ if $a_j(r)$ is odd. Then M contains B_r and it is distinct from B_r. Thus M is not an element of the sequence (B_m). In particular M does not belong to the family β_j defined in Section 9, j = 1,2,..., k(r)-1. On the other hand, if d is the distance from M to $R^n \sim \Omega$ we have that

$$d \geqslant \delta - \frac{\sqrt{n}}{2^{k(r)-1}} > \frac{\sqrt{n}}{2^{k(r)-2}} - \frac{\sqrt{n}}{2^{k(r)-1}} = \frac{\sqrt{n}}{2^{k(r)-1}}$$

and according to the definition of the class $\beta_{k(r)}$ it follows that M belongs to this class and we obtain a contradiction.

For every positive integer m let $g_m : R^n \longrightarrow R^n$ be the mapping defined by

$$g_m(x) = \left(\frac{2a_1(m)+1}{2^{k(m)+1}} + \frac{1}{2^{k(m)+1}}\left(1 + \frac{1}{4\sqrt{n}}\right) x_1, \ldots, \frac{2a_n(m)+1}{2^{k(m)+1}} + \right.$$

$$\left. + \frac{1}{2^{k(m)+1}}\left(1 + \frac{1}{4\sqrt{n}}\right) x_n\right)$$

for $x = (x_1,x_2,\ldots,x_n) \in R^n$. If

$$I = \{(x_1,x_2,\ldots,x_n) : -1 \leqslant x_j \leqslant 1, j= 1,2,\ldots,n\}$$

then g_m maps I onto A_m. Let h be an infinitely differentiable real function defined on R^n whose support is I and such that $h(x) > 0$ for x in $\mathring{I}$. If for every x of Ω we set

$$k_m(x) = \frac{h \circ g_m^{-1}(x)}{\Sigma\, h \circ g_j^{-1}(x)}, \quad m = 1,2,\ldots,$$

we have that $\{k_m : m = 1,2,\ldots\}$ is a partition of the unity of class C^∞ subordinate to the covering $\{\mathring{A}_m : m = 1,2,\ldots\}$.

(3) *Let* f *be an element of* $\mathcal{D}(H)$. *For every positive integer* r *and every multi-index* α *we have that*

$$\lim_m 2^{rk(m)} \sup \{|D^\alpha f \circ g_m(x)| : x \in I\} = 0$$

Proof. Let ε be an arbitrary positive number. We can find a non-void compact set L in Ω such that the modulus of the derivative of order r of $D^\alpha f$ in every direction in every point of $\Omega \sim L$ is bounded by

$$\frac{r!}{2} \left(\frac{1}{16\sqrt{n}}\right)^r \varepsilon.$$

If d is the distance from L to $R^n \sim \Omega$ we determine a positive integer p such that

$$\frac{\sqrt{n}}{2^{p-4}} < d.$$

Now we find a compact set P in Ω, $P \supset L$, such that if x belongs to $\Omega \sim P$ then

$$2^{rp} |D^\alpha f(x)| < \varepsilon.$$

Let q be a positive integer such that

$$A_m \cap P = \emptyset, \quad m > q$$

If $m > q$ and $z = (z_1, z_2, \ldots, z_n)$ belongs to A_m two cases can occur:

a) The length of the edge of B_m is larger or equal than $\frac{1}{2^p}$. The setting $x = g_m^{-1}(z)$ we have that

$$2^{rk(m)} |D^\alpha f \circ g_m(x)| = \left(\frac{1}{2^{k(m)+1}} \left(1 + \frac{1}{4\sqrt{n}}\right)\right)^{|\alpha|} 2^{rk(m)} |D^\alpha f(z)|$$

$$< \frac{1}{2^{k(m)|\alpha|}} 2^{rk(m)} |D^\alpha f(z)| < 2^{rp} |D^\alpha f(z)| < \varepsilon.$$

b) The length of the edge of B_m is less than $\frac{1}{2^p}$. Then $k(m) > p$ and applying (2) the distance from B_m to $R^n \sim \Omega$ is less or equal tah $\frac{\sqrt{n}}{2^{k(m)-2}}$. We find a point

$$\overset{\circ}{x} = (x_{10}, x_{20}, \dots, x_{no}) \in R^n \sim \Omega$$

whose distance to B_m is less than $\frac{\sqrt{n}}{2^{k(m)-3}}$. Then

$$||x^\circ - z|| \leqslant \frac{\sqrt{n}}{2^{k(m)-3}} + \frac{\sqrt{n}}{2^{k(m)}} + \frac{1}{2^{k(m)+3}} < \frac{\sqrt{n}}{2^{k(m)-4}} < \frac{\sqrt{n}}{2^{p-4}} < d$$

and consequently the distance of every point of the segment

$$\{x^\circ + w(z - x^\circ) : 0 \leqslant w \leqslant 1\}$$

to $R^n \sim \Omega$ is less than d and therefore this segment does not intersect L. We write $f = f_1 + if_2$ with f_1 and f_2 real functions and i the imaginary unity. Then

$$D^\alpha f_j(z) = \frac{1}{r!} \left[\frac{d^r}{dw^r} D^\alpha f_j(x_{10} + w(z_1 - x_{10}), \dots, x_{1n} + w(z_m - x_{no}))\right]_{w=\theta_j}$$

$$0 \leqslant \theta_j < 1, j = 1,2,$$

and therefore

$$|D^\alpha f(z)| \leqslant |D^\alpha f_1(z)| + |D^\alpha f_2(z)| \leqslant 2 \frac{||z - x^\circ||^r}{r!} \frac{r!}{2} \left(\frac{1}{16\sqrt{n}}\right)^r \varepsilon.$$

$$= ||z - x^\circ||^r \left(\frac{1}{16\sqrt{n}}\right)^r \varepsilon < \left(\frac{\sqrt{n}}{2^{k(m)-4}}\right)^r \left(\frac{1}{16\sqrt{n}}\right)^r \varepsilon = \frac{\varepsilon}{2^{rk(m)}}$$

and thus

$$2^{rk(m)} |D^\alpha f(z)| < \varepsilon.$$

If we set $x = g_m^{-1}(z)$ we have that

$$2^{rk(m)} |D^\alpha f \circ g_m^{(x)}| = \left(\frac{1}{2^{k(m)+1}} \left(1 + \frac{1}{4\sqrt{n}}\right)\right)^{|\alpha|} 2^{rk(m)} |D^\alpha f(z)|$$

$$\leqslant 2^{rk(m)} |D^\alpha f(z)| < \varepsilon$$

and the conclusion follows.

(4) *Let* f *be an element of* $\mathcal{D}(H)$.. *Given* $\varepsilon > 0$ *and a multi-index* α *there is*

a compact set L *of* Ω *such that*

$$|(fD^{\alpha}\Sigma h \circ g_j^{-1})(x)| < \varepsilon, \; x \in \Omega \sim L.$$

Proof. According to (3), there is a positive integer p such that if $m > p$ and if we set

$$b_m = \left(2^{k(m)+1}\frac{4\sqrt{n}}{4\sqrt{n}+1}\right)^{|\alpha|}4^n$$

we have that

$$b_m \sup\{|f \circ g_m(x)| : x \in I\}\sup\{|D^{\alpha}h(x)| : x \in I\} < \varepsilon$$

We take a compact set L in Ω with

$$A_m \subset L, \; m \leqslant p.$$

If z belongs to $\Omega \sim L$ there are positive integers $m_1, m_2, \ldots, m_q, q \leqslant 4^n$, with

$$z \notin \mathring{A}_m, \; m \neq m_j, \; m_j > p, j=1,2,\ldots,q,$$

according to 9.(16). Then

$$|(fD^{\alpha}\Sigma h \circ g_j^{-1})(z)| \leqslant \sum_{j=1}^{q} b_{m_j}4^{-n}|f(z)||(D^{\alpha}h) \circ g_m^{-1}(z)|$$

$$\leqslant \sum_{j=1}^{q} b_{m_j}4^{-n}\sup\{|f \circ g_j(x)| : x \in I\}\sup\{|D^{\alpha}h(x)| : x \in I\}$$

$$< \sum_{j=1}^{q} \varepsilon . 4^{-n} \leqslant \varepsilon.$$

(5) *Given a multi-index* γ *and an element* g *of* $\mathcal{D}(H)$ *we have that*

$$gD^{\gamma}\Sigma h \circ g_j^{-1}$$

belongs to $\mathcal{D}(H)$.

Proof. Given a multi-index λ we have that

$$D^{\lambda}(gD^{\gamma}\Sigma h \circ g_j^{-1}) = \sum_{\beta \leqslant \lambda} A_{\beta}D^{\beta}gD^{\lambda-\beta}(D^{\gamma}\Sigma h \circ g_j^{-1})$$

where A_{β} does not depend on the functions g and $D^{\gamma}\Sigma h \circ g^{-1}$.

Given $\varepsilon > 0$ we apply (4) for $f = D^{\beta}g$, $\alpha=\lambda-\beta+\gamma$, $\beta \leqslant \lambda$ to find a compact set L in Ω such that if x belongs to $\Omega \sim L$ then

$$\sum_{\beta \leqslant \lambda} |A_{\beta}D^{\beta}g(x)D^{\lambda-\beta+\gamma}\Sigma h \circ g_j^{-1}(x)| < \varepsilon$$

and therefore

$$|D^{\lambda}(gD^{\gamma}\Sigma h \circ g_j^{-1})(x)| < \varepsilon, \; x \in \Omega \sim L,$$

from where the conclusion follows.

We set

$$J = \{(x_1, x_2, \ldots, x_n) : -\frac{4\sqrt{n}}{4\sqrt{n}+1} < x_j < \frac{4\sqrt{n}}{4\sqrt{n}+1}, \; j = 1,2,\ldots,n\}.$$

The g_m maps J onto B_m, $m = 1,2,\ldots$

(6) *If* f *belongs to* $\mathcal{D}(H)$ *then*

(7) $$\frac{f}{\Sigma h \circ g_j^{-1}}$$

belongs to $\mathcal{D}(H)$.

Proof. Since J is a compact set contained in $\overset{\circ}{I}$ it follows that

$$\inf \{h(x) : x \in J\} = b > 0.$$

If x belongs to Ω we find a positive integer m such that x belongs to B_m. Then

(8) $$(\Sigma h \circ g_j^{-1})(x) \geq h \circ g_m^{-1}(x) = h(g_m^{-1}(x)) \geq b.$$

Given a multi-index λ we have that

(9) $$D^{\lambda} \frac{f}{\Sigma h \circ g_j^{-1}} = \sum_{\beta \leq \lambda} A_{\beta} D^{\lambda-\beta} f D^{\beta} \frac{1}{\Sigma h \circ g_j^{-1}}$$

where A_{β} does not depend on f and $\frac{1}{\Sigma h \circ g_j^{-1}}$

We have that $D^{\beta} \frac{1}{\Sigma h \circ g_j^{-1}}$ is a fraction whose numerator is a linear combination of terms of the form

$$D^{\gamma}(\Sigma h \circ g_j^{-1}) \, D^{\delta}(\Sigma h \circ g_j^{-1}) \ldots D^{\mu}(\Sigma h \circ g_j^{-1})$$

which dependes on β and whose denominator is a power of $\Sigma h \circ g_j^{-1}$ which depends on β again.

A reiterated aplication of (5) enables us to state that

$$D^{\lambda-\beta} f D^{\gamma}(\Sigma h \circ g_j^{-1}) D^{\delta}(\Sigma h \circ g_j^{-1}) \ldots D^{\mu}(\Sigma h \circ g_j^{-1})$$

belongs to $\mathcal{D}(H)$ and, according to (8), it follows that given $\varepsilon>0$ there is a compact set P in Ω such that, if x belongs to $\Omega \sim P$,

$$(10) \qquad \sum_{\beta<\lambda} |A_\beta D^{\lambda-\beta} f(x) D^\beta (\frac{1}{\Sigma h \circ g_j^{-1}})(x)| < \varepsilon$$

and therefore we obtain from (9) and (10)

$$|D^\lambda (\frac{f}{\Sigma h \circ g_j^{-1}})(x)| < \varepsilon, \; x \in \Omega \sim P,$$

from where it follows that (7) belongs to $\mathcal{D}(H)$.

(11) *If* f *belongs to* $\mathcal{D}(H)$, *given a positive integer* r *and a multi-index* α *then*

$$\lim_m 2^{rk(r)} \sup\{|D^\alpha((f \circ g_m)h)(x)| : x \in I\} = 0$$

Proof. We have that

$$D^\alpha((f \circ g_m) h) = \sum_{\beta<\alpha} A_\beta D^\beta (f \circ g_m) D^{\alpha-\beta} h$$

where A_β does not depend on the functions $f \circ g_m$ and h. Then

$$2^{rk(m)} \sup \{|D^\alpha((f \circ g_m)h)(x)| : x \in I\}$$

$$\leq (\sum_{\gamma\leq\alpha} \sup\{|D^\gamma h(x)| : x \in I\}) \sum_{\beta\leq\alpha} |A_\beta| 2^{rk(m)} \sup\{|D^\beta f \circ g_m(x)| : x \in I\}$$

and, according to (3), the conclusion follows.

We shall proof result (12) in order to apply it later.

(12) *If* f *belongs to* $\mathcal{D}(H)$, r *is a positive integer and* α *a multi-index then*

$$\lim_m 2^{rk(m)} \sup\{|D^\alpha(fk_m) \circ g_m(x)| : x \in I\} = 0$$

Proof. For every positive integer we have that

$$(fk_m) \circ g_m = (f \frac{h \circ g_m^{-1}}{\Sigma h \circ g_j^{-1}}) \circ g_m = (\frac{f}{\Sigma h \circ g_j^{-1}} \circ g_m) h.$$

According to (7),

$$\frac{f}{\Sigma h \circ g_j^{-1}} = g \in \mathcal{D}(H)$$

Now we apply (11) to obtain

$$\lim_m 2^{rk(m)} \sup\{|D^{\alpha}(fk_m) \circ g_m(x)| : x \in I\}$$
$$= \lim_m 2^{rk(m)} \sup\{|D^{\alpha}((g \circ g_m)h)|(x) : x \in I\} = 0$$

Now we denote by G the linear space over K of all sequences (f_m) of elements of $\mathcal{D}(I)$ such that, for every multi-index α and every positive integer r

$$\lim_m 2^{rk(m)} \sup\{|D^{\alpha} f_m(x)|: x \in I\} = 0$$

We provide G with the metrizable locally convex topology defined by the family of norms

$$P_{qr}((f_m)) = \sup_m 2^{rk(m)} \sum_{|\alpha| \leq q} \sup\{|D^{\alpha} f_m(x)|: x \in I\},$$
$$(f_m) \in G,\ q,r = 1,2,\ldots$$

Equivalently, G is the linear space over K of all sequences (f_m) of elements of $\mathcal{D}(I)$ such that, for every positive integer r,

$$(|f_m|_r) \in \lambda_o(a)$$

being $a = (2^{k(m)})$.

Since $\mathcal{D}(I)$ is isomorphic to s it results that G is isomorphic to the space $\lambda_o(a,s)$ defined in Section 3. According to 3.(13) the following result holds:

(13) G *is isomorphic to* $s \hat{\otimes} \lambda_o(a)$.

The Lebesgue measure of Ω in R^n coincides with the sum of the volume of the n-dimensional cubes $B_1, B_2, \ldots, B_m, \ldots$ Since Ω is bounded it has finite measure and consequently

$$\Sigma 2^{-nk(m)} < \infty$$

and therefore $\lambda_o(a)$ is nuclear, according 3.(3). Now we apply 4.(3) to obtain that $s \hat{\otimes} \lambda_o(a)$ is isomorphic to s. Therefore

(14) *The space* G *is isomorphic to* s.

For every (f_m) of G we set

$$T((f_m)) = \Sigma f_m \circ g_m^{-1}$$

Then $T : G \longrightarrow \mathcal{D}(H)$ is a linear mapping according to 9.(21).

(15) *T is continuous.*

Prof. Let (f_m) be an element of G and let α be a multi-index. Given a point z of Ω we find positive integers $m_1, m_2, \ldots, m_q$ with $q \leq 4^n$

$$z \notin \overset{\circ}{A}_m, \; m \neq m, \; j = 1,2,\ldots, q.$$

Then

$$|D^\alpha \Sigma f_m \circ g_m^{-1}(z)| = |\sum_{j=1}^{q} D^\alpha(f_{mj} \circ g_{mj}^{-1})(z)|$$

$$= |\sum_{j=1}^{q} (2^{k(m_j)+1} \frac{4\sqrt{n}}{4\sqrt{n}+1})^{|\alpha|} (D^\alpha f_{m_j}) \circ g_{m_j}^{-1}(z)|$$

$$\leq 4^n (\frac{8\sqrt{n}}{4\sqrt{n}+1})^{|\alpha|} \sup_m 2^{k(m)|\alpha|} \sup \{|D^\alpha f_m(x)| : x \in I\}$$

from where the continuity of T follows.

For every f of $\mathcal{D}(H)$ we set

$$Wf = (f k_m \circ g_m).$$

Then W is a linear mapping from $\mathcal{D}(H)$ into G, according to (12). Moreover

(16) $$T(Wf) = \Sigma(fk_m \circ g_m) \circ g_m^{-1} = f \Sigma k_m = f.$$

(17) *Given a multi-index* α *and a positive integer* r *there is a positive num ber* k *such that*

(18) $$2^{rk(m)} \sup \{|D^\alpha f \circ g_m(x)| : x \in I\} \leq k \sum_{|\beta| \leq |\alpha| + r} \sup \{|D^\alpha f(x)| : x \in H\}.$$

for every f *of* $\mathcal{D}(H)$ *and* $m = 1,2,\ldots$

Proof. Let x be any point of I. We set $z = g_m(x)$. Then z belongs to A_m. If $k(m) = 0$, then

(19) $$2^{rk(m)} |D^\alpha f \circ g_m(x)| = |D^\alpha f \circ g_m(x)| (\frac{1}{2} + \frac{1}{8\sqrt{n}})^{|\alpha|} |D^\alpha f(z)| \leq |D^\alpha f(z)| .$$

If $k(m)$ is larger than zero we apply (2) to obtain that the distance from B_m to $R^n \sim \Omega$ is less or equal than $\frac{\sqrt{n}}{2^{k(m)-2}}$. We take a point $x^\circ = (x_{11}, x_{20},\ldots,x_{no})$ in $R^n \sim \Omega$ whose distance to B_m is less than $\frac{\sqrt{n}}{2^{k(m)-3}}$. Proceeding as we did in the proof of (3) we have that

$$||x^\circ - z|| < \frac{\sqrt{n}}{2^{k(m)-4}}$$

and

$$|D^\alpha f_j(z)| = |\frac{1}{r!}(\frac{d^r}{dw^r} D^\alpha f_j(x_{1o} + w(z_1 - x_{1o}),\ldots,x_{no} + w(z_n - x_{no}))) \, w = \Theta_j|$$

and therefore there is a positive number Q (obtained in the calculation of the derivative above) which does not depend on f_j such that

$$|D^\alpha f_j(z)| \leq \frac{||z-x^\circ||^r}{r!} Q \sum_{|\beta| \leq |\alpha|+r} \sup\{|D^\beta f_j(y)| : y \in H\}$$

and consequently

$$2^{rk(m)} |D^\alpha f \circ g_m(x)| = \left(\frac{1}{2^{k(m)+1}}\left(1 + \frac{1}{4\sqrt{n}}\right)\right)^{|\alpha|} 2^{rk(m)} |D^\alpha f(z)|$$

$$\leq 2^{rk(m)} (|D^\alpha f_1(z)| + |D^\alpha f_2(z)|)$$

$$\leq 2^{rk(m)} \frac{2}{r!} ||z-x^\circ||^r Q \sum_{|\beta| \leq |\alpha|+r} \sup\{|D^\beta f(y)| : y \in H\}$$

$$\leq 2^{rk(m)} \frac{2}{r!} \left(\frac{\sqrt{n}}{2^{k(m)-4}}\right)^r Q \sum_{|\beta| \leq |\alpha|+r} \sup\{|D^\beta f(y)| : y \in H\}$$

$$\leq \frac{2}{r!} (16\sqrt{n})^r Q \sum_{|\beta| \leq |\alpha|+r} \sup\{|D^\beta f(y)| : y \in H\}$$

from where (18) is deduced having in mind (19) and taking

$$k = 1 + \frac{2}{r!}(16\sqrt{n})^r Q$$

(20) *W is an isomorphism from $\mathcal{D}(H)$ into G.*

Proof. Using (17) it is easy to prove that $W : \mathcal{D}(H) \longrightarrow G$ is a continuous linear mapping. Since T is continuous and (16) is verified the conclusion follows.

(21) $\mathcal{D}(H)$ *is isomorphic to a complemented subspace of* s.

Proof. $W \circ T$ is a continuous projection from G into itself such that $W \circ T(G) = W(\mathcal{D}(H))$. Since $\mathcal{D}(H)$ is isomorphic to $W(\mathcal{D}(H))$ and is isomorphic to s the conclusion follows.

Now we obtain the fundamental result of this section.

(22) $\mathcal{D}(H)$ *is isomorphic to* s.

Proof. It is an immediate consequence from (1), (21) and 2.(8).

Result (22) is due to VALDIVIA [18]. The methods exposed in this section can be found in VALDIVIA [19].

16. REPRESENTING THE SPACES $\mathcal{D}_{+\Gamma}$, $\mathcal{D}_{-\Gamma}$, $\mathcal{D}'_{+\Gamma}$ AND $\mathcal{D}'_{-\Gamma}$ Let Γ be a convex closed cone in the enclidean space R^n, $n > 1$, with vertex in the origin and with non-void interior. For every z of R^n we consider the cone $z + \Gamma$ having its vertex in the point z. Let G_z be the subspace of $\mathcal{E}(R^n)$ of all those functions having its support contained in $z + \Gamma$. It is obvious that G_z is closed and therefore a Fréchet space. We denote by $\mathcal{D}+$ the inductive limit of the family of Fréchet spaces

$$\{G_z : z \in R^n\}.$$

Let u be an interior point of Γ. Any cone of the form $z + \Gamma$ is contained in some cone

$$mu + \Gamma, \ m = -1, -2, \ldots$$

and therefore $\mathcal{D}_{+\Gamma}$ is the inductive limit of the family of Fréchet spaces

$$\{G_{mu} : m = -1, -2, \ldots\}.$$

Consequently, $\mathcal{D}_{+\Gamma}$ is an (LF)-space. $\mathcal{D}'_{+\Gamma}$ is the strong dual of $\mathcal{D}_{+\Gamma}$.

Let A be the family of all n-dimensional cubes of the form

(1) $$\{(x_1, x_2, \ldots, x_n) : p_j \leqslant x_j \leqslant p_j + 2, \ p_j \text{ integer } j=1,2,\ldots,n\}.$$

We take the vector u of Γ such that, if m is a positive integer larger than one, the subfamily A_m of A of all those elements contained in $-mu + \Gamma$ but not in $-(m-1)u + \Gamma$ is infinite.

Let

$$A_1 = \{A_{1r} : r = 1,2,\dots\}.$$

be the family of all those elements of A contained in $-u + \Gamma$. We write

$$A_m = \{A_{mr} : r = 1,2,\dots\},\ m = 2,3,\dots$$

and we set

$$E_{mr} = \mathcal{D}(A_{mr}),\ m,r = 1,2,\dots,$$

$$E_m = \prod_{r=1}^{\infty} E_{mr},\ m = 1,2,\dots,$$

$$E = \bigoplus_{m=1}^{\infty} E_m$$

Since E_{mr} is isomorphic to s it follows that E_m is isomorphic to s^N and consequently E is isomorphic to $(s^N)^{(N)}$.

The elements of E are the double sequences $\lambda = (\lambda_{mr})$ with

$$\lambda_{mr} \in E_{mr},\ m,r = 1,2,\dots,$$

and such that there is a positive integer $m(\lambda)$, depending on λ with

$$\lambda_{mr} = 0,\ m > m(\lambda),\ r = 1,2,\dots$$

We write

$$X\lambda = \Sigma\lambda_{mr}$$

(2) $X : E \longrightarrow \mathcal{D}_{+\Gamma}$ *is a continuous linear operator.*

Proof. If $\lambda = (\lambda_{mr})$ is an element of E there is a positive integer $m(\lambda)$ with

$$\lambda_{mr} = 0,\ m > m(\lambda),\ r = 1,2,\dots$$

Then $X\lambda$ is an element of $\mathcal{E}(R^n)$ whose support is contained in $-m(\lambda)u + \Gamma$. Thus X is a mapping, obviously linear, from E into $\mathcal{D}_{+\Gamma}$.

If $\lambda^k = (\lambda_{mr,k})$, $k = 1,2,\dots$, is a sequence in E converging to the origin there is a negative integer q such that

$$\operatorname{supp} \lambda_{mr,k} \subset qu + \Gamma,\ m,\ r,k = 1,2,\dots$$

For every pair of positive integers m and r, $\lambda_{mr,k}$, $k=1,2,\dots$, converges to the origin in E_{mr}. Since every compact subset of $q\,u + \Gamma$ can be covered by a finite number of n-dimensional cubes of the form A_{mr} it follows that

$$\lim_k X\lambda^k = 0$$

in G_{qu} and thus $X : E \longrightarrow \mathcal{D}_{+\Gamma}$ is sequentially continuous. Since E is an (LF)-space it follows that X is continuous.

Let

$$\{h_{mr} : m, r = 1,2,\ldots\}$$

be a partition of the unity of class C^∞ subordinated to the covering of R^n

$$\{\mathring{A}_{mr} : m,r = 1,2,\ldots\}$$

If f is any element of $\mathcal{D}_{+\Gamma}$ we set

$$Yf = (fh_{mr})$$

(3) *Y is an isomorphism from $\mathcal{D}_{+\Gamma}$ into* E.

Proof. It is immediate that if f belongs to $\mathcal{D}_{+\Gamma}$ then Yf belongs to E. Y is obviously linear. On the other hand, if (f_k) is a sequence in $\mathcal{D}_{+\Gamma}$ converging to the origin there is a negative integer q such that

$$\text{supp } f_k \subset qu + \Gamma$$

and therefore

$$Yf_k \in \bigoplus_{m=1}^{-q} E_m$$

and since for every pair of positive integers m and r the sequence

$$f_1h_{mr}, f_2h_{mr}, \ldots, f_kh_{mr}, \ldots$$

converges to the origin in E_{mr} it follows that (Yf_k) converges to the origin in E. Then Y is sequentially continuous. Since $D_{+\Gamma}$ is an (LF)-space we have that Y is continuous.

Finally, we have that

$$(X \circ Y)f = X((fh_{mr})) = \Sigma\, fh_{mr} = f\, \Sigma\, h_{mr} = f$$

and therefore Y is an isomorphism from $\mathcal{D}_{+\Gamma}$ into E.

(4) *$\mathcal{D}_{+\Gamma}$ is isomorphic to a complemented subspace of* $(s^N)^{(N)}$.

Proof. Y o X is a continuous projection from E into itself such that (Y o X)(E) = Y($\mathcal{D}_{+\Gamma}$). The subspace Y($\mathcal{D}_{+\Gamma}$) of E is isomorphic to $\mathcal{D}_{+\Gamma}$ and since E is isomorphic to $(s^N)^{(N)}$ it follows that $\mathcal{D}_{+\Gamma}$ is isomorphic to a complemented subspace of $(s^N)^{(N)}$.

We choose a double subsequence (B_{mr}) of (A_{mr}) satisfying

a) B_{1r} is contained in $-u+\Gamma$, $r = 1,2,\ldots$;

b) for every integer $m > 1$, B_{mr} is contained in $-mu+\Gamma$ but not contained in $-(m-1)u+\Gamma$;

c) if $(i,j) \neq (m,r)$ then $\overset{\circ}{B}_{ij} \cap \overset{\circ}{B}_{mr} = \emptyset$, $i,j,m,r = 1,2,\ldots$

For every pair of positive integers m and r, D_{mr} is a closed m-dimensional cube contained in $\overset{\circ}{B}_{mr}$. Let $T_{mr} : \mathcal{E}(D_{mr}) \longrightarrow \mathcal{D}(B_{mr})$ be a continuous linear extension operator.

We write

$$F_{mr} = \mathcal{E}(D_{mr}),\ mr = 1,2,\ldots,$$

$$F_m = \prod_{r=1}^{\infty} F_{mr},\ m = 1,2,\ldots,$$

$$F = \bigoplus_{m=1}^{\infty} F_m.$$

Since F_{mr} is isomorphic to s it follows that F_m is isomorphic to s^N and consequently F is isomorphic to $(s^N)(N)$.

The elements of F are the double sequences $\lambda = (\lambda_{mr})$ with

$$\lambda_{mr} \in F_{mr},\ m,r = 1,2,\ldots,$$

and such that there is a positive integer $m(\lambda)$ depending on λ with

$$\lambda_{mr} = 0,\ m > m(\lambda),\ r = 1,2,\ldots$$

We set g_{mr} for the restriction to D_{mr} of g, being g an element of $\mathcal{D}_{+\Gamma}$ and we write

$$Zg = (g_{mr})$$

(5) *Z is a continuous linear operator from $\mathcal{D}_{+\Gamma}$ into* F.

Proof. Obviously Z is a linear mapping from $\mathcal{D}_{+\Gamma}$ into F. On the other hand, if (f_k) is a sequence in $\mathcal{D}_{+\Gamma}$ converging to the origin there is a negative integer q such that f_k has its support contained in $q\,u + \Gamma$, k = 1, 2,..., and therefore

$$Z f_k \in \bigoplus_{m=1}^{-q} F_m, \quad k = 1,2,\dots$$

Since (f_k) converges to the origin in G_{qu} it is obvious that Z is continuos as a mapping from G_{qu} into $\bigoplus_{m=1}^{-q} F_m$ and, since $\mathcal{D}_{+\Gamma}$ is an (LF)-space, it follows that $Z : \mathcal{D}_{+\Gamma} \longrightarrow F$ is continuous.

For every $\lambda = (\lambda_{mr})$ of F we set

$$Wf = \Sigma\, T_{mr}(\lambda_{mr}).$$

(6) W *is an isomorphism from* F *into* $\mathcal{D}_{+\Gamma}$.

Proof. If $\lambda = (\lambda_{mr})$ is in F there is a positive integer $m(\lambda)$ with

$$\lambda_{mr} = 0, \; m > m(\lambda), \; r = 1,2,\dots$$

Then

$$W\lambda = \Sigma\, T_{mr}\lambda_{mr}$$

is an element of $E(R^n)$ whose support is contained in $-m(\lambda)u + \Gamma$. Thus W is a mapping, obviously linear, from F into $\mathcal{D}_{+\Gamma}$

If $\lambda^k = (\lambda_{mr}, k)$, k = 1,2,..., is a sequence in F converging to the origin there is a negative integer q such that

$$\operatorname{supp} T_{mr}\,\lambda_{mr},\, k \subset qu + \Gamma, \quad m,r,k = 1,2,\dots$$

For every pair of positive integers m and r, λ_{mr}, k, k = 1,2,..., converges to the origin in F_{mr} and therefore $T_{mr}\lambda_{mr}, k$, k = 1,2,..., converges to the origin in $\mathcal{D}(B_{mr})$. Since every compact subset of $qu + \Gamma$ intersects a finite number of cubes of the form B_{mr} it follows that

$$\lim_k W\lambda^k = 0$$

in G_{qu} and therefore W is sequentially continuous. Then $W : F \longrightarrow \mathcal{D}_{+\Gamma}$ is continuous since F is an (LF)-space.

Finally, if $\lambda = (\lambda_{mr})$ is in F we have that

$$(Z \circ W)\lambda = Z(\Sigma\, T_{mr}\, \lambda_{mr}) = (\lambda_{mr})$$

from where it follows that W is an isomorphism from f into $\mathcal{D}_{+\Gamma}$.

(7) *The space* $(s^N)^{(N)}$ *is isomorphic to a complemented subspace of* $\mathcal{D}_{+\Gamma}$.

Proof. W o Z is a continuous projection from $\mathcal{D}_{+\Gamma}$ into itself such that $(W \circ Z)(\mathcal{D}_{+\Gamma}) = W(F)$. Since the subspace W(F) of $\mathcal{D}_{+\Gamma}$ is isomorphic to F and this last space is isomorphic to $(s^N)^{(N)}$, it follows that $(s^N)^{(N)}$ is isomorphic to a complemented subspace of $\mathcal{D}_{+\Gamma}$.

(8) *The space* $\mathcal{D}_{+\Gamma}$ *is isomorphic to* $(s^N)^{(N)}$.

Proof. It is an obvious consequence of (4), (7) and 2.(2).

(9) *The space* $\mathcal{D}'_{+\Gamma}$ *is isomorphic to* $(s'^{(N)})^N$.

Proof. See result (8)

If Γ_1 is the symmetric cone of Γ we define $\mathcal{D}_{-\Gamma}$ as $\mathcal{D}_{+\Gamma_1}$. $\mathcal{D}'_{-\Gamma}$ is the strong dual of $\mathcal{D}_{-\Gamma}$. Now the following results are clear:

(10) *The space* $\mathcal{D}_{-\Gamma}$ *is isomorphic to* $(s^N)^{(N)}$.

(11) *The space* $\mathcal{D}'_{-\Gamma}$ *is isomorphic to* $(S^{(N)})^N$.

More general results can be found in VALDIVIA [20].

17. A REPRESENTATION OF THE SPACE $\mathcal{S}(R^n)$. We represent by $\mathcal{S}(R^n)$ or by $\mathcal{S}$ the linear subspace of $\mathcal{E}(R^n)$ of all those functions f such that for every results index α and every positive integer r

$$\lim_{||x|| \to \infty} ||x||^r\, |D^\alpha f(x)| = 0.$$

For every pair of positive integers p and r we set

$$|f|_{p,r} = \sum_{|\alpha| \leq p} \sup\{|(1+||x||^2)^r D^\alpha f(x)| : x \in R^n\}.$$

Then $|.|_{p,r}$, p,r = 1,2,..., is a system of norms defining a metrizable locally convex topology on S. We shall suppose S endowed with this topology

(1) *S is a Fréchet space.*

Proof. Let (f_m) be a Cauchy sequence in S which is a Cauchy sequence in $E(R^n)$ and therefore converges to the function f in this space. We fix positive integers p and r. Given $\varepsilon > 0$ there is a positive integer q such that

$$|f_m - f_k|_{p,r} < \frac{\varepsilon}{(p+1)^m}, \quad m,k \geq q.$$

Then if x belongs to R^n, $m > q$ and $|\alpha| \leq p$, we have that

$$||x||^r \; |D^\alpha f_m(x) - D^\alpha f_q(x)| < \frac{\varepsilon}{(p+1)^n}$$

from where it follows

$$||x||^r \; |D^\alpha f\;(x) - D^\alpha f_q(x)| \leq \frac{\varepsilon}{(p+1)^n}.$$

We find $h > 0$ such that

$$||x||^r \; |D^\alpha f_q(x)| < \frac{\varepsilon}{2}, \quad ||x|| > h.$$

Consequently if $||x|| > h$ then

$$||x||^r \; |D^\alpha f(x)| \leq ||x||^r \; |D^\alpha f(x) - D^\alpha f_q(x)| + ||x||^r \; |D^\alpha f_q(x)| < \varepsilon$$

and therefore f belongs to S. Finally for every x of R^n and positive integers m and k larger than q

$$(1+||x||^2)^r \; |D^\alpha f_m(x) - D^\alpha f_k(x)| < |f_m - f_k|_{p,r} < \frac{\varepsilon}{(p+1)^n}$$

and therefore

$$(1+||x||^2)^r \; |D^\alpha f_m(x) - D^\alpha f(x)| \leq \frac{\varepsilon}{(p+1)^n}.$$

Consequently

$$|f_m - f|_{p,r} = \sum_{|\alpha| \leq p} \sup \{|(1+||x||^2)^r D^\alpha (f_m(x) - f(x)) : x \in R^n\} < \varepsilon$$

for $m > q$ and therefore the sequence (f_m) converges to f in S. This completes the proof.

We set

$$M_m = \{(x_1,x_2,\ldots,x_n) : -m \leqslant x_j \leqslant m, j = 1,2,\ldots,n\}, \ m = 1,2,\ldots$$

Let A_m be the family of all the cubes of R^n contained in M_m of the form

$$\{(x_1,x_2,\ldots,x_n) : a_j \leqslant x_j \leqslant a_j + 1, \ a_j \text{ integer}, \ j = 1,2,\ldots,n\}.$$

By recurrence we construct a sequence (B_m) of elements of $A = \bigcup_{m=1}^{\infty} A_m$ in the following way: we order the elements of A_1; given a positive integer m and an ordering of A_m, we ordered the cubes of $A_{m+1} \sim A_m$ and we suppose them posterior to the cubes of A_m. Since there are $(2m)^m$ cubes in A_m the first element of $A_{m+1} \sim A_m$ is $B_{(2m)^n+1}$ and the last one is $B_{(2m+2)^n}$. For every positive integer m we write

$$B_m = \{(x_1,x_2,\ldots,x_n) : a_j^{(m)} \leqslant x_j \leqslant a_j^{(m)} + 1, \ j = 1,2,\ldots,n\}.$$

We consider (B_m) as the sequence we used in Section 9 for the case $\Omega = R^n$. Let (A_m) be the sequence of cubes constructed in Section 9. We denote by $g_m: R^n \longrightarrow R^n$ the mapping defined in Section 15 by setting $k(m) = 0$, $m = 1,2,\ldots$ If

$$I = \{(x_1,x_2,\ldots,x_n) : -1 \leqslant x_j \leqslant 1, \ j = 1,2,\ldots,n\}$$

then g_m maps I onto A_m. The functions h and k_m are defined as in Section 15. Then $\{k_m : m = 1,2,\ldots\}$ is a partition of the unity of class C^∞ subordinated to the covering $\{\overset{\circ}{A}_m : m = 1,2,\ldots\}$ of R^n.

We denoted by F the linear space over K of the sequences (f_m) of elements of $\mathcal{D}(I)$ such that for every multi-index α and every positive integer r

$$\lim m^r \sup \{|D^\alpha f_m(x)| : x \in I\} = 0$$

We suppose F endowed with the metrizable locally convex topology defined by the family of norms

$$||(f_m)||_{pr} = \sup m^r \sum_{|\alpha| \leqslant p} \sup \{|D^\alpha f_m(x)| : x \in I\}$$

$$= \sup m^r |f_m|_p, \ (f_m) \in F, p,r = 1,2,\ldots$$

Obviously F is the linear space over K of the sequences (f_m) of $\mathcal{D}(I)$ such that for every positive integer r

$$(|f_m|_r) \in \lambda_0(a) \text{ with } a = (m).$$

Since $\mathcal{D}(I)$ is isomorphic to s it follows that F is isomorphic to the space $\lambda_0(a,s)$ defined in Section 3.

(2) *F is isomorphic to s.*

Proof. Since $\Sigma \frac{1}{m^2}$ is convergent we apply Chapter Two, 4, §4.(3) to obtain that $\lambda_0(a)$ is nuclear. The conclusion follows according to 3.(13) and 4.(3).

For every $(f_m) \in F$ we set

$$T((f_m)) = \Sigma f_m \circ g_m^{-1}$$

(3) *T is a linear mapping from F into S.*

Proof. It is obvious that if (f_m) belongs to F then $T\,(f_m)$ is well defined and belongs to $\mathcal{E}(R^n)$. Given a positive integer r and $\varepsilon > 0$ there is an integer $q > 1$ such that

$$n^r \sum_{m=q}^{\infty} m^r \, |f_m|_r < \varepsilon.$$

If x belongs to R^n and $\|x\| > nq$ there are positive integers t and k such that

(4) $$x \in B_t \in A_{k+1} \sim A_k$$

and therefore

(5) $$(2k)^n + 1 \leq t \leq (2k+2)^n$$

We have that $k \geq q$. If $A_{p_1}, A_{p_2}, \ldots, A_{p_j}$ are all the cubes of the form A_m to which x belongs then

(6) $$k+1 \leq (2(k-1))^n + 1 \leq p_i \leq (2(k+1)+2)^n, \; i = 1,2,\ldots,j,$$

and therefore if α is a multi-index with $|\alpha| \leq r$ we have that

$$\|x\|^r \, |D^\alpha \, \Sigma(f_m \circ g_m^{-1})(x)| \leq \|x\|^r \sum_{i=1}^{j} |D^\alpha (f_{p_i} \circ g_{p_i}^{-1})(x)|$$

$$\leq ((k+1)\sqrt{n})^r \sum_{i=1}^{j} \left(\frac{1}{2}+\frac{1}{8\sqrt{n}}\right)^{-|\alpha|} \sup\{|D^\alpha f_{p_i}(u)| : u \in I\}$$

$$\leq (\sqrt{n})^r \left(\frac{1}{2}+\frac{1}{8\sqrt{n}}\right)^{-|\alpha|} \sum_{i=1}^{j} p_i^{\,r} |f_{p_i}|_r \leq n^r \left(\frac{1}{2}+\frac{1}{8\sqrt{n}}\right)^{-|\alpha|}$$

$$\sum_{m=q}^{\infty} m^r |f_m|_r < \varepsilon$$

and therefore $T(ff_m))$ belongs to S. Obviously T is linear.

(7) $T : F \longrightarrow S$ *is continuous.*

Proof. Fix an integer $q > 1$. Let (f_m) be any element of F and take positive integers p and r and a point x of R^n. Let α be a multi-index with $|\alpha| \leq p$. If $||x|| \leq nq$ we have that

$$(1+||x||^2)^r |D^\alpha \Sigma f_m \circ g_m^{-1}(x)| \leq (1+n^2q^2) \left(\frac{1}{2}+\frac{1}{8\sqrt{n}}\right)^{-|\alpha|}$$

$$\Sigma \sup\{|D^\alpha f_m(u) : u \in I\} \leq (2n^2q^2)^r \left(\frac{1}{2}+\frac{1}{8\sqrt{n}}\right)^{-|\alpha|} \Sigma\, m^{2r} |f_m|_p.$$

If $||x|| > nq$ there are positive integers t and k such that (4) holds and therefore (5) is verified. If $A_{p_1}, A_{p_2}, \ldots, A_{p_j}$ are all the cubes of the form A_m to which x belongs then (6) holds. Consequently

$$(1+||x||^2)^r |D^\alpha \Sigma f_m \circ g_m^{-1}(x)| \leq (1+||x||^2)^r \sum_{i=1}^{j} |D^\alpha f_{p_i} \circ g_{p_i}^{-1}(x)|$$

$$\leq (1+(k+1)^2 n)^r \left(\frac{1}{2}+\frac{1}{8\sqrt{n}}\right)^{-|\alpha|} \sum_{i=1}^{j} \sup\{|D^\alpha f_{p_i}(u) : u \in I\}$$

$$\leq (2n)^r \left(\frac{1}{2}+\frac{1}{8\sqrt{n}}\right)^{-|\alpha|} \sum_{i=1}^{j} p_i^{\,2r} |f_{p_i}|_p \leq (2n^2q^2)^r \left(\frac{1}{2}+\frac{1}{8\sqrt{n}}\right)^{-|\alpha|}$$

$$\Sigma m^2 |f_m|_p$$

We set $H = (2n^2q^2)^r \left(\frac{1}{2}+\frac{1}{8\sqrt{n}}\right)^{-p}$. Then we have that

$$|T((f_m))|_{p,r} = \sum_{|\alpha| \leq p} \sup\{|(1+||x||^2)^r D^\alpha \Sigma f_m \circ g_m^{-1}(x)| : x \in R^n\}$$

$$\leq \sum_{|\alpha| \leq p} H\, \Sigma m^{2r} |f_m|_p \leq (p+1)^n H \;\; \Sigma m^{2r} |f_m|_p$$

$$\leq (p+1)^n H \left(\Sigma \frac{1}{m^2}\right) \sup m^{2r+2} |f_m|_p = (p+1)^n H \left(\Sigma \frac{1}{m^2}\right) ||(f_m)||_{p(2r+2)}$$

from where the continuity of T follows.

(8) *Given a multi-index* α, $D^\alpha \Sigma h \circ g_m^{-1}$ *is bounded on* R^n.

Proof. Let x be any point of R^n. Let $A_{p1}, A_{p2}, \ldots, A_{pr}$ the cubes of the form A_m to which x belongs. It is immediate that $r < 3^m$ and therefore

$$|D^\alpha h \circ g_m^{-1}(x)| = |\sum_{j=1}^{r} D^\alpha h \circ g_m^{-1}(x)| \leqslant 3^n \left(\frac{1}{2} + \frac{1}{8\sqrt{n}}\right)^{-|\alpha|}$$

$$\sup \{|D^\alpha f(u)| : u \in R^n\}$$

from where the conclusion follows.

We set

$$J = \{(x_1, x_2, \ldots, x_n) : -\frac{4\sqrt{n}}{4\sqrt{n+1}} \leqslant x_j \leqslant \frac{4\sqrt{n}}{4\sqrt{n}+1}, \; j = 1,2,\ldots,n\}$$

Then g_m maps J onto B_m, $m = 1,2,\ldots$

(9) *Given the multi-index* α, $D^\alpha \dfrac{1}{\Sigma h \circ g_m^{-1}}$ *is bounded on* R^n.

Proof. Since J is a compact set contained in the interior of I we have that

$$\inf \{h(x) : x \in J\} = b > 0.$$

If x belongs to R^n we find a positive integer r such that x belongs to B_r. Then

(10) $$|\Sigma h \circ g_m^{-1}(x)| \geqslant |h \circ g_r^{-1}(x)| = |h(g_r^{-1}(x))| \geqslant b$$

We have that $D^\alpha \dfrac{1}{\Sigma h \circ g_m^{-1}}$ is a ratio whose numerator is a linear combination of termus of the form

$$D^\gamma(\Sigma h \circ g_m^{-1})\, D^\delta(\Sigma h \circ g_m^{-1}) \ldots D^\mu(\Sigma h \circ g_m^{-1})$$

where $\gamma, \delta, \ldots, \mu$ are multi-índices and whose denominator is a positive integer power of $\Sigma h \circ g_m^{-1}$. According to (8) and (10) the conclusion follows easily.

(11) *If* f *belongs to* S *then*

$$\frac{f}{\Sigma\ h \circ g_m^{-1}} \in S.$$

Proof. Given a multi-index α we have that

$$(12) \qquad D^{\alpha} \frac{f}{\Sigma\ f \circ g_m^{-1}} = \sum_{\beta \leq \alpha} A_{\alpha\beta}\, D^{\beta} \frac{1}{\Sigma\ h \circ g_m^{-1}} D^{\alpha-\beta} f$$

where $A_{\alpha\beta}$ does not depend on $\frac{1}{\Sigma\ h \circ g_m^{-1}}$ and on f. According to (10) we find a positive number H such that

$$\left|D^{\beta} \frac{1}{\Sigma\ h \circ g_m^{-1}(x)}\right| \leq H,\ \beta \leq \alpha,\ x \in R^n.$$

Then, if r is any positive integer,

$$||x||^r \left|D^{\alpha} \frac{f(x)}{\Sigma\ h \circ g_m^{-1}(x)}\right| = ||x||^r \left|\sum_{\beta<\alpha} A_{\alpha\beta}\, D^{\beta} \frac{1}{\Sigma\ h \circ g_m^{-1}(x)} D^{\alpha-\beta} f(x)\right|$$

$$\leq H \sum_{\beta \leq \alpha} |A_{\alpha\beta}| . ||x||^r\ |D^{\alpha-\beta} f(x)|$$

and since

$$\lim_{||x|| \to \infty} ||x||^r\ |D^{\alpha-\beta} f(x)| = 0$$

it follows that

$$\lim_{||x|| \to \infty} ||x||^r \left|D^{\alpha} \frac{f(x)}{\Sigma\ h \circ g_m^{-1}(x)}\right| = 0$$

from where the conclusion follows.

For every f of S we set $Xf = \frac{f}{\Sigma\ h \circ g_m^{-1}}$.

(13) X *is a continuous linear mapping from* S *into* S.

Proof. Obviously X is linear. Let p and r be two positive integers. According to (9) there is H > 0 such that

$$\left|D^{\beta} \frac{1}{\Sigma\ h \circ g_m^{-1}(x)}\right| < H,\ |\beta| \leq p,\ x \in R^n.$$

If f is any element of S we recall (12) to obtain

$$|Xf|_{pr} = \sum_{|\alpha| \leq p} \sup\{|(1+||x||^2)^r\ |D^{\alpha} f(x)| : x \in R^n\}$$

$$= \sum_{|\alpha|\leq p} \sup\{ (1+||x||^2)^r | \sum_{\beta \leq \alpha} |A_{\alpha\beta} D^{\beta} \frac{1}{\Sigma\, h \circ g_m^{-1}(x)} D^{\alpha-\beta} f(x)|$$

$$: x \in R^n\} \leq H \sum_{|\alpha| \leq p} \sum_{\beta\leq\alpha} |A_{\alpha\beta}| \sup\{(1+||x||^2)^r | D^{\alpha-\beta} f(x)|$$

$$: x \in R^n \}$$

and therefore there is a positive number L which does not depend on f such that

(14) $$|Xf|_{pr} \leq L \sum_{|\alpha|\leq p} \sup\{ (1+||x||^2)^r | D^{\alpha} f(x)| : x \in R^n\} = L|f|_{pr}$$

from where the continuity of X follows.

(15) *If* (f_m) *is a sequence of elements of* E(I) *such that*

$$\lim m^r \sup \{|D^{\alpha} f_m(x)| : x \in I\} = 0$$

for every positive integer r *and multi-index* α, *then* (hf_m) *belongs to* F.

Proof. Clearly hf_m belongs to $\mathcal{D}(I)$, m = 1,2,... Given a multi-index α there is a positive number k such that

$$\sup_{\beta<\alpha} \sup \{|D^{\alpha-\beta} h(x)| : x \in I\} < k$$

On the other hand

$$D^{\alpha}(h f_m) = \sum_{\beta<\alpha} A_{\beta} D^{\alpha-\beta} h D^{\beta} f_m$$

where A_{β} does not depend on h and f_m. Then

$$\sup \{|D^{\alpha} h(x) f_m(x)| : x \in I\} < \sum_{\beta<\alpha} |A_{\beta}| k \sup \{|D^{\beta} f_m(x)| : x \in I\}$$

from where it follows that

$$\lim m^r \sup \{|D^{\alpha} h(x) f_m(x)| : x \in I\} < k \sum_{\beta<\alpha} |A_{\beta}| \lim m^r \sup|\{D^{\beta} f_m(x) : x \in I\} = 0$$

for every positive integer r. The proof is complete.

(16) *Let* f *be an element of* S. *If* f_m *denotes the restriction of* $f \circ g_m$ *to* I, m = 1,2,..., *then*

$$\lim m^r \sup \{|D^{\alpha} f_m(x)| : x \in I\} = 0$$

for every positive integer r *and multi-index* α.

Proof. For every positive integer m there is a point x_m in I such that

$$\sup \{|D^\alpha f_m(x)| : x \in I\} = |D^\alpha f_m(x_m)|.$$

We set $z_m = g_m(x_m)$. Since z_m belongs to B_m we have that $\lim ||z_m|| = \infty$ and therefore

$$(17) \qquad \lim ||z_m||^{rn} |D^\alpha f_m(x_m)| = \lim ||z_m||^{rn} |D^\alpha f(z_m)| = 0.$$

If $m > 4^n+1$ there is a positive integer $k(m) > 1$ such that

$$(18) \qquad 2k(m)^n+1 \leq m \leq (2k(m)+2)^n$$

Then B_m belongs to $A_{k(m)+1} \sim A_{k(m)}$ and therefore $z_m \notin M_{k(m)-1}$ from where it follows that $||z_m|| > k(m) - 1 > 1$. Consequently

$$(19) \qquad m^r \sup \{|D^\alpha f_m(x)| : x \in I\} \leq (2km(m)+2)^{rn} |D^\alpha f_m(x_m)|$$

$$\leq 2^{rn}(||z_m||\pm 2)^{rn} |D^\alpha f(z_m)| \leq 6^{rn}||z_m||^{rn} |D^\alpha f(z_m)|$$

according to (18). We apply (17) to reach the conclusion.

For every f of S we set $Zf = ((kmf) \circ g_m)$.

(20) Z *is an isomorphism from* S *into* F.

Proof. Let f be any element of S. According to (11) Xf belongs to S. For every positive integer m we have that

$$(21) \qquad (k_m f) \circ g_m = \left(\frac{h \circ g_m^{-1}}{\Sigma\, h \circ g_m^{-1}} f\right) \circ g_m$$

$$= ((h \circ g_m^{-1})Xf) \circ g_m = h\,((Xf) \circ g_m)$$

and therefore we apply (13), (15) and (16) to obtain that Zf belongs to F. Obviously Z is linear and since

$$T(Zf) = \Sigma\, ((k_m f) \circ g_m) \circ g_m^{-1} = \Sigma\, k_m f = f$$

it follows that Z is open and injective. Finally since $X : S \longrightarrow S$ is a linear continuous it is enough to show that the mapping from S into F defined by

$$f \longrightarrow (h(f \circ g_m)),\ f \in S,$$

is continuous (see (21)). For every multi-index α and f belonging to S we have that

$$D^{\alpha} h\ (f \circ g_m) = \sum_{\beta \leq \alpha} A_{\alpha\beta}\ D^{\alpha-\beta}\ h\ D^{\alpha}\ (f \circ g_m)$$

where $A_{\alpha\beta}$ does not depend on f. We fix two positive integers p and r. We take $m > 4^n + 1$. Applying (19) it follows that

$$||(h(f \circ g_m))||_{pr} = \sup\ m^r \sum_{|\alpha| \leq p} \sup\ \{|D^{\alpha} h\ (f \circ g_m)(x)| : x$$

$$\in I\} \leq \sup m^r \sum_{|\alpha| \leq p} \sum_{\beta \leq \alpha} |A_{\alpha\beta}|\ \sup\ \{|D^{\alpha-\beta} h(x)| : x$$

$$\in I\}\ \sup\ \{|D^{\alpha} f \circ g_m(x)| : x \in I\} \leq 6^{rn} \sum_{|\alpha| \leq p} \sup\{||z||^{rn} |D^{\alpha} f(z)| : z$$

$$\in R^n\} \sum_{\beta \leq \alpha} |A_{\alpha\beta}| \sup\ \{|D^{\alpha-\beta}\ h(x)| : x \in I\}$$

and therefore there is a positive integer k such that

$$||h(f \circ g_m)||_{pr} \leq k \sum_{|\alpha| \leq p} \sup\{||x||^{rn}\ |D^{\alpha} f(x)| : x \in R^n\}$$

$$\leq k \sum_{|\alpha| \leq p} \sup\ \{(1+||x||^2)^{rn}\ |D^{\alpha} f(x)| : x \in R^n\} = k\ |f|_{p(rn)}$$

from where the continuity of Z follows.

(22) *S isomorphic to a complemented subspace of* F.

Proof. Clearly $Z \circ T$ is a continuous projection from F into itself such that $(Z \circ T)\ (F) = Z(S)$ and therefore $Z(S)$ is a complemented subspace of F. We apply (20) to obtain that $Z(S)$ is isomorphic to S.

(23) *s is isomorphic to a complemented subspace of S.*

Proof. Let $Y : E(J) \longrightarrow \mathcal{D}(I)$ be a continuous linear extension operator. Then Y is an extension operator from $E(J)$ into S. Since $E(J)$ is isomorphic to s we have that $Y(E(J))$ is a subspace of S isomorphic to s. Finally the subspace of S of all those functions vanishing in J is a topological complement of $Y(E(J))$. The proof is complete.

Now we give the fundamental result of this section.

(24) *S is isomorphic to* s.

Proof. It is an immediate consequence of (22), (23) and 2.(8).

A proof of (24) can be found in SCHWARTZ [1], Chap. VII, Ex.7, using the Fourier transform.

18. REPRESENTATIONS OF THE SPACE $\overset{o}{B}(\Omega)$. Let Ω be a non-void open subset of R^n. We denote by $\overset{o}{B}(\Omega)$ the linear space over K of the K-valued function f defined on Ω which are infinitely differentiable and such that given $\varepsilon > 0$ and a multi-index α there is a compact set H in Ω depending of f and verifying $|D^\alpha f(x)| < \varepsilon$, $x \in \Omega \sim H$. For every positive integer we set

$$|f|_r = \sum_{|\alpha| \leq r} \sup \{|D^\alpha f(x)| : x \in \Omega\}$$

Then $|.|_r$, r = 1,2,..., is a system of norms on $\overset{o}{B}(\Omega)$ defining a metrizable locally convex topology. We suppose $\overset{o}{B}(\Omega)$ endowed with this topology. One proves easily that $\overset{o}{B}(\Omega)$ is a Fréchet space. If $\Omega = R^n$, $\overset{o}{B}(\Omega)$ coincides with the space $\overset{o}{B}$(cf. SCHWARTZ [1], Chap. VI, §8).

If f belongs to $\overset{o}{B}(\Omega)$ let g be the function defined on R^n which coincides with f in Ω and vanishes in $R^n \sim \Omega$. Then g is of class C^∞ in R^n and all its partial derivatives of all orders vanish in $R^n \sim \Omega$. If we suppose all functions of $\overset{o}{B}(\Omega)$ extended as we mentioned above it follows that $\overset{o}{B}(\Omega)$ is the subspace of $\overset{o}{B}$ of all those functions vanishing in $R^n \sim \Omega$ as well as their derivatives of all orders. In what follows we suppose that $\overset{o}{B}(\Omega)$ is this subspace of $\overset{o}{B}$.

Let (A_m) and (B_m) the sequence of n-dimensional cubes defined in Section 9. Let $g_m: R^n \longrightarrow R^n$ be the function defined in Section 15. If I and J are the n-dimensional cubes defined in Section 17 then g_m maps I onto A_m and J onto B_m. The functions h and k_m are defined as in Section 15. Then $\{k_m : m = 1,2,...\}$ is a partition of the unity of class C^∞ subordinated to the covering $\{\overset{\circ}{A}_m : m = 1,2,...\}$ of Ω.

Let G be the Fréchet space introduced in Section 15. Then G is isomorphic to $s \hat{\otimes} \lambda_0(a)$ with $a = (2^{k(m)})$. In that section k(m) is distinct from zero for an infinity of values of m but here k(m) is equal to zero, m = 1,2,..., when $\Omega = R^n$. For every (f_m) of G we set $T((f_m)) = \Sigma f_m \circ g_m^{-1}$.

Then T is a mapping from G into $\mathcal{B}(\Omega)$ according to 9.21. Obviously T is linear. If we change in Section 15 the interior of H by any open set Ω in R^n, $\Omega \neq \emptyset$ all the results given there are still valid with exception of 15.(14), 15.(21) and 15.(22). In particular the following result holds:

(1) *If Ω is distinct from R^n, $\overset{o}{\mathcal{B}}(\Omega)$ is isomorphic to a complemented subspace of* $s \hat{\otimes} \lambda_0(a)$.

(2) *If Ω coincides with R^n then T is continuous.*

Proof. Let (f_m) be an element of G and let α be any multi-index. We take x in R^n. Let $A_{m_1}, A_{m_2}, \ldots, A_{m_r}$ be the cubes of the form A_m to which x belongs. Then $r < 4^n$ and therefore

$$|D^\alpha \Sigma f_m \circ g_m^{-1}(x)| = |D^\alpha \sum_{j=1}^{r} f_{m_j} \circ g_{m_j}^{-1}(x)|$$

$$\leq 4^n \left(\frac{1}{2} + \frac{1}{8\sqrt{n}}\right)^{-|\alpha|} \sup_m \sup \{|D f_m(u)| : u \in I\}$$

from where the continuity of T follows.

(3) *If f belongs to $\overset{o}{B}$ then $f/\Sigma\, h \circ g_m^{-1}$ belongs to $\overset{o}{B}$.*

Proof. Given a multi-index α we have that

$$D^\alpha \frac{f}{\Sigma\, h \circ g_m^{-1}} = \sum_{\beta < \alpha} A_{\alpha\beta} D^\beta \frac{1}{\Sigma\, h \circ g_m^{-1}} D^{\alpha-\beta} f$$

where $A_{\alpha\beta}$ does not depend on $\dfrac{1}{\Sigma\, h \circ g_m^{-1}}$, and f. According to 17.(10) we find a positive number k such that

$$|D^\beta \frac{1}{\Sigma\, h \circ g_m^{-1}(x)}| \leq k, \; \beta \leq \alpha, \; x \in R^n$$

and therefore

(4) $$|D^\alpha \frac{f(x)}{\Sigma\, h \circ g_m^{-1}(x)}| \leq k \sum_{\beta \leq \alpha} |A_{\alpha\beta}|\,|D^{\alpha-\beta} f(x)|$$

The conclusion follows from (4).

For every f of $\overset{o}{B}$ we set $Xf = \dfrac{f}{\Sigma\, h \circ g_m^{-1}}$.

(5) *X is a continuous linear mapping from* $\overset{o}{B}$ *into* $\overset{o}{B}$.

Proof. According to (3), X is a mapping from $\overset{o}{B}$ into $\overset{o}{B}$ which is obviously linear. The continuity of X follows immediately from (4).

(6) *If* Ω *coincides with* R^n *and if* (f_m) *is a sequence of elements of* $E(I)$ *such that, for every multi-index* α,

$$\lim \sup \{|D^{\alpha}f_m(x)| : x \in I\} = 0$$

then (hf_m) *belongs to* G.

Proof. It is obvious that hf_m belongs to $\mathcal{D}(I)$, m = 1,2,... Given a multi-index α there is a positive number k such that

$$\sup_{\beta \leq \alpha} \sup \{|D^{\alpha-\beta}h(x)| : x \in I\} < k$$

On the other hand,

$$D^{\alpha}(hf_m) = \sum_{\beta \leq \alpha} A_{\beta} D^{\alpha-\beta}h \, D^{\beta}f_m$$

where A_{β} does not depend on h and f. Then

$$\sup \{|D^{\alpha}h(x)f_m(x)| : x \in I\}$$

$$\leq \sum_{\beta \leq \alpha} |A_{\beta}| k \sup \{|D^{\beta}f_m(x)| : x \in I\}$$

from where

$$\lim \sup \{|D^{\alpha}h(x)f_m(x)| : x \in I\} = 0$$

and the conclusion follows.

(7) *Let* f *be an element of* $\overset{o}{B}$. *If* f_m *is the restriction of* $f \circ g_m$ *to* I, m = 1,2,..., *then*

$$\lim \sup \{|D^{\alpha}f_m(x)| : x \in I\} = 0$$

for every multi-index α.

Proof. Given $\varepsilon > 0$ we find a compact subset H of R^n such that

$$|D^{\alpha}f(x)| < \varepsilon, \ x \in R^n \sim H.$$

We find a positive integer q such that $A_m \cap H = \emptyset$, m > q. Then

$$\sup \{|D^{\alpha}f_m(x)| : x \in I\}$$

$$= \left(\frac{1}{2} + \frac{1}{8\sqrt{n}}\right)^{|\alpha|} \sup \{|D^{\alpha} f(u) : u \in A_m\} < \varepsilon$$

for those values of m.

For every f of $\overset{o}{B}$ we set $Zf = ((k_m f) \circ g_m)$.

(8) *If* Ω *coincides with* R^n *then* Z *is an isomorphism from* $\overset{o}{B}$ *into* G.

Proof. Let f be any element of $\overset{o}{B}$. We apply (3) to obtain that Xf belongs to $\overset{o}{B}$. For every positive integer m we have that

(9) $(k_m f) \circ g_m = h((Xf) \circ g_m)$

and therefore Zf belongs to G applying (5), (6) and (7). Clearly Z is linear. Since

$$T(Zf) = \Sigma((k_m f) \circ g_m) \circ g_m^{-1} = \Sigma\, k_m f = f$$

it follows that $X : \overset{o}{B} \longrightarrow G$ is open and injective.

Finally, since $X : \overset{o}{B} \longrightarrow \overset{o}{B}$ is a continuous linear operator it is enough to show that the mapping

$$f \longrightarrow (h(f \circ g_m)),\ f \in \overset{o}{B}$$

from $\overset{o}{B}$ into G is continuous according to (9). Given a multi-index α we find a positive number k such that

$$|D^{\alpha-\beta} h(x)| < k,\ x \in I,\ \beta \leq \alpha$$

For every f of $\overset{o}{B}$ we have that

$$D^{\alpha} h\, (f \circ g_m) = \sum_{\beta \leq \alpha} A_{\alpha\beta}\, D^{\alpha-\beta} h\, D^{\alpha}(f \circ g_m)$$

where $A_{\alpha\beta}$ does not depend of $\overset{o}{B}$. Then given any point x of R^n it follows that

$$|D^{\alpha} h(x)\ (f \circ g_m)(x)| \leq \sum_{\beta \leq \alpha} |A_{\alpha\beta}|\ k \sup \{|D^{\alpha}(f \circ g_m)(u)| : u$$

$$\in I\} = k \left(\frac{1}{2} + \frac{1}{8\sqrt{n}}\right)^{|\alpha|} \sum_{\beta \leq \alpha} |A_{\alpha\beta}| \sup \{|D^{\alpha} f(u)| : u \in R\}$$

which proves the continuity of Z.

(10) *If* Ω *coincides with* R^n *then* $\overset{o}{B}$ *is isomorphic to a complemented subspace of* $s \hat{\otimes} \lambda_o(a)$.

Proof. It is immediate that Z o T is a continuos projection from G into itself such that $Z(\overset{o}{B}) = Z \circ T(G)$ and therefore $Z(\overset{o}{B})$ is a complemented subspace of G. Now we apply (8) to obtain that $Z(\overset{o}{B})$ is isomorphic to $\overset{o}{B}$. The conclusion follows having in mind that G is isomorphic to $s \hat{\otimes} \lambda_o(a)$.

Suppose now that Ω is a non-void open subset of R^n. We denote by L the cube

$$\{(x_1,x_2,\ldots,x_n) : -\frac{1}{5} \leq x_j \leq \frac{1}{5},\ j = 1,2,\ldots,n\}.$$

Then L is contained in the interior of J. Denote by E the linear space over K of all sequences (f_m) of element of $E(L)$ such that

$$\lim 2^{rk(m)} \sup \{|D^\alpha f_m(x)| : x \in L\} = 0$$

for every multi-index α and every positive integer r. We provide E with the metrizable locally convex topology defined by the family of norms

$$P_{qr}((f_m)) = \sup_m 2^{rk(m)} \sum_{|\alpha| \leq q} \sup \{|D^\alpha f_m(x)| : x \in L\},$$

$$(f_m) \in t,\ q,r = 1,2,\ldots$$

Since L is isomorphic to s it follows that E is isomorphic to the space $\lambda_o(a,s)$ with $a = (2^{k(m)})$ defined in Section 3. According to 3.(13) we have the following result:

(11) E *is isomorphic to* $s \hat{\otimes} \lambda_o(a)$.

Let $U : E(L) \longrightarrow D(J)$ be a continuous linear extension operator. Given a multi-index α there is a positive integer q and a positive number M such that

(12) $$\sup \{|D^\alpha Uf(x)| : x \in J\} \leq M \sum_{|\beta| \leq q} \sup \{|D^\beta f(x)| : x \in L\},\ f \in E(L)$$

For every (f_m) of E we set $Y(f_m) = (Uf_m)$. It follows that (12) that $Y : E \longrightarrow G$ is an isomorphism.

(13) *If* f *belongs to* $\overset{o}{B}(\Omega)$ *then*

$$\lim 2^{rk(m)} \sup \{|D^\alpha f \circ g_m(x)| : x \in I\} = 0$$

for every positive integer r *and for every multi-index* α.

Proof. If $\Omega \neq R^n$ see 15.(3) for the proof substituting $\mathcal{D}(H)$ by $\overset{o}{B}(\Omega)$, If Ω coincides with R^n the result is contained in (7).

(14) *Given a multi-index* α *and a positive integer* r *there is a positive number* k *such that*

$$(15)\qquad 2^{rk(m)} \sup \{|D^\alpha f \circ g_m(x)| : x \in I\} \leq k \sum_{|\beta| \leq |\alpha|+r} \sup\{|D^\beta f(x)| : x \in \Omega\}$$

for every f *of* $\overset{o}{B}(\Omega)$ *and* $m = 1,2,\ldots$

Proof. If $\Omega \neq R^n$ see 15.(17) for the proof substituting $\overset{o}{H}$ by Ω. If Ω coincides with R^n we have that

$$2^{rk(m)} \sup \{|D^\alpha f \circ g_m(x)| : x \in I\}$$

$$= \left(\frac{1}{2} + \frac{1}{8\sqrt{n}}\right)^{|\alpha|} \sup \{|D^\alpha f(u)| : u \in \Omega\} \leq \sum_{\beta \leq |\alpha|+r} \sup \{|D^\beta f(x)| : x \in \Omega\}$$

For every f of $\overset{o}{B}(\Omega)$ we set $V(f) = (V_m f)$ being

$$V_m f(x) = f \circ g_m(x),\ x \in L,\ m = 1,2,\ldots$$

According to (13) V is a mapping from $\overset{o}{B}(\Omega)$ into E. Obviously V is linear.

(16) $V : \overset{o}{B}(\Omega) \longrightarrow E$ *is continuous.*

Proof. Given the positive integers q and r we apply (14) to obtain a number $k > 0$ such that, for every f of $\overset{o}{B}(\Omega)$,

$$p_{qr}(Vf) = \sup_m 2^{rk(m)} \sum_{|\alpha| \leq q} \sup \{|D^\alpha V_m f(x)| : x \in L\}$$

$$\leq \sup_m 2^{rk(m)} \sum_{|\alpha| \leq q} \sup \{|D^\alpha f \circ g_m(x)| : x \in I\}$$

$$\leq k \sum_{|\alpha| \leq q} \sum_{|\beta| \leq |\alpha|+r} \sup \{|D^\beta f(x)| : x \in \Omega\}$$

$$\leq k \sum_{|\alpha|\leq q} |f|_{q+r} \leq k(q+1)^n |f|_{q+r}$$

from where the continuity of V follows.

(17) $\overset{o}{B}(\Omega)$ *has a complemented subspace isomorphic to* s $\hat{\otimes}$ $\lambda_0(a)$.

Proof. It is immediate that T o Y o V is a continuous projection from $\overset{o}{B}(\Omega)$ into itself such that T o Y o V $(\overset{o}{B}(\Omega))$ = T(Y(E)). Therefore T(Y(E)) is a complemented subspace of $\overset{o}{B}(\Omega)$. If S is the restriction of T to Y(E) we have that S is an injective continuous linear mapping from the Fréchet space Y(E) onto the Fréchet space T(Y(E)). We apply the open mapping theorem to conclude that Y(E) is isomorphic to T(Y(E)). Since Y(E)is isomorphic to E and E is isomorphic to s $\hat{\otimes}$ $\lambda_0(a)$ we are done.

The fundamental results of this section are (18), (19) and (20).

(18) $\overset{o}{B}(\Omega)$ *is isomorphic to* s $\hat{\otimes}$ $\lambda_0(a)$.

Proof. It is immediate consequence of (1), (10), (17), and 2.(7).

(19) *If* $\overset{o}{B}(\Omega)$ *is nuclear then it is isomorphic to* s.

Proof. It is a straightforward consequence of (18) and 4.(3).

(20) *If* $\overset{o}{B}(\Omega)$ *is not reflexive then it is isomorphic to* s $\hat{\otimes}$ c_0.

Proof. See (18) and 5.(2).

Results (18), (19) and (20) have been obtained by VALDIVIA [19]. Result (19) has been obtained independently by P. DIEROLF and J. VOIGHT [1] using a different method. During the preparation of this section we received a preprint by D. VOGT [1] where result (20) can also be found.

19. REFLEXIVITY AND NUCLEARITY OF $\overset{o}{B}(\Omega)$. We continue with the same notations used in the former section. We represent by ϕ the function defined on Ω such that if $\Omega = R^n$,

$$\phi(x) = +\infty, \ x \in R^n$$

and if $\Omega \neq R^n$

$$\phi(x) = \inf\{||x-y|| : y \in R^n \sim \Omega\}.$$

Ω is said to be quasi-bounded if for every $\varepsilon > 0$ the set

$$H_\varepsilon = \{x \in R^n : \phi(x) \geq \varepsilon\}$$

is compact.

(1) *$\overset{o}{B}(\Omega)$ is reflexive if and only if Ω is quasi-bounded.*

Proof. If Ω is quasi bounded, given $\varepsilon > 0$, there is a positive integer p such that $\bigcup_{m=1}^{p} \overset{o}{A}_m \supset H_\varepsilon$. For every positive integer m, $\overset{o}{A}_m$ intersects at most 4^n cubes of the form $\overset{o}{A}_r$ according 9.(14). Consequently there is a positive integer q such that $B_m \cap H_\varepsilon = \emptyset$, $m > q$. Then $2^{-k(m)} < \varepsilon$, $m > q$. According to 3.(2), $\lambda_0(a)$ is a Schwartz space and therefore it is reflexive (cf. SCHAEFER [1], Chapter IV, §9). According to 18.(18), $\overset{o}{B}(\Omega)$ is isomorphic to $s \hat{\otimes} \lambda_0(a)$ and thus $\overset{o}{B}(\Omega)$ is reflexive.

Suppose that $\overset{o}{B}(\Omega)$ is reflexive. Since $\lambda_0(a)$ is isomorphic to a closed subspace of $s \hat{\otimes} \lambda_0(a)$ it follows that $\lambda_0(a)$ is reflexive. On the other hand, $\lambda_0(a)$ is an echelon space of order zero and therefore, according to 3.(1) and Chapter Two, §4, 2.(1), $\lambda_0(a)$ is a Schwartz space. We apply 3.(2) to obtain that $(2^{(km)})$ diverges to infinite. Then Ω is distinct from R^n. Given $\varepsilon > 0$ we find a positive integer q such that $\sqrt{n} < \varepsilon . 2^{q-1}$. If $z = (z_1, z_2, \ldots, z_n)$ is any point of H_ε we take the integer a_j such that

$$\frac{a_j}{2^q} \leq z_j \leq \frac{a_j + 1}{2^q}, \quad j = 1,2,\ldots, n,$$

and we set Q for the cube

$$\{(x_1, x_2, \ldots, x_n) : \frac{a_j}{2^q} \leq x_j \leq \frac{a_j + 1}{2^q}, \quad j = 1,2,\ldots, n\}.$$

Let y be any point of Q. Let d and δ be the distances of z and y to $R^n \sim \Omega$ respectively. We have that

$$\delta \geq d - ||y-z|| \geq \varepsilon - \frac{\sqrt{n}}{2^q} > \frac{\sqrt{n}}{2^q}$$

and therefore Q is contained in Ω. Thus z belongs to a cube of the sequence (B_m) whose diameter is larger or equal than $\frac{\sqrt{n}}{2^q}$. Then there is a positive integer r such that $\bigcup_{m=1}^{r} B_m \supset H_\varepsilon$ and therefore H_ε is compact and we are done.

(2) *If $\overset{o}{B}(\Omega)$ is reflexive then it is a Schwartz space.*

Proof. See the proof of (1).

An open set Ω is quasi-integrable if there is a positive integer p such that ϕ belongs to $L^p(R^n)$.

(3) $\overset{o}{\mathcal{B}}(\Omega)$ *is nuclear if and only if* Ω *is quasi-integrable.*

Proof. First we suppose that Ω is quasi integrable. We find a positive integer p such that ϕ belongs to $L^p(R^n)$. Then

$$\Sigma 2^{-(p+n)k(m)} = \Sigma \int_{Bm} 2^{-pk(m)} dx$$
$$\leq \left(\frac{1}{\sqrt{n}}\right)^p \Sigma \int_{B_m} \phi^p dx = \left(\frac{1}{\sqrt{n}}\right)^p \int_{R^n} \phi^p dx < \infty$$

and $\lambda_o(a)$ is nuclear, according to 3.(3). We apply 4.(3) to obtain that $s \hat{\otimes} \lambda_o(a)$ is nuclear. Therefore $\overset{o}{\mathcal{B}}(\Omega)$ is nuclear.

Suppose that $\overset{o}{\mathcal{B}}(\Omega)$ is nuclear. According to 3.(3) there is a positive integer p such that $\Sigma\, 2^{-pk(m)} < \infty$. We find a positive integer q such that $k(m) > 0$ for $m > q$. We apply 15.(2) to obtain that the distance from B_m to $R^n \sim \Omega$ is less or equal than $\frac{\sqrt{n}}{2^{k(m)-2}}$ for $m > q$. Then

$$\int_{R^n} \phi^p dx = \sum_{m=1}^{q} \int_{B_m} \phi^p dx + \sum_{m=q+1}^{\infty} \int_{B_m} \phi^p dx$$

$$\leq \sum_{m=1}^{q} \int_{B_m} \phi^p dx + \sum_{m=q+1}^{\infty} \int_{B_m} \left(\frac{\sqrt{n}}{2^{k(m)}} + \frac{\sqrt{n}}{2^{k(m)-2}}\right)^p dx$$
$$\leq \sum_{m=1}^{q} \int_{B_m} \phi^p dx + (5\sqrt{n})^p \sum_{m=q+1}^{\infty} 2^{-(p+m)k(m)} < \infty$$

and the proof is complete.

Results (1) and (2) are due to P. DIEROLF [1] where different methods of proof can be found. Result (3) has been obtained by VALDIVIA [19]. Independently, a characterization of the nuclearity of $\overset{o}{\mathcal{B}}(\Omega)$ analogous to (3) has been obtained by P. DIEROLF and J. VOIGT [1].

20. ON CERTAIN SEQUENCES IN LOCALLY CONVEX SPACES. A subset M of a locally convex space G has the property of the interchangeable double limits (p.i.d.l.) if for every sequence (x_m) in M and every sequence (u_m) in P, being P any weakly compact absolutely convex subset of G', the existence of both double limits.

$$\lim_m \lim_n \langle x_m, u_n\rangle \text{ and } \lim_n \lim_m \langle x_m, u_n\rangle$$

implies their equality

(1) *If* M *a weakly relatively countably compact subset of a locally convex space* G *then* M *has* (p.i.d.1).

Proof. Let P be a compact absolutely convex subset of $G'[\sigma(G',G)]$. We take any sequence (x_m) and (u_m) in M and P respectively such that

$$a = \lim_m \lim_n \langle x_m, u_n\rangle \quad \text{and} \quad b = \lim_n \lim_m \langle x_m, u_n\rangle$$

exist. Let x be a weakly adherent point of the sequence (x_m) and let u be a weakly adherent point of the sequence (u_m). Then

$$\lim_n \langle x_m, u_n\rangle = \langle x_m, u\rangle \text{ and } \lim_m \langle x_m, u_n\rangle = \langle x, u_n\rangle$$

and therefore

$$a = \lim_m \langle x_m, u\rangle = \langle x, u\rangle = \lim_n \langle x, u_n\rangle = b.$$

(2) *Let* M *be a weakly relatively countably compact subset of the locally convex space* G. *If* H *is the absolutely convex hull of* M *then* H *has* (p.i.d.1).

Proof. Let L be the balanced hull of M. It is immediate that L is weakly relatively countably compact and therefore has (p.i.d.1), H coincides with the convex hull of L and therefore H has (p.i.d.1) (cf. KÖTHE [1], Chapter 5, §24, Section 6).

In what follows E is a locally convex space and F a subspace of E' which has the following property: If u belongs to E' there is in F an E-equicontinuous subset, which is $\sigma(E',E)$-relatively countably compact, such that u belongs to the $\sigma(E',E)$-closed absolutely convex hull of A.

(3) *If* (x_m) *is a bounded sequence in* E *converging to the origin for the topology* $\sigma(E,F)$, *then* (x_m) *converges to the origin for the topology* $\sigma(E,E')$.

Proof. First we suppose E complete. Then, for every element (b_m) of ℓ^1, the series $\Sigma\, b_m x_m$ converges in E. Let L be the subset of E.

$$\{ \Sigma\, a_m x_m : a_m \in K,\ m = 1,2,\ldots, \Sigma\, |a_m| \leq 1\}.$$

We set $T((b_m)) = \Sigma\, b_m x_m$, $(b_m) \in \ell^1$. Then T is a linear mapping from ℓ^1 into E which maps the closed unit ball B of ℓ^1 in L. Let $\{(b_m(j)) : j$

$\in J, \geq\}$ be a net in ℓ^1 $\sigma(\ell^1,c_0)$-convergent to the origin. If v is any element of F we have that $(\langle x_m,v\rangle)$ belongs to c_0. Therefore

$$\lim \{\langle \Sigma b_m^{(j)} x_m,v\rangle : j \in J, \geq\} = \lim \{\Sigma b_m^{(j)} \langle x_m,v\rangle : j \in J,\geq\} = 0$$

and consequently $T : \ell^1 [\sigma(\ell^1,c_0)] \longrightarrow E [\sigma(E,F)]$ is continuous. Since B is $\sigma(\ell^1,c_0)$-compact, then L is $\sigma(E,F)$-compact. Let P be the linear hull of L. Let Q be the subspace of E' orthogonal to P. We set $X : E' \longrightarrow E'/Q$ the canonical surjection. We suppose the existence of an element u of E' such that $(\langle x_m,u\rangle)$ does not converge to the origin. Since the sequence (x_m) is bounded in E we extract from it a subsequence (z_m) such that $\lim \langle z_m,u\rangle = b \neq 0$.

We find a subset A of F which is E-equicontinuous and countably compact for the topology $\sigma(E',E)$ such that if D is the absolutely convex hull of A then u belongs to $\sigma(E',E)$-closure $\bar{D}$ of D.

If $e_m = (w_n)$ is the element of ℓ^1 such that $w_n = 0$, $n \neq m$, $w_m = 1$, we have that $\{e_1,e_2,\dots,e_m,\dots\}$ is total in ℓ^1 and therefore $\{x_1,x_2,\dots,x_n,\dots\}$ is total in P for the Banach space topology defined by the gauge of L. Since L is bounded in E, it follows that P is $\sigma(P,E')$- separable. Consequently $X(\bar{D})$ is a metrizable compact set of $(E'/Q)[\sigma(E'/Q,P)]$. Therefore there is a sequence (u_m) in D such that $X u_m$ converges to $X u$ for the topology $\sigma(E'/Q,P)$. Then

$$\lim_m \langle z_m,u_n\rangle = 0, \; n = 1,2,\dots,$$

$$\lim_n \langle z_m,u_n\rangle = \lim_n \langle z_m,Xu_n\rangle = \langle z_m,Xu\rangle , \; m = 1,2,\dots,$$

and therefore

$$\lim_n \lim_m \langle z_m,u_n\rangle = 0 \text{ and } \lim_m \lim_n \langle z_m,u_n\rangle = b$$

We apply (2) for $G = F [\sigma(F,E)]$, $A = M$ and $D = H$. Since z_m belongs to L, $m = 1,2,\dots$, and this set is absolutely convex and $\sigma(E,F)$-compact it follows that $b = 0$ which is a contradiction.

Now we suppose E not complete. Let S be the completion of E. If M is a $\sigma(E',E)$-closed E-equicontinuous subset then M is S-equicontinuos and $\sigma(E',S)$-closed and therefore $\sigma(E',S)$-compact; thus $\sigma(E',S)$ and $\sigma(E',S)$ coincide on M. Given an element u of E' there is a $\sigma(E',E)$-relatively countably compact E-equicontinuous subset A of F such that u belongs to the closed absolutely convex hull of A in $E'[\sigma(E', E)]$. Then A is $\sigma(E', S)$-relatively countably compact S-equicontinuous set such that u belongs to the closed absolutely convex hull of A in $E'[\sigma(E', S)]$. If (x_m) is a bounded sequence in E which is $\sigma(E, F)$-convergent to the origin, then (x_m) is a bounded sequence of S. Obviously F is dense in $E'[\sigma(E', S)]$. We apply the arguments used above to obtain that (x_m) $\sigma(E, E')$-converges to the origin

(4) *If* (x_m) *is a bounded sequence of* E *which is a Cauchy sequence in* $E[\sigma(E, F)]$, *then* (x_m) *is a Cauchy sequence in* $E[\sigma(E, E')]$.

Proof. We suppose the property false. Then there is an element u of E' such that $(\langle x_m, u\rangle)$ is not a Cauchy sequence in K. Therefore there is $\varepsilon > 0$ and a sequence of positive integers $m_1 < m_2 < \ldots < m_r < \ldots$ such that

$$|\langle x_{n_{2r-1}} - x_{n_{2r}}, u\rangle| > \varepsilon, \; r = 1,2,\ldots$$

Consequently the sequence $(x_{n_{2r-1}} - x_{n_{2r}})$, which is obviously bounded in E, converges to the origin for the topology $\sigma(E, F)$ and does not converge to the origin in $E[\sigma(E, E')]$; a fact which contradicts (3). The proof is complete.

Let Ω be a non-void open subset of the euclidean space R^n. For every x of R^n and for every multi-index α we set δ_x^α to denote the derivative of order α of the measure δ of Dirac concentratect in x, i.e., if f belongs to $\mathcal{E}(R^n)$, then

$$\langle f, \delta_x^\alpha\rangle = (-1)^{|\alpha|} D^\alpha f(x).$$

It is obvious that δ_x^α is a continuous linear form on $\overset{o}{\mathcal{B}}(\Omega)$. We denote by $\overset{o}{\mathcal{B}}'(\Omega)$ the topological dual of $\overset{o}{\mathcal{B}}(\Omega)$ with the strong topology. We set

$$A_\alpha = \{\delta_x^\alpha : x \in \Omega\} \cup \{0\}$$

where 0 is the origin of $\overset{o}{\mathcal{B}}'(\Omega)$. For every positive integer m we set

$$A_m = \cup \{A_\alpha : |\alpha| \leq m\}.$$

(5) *For every positive integer* m, A_m *is a compact subset of* $\overset{o}{\mathcal{B}}(\Omega)$ $[\sigma(\overset{o}{\mathcal{B}}'(\Omega), \overset{o}{\mathcal{B}}(\Omega))]$

Proof. According to the definition of A_m, it is enough to prove that, given a multi-index α, A_α is $\sigma(\overset{o}{B}'(\Omega), \overset{o}{B}(\Omega))$-compact. We take in A_α the net

(6) $\{\delta^\alpha_{x_j} : j \in J, \geq\}$

Since Ω is locally compact we take its Alexandroff compactification Z by adjoining the point ∞. We extract from the net $\{x_j : j \in J, \geq\}$ a subnet

(7) $\{z_i : i \in I, \geq\}$

which converges to an element z of Z. Then

(8) $\{\delta^\alpha_{z_i} : i \in I, \geq\}$

is a subnet of (6). Let f be any element of $\overset{o}{B}(\Omega)$. If z is distinct from ∞ the net (7) converges to z in Ω and therefore

$$\lim \{<f,\delta^\alpha_{z_i}> : i \in I,\geq\} = \lim \{(-1)^{|\alpha|}D^\alpha f(z_i) : i \in I, \geq\}$$

$$= (-1)^{|\alpha|}D^\alpha f(z) = <f,\delta^\alpha_z>$$

and thus the net (8) converges to δ^α_z in A_α for the topology $\sigma(\overset{o}{B}'(\Omega), \overset{o}{B}(\Omega))$. If $z = \infty$, given $\varepsilon > 0$, we find a compact Y in Ω such that

$$|D^\alpha f(x)| < \varepsilon, \ x \in \Omega \sim Y$$

Then

$$\{y \in Z : y \in \Omega \sim Y\} \cup \{\infty\}$$

is a neighbourhood of ∞ in Z and therefore there is and index h in I such that $z_i \in \Omega \sim Y$, $i > h$. Consequently

$$|<f,\delta^\alpha_{z_i}>| = |D^\alpha f(z_i)| < \varepsilon$$

if $i > h$, and therefore the net (8) $\sigma(\overset{o}{B}'(\Omega), \overset{o}{B}(\Omega))$-converges to the origin in $\overset{o}{B}'(\Omega)$ and we are done.

(9) *If* u *is any element of* $\overset{o}{B}'(\Omega)$ *there is a positive integer* m *such that* u *belongs to the closed absolutely convex hull of* m A_m *in* $\overset{o}{B}'(\Omega)$ $[\sigma(\overset{o}{B}'(\Omega), \overset{o}{B}(\Omega))]$.

Proof. Since u is continuous in $\overset{o}{B}(\Omega)$ there is a positive integer and $\varepsilon > 0$ such that

(10) $|<f,u>| < 1, \ f \in \overset{o}{B}(\Omega), \ |f|_r < \varepsilon$

Let m be a positive integer with

$$r < m, \ (r+1)^n < m\,\varepsilon.$$

Let U be the polar set of m A_m in $\overset{o}{\mathcal{B}}(\Omega)$. If V is the polar set of U in $\overset{o}{\mathcal{B}}'(\Omega)$ then V coincides with the closed absolutely convex hull of m A_m in $\overset{o}{\mathcal{B}}'(\Omega)$ $[\sigma\,(\overset{o}{\mathcal{B}}'(\Omega), \overset{o}{\mathcal{B}}(\Omega))]$. Now we take any element f of U. If the multi-index α verifies $|\alpha| \leq m$ and if x is any point of Ω we have that m δ_x^α belongs to m A_m and therefore

$$|\langle f, m\delta_x^\alpha\rangle| = m\,|D^\alpha f(x)| \leq 1.$$

Since the number of multi-index α with $|\alpha| \leq r$ is less or equal than $(r+1)^n$ it follows that

$$|f|_r = \sum_{|\alpha| \leq r} \sup\{|D^\alpha f(x)| : x \in \Omega\} \leq \frac{(r+1)^n}{m} < \varepsilon$$

and observing (10), we have that $|\langle f,u\rangle| < 1$; thus u belongs to V.

Let F be the linear hull of $\cup\{A_m : m = 1,2,\ldots\}$ in $\overset{o}{\mathcal{B}}'(\Omega)$.

(11) *Let* (f_m) *be a bounded sequence of* $\overset{o}{\mathcal{B}}(\Omega)$. *If for every multi-index* α *and for every* x *of* Ω, *the sequence* $(D^\alpha f_m(x))$ *converges to zero then* (f_m) *converges weakly to the origin in* $\overset{o}{\mathcal{B}}(\Omega)$.

Proof. For every V of F the sequence $(\langle f_m, v\rangle)$ converges to zero and the conclusion follows easily from (3) by observing that (9) is verified.

(12) *Let* (f_m) *be a bounded sequence of* $\overset{o}{\mathcal{B}}(\Omega)$. *If, for every multi-index* α *and* x *in* Ω, *the sequence* $(D^\alpha f_m(x))$ *is a Cauchy sequence in* K, *then* (f_m) *is a weak Cauchy sequence in* $\overset{o}{\mathcal{B}}(\Omega)$.

Proof. For every v of F, the sequence $(\langle f_m, v\rangle)$ is a Cauchy sequence in K. Since (9) is verified the conclusion follows easily from (4).

Let $\lambda_0(a)$ be the Fréchet space used in Section 18. Then $\overset{o}{\mathcal{B}}(\Omega)$ is isomorphic to $s \hat{\otimes} \lambda_0(a)$, according to 18.(17). If $\lambda_0'(a)$ is the strong dual of $\lambda_0(a)$ we have that $\overset{o}{\mathcal{B}}'(\Omega)$ is isomorphic to $s' \hat{\otimes} \lambda_0'(a)$ (cf. SCHAEFER [1], Chapter IV, 99). Since s' and $\lambda_0'(a)$ are separable, it is immediate that $\overset{o}{\mathcal{B}}'(\Omega)$ is separable. If A is a bounded subset of the strong bidual $\overset{o}{\mathcal{B}}''(\Omega)$ of $\overset{o}{\mathcal{B}}(\Omega)$ we have that A is $\sigma(\overset{o}{\mathcal{B}}''(\Omega), \overset{o}{\mathcal{B}}'(\Omega))$-relatively compact and, since $\overset{o}{\mathcal{B}}'(\Omega)$ is separable, A is $\sigma(\overset{o}{\mathcal{B}}''(\Omega), \overset{o}{\mathcal{B}}'(\Omega))$-metrizable. Consequently every element of $\overset{o}{\mathcal{B}}''(\Omega)$ is the $\sigma(\overset{o}{\mathcal{B}}''(\Omega), \overset{o}{\mathcal{B}}'(\Omega))$-limit of a sequence contained in $\overset{o}{\mathcal{B}}(\Omega)$.

(13) F *is a dense subspace of* $\overset{o}{\mathcal{B}}'(\Omega)$.

Proof. Suppose F not dense in $\overset{o}{\mathcal{B}}'(\Omega)$. Then there are elements u of $\overset{o}{\mathcal{B}}'(\Omega)$ and w of $\overset{o}{\mathcal{B}}''(\Omega)$ such that

$$\langle w,u\rangle = 1, \ \langle w,v\rangle = 0, \text{ for every } v \text{ of } F.$$

Let (f_m) be a sequence in $\overset{o}{\mathcal{B}}(\Omega)$ converging to w in $\overset{o}{\mathcal{B}}''(\Omega)$ $[\sigma(\overset{o}{\mathcal{B}}''(\Omega), \overset{o}{\mathcal{B}}'(\Omega))]$. Then (f_m) is bounded in $\overset{o}{\mathcal{B}}(\Omega)$ and, for every v of F,

$$\lim \langle f_m,v\rangle = \langle w,v\rangle = 0$$

We apply (1) to obtain

$$\lim \langle f_m,u\rangle = \langle w,u\rangle = 0.$$

This is a contradiction.

21. A REPRESENTATION OF THE SPACE $\mathcal{B}_1(\Omega)$. We denote by $\mathcal{B}(R^n)$, or shortly $\mathcal{B}$, the linear space over K of all K-valued functions f defined on R^n which are infinitely differentiable and such that

$$|D^\alpha f(x)| < h, \ x \in R^n,$$

h being a positive number depending of f and the multi-index α. For every positive integer r we set

$$|f|_r = \sum_{|\alpha|\leq r} \sup \{|D^\alpha f(x)| : x \in R^n\}$$

Then $|.|_r$, r = 1,2,..., is a system of norms on $\mathcal{B}$ defining a metrizable locally convex topology. We suppose $\mathcal{B}$ endowed with this topology. It is easy to prove that $\mathcal{B}$ is complete. Thus $\mathcal{B}$ is a Fréchet space.

Let Ω be a non-void open subset of R^n. We denote by $\mathcal{B}_1(\Omega)$ the subspace of $\mathcal{B}$ of all those functions vanishing in $R^n \sim \Omega$ as well as their partial derivatives of all orders. It is obvious that $\mathcal{B}_1(\Omega)$ is a closed subspace of $\mathcal{B}_1(\Omega)$.

We fix a function g of $\mathcal{D}(R^n)$ taking the value one on a neighbourhood of the origin in R^n. For every positive integer m let g_m the element of $\mathcal{D}(R^n)$ such that, if $x = (x_1,x_2,\ldots,x_m)$ belongs to R^n,

$$g_m(x) = g\left(\frac{1}{m}x_1, \frac{1}{m}x_2, \ldots, \frac{1}{m}x_n\right).$$

(1) *If* f *is any element of* $B_1(\Omega)$ *then*

(2) $$\{fg_1, fg_2, \ldots, fg_m, \ldots\}$$
is a bounded subset of $\overset{o}{B}(\Omega)$.

Proof. For every positive integer m, it is obvious that fg_m belongs to $\mathcal{D}(R^n)$ and vanishes in $R^n \sim \Omega$ as well as all its partial derivatives of all orders. Therefore fg_m belongs to $\overset{o}{B}(\Omega)$.

Given the multi-index α, we have that

$$D^\alpha(fg_m) = \sum_{\beta \leq \alpha} A_\beta D^\beta f D^{\alpha-\beta} g_m$$

where A_β does not depend on f and g_m. Therefore, if $x = (x_1, x_2, \ldots, x_n)$ belongs to R^n, we have that

$$|D^\alpha(fg_m)(x)| \leq \sum_{\beta \leq \alpha} |A_\beta D^\beta f(x)| \cdot \left| D^{\alpha-\beta} g\left(\frac{1}{m}x_1, \frac{1}{m}x_2, \ldots, \frac{1}{m}x_n\right)\right|$$

$$\leq \sup_{\beta \leq \alpha} |A_\beta D^\beta f(x)| \sum_{\beta \leq \alpha} \left(\frac{1}{m}\right)^{|\alpha-\beta|} \left|D^{\alpha-\beta} g\left(\frac{1}{m}x_1, \frac{1}{m}x_2, \ldots, \frac{1}{m}x_n\right)\right|$$

$$\leq |f|_{|\alpha|} \sup_{\beta \leq \alpha} |A_\beta| \sum_{\beta \leq \alpha} \sup\{|D^{\alpha-\beta} g(y)| : y \in R^n\} = h < +\infty$$

where h does not depend on m. We deduce that the set (2) is bounded in $\overset{o}{B}(\Omega)$.

(3) *If* f *is any element of* $B_1(\Omega)$, *the sequence* (fg_m) *converges to* f *for the topology on* $B_1(\Omega)$ *induced by* $E(R^n)$.

Proof. In R^n let M be an open ball of center the origin such that $g = 1$ in M. Given a compact subset H of R^n we find a positive integer p such that H is contained in pM. Then $g_m = 1$ in a neighbourhood of H for $m > p$. Consequently $D^\alpha(fg_m)(x) = D^\alpha f(x)$. The conclusion follows.

Let F be the linear space introduced in the former section provided with the topology induced by $\overset{o}{B}'(\Omega)$.

(4) *If* f *is any element of* $B_1(\Omega)$ *then the sequence* (fg_m) *converges in* $\overset{o}{B}''(\Omega)$ $[\sigma(\overset{o}{B}''(\Omega), \overset{o}{B}'(\Omega))]$ *to an element* Tf *which coincides with* f *in* F.

Proof. According to (2) and (3), the sequence (fg_m) is bounded in $\overset{o}{B}(\Omega)$; this sequence is a Cauchy sequence for the topology on $\sigma(\overset{o}{B}(\Omega), F)$. We apply 20.(12) to obtain that (fg_m) is a weak Cauchy sequence in $\overset{o}{B}(\Omega)$ and therefore converges in $\overset{o}{B}''(\Omega)$ $[\sigma(\overset{o}{B}''(\Omega), \overset{o}{B}'(\Omega))]$ to an element Tf.

Given any element δ_x^α of A_α we have that

$$<Tf, \delta_x^\alpha> = \lim <fg_m, \delta_x^\alpha> = \lim(-1)^{|\alpha|} D^\alpha (fg_m(x)$$

$$= (-1)^{|\alpha|} D^\alpha f(x) = <f, \delta_x^\alpha>$$

and therefore f coincides with Tf in F.

(5) $T : B_1(\Omega) \longrightarrow \overset{o}{B}''(\Omega)$ *is an algebraic isomorphism.*

Proof. Obviously T is linear and injective. Let w be any element of $\overset{o}{B}''(\Omega)$. We find a sequence (f_m) in $\overset{o}{B}(\Omega)$ converging to w in $\overset{o}{B}''(\Omega)$ $[\sigma(\overset{o}{B}''(\Omega), \overset{o}{B}'(\Omega))]$. Since

(6) $$\{f_1, f_2, \ldots, f_m, \ldots\}$$

is a bounded subset of $\overset{o}{B}(\Omega)$, (6) is a bounded set of $E(R^n)$ and, since this last space is Montel, we can extract from (f_m) a subsequence converging to f in $E(R^n)$. For every x of R^n and every multi-index α, it is obvious that $(D^\alpha f_m(x)$ converges to $D^\alpha f(x)$. On the other hand, there is $h>0$ such that

$$\sup \{|D^\alpha f_m(z)| : z \in R^n, m = 1,2,\ldots\} < h$$

and therefore

$$\sup \{|D^\alpha f(x)| : z \in R^m\} \leq h$$

from where it follows that f belongs to $B_1(\Omega)$. Finally given any element δ_x^α of A_α we have that

$$<w, \delta_x^\alpha> = \lim <f_m, \delta_x^\alpha> = \lim (-1)^{|\alpha|} D^\alpha f_m(x)$$

$$= (-1)^{|\alpha|} D^\alpha f(x) = <f, \delta_x^\alpha>$$

and consequently w coincides with f in F. We apply now 20.(13) to obtain that w equals Tf and the proof is complete.

(7) T^{-1} *is sequentially continuous.*

Proof. Let (f_m) be a sequence in $B_\Lambda(\Omega)$ such that (Tf_m) converges to the origin in $B''_1(\Omega)$. Given the multi-index α, A_α is a subset of $\overset{o}{B}'(\Omega)$ which is an equicontinuous in $\overset{o}{B}''(\Omega)$ and therefore

$$0 = \lim_m \sup \{|<Tf_m, \delta_x^\alpha>| : x \in \Omega\}$$

$$= \lim_m \sup \{|\langle f_m \, \delta_x^{\alpha}\rangle| : x \in \Omega\} = \lim_m \sup \{|D^{\alpha} f_m(x)| : x \in \Omega\}$$

from where the conclusion follows.

(8) $T : B_1(\Omega)_{o} \longrightarrow \overset{o}{B}''(\Omega)$ *is an isomorphism.*

Proof. $\overset{o}{B}''(\Omega)$ is the strong bidual of the Fréchet space $\overset{o}{B}(\Omega)$ and therefore a Fréchet space. Since $B_1(\Omega)$ is a Fréchet space, it follows that T^{-1} is continuous, according to (7). Finally we apply the closed graph theorem to obtain that T is an isomorphism.

In what follows we identify f with Tf, $f \in B_1(\Omega)$.

(9) $B_1(\Omega)$ *coincides with the strong bidual of* $\overset{o}{B}(\Omega)$.

Proof. It is an immediate consequence of (8).

Let $\lambda_o(a)$ be the space introduced in Section 18. According to Chapter Two, §4, 1.(10) the strong bidual of $\lambda_o(a)$ coincides with $\lambda_\infty(a)$ and therefore the strong bidual of $s \hat{\otimes} \lambda_o(a)$ is isomorphic to $s \hat{\otimes} \lambda_\infty(a)$ (cf. SCHAEFER [1], Chapter IV, 9.(9). Since $\overset{o}{B}(\Omega)$ is isomorphic to $s \hat{\otimes} \lambda_o(a)$ results (10), (11), (12), (13), (14), (15) and (16) are now immediate.

(10) $B_1(\Omega)$ *is isomorphic to* $s \hat{\otimes} \lambda_\infty(a)$.

(11) $B_1(\Omega)$ *is a Schwartz space if it is reflexive.*

(12) $B_1(\Omega)$ *is reflexive if and only if* Ω *is quasi-bounded.*

(13) $B_1(\Omega)$ *is nuclear if and only if* Ω *is quasi-integrable.*

(14) *If* $B_1(\Omega)$ *is nuclear, then it is isomorphic to* s.

(15) *If* $B_1(\Omega)$ *is not reflexive, then it is isomorphic to* $s \hat{\otimes} \ell^\infty$

(16) B *is isomorphic to* $s \hat{\otimes} \ell^\infty$

(17) F *has its Mackey topology.*

Proof. Let U be a closed absolutely convex neighbourhood of the origin in $F[\mu(F, \overset{o}{B}''(\Omega))]$. Let $\bar{U}$ be the closure of U in $\overset{o}{B}'(\Omega)$. If V denotes the polar set of U in $\overset{o}{B}''(\Omega)$ we have that V is absolutely convex and $\sigma(B''(\Omega),F)$-compact. Consequently if G denotes the linear hull of V normed with the gauge of V, then G is a Banach space.

If $S : G \longrightarrow \overset{o}{B}''[\sigma(\overset{o}{B}''(\Omega),F)]$ is the canonical injection, it is obvious that S is continuous and therefore has closed graph in $G \times \overset{o}{B}''(\Omega)$. We apply the closed graph theorem to obtain that $S : G \longrightarrow B''(\Omega)$ is continuous. Consequently $S(V) = V$ is a bounded set of $\overset{o}{\mathcal{B}}''(\Omega)$. The space $\overset{o}{\mathcal{B}}'(\Omega)$ is separable and therefore barrelled (cf. KOTHE [1], Chapter Six, §29, Section 3) from where it follows that V is an equicontinuous set on $\overset{o}{\mathcal{B}}'(\Omega)$. Then $\bar{U}$ is a neighbourhood of the origin in $\overset{o}{\mathcal{B}}'(\Omega)$ and the proof is complete.

For every positive integer m, let B_m be the closed absolutely convex hull of m A_m in $\overset{o}{\mathcal{B}}'(\Omega)$ $[\sigma(\overset{o}{\mathcal{B}}'(\Omega), \overset{o}{\mathcal{B}}(\Omega))]$. Then $\{B_1, B_2, \ldots, B_m, \ldots\}$ is a fundamental system of bounded sets in $\overset{o}{\mathcal{B}}'(\Omega)$. We set E_m to denote the linear hull of B_m endowed with the norm derived from the gauge of B_m.

(18) *Let* M *be a compact subset of* R^m. *Given a multi-index* α *there is a positive integer* m *such that the set*

(19) $\{\delta_x^\alpha : x \in M\}$

is a compact subset of the Banach space E_m.

Proof. Let

(20) $\{\delta_{x_j}^\alpha : j \in J, \geq\}$

be a net in the set (19). Then

(21) $\{x_j : j \in J, \geq\}$

is a net in M. Since m is a compact subset of R^n, there is a subnet $\{z_i : i \in I, \geq\}$ of (21) converging in M to a point z. Then given any element f of $\overset{o}{\mathcal{B}}(\Omega)$ we have that

$$<f, \delta_z^\alpha> = (-1)^{|\alpha|} D^\alpha f(z) = \lim (-1)^{|\alpha|} \{D^\alpha f(z_i) : i \in I, \geq\}$$

$$= \lim \{<f, \delta_{z_i}^\alpha> : i \in I, \geq\}$$

and therefore the net $\{\delta_{z_i}^\alpha : i \in I, \geq\}$, which is a subnet of (20) converges to the element $\delta_{z_o}^\alpha$ of (19) for the topology $\sigma(\overset{o}{\mathcal{B}}'(\Omega), \overset{o}{\mathcal{B}}(\Omega))$. Consequently the set (19) is $\sigma(\mathcal{B}''(\Omega), \overset{o}{\mathcal{B}}(\Omega))$-compact.

Let T be the topology on $\overset{o}{\mathcal{B}}(\Omega)$ induced by $\mathcal{E}(R^n)$. If G denotes the topological dual of $\overset{o}{\mathcal{B}}(\Omega)[T]$ it is immediate that $\overset{o}{\mathcal{B}}'(\Omega) \supset G \supset F$. Obviously the set (19) is T-equicontinuous and, since $\overset{o}{\mathcal{B}}(\Omega)$ $[T]$ is a Schwartz space, there is in G an absolutely convex and $\sigma(G, \overset{o}{\mathcal{B}}(\Omega))$-closed there in G an absolutely convex $\sigma(G, \overset{o}{\mathcal{B}}(\Omega))$-closed T-equicontinuous subset Q such that (19) is a compact subset of G_Q. Obviously Q is $\sigma(\mathcal{B}'(\Omega), \mathcal{B}(\Omega))$-bounded and, accordingly there is a positive integer m with $B_m \supset Q$. Therefore (19) is a compact set in E_m.

(22) *The space F is bornological.*

Proof. For every positive integer m let F_m be the subspace $E_m \cap F$ of E_m. Let $\mathcal{U}$ be the topology on $\mathcal{B}_1(\Omega)$ of the uniform convergence on every sequence of F converging to the origin in some space F_m, m = 1,2,... Let

(23) $\{f_j : j \in J, \geq\}$

be a Cauchy net in $\mathcal{B}_1(\Omega)$ $[\mathcal{U}]$. Let M be a compact subset of R^n. Given a multi-index α, we set M_α to denote the set $\{\delta_x^\alpha : x \in M\}$. According to (18) there is a positive integer m such that M_α is a compact subset of F_m. We find in F_m a sequence converging to the origin such that M_α is in the closed absolutely convex hull in F_m of this sequence (cf. KÖTHE [1], Chapter Four, §21, Section 10). Consequently (23) is a Cauchy net in $\mathcal{E}(R^n)$ and therefore there exits the limit f of (23) in $\mathcal{E}(R^n)$. Since all the functions of the net (23) vanish in $R^n \sim \Omega$ as well as all their partial derivatives of all orders, it follows that f vanishes in $R^n \sim \Omega$ as well as all its partial derivatives of all orders. Now we suppose that f does not belongs to $\mathcal{B}$. Then there is a multi-index α and a sequence (x_r) in R^n such that

$$|D^\alpha f(x_r)| > r^2, \quad r = 1,2,\ldots$$

Since

(24) $\{\delta_{x_r}^\alpha : r = 1,2,\ldots\}$

is a bounded subset of $\overset{o}{\mathcal{B}}'(\Omega)$ there is a positive integer m such that (24) is a bounded subset of F_m. Then the sequence $(\frac{1}{r}\,\delta_{x_r}^\alpha)$ converges to the origin in F_m and therefore there is an index i in J such that

$$\frac{1}{r}\,|D^\alpha f_i(x_r)-D^\alpha f_j(x_r)| = |\langle f_i-f_j, \frac{1}{r}\,\delta_{x_r}^\alpha\rangle| < 1$$

$$j \in J,\ j \geq i,\ r = 1,2,\ldots,$$

and accordingly

$$|D^\alpha f_i(x_r)-D^\alpha f(x_r)| \leq r, \quad r = 1,2,\ldots,$$

from where it follows that

$$r^2 < |D^\alpha f(x_r)| \leq |D^\alpha f(x_r) - D^\alpha f_i(x_r)| + |D^\alpha f_i(x_r)|$$

$$\leq r + |D^\alpha f_i(x_r)|$$

for r = 1,2,... and therefore $D^\alpha f_i$ is not bounded on R^n; that is a contradiction. Thus f belongs to $\mathcal{B}_1(\Omega)$.

Finally (23) is a Cauchy net in $\mathcal{B}_1(\Omega)$ $[\mathcal{U}]$ converging to f in $\mathcal{B}_1(\Omega)$ $[\sigma(\mathcal{B}_1(\Omega), \overset{o}{\mathcal{B}}'(\Omega))]$ and therefore (23) converges to f in $\mathcal{B}_1(\Omega)$ $[\mathcal{U}]$. According to (17) F is a Mackey space and this F is bornological (cf. KÖTHE [1], Chapter Six, §28, Section 5).

Result (9) can be found in P. DIEROLF [1]. The proof of this result appears in P. DIEROLF and J. VOIGT [1] and it is different from the proof given here.

22. SPACES OF FUNCTIONS DEFINED ON A C^∞-DIFFERENTIABLE MANIFOLD. Given a topological space V let f be a K-valued continuous function defined on V. The support of f is the closure of $\{x \in V : f(x) \neq 0\}$. Given a positive integer n, we suppose now V an n-dimensional C^∞-differentiable manifold. Let (U_i, ϕ_i), $i \in I$, an atlas in V. A K-valued function defined on V is said to be infinitely differentiable if for every $i \in I$ the function $f \circ \phi_i^{-1}$ is infinitely differentiable in the open set of R^n $\phi_i(U_i)$. We write $\mathcal{E}(V)$ to denote the linear space over K of all K-valued functions defined on V which are infinitely differentiable. Given a positive integer r and $i \in I$, we set

$$|f|_{r,L,i} = \sum_{|\alpha| \leq r} \sup \{|D^\alpha(f \circ \phi_i^{-1})(x)| : x \in \phi_i(L)\}$$

for every compact subset of V contained in U_i. Then $|.|_{r,L,i}$ is a seminorm on $\mathcal{E}(V)$. We suppose $\mathcal{E}(V)$ endowed with the locally convex topology defined by $|.|_{r,L,i}$, $r \in N$, L compact contained in U_i, $i \in I$. It is not difficult to prove that $\mathcal{E}(V)$ is complete. Moreover, if we use an atlas equivalent to (U_i, ϕ_i), $i \in I$, to define $\mathcal{E}(V)$ it is easy to see that the same locally convex space is obtained.

Let H be a compact subset of V. We denote by $\mathcal{D}(H)$ the subspace of $\mathcal{E}(V)$ of all those functions which have their support contained in H. If M denotes the closure of the interior $\overset{o}{H}$ of H it is obvious that $\mathcal{D}(H)$ coincides with $\mathcal{D}(M)$.

Let $\mathcal{D}(V)$ be the linear subspace of $\mathcal{E}(V)$ of all those functions with compact support. Let $\mathcal{H}$ be the family of all compact subset of V. We suppose $\mathcal{D}(V)$ endowed with the locally convex topology such that $\mathcal{D}(V)$ is the inductive limit of the family of locally convex spaces $\{\mathcal{D}(H) : H \in \mathcal{H}\}$.

Given $i \in I$, a subset Q of V is a cube in U_i if Q is contained in V and there are real numbers $a_j < b_j$, $j = 1,2,\ldots,n$ such that

$$\phi_i(Q) = \{(x_1,x_2,\ldots,x_n) : a_j \leq x_j \leq b_j,\ j = 1,2,\ldots, n\}.$$

Let H be a compact subset of V with non-void interior. Let M be the closure of $\overset{\circ}{H}$. For every z of M we find i_z in I and a cube Q_z in U_{i_z} such that z belongs to $\overset{\circ}{Q}_z$. Since M is compact we find the points $z_1,z_2,\ldots,z_m$ in M with $\bigcup\{\overset{\circ}{Q}_{z_j} : j = 1,2,\ldots, m\}\supset M$. We set

$$Q_{z_j} \cap M = P_j,\ \phi_{i_{z_j}} = \psi_j,\ U_{z_j} = V_j,\ j = 1,2,\ldots,m.$$

(1) *If* j *is an integer with* $1 \leq j \leq r$ *then* $\mathcal{D}(P_j)$ *is isomorphic to* s.

Proof. If f is any element of $\mathcal{D}(P_j)$ we set Xf to denote the element of $\mathcal{E}(R^n)$ such that

$$Xf(x) = f \circ \psi_j^{-1}(x),\ x \in \psi_j(V_j),\ Xf(x) = 0,\ x \in R^n \sim \psi_j(V_j).$$

It is obvious that X is an injective linear mapping from $\mathcal{D}(P_j)$ in $\mathcal{D}(\psi_j(P_j))$. If g is any element of $\mathcal{D}(\psi_j(P_j))$, we set Yg to denote the element of $\mathcal{D}(\psi_j(P_j))$, we set Yg to denote the element of $\mathcal{D}(P_j)$ with

$$Yg(z) = g \circ \psi_j(z),\ z \in V_j,\quad Yg(z) = 0,\ z \in V \sim V_j.$$

Then X(Yg) = g and therefore X is an algebraic isomorphis from $\mathcal{D}(P_j)$ onto $\mathcal{D}(\psi_j(P_j))$. Given a positive integer r we have that

$$|Xf|_r = \sum_{|\alpha| \leq r} \sup \{|D\ Xf(x)| : x \in R^n\}$$

$$= \sum_{|\alpha|\leq r} \sup \{|D^\alpha f \circ \psi_j^{-1}(x)| : x \in \psi_j(P_j)\} = |f|_{r,P_j,\ i_{z_j}}$$

for every f of $\mathcal{D}(P_j)$ from where it follows that X is continuous. It is not difficult to show that $\mathcal{D}(P_j)$ is a Fréchet space. Therefore X is an isomorphism. Finally the interior of P_j is non-void and therefore $\psi_j(P_j)$ has non-void interior. According to 15.(22), $\mathcal{D}(\psi_j(P_j))$ is isomorphic to s and the conclusion follows.

If $(f_1,f_2,\ldots f_m)$ is any element of $\prod_{j=1}^{m} \mathcal{D}(P_j)$ we write

$$Z((f_1,f_2,\ldots,f_m) = \sum_{j=1}^{m} f_j$$

Then $Z : \prod_{j=1}^{m} \mathcal{D}(P_j) \longrightarrow \mathcal{D}(H)$ is linear and continuous.

Given an integer j, $1 \leqslant j \leqslant m$, we take in $\mathcal{D}(\psi_j(Q_{zj}))$ an element k_j whit $k_j(x) > 0$ when x belongs to the interior of $\psi_j(Q_{zj})$. If we set

$$g_j(z) = k_j \circ \psi_j(z),\ z \in V_j, g_j(z) = 0,\ z \in V \sim V_j,$$

we have that g_j belongs to $\mathcal{D}(Q_{zj})$ and $g_j(z) > 0$ when z belongs to the interior of Q_{zj}. If g is any element of $\mathcal{D}(H)$ we write $T_j q$ to denote the vector of $\mathcal{D}(P_j)$ such that

$$T_j g(z) = \frac{g(z)g_j(z)}{\sum_{r=1}^{m} g_r(z)},\ z \in M,\ T_j g(z) = 0,\ z \in V \sim M.$$

$T_j : \mathcal{D}(H) \longrightarrow \mathcal{D}(P_j)$ is obviously linear. Since

$$\sum_{j=1}^{m} g_j(z) > 0,\ z \in M,$$

and since M is compact, it is not difficult to show that T_j is continuous. If

$$Tg = (T_1 g,\ T_2 g,\ldots,T_m g),$$

$T : \mathcal{D}(H) \longrightarrow \prod_{j=1}^{m} \mathcal{D}(P_j)$ is linear and continuous.

(2) T *is an isomorphism from* $\mathcal{D}(H)$ *into* $\prod_{j=1}^{m} \mathcal{D}(P_j)$.

Proof. If g is any element of $\mathcal{D}(H)$ then $Z(Tg) = \sum_{j=1}^{m} T_j g$. If z belongs to M we have that

$$Z(Tg)(z) = \sum_{j=1}^{m} \frac{g(z)g_j(z)}{\sum_{r=1}^{m} g_r(z)} = g(z),$$

and if z does not belong to M

$$Z(Tg)(z) = 0 = g(z)$$

Therefore $Z(Tg) = g$ and the conclusion follows.

(3) *$\mathcal{D}(H)$ is isomorphic to a complemented subspace of* s.

Proof. T o Z is a continuous projection from $\prod_{j=1}^{m} \mathcal{D}(P_j)$ into itself such that T o Z $(\prod_{j=1}^{m} \mathcal{D}(P_j)) = T(\mathcal{D}(H))$. According to (2), $T(\mathcal{D}(H))$ is isomorphic to $\mathcal{D}(H)$ and therefore $\mathcal{D}(H)$ is isomorphic to a complemented subspace of $\prod_{j=1}^{m} \mathcal{D}(P_j)$. Finally, we apply (1) and 1.(5) to obtain that $\prod_{j=1}^{m} \mathcal{D}(P_j)$ is isomorphic to s. The proof is complete.

(4) s *is isomorphic to a complemented subspace of* $\mathcal{D}(H)$.

Proof. Let z be an interior point of H. We find $i \in I$ such that z belongs to U_i. Let Q be a cube in U_i contained in $\overset{\circ}{H}$. Let W be a continuous linear extension operator from $\mathcal{E}(\phi_i(Q))$ into $\mathcal{D}(\phi_i(U_i \cap \overset{\circ}{H}))$. The mapping $J:\mathcal{E}(\phi_i((Q)) \longrightarrow \mathcal{D}(H)$ such that

$$Jf(x) = (Wf) \text{ o } \phi_i(x), \; x \in U_i,$$

$$Jf(x) = 0, \; x \in V \sim U_i$$

for every f of $\mathcal{E}(\phi_i(Q))$ is an isomorphism from $\mathcal{E}(\phi_i(Q))$ into $\mathcal{D}(H)$. On the other hand, the subspace of $\mathcal{D}(H)$ whose functions vanish in Q is a (H) topological complement of $J(\mathcal{E}(\phi_i(Q)))$ in that space. Finally the conclusion follows since $\mathcal{E}(\phi_i(Q))$ is isomorphic to s.

Now we arrive to one of the fundamental results of this section:

(5) *Let* V *be an* n*-dimensional* C^∞*-differentiable manifold. Let* H *be a compact subset of* V *with non-void interior. Then* $\mathcal{D}(H)$ *is isomorphic to* s.

Proof. It is an immediate consequence from (3), (4) and 2.(8).

(6) *If* V *is an* n*-dimensional compact* C^∞*-differentiable manifold then* $\mathcal{E}(V)$ *is isomorphic to* s.

Proof. Since $\mathcal{D}(V) = \mathcal{E}(V)$, the result is an immediate consequence from (5).

In what follows we suppose V non compact and countable at infinity.

If we follow an analogous method to the one used in 8.(5), we can extract from the atlas (U_i, ϕ_i), $i \in I$, a sequence of charts

(7) $$(W_1, \alpha_1), (W_2, \alpha_2), \ldots, (W_r, \alpha_r), \ldots$$

such that in W_r there is a cube Q_r such that

(8) $$\{\overset{\circ}{Q}_r : r = 1,2,\ldots\}$$

is a locally finite covering of V.

For every positive integer r let h_r be an element of $\mathcal{D}(\alpha_r(Q_r))$ such that $h_r(x) > 0$ for every x of the interior of $\alpha_r(Q_r)$. Let λ_r be the function defined on V such that

$$\lambda_r(z) = h_r \circ \alpha_r(z), \; z \in W_r,$$

$$\lambda_r(z) = 0, \; z \in V \sim W_r$$

We set

$$k_r = \frac{\lambda_r}{\Sigma \lambda_j}, \; r = 1,2,\ldots$$

(9) $\mathcal{E}(V)$ *is isomorphic to a complemented subspace of* s^N.

Proof. We write E to denote $\prod_{r=1}^{\infty} \mathcal{D}(Q_r)$. Let λ be the mapping from E into $\mathcal{E}(V)$ defined by

$\lambda(f_1, f_2, \ldots, f_r, \ldots) = \Sigma f_r, f_r \in \mathcal{D}(Q_r)$, $r = 1,2,\ldots$ It is not difficult to check that λ is linear and continuous. If μ is the mapping from $\mathcal{E}(V)$ into E defined by

$$\mu(f) = (fk_1, fk_2, \ldots, fk_r \ldots), \; f \in \mathcal{E}(V)$$

then μ is linear and continuous. Moreover

$$\lambda \circ \mu(f) = \Sigma f k_r = f$$

and therefore μ is an isomorphism from $\mathcal{E}(V)$ into E. Consequently $\mu(\mathcal{E}(V))$ is a subspace of E isomorphic to $\mathcal{E}(V)$. It is immediate that $\mu \circ \lambda$ is a continuous projection from E into itself such that $\mu \circ \lambda(E) = \mu(\mathcal{E}(V))$. Therefore $\mu(\mathcal{E}(V))$ is a complementd subspace of E. According to (5), E is isomorphic to s^N, Now the conclusion follows.

Since the covering (8) is locally finite we can find a sub-sequence of (7):

$$(W_{r_1},\alpha_{r_1}),\ (W_{r_2},\alpha_{r_2}),\dots,(W_{r_m},\alpha_{r_m}),\dots$$

such that the set $Q_{r_1}, Q_{r_2},\dots,Q_{r_m},\dots$ are pairwise disjoint. We set

$$B_j = W_{r_j}, \beta_j = \alpha_{r_j},\ H_j = Q_{r_j},\ j = 1,2,\dots$$

Let L_j be a cube in B_j contained in the interior of H_j, $j = 1,2,\dots$

(10) s^N *is isomorphic to a complemented subspace of* $E(V)$.

Proof. For every positive integer r let X_r be a continuous linear extension operator from $E(\beta_r(L_r))$ into $\mathcal{D}(\beta_r(H_r))$. If f belongs to $E(\beta_r(L_r))$ we set

$$Z_r f(x) = (X_r\ f) \circ \beta_r(x),\ x \in H_r,$$

$$Z_r f(x) = 0,\ x \in V\sim H_r$$

If f_r belongs to $E(\beta_r(L_r))$, $r = 1,2,\dots$, we set

$$\rho(f_1,f_2,\dots,f_r,\dots) = \Sigma\ Z_r f_r$$

We write F to denote $\prod_{r=1}^{\infty} E(\beta_r(L_r))$. It is immediate that ρ is an isomorphism from F into $E(V)$. Since $E(\beta_r(L_r))$ is isomorphic to s it follows that $\rho(F)$ is isomorphic to s^N. Finally the subspace of $E(V)$ of all those functions vanishing in $L_1 \cup L_2 \cup \dots \cup L_r \cup \dots$. is a topological complement of $\rho(F)$ in $E(V)$. The proof is complete.

Next we have another fundamental result of this section.

(11) *Let* V *be an* n-*dimensional* C^∞-*differentiable manifold. If* V *is not compact and countable at infinity the* $E(V)$ *is isomorphic to* s^N.

Proof. It is an immediate consequence from (9), (10) and 2.(1).

(12) $\mathcal{D}(V)$ *is isomorphic to a complemented subspace of* $s^{(N)}$.

Proof. We set G to denote $\bigoplus_{r=1}^{\infty} \mathcal{D}(Q_r)$. If γ is the mapping from G into $\mathcal{D}(V)$ defined by

$$\gamma(f_1,f_2,\dots,f_r\dots) = \Sigma\ f_r, (f_1,f_2,\dots,f_r\dots) \in G$$

it is not difficult to chech that γ is linear and continuous. If δ is the mapping from $\mathcal{D}(V)$ into G such that

$$\delta(f) = (fk_1, fk_2, \ldots, fk_r, \ldots), \ f \in \mathcal{D}(V),$$

then δ is linear and continuous. Moreover

$$\gamma \circ \delta(f) = \Sigma\, fk_r = f$$

and therefore δ is an isomorphism from $\mathcal{D}(V)$ into G. Consequently $\delta(\mathcal{D}(V))$ is a subspace of G isomorphic to $\mathcal{D}(V)$.

It is immediate that $\delta \circ \gamma$ is a continuous projection from G into itself such that $\delta \circ \gamma(G) = \delta(\mathcal{D}(V))$. Consequently $\delta(\mathcal{D}(V))$ is a complemented subspace of G. Since this last space is isomorphic to $s^{(N)}$ the conclusion follows.

(13) $s^{(N)}$ *is isomorphic to a complemented subspace of* $\mathcal{D}(V)$.

Proof. We write L to denote $\bigoplus_{r=1}^{\infty} E(\beta_r(L_r))$. If Z_r is the mapping defined in the proof of (10) we set

$$\alpha(f_1, f_2, \ldots, f_r \ldots) = \Sigma\, Z_r f_r, (f_1, f_2, \ldots, f_r, \ldots) \in L.$$

It is immediate that α is an isomorphism from L into $\mathcal{D}(V)$. Consequently $\alpha(L)$ is isomorphic to $s^{(N)}$. Finally the subspace of $\mathcal{D}(V)$ of all those functions vanishing on $L_1 \cup L_2 \cup \ldots \cup L_r \cup \ldots$ is a topological complement of $\alpha(L)$ in $\mathcal{D}(V)$. The proof is complete.

We arrive to the last fundamental result of this section:

(14) *Let V be an n-dimensional C^∞-differentiable manifold. If V is not compact and countable at inifinity, then $\mathcal{D}(V)$ is isomorphic to $s^{(N)}$.*

Proof. It is an immediate consequence from (12), (13) and 2.(2).

Result (5) can be found in VALDIVIA [18]. Result (6) with some additional conditions appears in OGRODSKA [1]. Results (11) and (14) can be found in VALDIVIA [27].

23. SPACES OF FUNCTIONS DEFINED ON A C^∞-DIFFERENTIABLE MANIFOLD WITH BOUNDARY. Given a positive integer n we set R^n_+ to denote the subspace of R^n :

$\{(x_1,x_2,\ldots,x_n) : x_1 \geqslant 0\}$. If L is a compact set of R^n_+ we denote by $\mathcal{D}_+(L)$ the linear space over K of all K-valued functions defined on R^n_+ which are infinitely differentiable and with support contained in L. Given a positive integer r we set

$$|f|_r = \sum_{|\alpha|\leqslant r} \sup \{|D^\alpha f(x)| : x \in L\}, \quad f \in \mathcal{D}_+(L)$$

Then $|.|_r$, $r = 1,2,\ldots$, is a family of norms on $\mathcal{D}_+(L)$ defining a metrizable locally convex topology. We suppose $\mathcal{D}_+(L)$ endowed with this topology.

We suppose now that L has non-void interior. We find a positive number b such that

$$\sup \{||x|| : x \in L\} < b.$$

We set

$$A = \{(x_1,x_2,\ldots,x_n) : -b \leqslant x_j \leqslant b,\ j = 1,2,\ldots,n\}$$

$$B = \{(x_1,x_2,\ldots,x_n) : 0 \leqslant x_1 \leqslant b,\ -b \leqslant x_j \leqslant b,\ j = 2,3,\ldots,n\}$$

$$D = \{(x_1,x_2,\ldots,x_n) : -b \leqslant x_1 \leqslant 0,\ -b \leqslant x_j \leqslant b,\ j = 2,3,\ldots,n\}$$

Let g be an element of $\mathcal{D}(A)$ such that $g(x) = 1$, $x \in L$. Let $\nu : \mathcal{E}(B) \longrightarrow \mathcal{E}(A)$ a continuous linear extension operator. We set

$$\beta f = g\,\nu f, \quad f \in \mathcal{D}_+(L)$$

Then β is a continuous linear extension operator from $\mathcal{D}_+(L)$ into $\mathcal{D}(M)$ with $M = D \cup L$.

(1) *$\mathcal{D}_+(L)$ is isomorphic to a complemented subspace of* s.

Proof. $\beta(\mathcal{D}_+(L))$ is a subspace of $\mathcal{D}(M)$ isomorphic to $\mathcal{D}_+(L)$. It is obvious that the subspace of $\mathcal{D}(M)$ of all those functions vanishing in L is a topological complement of $\beta(\mathcal{D}_+(L))$ in $\mathcal{D}(M)$. According to 15.(22), $\mathcal{D}(M)$ is isomorphic to s and the conclusion follows.

(2) s *is isomorphic to a complemented subspace of $\mathcal{D}_+(L)$.*

Proof. It is analogous to the proof of 15.(1) taking the cube B in the interior of L in R^n.

(3) *$\mathcal{D}_+(L)$ is isomorphic to* s.

Proof. It is an immediate consequence from (1), (2) and 2.(8).

Let V be an n-dimensional C^∞-differentiable manifold with boundary. Let (U_i, ϕ_i), $i \in I$, be an atlas in V such that $\phi_i(U_i)$ is an open set in R^n_+ $i \in I$. If z belongs to the boundary of V, then $\phi_i(z)$ has its first coordinate nulle. As in Section 22 we define the space $E(V)$ by taking R^n_+ instead of R^n. If H is a compact subset of V, then $\mathcal{D}(H)$ is the subspace of $E(V)$ whose functions have support contained in H. $\mathcal{D}(V)$ is defined as in the former section as well as the notion of a cube in U_i, $I \in I$.

We shall use the same notations as in Section 22.

(4) *If* j *is an integer with* $1 \leq j \leq r$, *then* $\mathcal{D}(P_j)$ *is isomorphic to* s.

Proof. See 22.(1) taking $\mathcal{D}_+(\psi_j(P_j))$ instead of $\mathcal{D}(\psi_j(P_j))$ and apply (3).

Given an integer j, $1 \leq j \leq r$, we take an element k_j in $\mathcal{D}_+(\psi_j(Qz_j))$ such that $k_j(x) > 0$ when x belongs to the interior of $\psi_j(Qz_j)$ in R^n. Now we define g_j, T_j and T, $j = 1,2,\ldots, r$, as we did in the former section.

Result (5) and (6) can be proved as in 22.(2) and 22.(3) respectively

(5) T *is an isomorphism from* $\mathcal{D}(H)$ *into* $\prod_{j=1}^{r} \mathcal{D}(P_j)$.

(6) $\mathcal{D}$ (H) *is isomorphic to a complemented subspace of* s.

(7) s *is isomorphic to a complemented subspace of* $\mathcal{D}(H)$.

Proof. See 22.(4) and take z such that it does not belong to the boundary P of V and take $\mathcal{D}(\phi_i(U_i \cap \overset{\circ}{H} \sim P))$ instead of $\mathcal{D}(H)$.

We arrive to one of the fundamental results of this section:

(8) *Let* V *be an* n-*dimensional* C^∞-*differentiable manifold with boundary. Let* H *be a compact subset of* V *whose interior is non-void. Then* D(H) *is isomorphic to* s.

Proof. It is an immediate consequence from (6), (7) and 2.(8).

(9) *If* V *is an* n-*dimensional compact* C^∞-*differentiable manifold with boundary, then* E(V) *i isomorphic to* s.

Prof. Since $E(V) = \mathcal{D}(V)$, it is an immediate consequence of (8).

In what follows we suppose V countable at infinity and non compact. We extract from the atlas (U_i,ϕ_i), $i \in I$, a sequence of charts

$$(W_1,\alpha_1),\ (W_2,\alpha_2),\dots,(W_r,\alpha_r),\dots$$

such that there is in W_r a cube Q_r such that $\{\overset{\circ}{Q}_r : r = 1,2,\dots\}$ is a locally finite covering of V (see the proof of 8.(5)).

For every positive integer r let h_r be an element of $\mathcal{D}_+(\alpha_r(Q_r))$ such that $h_r(x) > 0$ for every x belonging to the interior of $\alpha_r(Q_r)$ in R^n_+. We define λ_r and k_r as we did in Section 22.

The proof of (10) is the same as 22.(9).

(10) *$E(V)$ is isomorphic to a complemented subspace of s^N.*

We define V_j, β_j and H_j as we did in former section. Then result (11) can be proven as 22.(11) taking $\mathcal{D}_+(\beta_r(H_r))$ instead of $\mathcal{D}(\beta_r(H_r))$.

(11) *s^N is isomorphic to a complemented subspace of $E(V)$.*

Now we arrive to another fundamental result of this section:

(12) *Let V be an n-dimensional C^∞-differentiable manifold with boundary. If V countable at infinity and non compact, then $E(V)$ is isomorphic to s^N.*

Proof. It is an immediate consequence from (10), (11) and 2.(1).

Results (13) and (14) can be proven as 22.(12) and 22.(13) respectively.

(13) *$\mathcal{D}(V)$ is isomorphic to a complemented subspace of $s^{(N)}$.*

(14) *$s^{(N)}$ is isomorphic to a complemented subspace of $\mathcal{D}(V)$*

The last fundamental result of this section is the following:

(15) Let V *be an n-dimensional C^∞-differentiable manifold with boundary. If V is countable at infinite and non compact, then $\mathcal{D}(V)$ is isomorphic to $s^{(N)}$.*

Proof. It is an immediate consequence from (13), (14) and 2.(2).

Results (8), (9), (12) and (15) appear here for the first time.

24. OTHER RESULTS. Let $\mathcal{O}_M$ be the linear spacer over K of all the K-valued functions defined on R^n which are infinitely differentiable such that $D^\alpha f$

is a solwly increasing function for every multi - index α, i.e., for every g of $\mathcal{S}$, g $D^\alpha f$ is an element of $\mathcal{B}$. A system of seminorms describing the topology of $\mathcal{O}_M$ is

(1) $\{ p_{g,r} : g \in \mathcal{S}, \quad r = 1,2, \ldots \}$

with

$$p_{g,r}(f) = \sum_{|\alpha| \leq r} \sup \{ |g(x) D^\alpha f(x)| : x \in R^n \} , \quad f \in \mathcal{O}_M.$$

It can be proven that $\mathcal{O}_M$ is isomorphic to s $\hat{\otimes}$ s' (cf. VALDIVIA [21]).

The space $\mathcal{D}_{L^p}$, $1 \leq p < \infty$, is the linear space over K of all K- valued functions f defined on R^n which are infinitely differentiable and such that $D^\alpha f$ belongs to $L^p(R^n)$ for every multi - index α. We suppose $\mathcal{D}_{L^p}$ provided with its ordinary topology of Fréchet space. Then a sequence (f_m) of $\mathcal{D}_{L^p}$ converges to the origin if and only if $(D^\alpha f_m)$ converges to the origin in $L^p(R^n)$ for every multi - index α (cf. SCHWARTZ [1], p. 199). It can be shown that $\mathcal{D}_{L^p}$ is isomorphic to s $\hat{\otimes}$ ℓ_p (cf. VALDIVIA [22]).

§ 2. SPACES OF C^m- DIFFERENTIABLE FUNCTIONS

1. A PROPERTY OF COMPLEMENTATION. If $z = (z_m)$ is any sequence in K we set

$$||z|| = \sup \{ |z_m| : m = 1,2,\ldots \}$$

Then $||\cdot||$ is the norm of c_0. Let b_0 be the linear space over K of the double sequences $x = (x_{ij})$ with

$$\lim_{i+j\to\infty} x_{ij} = 0.$$

If for any double sequence $u = (u_{ij})$ we set

$$||u|| = \sup \{ |u_{ij}| : i, j = 1,2,\ldots \}$$

then it is obvious that $||.||$ is a norm on b_0 providing structure of Banach space. We suppose b_0 endowed with this topology.

Let X and Y be the mappings defined in § 1, Section 1. We set T to denote the restriction of X to b_0 and S for the restriction of Y to c_0. It is obvious that $T(b_0)$ is contained in c_0 and that $S(c_0)$ is contained in b_0. Moreover, if x belongs to b_0 and z to c_0 we have that $||Tx|| = ||x||$, $||Sz|| = ||z||$ and therefore T is an isomorphism from b_0 onto c_0.

If $y = (y_m)$ and $z = (z_m)$ are any two elements of c_0 we set $B(y, z) = (y_m z_n)$. Then

$$||B(y, z)|| = \sup \{|y_m z_n| : m, n = 1,2,\ldots\}$$

$$\leq \sup \{|y_m| : m = 1,2,\ldots\} \sup \{|z_n| : n = 1,2,\ldots\} = ||y||.||z||$$

and therefore $B : c_0 \times c_0 \longrightarrow b_0$ is a continuous linear mapping.

Let $\psi: c_0 \times c_0 \longrightarrow c_0 \otimes c_0$ the canonical bilineal mapping. Then there is a continuous linear mapping $Z : c_0 \otimes c_0 \longrightarrow b_0$ such that $Z \circ \psi = B$ (cf. SCHAEFER [1], Chapter III, 6.1)

(1) Z *is an isomorphism from* $c_0 \otimes_\varepsilon c_0$ *into* b_0.

Proof. Let z be an element of $c_0 \otimes c_0$. Then

$$z = \sum_{j=1}^{p} x^j \otimes y^j, \; x^j = (x_m^{(j)}), \; y^j = (y_m^{(j)}) \in c_0, \; j = 1,2,\ldots, p$$

We set $||.||$ and $|.|$ to denote the norms of $c_0 \otimes_\varepsilon c_0$ and ℓ^1 respectively. Then

$$||z|| = \sup \{|\sum_{j=1}^{p} <x^j, u> <y^j, v>| : u, v \in \ell^1, |u| \leq 1, |v| \leq 1\}$$

$$= \sup \{|\sum_{j=1}^{p} (\sum_m x_m^{(j)} u_m) (\sum_n y_n^{(j)} v_n) : \Sigma|u_m| \leq 1, \Sigma| v_n | \leq 1\}$$

$$= \sup \{|\sum_{m,n} (\sum_{j=1}^{p} x_m^{(j)} y_n^{(j)}) u_m v_n| : \Sigma|u_m| \leq 1, \Sigma|v_n| \leq 1$$

$$= \sup \{| \sum_{m,n} (\sum_{j=1}^{p} x_m^{(j)} y_n^{(j)}) w_{mn}| : \Sigma| w_{mn}| \leq 1\}$$

$$= \sup \{|<B(x,y), w>| : w = (w_{mn}), \Sigma|w_{mn}| \leq 1 \}$$

$$= ||B(x, y)|| = ||\sum_{j=1}^{p} B(x^j, y^j)|| = ||\sum_{j=1}^{p} Z(x^j \otimes y^j)||$$

$$=||Z(\sum_{j=1}^{p} x^j \otimes y^j)|| = ||Z(z)||$$

and the conclusion follows.

(2) $c_0 \hat{\otimes}_\varepsilon c_0$ *is isomorphic to* c_0.

Proof. Let $\{e_m : m = 1,2,\ldots\}$ be the unit vector of c_0. Then $Z(e_m \otimes e_n)$ is the element of b_0 with nulle coordinates save the coordinate (n, m) whose value is one. Consequently $Z(c_0 \otimes c_0)$ is dense in b_0. Since $c_0 \hat{\otimes}_\varepsilon c_0$ and b_0 are Banach spaces, Z can be extended to an isomorphism from $c_0 \hat{\otimes}_\varepsilon c_0$ onto b_0. Since T is an isomorphism from b_0 onto c_0 it follows that $c_0 \hat{\otimes}_\varepsilon c_0$ is isomorphic to c_0.

If b belongs to K and if (a_m) belongs to c_0, we set $U(b,(a_m)) = (b_m)$ with $b_1 = b$, $b_{m+1} = a_m$, $m = 1,2,\ldots$ It is not difficult to prove that U is an isomorphism from $K \times c_0$ onto c_0.

If (a_m) and (b_m) are in c_0 we set $V((a_m), (b_m)) = (c_m)$ with $c_{2m-1} = a_m$, $c_{2m} = b_m$, $m = 1,2,\ldots$ It is easy to check that V is an isomorphism from $c_0 \times c_0$ onto c_0. Results (3) and (4) are now obvious.

(3) *The topological product* $K \times c_0$ *is isomorphic to* c_0.

(4) *The topological product* $c_0 \times c_0$ *is isomorphic to* c_0.

Now we arrive to the fundamental result of this section:

(5) *Let* F *be a locally convex space. Let* G *be a complemented subspace of* $c_0 \hat{\otimes}_\varepsilon F$. *Let* H *be a complemented subspace of* G. *If* H *is isomorphic to* $c_0 \hat{\otimes}_\varepsilon F$, *then* G *is isomorphic to* $c_0 \hat{\otimes}_\varepsilon F$.

Proof. Since c_0 verifies (2), (3) and (4), it is enough to apply §1, 2.(9) for $c_0 = E$ to obtain the conclusion.

2. THE SPACES $C^m(H)$ AND $\mathcal{D}^m(H)$. In what follows m denotes a non negative integer. Let H be a compact subset of the n-dimensional euclidean space R^n. If H coincides with the closure of its interior $\overset{\circ}{H}$, then $C^m(H)$ is the linear space over K of all K-valued functions f defined on H which have con-

tinuous partial derivatives of order α in $\overset{\circ}{H}$, α being a multi-index with $|\alpha| \leq m$, and such that $D^{\alpha}f$ can be continuously extended to H from $\overset{\circ}{H}$.

Given the real numbers $a_j < b_j$, $j = 1,2,\ldots,n$, we set

$$L = \{(x_1,x_2,\ldots,x_n) : a_j \leq x_j \leq b_j,\ j = 1,2,\ldots, n\}.$$

Let E be the linear space over K of all those K-valued functions f defined on L which admit continuous partial derivatives of order α in L, with $|\alpha| \leq m$. The derivatives on the boundary of L are defined laterally. The following result can be proved as in §1,8.(1):

(1) *The linear space* E *coincides with* $C^m(L)$.

We suppose now that the closure of $\overset{\circ}{H}$ coincides with H. If f belongs to $C^m(H)$ and if α is a multi-index with $|\alpha| \leq m$, we denote by $D^{\alpha}f$ the extension to H of the partial derivative of order α of f in $\overset{\circ}{H}$. We set

$$|f| = \sum_{|\alpha| \leq m} \sup\{|D^{\alpha}f(x)| : x \in H\}.$$

Then $|.|$ is a norm on $C^m(H)$. We suppose $C^m(H)$ endowed with the topology derived from this norm. Analogously to §1,8.(3) we have the following result:

(2) $C^m(H)$ *is a Banach space.*

Given any compact subset H of R^n we denote by $\mathcal{D}^m(H)$ the linear space over K of all those K-valued functions defined on R^n whose supports are contained in H and which admit continuous partial derivatives of order α, with $|\alpha| \leq m$. If f belongs to $\mathcal{D}^m(H)$ we set

$$|f| = \sum_{|\alpha| \leq m} \sup\{|D^{\alpha}f(x)| : x \in R^n\} = \sum_{|\alpha| \quad m} \sup\{|D^{\alpha}f(x)| : x \in H\}.$$

Then $|.|$ is a norm on $\mathcal{D}^m(H)$. We suppose this space endowed with the topology derived from $|.|$. If M is the closure of $\overset{\circ}{H}$ in R^n it is obvious that $\mathcal{D}^m(H)$ coincides with $\mathcal{D}^m(M)$. If $\overset{\circ}{H} \neq \emptyset$ and if f belongs to $\mathcal{D}^m(H)$ we set Tf to denote the restriction of f to M. Then $T : \mathcal{D}^m(H) \longrightarrow C(M)$ is a linear mapping with $|Tf| = |f|$. Obviously $T(\mathcal{D}^m(H))$ is a closed subspace of $C^m(M)$ and therefore

(3) $\mathcal{D}^m(H)$ *is a Banach space.*

Sometimes we identify $T(\mathcal{D}^m(H))$ with $\mathcal{D}^m(H)$ when we deal with isomorphism problems.

3. LINEAR EXTENSION OPERATOR. Let A,B and D be the n-dimensional intervals defined in §1, Section 10. Let E be the subspace of $C^m(A)$ of all those functions f such that $D^\alpha f$ vanishes in the face of A:

$$\{(x_1,x_2,\ldots,x_n) : x_1 = a,\ a_j < x_j < b_j,\ j = 2,3,\ldots,n\}$$

for every multi-index α with $|\alpha| \leq m$. Let us suppose the functions f of E extended to the set M defined in §1, Section 10 by setting $f(x) = 0$ when $x = (x_1,x_2,\ldots,x_n)$ is a point of M with $x_1 < a_1$. Obviously $D^\alpha f$, $|\alpha| < m$, exists and is continuous on M.

If f belongs to E we set Xf to denote the function defined on B in §1,10.(1). The proofs of §1,10.(2) and §1,10.(4) suggest

(1) *If* f *belongs to* E *then* Xf *belongs to* $C^m(B)$.

(2) $X : E \longrightarrow C^m(B)$ *is a continuous linear operator.*

Now we suppose that P and Q are two n-dimensional compact intervals such that P is contained in the interior of Q. Using the methods developed in §1, Section 10 and Section 11 we obtain analogous results to §1, 10.(10) and §1,11.(14) respectively:

(3) *There is a continuous linear extension operator from* $C^m(P)$ *into* $C^m(Q)$.

(4) $\mathcal{D}^m(Q)$ *is isomorphic to a complemented subspace of* $C^m(Q)$.

WHITNEY [1] defined the functions of class C^m in any closed subset M of R^n to prove that every function of class C^m in M can be extended to a function of class C^m in R^n. The corresponding extension operator is linear. HESTENES [1] modifies the proof of WHITNEY and provides another method of extending to R^n the function of class C^m defined on M when the boundary of M satisfies suitable conditions. This method is a generalization of the extension procedure used by LICHTENSTEIN [1].

4. THE SPACE $c_0(E)$. Let E be a locally convex space over the field K. We denote by $\mathcal{P}$ the family of all continuous seminorms on E. If $x = (x_r)$ is

any sequence in E and if p belongs to P we set

$$|x|_p = \sup \{p(x_r) : r = 1,2,..\} .$$

$c_o(E)$ is the linear space over K of all sequences $x = (x_r)$ in E with

$$\lim p \ (x_r) = 0, \ p \in P$$

It is immediate that $|.|_p, p \in P$, is a system of seminorms on $c_o(E)$ defining a locally convex topology. We suppose $c_o(E)$ endowed with this topology.

(1) *If* E *is complete, then* $c_o(E)$ *is complete.*

Proof. Let

(2) $x^j = \{(x_r{}^j) : j \in J, \geqslant\}$

be a Cauchy net in $c_o(E)$. If p belongs to P and if n is a positive integer we have that

(3) $p(x_n{}^i - x_n{}^j) \leqslant |x^i - x^j|_p$, $i,j \in J$,

and therefore $\{x_n{}^j : j \in J, \geqslant\}$ is a Cauchy net in E. Since E is complete, this net converges in E to a point x_n. Given $\varepsilon > 0$ we find $h > J$ such that

(4) $$|x^j - x^h|_p < \frac{\varepsilon}{2} , \ j > h.$$

A positive integer q can be found such that

$$p(x_n{}^h) < \frac{\varepsilon}{2} , \ n > q.$$

According to (3) and (4)

(5) $$p(x_n - x_n{}^h) \leqslant \frac{\varepsilon}{2} , \ n = 1,2,...$$

Then

$$p(x_n) \leqslant p(x_n{}^h) + p(x_n - x_n{}^h) < \varepsilon$$

if $n > q$ and therefore $x = (x_r)$ belongs to $c_o(E)$. By (5) we have that $|x-x^h|_p \leqslant \frac{\varepsilon}{2}$. Finally, if j belongs to J, $j > h$, it follows that

$$|x-x^j|_p \leqslant |x-x^h|_p + |x^h-x^j|_p < \varepsilon$$

Consequently the net (2) converges to x in $c_o(E)$. The proof is complete.

Let z be an element of $c_o \otimes E$. Then

(6) $$z = \sum_{j=1}^{r} x^j \otimes y^j,\ x^j = (x_m^j) \in c_o,\ y^j \in E.$$

We set $Tz = (\sum_{j=1}^{r} x_m^{(j)} y^j)$. It is easy to check that Tz belongs to $c_o(E)$ and that it does not depend on the particular representation of z. Therefore T is a mapping from $c_o \otimes E$ into $c_o(E)$ which is obviously linear,

(7) T *is an isomorphism from* $c_o \otimes_\varepsilon E$ *into* $c_o(E)$.

Proof. Let p be and element of $\mathcal{P}$. Let V be the polar set in E' of $\{x \in E : p(x) \leqslant 1\}$. If U denotes the unit ball of ℓ^1, given the element (6) of $c_o \otimes E$ we have that

$$|z|_p = \sup \{|\sum_{j=1}^{r} \langle x^j,u\rangle \langle y^j,v\rangle| : u \in U,\ v \in V\}$$

$$= \sup \{|\sum_{j=1}^{r} \langle y^j,v\rangle \sum_m x_m^{(j)} u_m| : \Sigma|u_m| \leqslant 1,\ v \in V\}$$

$$= \sup \{\sum_m |\langle \sum_{j=1}^{r} x_m^{(j)} y^j, v\rangle u_m| : \Sigma|u_m| \leqslant 1,\ v \in V\}$$

$$= \sup \{\sum_m p(\sum_{j=1}^{r} x_m^{(j)} y^j)\ |u_m| : \Sigma|u_m| \leqslant 1\}$$

$$= \sup \{p(\sum_{j=1}^{r} x_m^{(j)} y^j) : m = 1,2,\dots\} = |Tz|_p$$

and the conclusion follows.

(8) *If* E *is complete then* $c_o \hat{\otimes}_\varepsilon E$ *is isomorphic to* $c_o(E)$.

Proof. If $x = (x_r)$ is any element of $c_o(E)$ we set $x\{r\}$ to denote the sequence $x_1, x_2, \dots, x_{r-1}, 0, 0, \dots$ Given an element p of $\mathcal{P}$ we have that

$$\lim_r |x - x\{r\}|_p = \lim_r \sup \{p(x_m) : m \geqslant r\} = 0$$

and therefore the set

$$\{x\{r\} : x \in c_o(E),\ r = 1,2,\dots\}$$

is dense in $c_0(E)$. Since

$$T\left(\sum_{j=1}^{r-1} e_i \otimes x_j\right) = x\{r\},\ r = 1,2,\ldots,$$

it follows that $T(c_0 \otimes E)$ is dense in $c_0(E)$. By (1), $c_0(E)$ is complete and accordingly T can be extended to an isomorphism from $c_0 \hat{\otimes} E$ onto $c_0(E)$. The proof is complete.

Now take a sequence (a_r) of positive numbers. If $x = (x_r)$ is any sequence of E and if p belons to P we set

$$||x||_p = \sup\{a_r p(x_r) : r = 1,2,\ldots\}.$$

Let F be the linear space over K of all sequences $x = (x_r)$ in E with

$$\lim a_r p(x_r) = 0,\ p \in P$$

It is immediate that $||.||_p$, $p \in P$, is a system of seminorms on F defining a locally convex topology. We suppose F endowed with this topology.

(9) F *is isomorphic to* $c_0 \hat{\otimes}_\varepsilon E$.

Proof. According to (8) it is enough to show that F is isomorphic to $c_0(E)$. For every element $x = (x_r)$ of $c_0(E)$ we set $Sx = (\frac{1}{a_r} x_r)$. It is obvious that $S : c_0(E) \longrightarrow F$ is a linear mapping. On the other hand,

$$||Sx||_p = \sup\{a_r p(\frac{1}{a_r} x_r) : r = 1,2,\ldots\} = |x|_p$$

if p belongs to P and therefore S is an isomorphism from $c_0(E)$ into F. If $z = (z_r)$ belongs to F it is obvious that $y = (a_r z_r)$ belongs to $c_0(E)$ and $Sy = z$. Now the proof is complete.

5. A REPRESENTATION OF THE SPACE $C_0^m(\Omega)$. Let Ω be a non-void open subset of R^n. $C_0^m(\Omega)$ is the linear space over K of all K-valued functions f defined on Ω which admit continuous partial derivatives of order α with $|\alpha| \leq m$ and such that for every $\varepsilon > 0$ there is a compact subset H of Ω, depending on f, such that

$$|D^\alpha f(x)| < \varepsilon,\ |\alpha| \leq m,\ x \in \Omega \sim H.$$

We set

$$|f| = \sum_{|\alpha|\leq m} \sup\{|D^{\alpha}f(x)| : x \in \Omega\}$$

Then $|.|$ is a norm on $C_0^m(\Omega)$. We suppose $C_0^m(\Omega)$ endowed with the topology derived from the norm. One shows easyly that $C_0^m(\Omega)$ is a Banach space.

If f belongs to $C_0^m(\Omega)$ let g be the function defined on R^n which vanishes in $R^n \sim \Omega$ and coincides with f in Ω. Then g is a function of class C^m and $D^{\alpha}g$ vanishes in $R^n \sim \Omega$ for $|\alpha|\leq m$. All the functions of $C_0^m(\Omega)$ are supposed to be extended in that way. Then this space is a closed subspace of $C_0^m(R^n)$.

Let A_r and B_r be the sequences of n-dimensional cubes introduced in §1, Section 9. We denote by g_r the mapping from R^n into R^n defined in §1, Section 15 and by I the n-dimensional cube

$$\{(x_1,x_2,\ldots,x_n) = -1 \leq x_j \leq 1,\ j = 1,2,\ldots,n\}.$$

Then g_r maps I onto A_r, $r = 1,2,\ldots$ Let h be the function defined in §1, Section 15 and let $\{k_r : r = 1,2,\ldots\}$ be the partition of the unity of class C^{∞} used in §1, Section 18. We denote by G the linear space over K of all the sequence (f_r) of $\mathcal{D}^m(I)$ such that

$$\lim 2^{mk(r)} \sum_{|\alpha|\leq m} \sup\{|D^{\alpha}f_r(x)| : x \in I\} = 0.$$

We provide G with the topology derived from the norm

$$|(f_r)| = \sup_r 2^{mk(r)} \sum_{|\alpha|\leq m} \sup\{|D^{\alpha}f_r(x)| : x \in I\},\ (f_r) \in G$$

(1) G *is isomorphic to* $c_0 \hat{\otimes}_{\varepsilon} \mathcal{D}^m(I)$.

Proof. If we set $\mathcal{D}^m(I)$ for E and $a_r = 2^{mk(r)}$, $r = 1,2,\ldots$, in the former section, then F coincides with G and the conclusion follows applying 4.(9).

For every (f_r) of G, we set $T((f_r)) = \Sigma f_r \circ g_r^{-1}$. As in §1,9.(20) we can prove that $T((f_r))$ belongs to $C_0^m(\Omega)$ with α such that $|\alpha|\leq m$. Obviously T is linear.

(2) $T : G \longrightarrow C_0^m(\Omega)$ *is continuous.*

Proof. If Ω is distinct from R^n, see §1, 15.(15) and if Ω coincides with R^n see §1, 18.(2) taking $|\alpha|\leq m$.

(3) *If* $\Omega \neq R^n$ *and if* f *belongs to* $C_o^m(\Omega)$, *then*

$$\lim_r 2^{mk(r)} \sup \{|D^\alpha f \circ g_r(x)| : x \in I\} = 0$$

for every multi-index α *with* $|\alpha| \leq m$.

Proof. Let ε be a positive number. We find a non-void compact subset L in Ω such that the absolute value of the partial derivative of $D^\alpha f$ of order $h = m-|\alpha|$ in every direction and in every point in $\Omega \sim L$ in less or equal than

$$\frac{h!}{2}\left(\frac{1}{16\sqrt{n}}\right)^h \varepsilon$$

If d is the distance from L to $R^n \sim \Omega$ we determine a positive integer p such that $\sqrt{n} < 2^{p-4}d$. Now select a compact subset P in Ω, $P \supset L$, such that if x belongs to $\Omega \sim P$ then $2^{hp}\,|D^\alpha f(x)| < \varepsilon$. Let q be a positive integer with $A_r \cap P = \emptyset$, $r > q$. If $r > q$ and $z = (z_1, z_2, \ldots, z_n)$ belongs to A_r the following two cases can occur:

a) The length of the edge of B_r is larger or equal than $\frac{1}{2^p}$. Then, setting $x = g_r^{-1}(z)$, we have that

$$2^{mk(r)}\,|D^\alpha f \circ g_r(x)| = \left(\frac{1}{2^{k(r)+1}}\left(1+\frac{1}{4\sqrt{n}}\right)\right)^{|\alpha|} 2^{mk(r)}\,|D^\alpha f(z)|$$

$$\leq \frac{1}{2^{k(r)|\alpha|}}\, 2^{mk(r)}\,|D^\alpha f(z)| = 2^{hk(r)}\,|D^\alpha f(z)| \leq 2^{hp}\,|D^\alpha f(z)| < \varepsilon$$

b) The length of the edge of B_r is less than $\frac{1}{2^p}$. Then $k(r) > p$ and, applying §1, 15.(2), the distance from B_r to $R^n \sim \Omega$ is less or equal than $\frac{\sqrt{n}}{2^{k(r)-2}}$. We find a point

$$x^o = (x_{1o}, x_{2o}, \ldots, x_{no}) \in R^n \sim \Omega$$

whose distance to B_r is less than $\frac{\sqrt{n}}{2^{k(r)-3}}$. Then

$$||x^o - z|| \leq \frac{\sqrt{n}}{2^{k(r)-3}} + \frac{\sqrt{n}}{2^{k(r)}} + \frac{1}{2^{k(r)+3}} < \frac{\sqrt{n}}{2^{k(r)-4}} < \frac{\sqrt{n}}{2^{p-4}} < d$$

and consequently the distance from every point of the segment $\{x^o + w(z-x^o) : 0 \leq w \leq 1\}$ to $R^n \sim \Omega$ is less than d and therefore this segment does not meet L. We write $f = f_1 + i\, f_2$ with f_1 and f_2 real functions, i being the imaginary unity. Then

$$D^{\alpha}f_j(z) = \frac{1}{h!}\left[\frac{d^h}{dw^h} D^{\alpha}f_j(x_{10}+w(z_1-x_{10}),\dots,x_{no}+w(z_n-x_{no}))\right]_{w=\theta_j}$$

$$0 \leqslant \theta_j < 1,\ j = 1,2$$

and therefore

$$|D^{\alpha}f(z)| \leqslant |D^{\alpha}f_1(z)| + |D^{\alpha}f_2(z)| \leqslant 2\,\frac{\|z-x^0\|^h}{h!}\,\frac{h!}{2}\left(\frac{1}{16\sqrt{n}}\right)^h \varepsilon$$

$$= \|z-x^0\|^h\left(\frac{1}{16\sqrt{n}}\right)^h\varepsilon < \left(\frac{\sqrt{n}}{2^{k(r)-4}}\right)^h\left(\frac{1}{16\sqrt{n}}\right)^h \varepsilon = \frac{\varepsilon}{2^{hk(r)}})$$

consequently

$$2^{hk(r)}\,|D^{\alpha}f(z)| < \varepsilon$$

If we set $x = g_r^{-1}(z)$ we have that

$$2^{mk(r)}\,|D^{\alpha}f \circ g_r(x)| = \left(\frac{1}{2^{k(r)+1}}\left(1+\frac{1}{4\sqrt{n}}\right)\right)^{|\alpha|} 2^{mk(r)}\,|D^{\alpha}f(z)|$$

$$\leqslant \frac{1}{2^{k(r)|\alpha|}}\,2^{mk(r)}\,|D^{\alpha}f(z)| = 2^{hk(r)}\,|D^{\alpha}f(z)| < \varepsilon$$

and the conclusion follows.

(4) *Let Ω be a non-void open subset of R^n distinct from R^n. Let f be an element of $C_0^m(\Omega)$. Let α and β be two multi-indices such that $|\alpha| + |\beta| \leqslant m$. Given $\varepsilon > 0$ there is a compact subset L of Ω such that*

$$|D^{\alpha}f\, D^{\beta}(\Sigma\, h \circ g_j^{-1})(x)| < \varepsilon,\ x \in \Omega \sim L.$$

Proof. Given a positive integer r we set

$$b_r = \left(2^{k(r)+1}\,\frac{4\sqrt{n}}{4\sqrt{n}+1}\right)^{|\beta|} 4^n$$

Since f belongs to $C_0^m(\Omega)$ it follows that $D^{\alpha}f$ belongs to $C_0^{m-|\alpha|}(\Omega)$ and, according to (3), there is a positive integer p such that

$$b_r \sup\{|(D^{\alpha}f \circ g_r)(x)| : x \in I\} \sup\{|D^{\beta}h(x)| : x \in I\} < \varepsilon.$$

We take a compact set L in Ω with

$$A_r \subset L,\ r = 1,2,\dots, p.$$

If z belongs to $\Omega \sim L$ there are positive integers $r_1, r_2, \ldots, r_q$, $q \leq 4^n$ with

$$z \notin \mathring{A}_r, \quad r \neq r_j, \quad r_j > p, \quad j = 1,2, \ldots, q,$$

according to § 1, 9 (16). Then

$$|D^\alpha f(z) D^\beta(\Sigma\, h \circ g_j^{-1}(z)| \leq \sum_{j=1}^{q} b_{r_j} 4^{-n} |D^\alpha f(z)| . |(D^\beta g) \circ g_{r_j}^{-1}(z)|$$

$$\leq \sum_{j=1}^{q} b_{r_j} 4^{-n} \sup \{|D^\alpha f(g_{r_j}(x))| : x \in I\} \sup \{ |D^\alpha h(x)| : x \in I\} < \varepsilon$$

(5) *Let* Ω *be a non-void open subset of* R^n *distinct from* R^n. *Let* g *be an element of* $C_0^m(\Omega)$. *If* α *and* β *are two multi-indices such that* $|\alpha| + |\beta| \leq m$ *then* $D^\alpha g\, D^\beta\, \Sigma\, h \circ g_j^{-1}$ *belongs to* $C_0^{m-|\alpha|-|\beta|}(\Omega)$.

Proof. Given a multi-index λ with $|\lambda| \leq m-|\alpha|-|\beta|$ we have that

$$D^\lambda(D^\alpha g\, D^\beta\, \Sigma\, h \circ g_j^{-1}) = \sum_{\gamma \leq \lambda} A_\gamma D^{\alpha+\gamma} g\, D^{\lambda-\gamma+\beta} \Sigma\, h \circ g_j^{-1}$$

where A_γ does not depend on the funcions g and $D^\beta\, \Sigma\, h \circ g_j^{-1}$.

Since $|\alpha + \gamma| + |\lambda - \gamma + \beta| \leq m$, given $\varepsilon > 0$, we apply (4) with g=f to find a compact set L in Ω such that, if x belongs to $\Omega \sim L$,

$$\sum_{\gamma \leq \lambda} |A_\gamma D^{\alpha+\gamma} g(x) D^{\lambda-\gamma+\beta}\, \Sigma h \circ g_j^{-1}(x)| < \varepsilon$$

and therefore

$$|D^\lambda (D^\alpha g\, D^\beta\, \Sigma\, h \circ g_j^{-1}(x)| < \varepsilon, \quad x \in \Omega \sim L$$

from where the conclusion follows.

We set $J = \{(x_1, x_2, \ldots, x_n) : -\dfrac{4\sqrt{n}}{4\sqrt{n}+1} \leq x_j \leq \dfrac{4\sqrt{n}}{4\sqrt{n}+1},\ j = 1,2,\ldots,n\}$.

Then g_r maps onto B_r, r= 1,2,..

(6) *Let* Ω *be a non-void open subset of* R^n *distinct from* R^n. *If* f *belongs to* $C_0^m(\Omega)$, *then*

$$\frac{f}{\Sigma\, h \circ g_j^{-1}} \in C_0^m(\Omega)$$

Proof. Since J is a compact set contained in $\overset{\circ}{I}$ it follows that

$$\inf \{h(x) : x \in J\} = b > 0$$

Then

(7) $$(\Sigma h \circ g_j^{-1})(x) \geq b \text{ for every } x \in \Omega$$

Given a multi-index λ with $|\lambda| \leq m$ we have that $D^{\lambda} \frac{f}{\Sigma h \circ g_j^{-1}}$ is a ratio whose denominator is a power of $\Sigma h \circ g_j^{-1}$ and whose numerator is a linear combination of termus of the form

(8) $$D^{\gamma} f \, D^{\delta}(\Sigma h \circ g_j^{-1}) \ldots D^{\mu}(\Sigma h \circ g_j^{-1})$$

with $|\gamma| + |\delta| + \ldots + |\mu| \leq m$.

A reiterated application of (5) enables us to state that (8) is an element of $C_0^{0}(\Omega)$. Then, according to (7), it follows that given $\varepsilon > 0$ there is a compact set P in Ω such that, if x belongs to $\Omega \sim P$,

$$|D^{\lambda}(\frac{f}{\Sigma h \circ g_j^{-1}})(x)| < \varepsilon$$

from where the conclusion follows.

(9) *If* $\Omega \neq R^n$ *and if* f *belongs to* $C_0^m(\Omega)$, *given a multi-index* α *with* $|\alpha| \leq m$ *then*

$$\lim_r 2^{mk(r)} \sup\{|D^{\alpha}((f \circ g_r)h)(x)| : x \in I\} = 0$$

Proof. The same proof of §1, 15.(11) can be applied with slight modifications, using (3) instead of §1.15(3).

The next result will be applied later on.

(10) *If* $\Omega \neq R^n$ *and if* f *belongs to* $C_0^m(\Omega)$, *given a multi-index* α *wiht* $|\alpha| \leq m$, *then*

$$\lim_r 2^{mk(r)} \sup \{|D^{\alpha}(fk_r) \circ g_r(x)| : x \in I\} = 0$$

Proof. See §1, 15.(12).

For every f of $C_0^m(\Omega)$ we set $Wf = (fk_r \circ g_r)$. According to (10), W is a mapping from $C_0^m(\Omega)$ into G, $\Omega \neq R^n$. Obviously , W is linear. Moreover

(11) $T(Wf) = \Sigma(fk_r \circ g_r) \circ g_r^{-1} = f\,\Sigma k_r = f$

(12) *If* $\Omega \neq R^n$, *given a multi-index* α *with* $|\alpha|\leqslant m$, *there is a positive number* k *such that*

(13) $$2^{mk(r)} \sup\{|D^\alpha f \circ g_r(x)| : x \in I\}$$

$$\leqslant k \sum_{|\beta|\leqslant m} \sup\{|D^\beta f(x)| : x \in \Omega\}$$

for every f *of* $C_0^m(\Omega)$ *and* $r = 1, 2, \ldots$

Proof. Let x be any point of I. Given any positive number r we set $z = g_r(x)$. Then z belongs to A_r. If $k(r) = 0$ we have that

(14) $$2^{mk(r)} |D^\alpha f \circ g_r(x)| = \left(\frac{1}{2} + \frac{1}{8\sqrt{n}}\right)^{|\alpha|} |D^\alpha f(z)| \leqslant |D^\alpha f(z)|.$$

If $k(r)$ is larger than zero, the distance from B_r to $R^n \sim \Omega$ is less or equal than $\frac{\sqrt{n}}{2^{k(r)-2}}$ according to §1, 15.(2). We take a point $x^0 = (x_{10}, x_{20}, \ldots, x_{n0})$ in $R^n \sim \Omega$ whose distance to B_r is less than $\frac{\sqrt{n}}{2^{k(r)-3}}$. We set $h = m - |\alpha|$. Proceeding as we did in the proof of (3) we have that

$$||x^0 - z|| < \frac{\sqrt{n}}{2^{k(r)-4}}$$

and

$$|D^\alpha f_j(z)| = \left|\frac{1}{h!}\left(\frac{d^h}{dw^h} D^\alpha f_j(x_{10}+w(z_1-x_{10}), \ldots, x_{no}+w(z_n-x_{no}))\right)_{w=\theta_j}\right|$$

$$0 < \theta_j \leqslant 1, \; j = 1, 2,$$

and therefore there exists a positive number Q (obtained calculating the former derivative) which does not depend on f_j and on r such that

$$|D^\alpha f_j(z)| \leqslant \frac{||z-x^0||^h}{h!} Q \sum_{|\beta|\leqslant m} \sup\{|D^\beta f_j(y)| : y \in \Omega\}$$

and consequently

$$2^{mk(r)} |D^\alpha f \circ g_r(x)| = \left(\frac{1}{2^{k(r)+1}}\left(1 + \frac{1}{4\sqrt{n}}\right)\right)^{|\alpha|} 2^{mk(r)} |D^\alpha f(z)|$$

$$\leqslant \frac{1}{2^{k(r)|\alpha|}} \; 2^{mk(r)} \; (|D^{\alpha}f_1(z)| + |D^{\alpha}f_2(z)|)$$

$$\leqslant 2^{hk(r)} \; \frac{2}{h!} \; ||z-x^o||^h \; Q \sum_{|\beta|\leqslant m} \sup \{|D^{\beta}f(y)| : y \in \Omega\}$$

$$\leqslant 2^{hk(r)} \; \frac{2}{h!} \; \left(\frac{\sqrt{n}}{2^{k(r)-4}}\right)^h Q \sum_{|\beta|\leqslant m} \sup \{|D^{\beta}f(y)| : y \in \Omega\}$$

$$\leqslant 2(16\sqrt{n})^h \; Q \sum_{|\alpha|\leqslant m} \sup \{|D^{\beta}f(y)| : y \in \Omega\}$$

from where it follows (11), having in mind (12) and taking $k = 1+2(16\sqrt{n})^m Q$.

For every f of $C_o^m(\Omega)$ we set

$$W_1 f = \frac{f}{\Sigma \; h \circ g_j^{-1}}, \; W_2 f = ((f \circ g_r)h).$$

If Ω is distinct from R^n we have that W_1 is a mapping from $C_o^m(\Omega)$ into $C_o^m(\Omega)$, according to (6). Obviously W_1 is linear. The proof of (6) shows that W_1 is continuous.

If Ω is distinct from R^n we have that W_2 is a mapping from $C_o^m(\Omega)$ into G, according to (9). Obviously W_2 is linear. If we observe (12), it is not difficult to show that W_2 is continuous.

(15) *If Ω is distinct from R^n, W is an isomorphism from $C_o^m(\Omega)$ into G.*

Proof. Since $W = W_2 \circ W_1$ we have that W is continuous. On the other hand, T is continuous. We apply (11) to reach the conclusion.

(16) *If Ω is distinct from R^n, then $C_o^m(\Omega)$ is isomorphic to a complemented subspace of $c_o \hat{\otimes}_{\varepsilon} \mathcal{D}^m(I)$.*

Proof. $W \circ T$ is a continuous projection from G into itself such that $W \circ T(G) = W(C_o^m(\Omega))$. Since $W(C_o^m(\Omega))$ is isomorphic to $C_o^m(\Omega)$ and, since G is isomorphic to $c_o \hat{\otimes}_{\varepsilon} \mathcal{D}^m(I)$, the conclusion follows.

(17) *If $\Omega = R^n$, T is continuous.*

Proof. See §1,18(2) and take $|\alpha| \leqslant m$.

(18) *If* f *belongs to* $C_o^m(R^n)$, *then*

$$\frac{f}{\Sigma\, h \circ g_j^{-1}} \in C_0^m(R^n).$$

Proof. See §1,18.(3) and take $|\alpha|\leq m$.

(19) *If* $\Omega = R^n$ *and if* (f_r) *is a sequence of elements of* $C^m(I)$ *such that*

$$\lim \sup \{|D^\alpha f_r(x)| : x \in I\} = 0$$

for every multi-index α *with* $|\alpha|\leq m$, *then* (hf_m) *belongs to* G.

Proof. See §1,18.(6)

(20) *Let* f *be an element of* $C_0^m(\Omega)$. *If* f_r *is the restriction of* $f \circ g_r$ *to* I, r = 1,2,.... *then*

$$\lim \sup \{|D^\alpha f_r(x)| : x \in I\} = 0$$

for every multi-index α *with* $|\alpha|\leq m$.

Proof. See §1,18.(7).

(21) *If* $\Omega = R^n$, *then* W *is an isomorphism from* $C_0^m(R^n)$ *into* G.

Proof. See § 1,18.(8).

(22) $C_0^m(R^n)$ *is isomorphic to a complemented subspace of* $c_0 \hat{\otimes}_\varepsilon \mathcal{D}^m(I)$

Proof. See (14).

We set L for the n-dimensional cube

$$\{(x_1,x_2,\ldots,x_n) : -\tfrac{1}{5}\leq x_j \leq \tfrac{1}{5},\ j = 1,2,\ldots,n\}.$$

L is contained in J. We denote by E the linear space over K of the sequences (f_r) of $C^m(L)$ with

$$\lim_r 2^{mk(r)} \sum_{|\alpha|\leq m} \sup \{|D^\alpha f_r(x)| : x \in L\} = 0.$$

We endowe E with the topology derived from the norm

$$|(f_r)| = \sup_r 2^{mk(r)} \sum_{|\alpha|\leq m} \sup \{|D^\alpha f_r(x)| : x \in L\},\ (f_r) \in E.$$

According to 3.(3) there is a continuous linear extension operator Z: $C^m(L) \longrightarrow C^m(J)$. If ψ is an element of $\mathcal{D}(J)$ which takes the value one on L, we set $Uf = (Zf)\psi$. Then U is a continuous linear extension operator from $C^m(L)$ into $\mathcal{D}^m(J)$. Therefore there exists a number $M > 0$ such that

(23) $$\sum_{|\alpha|\leq m} \sup\{|D^\alpha Uf(x)| : x \in J\} \leq M \sum_{|\beta|\leq m} \sup\{|D^\beta f(x)| : x \in L\}.$$

for every f of $C^m(L)$.

For every (f_r) of E we set $Y((f_r)) = (Uf_r)$. Y is an isomorphism from E into G, according to (23).

(25) *Given a multi-index* α *there is a positive number* k *such that*

$$2^{mk(r)} \{|D^\alpha f \circ g_r(x)| : x \in I\} \leq k \sum_{|\alpha|\leq m} \sup\{|D^\alpha f(x)| : x \in \Omega\}$$

for every f *of* $C_0^m(\Omega)$ *and* $r = 1,2,\ldots$

Proof. The case $\Omega \neq R^n$ has been proved in (12). If $\Omega = R^n$ we have that

$$2^{mk(r)} \sup\{|D^\alpha f \circ g_r(x)| : x \in I\}$$

$$= \left(\frac{1}{2} + \frac{1}{8\sqrt{n}}\right)^{|\alpha|} \sup\{|D^\alpha f(u)| : u \in R^n\} \leq \sum_{|\beta|\leq m} \sup\{|D^\beta f(x)| : x \in R^n\}.$$

For every f of $C_0^m(\Omega)$ we set $Vf = (V_r f)$ where

$$V_r f(x) = f \circ g_r(x),\ x \in L,\ r = 1,2,\ldots$$

According to (3) and (18), V is a mapping from $C_0^m(\Omega)$ into E. Obviously V is linear.

(26) $V : C_0^m(\Omega) \longrightarrow E$ *is continuous.*

Proof. We apply (25) to obtain $k > 0$ such that

$$|(f_r)| = \sup_r 2^{mk(r)} \sum_{|\alpha|\leq m} \sup\{|D^\alpha V_r f(x)| : x \in L\}$$

$$\leq \sup_r 2^{mk(r)} \sum_{|\alpha|\leq m} \sup\{|D^\alpha f \circ g_r(x)| : x \in I\}$$

$$\leq k \sum_{|\alpha|\leq m} \sup\{|D^\alpha f(x)| : x \in \Omega\} = k\,|f|$$

for every f of $C_0^m(\Omega)$. The continuity of V follows.

(27) $C_0^m(\Omega)$ *has a complemented subspace isomorphic to* E.

Proof. It is immediate that $T \circ Y \circ V$ is a continuous projection from $C_0^m(\Omega)$ into itself such that

$$T \circ Y \circ V\,(C_0^m(\Omega)) = T\,(Y(E)).$$

Therefore $T\,(Y(E))$ is a complemented subspace of $C_0^m(\Omega)$.

Let T_1 be the restriction of T to $Y(E)$. We have that T_1 is an injective cotinuous mapping from the Fréchet space $Y(E)$ onto the Fréchet space $T(Y(E))$. We apply the open mapping theorem to obtain that $Y(E)$ is isomorphic to $T(Y(E))$. Since $Y(E)$ is isomorphic to E the proof is finished.

We consider the mapping U defined above as an extension operator from $C^m(L)$ into $\mathcal{D}_m(I)$. Then $C^m(L)$ is isomorphic to $U(C^m(L))$. We set $E_2 = U(C^m(L))$. Since the subspace E_1 of $\mathcal{D}^m(I)$ of all those functions vanishing in L is a topological complement of E_2 in $\mathcal{D}^m(I)$ and since E_2 is isomorphic to $C^m(L)$, we have that

$$c_0 \hat{\otimes}_\varepsilon \mathcal{D}^m(I) \simeq c_0 \hat{\otimes}_\varepsilon (E_1 \times E_2) \simeq (c_0 \hat{\otimes}_\varepsilon E_1) \times (c_0 \hat{\otimes}_\varepsilon E_2)$$

$$\simeq (c_0 \hat{\otimes}_\varepsilon E_1) \times (c_0 \hat{\otimes}_\varepsilon C^m(L))$$

and therefore we obtain the following reults:

(28) $c_0 \hat{\otimes}_\varepsilon C^m(L)$ *is isomorphic to a complemented subspace of* $c_0 \hat{\otimes}_\varepsilon \mathcal{D}^m(I)$.

According to 3.(4), $\mathcal{D}^m(I)$ is isomorphic to a complemented subspace of $C^m(L)$. It is not difficult to check that $C^m(L)$ is isomorphic to $\mathcal{D}^m(I)$ using an affine mapping from R^n onto R^n which maps L onto I. Therefore $C^m(L)$ is isomorphic to the product $E_1 \times E_2$ of Banach spaces, F_1 being isomorphic to $\mathcal{D}^m(I)$. Then

$$c_0 \hat{\otimes}_\varepsilon C^m(L) \simeq c_0 \hat{\otimes}_\varepsilon (F_1 \times F_2) \simeq (c_0 \hat{\otimes}_\varepsilon F_1) \times (c_0 \hat{\otimes}_\varepsilon F_2)$$

$$\simeq (c_0 \hat{\otimes}_\varepsilon \mathcal{D}^m(I)) \times (c_0 \hat{\otimes}_\varepsilon F_2)$$

and therefore we obtain the following result:

(29) $c_0 \hat{\otimes}_\varepsilon \mathcal{D}^m(I)$ *is isomorphic to a complemented subspace of* $c_0 \hat{\otimes}_\varepsilon C^m(L)$.

(30) $c_0 \hat{\otimes}_\varepsilon C^m(L)$ *is isomorphic to* E.

Proof. It is obvious that E coincides with $c_0(C^m(L))$. It is enough to apply 4.(9) to obtain the conclusion.

(31) E *is isomorphic to* $c_0 \hat{\otimes}_\varepsilon \mathcal{D}^m(I)$.

Proof. It is enough to apply (30), (28), (29) and 1.(5).

(32) $C_0^m(\Omega)$ *has a complemented subspace isomorphic to* $c_0 \hat{\otimes}_\varepsilon \mathcal{D}^m(I)$.

Proof. It is an immediate consequence from (27) and (31).

(33) $C_0^m(\Omega)$ *is isomorphic to* $c_0 \hat{\otimes}_\varepsilon \mathcal{D}^m(I)$.

Proof. It is enough to apply (16), (22), (32) and 1.(5).

Now we have the fundamental result of this section:

(34) $C_0^m(\Omega)$ *is isomorphic to* $C^m(I)$.

Proof. We apply (33) to obtain that $\mathcal{D}^m(I)$, which coincides with $C_0^m(\mathring{I})$, is isomorphic to $c_0 \hat{\otimes}_\varepsilon \mathcal{D}^m(I)$. According to 3.(4), $c_0 \hat{\otimes}_\varepsilon \mathcal{D}^m(I)$ is isomorphic to a complemented subspace of $C^m(I)$. By (30), $C^m(I)$, which is isomorphic to $C^m(L)$, is isomorphic to a complemented subspace of E. By (31), $C^m(I)$ is isomorphic to a complemented subspace of $c_0 \hat{\otimes}_\varepsilon \mathcal{D}^m(I)$. We apply 1.(5) to obtain that

$$C^m(I) \simeq c_0 \hat{\otimes}_\varepsilon \mathcal{D}^m(I).$$

The conclusion follows from (33).

(35) *Let* H *be a compact subset of* R^n *with non-void interior. Then* $\mathcal{D}^m(H)$ *is isomorphic to* $C^m(I)$.

Proof. Since $C_0^m(\mathring{H})$ coincides with $\mathcal{D}^m(H)$, the result is an immediate consequence from (30).

Result (34) appears here for the first time.

6. SPACES OF FUNCTIONS DEFINED ON A C^m-DIFFERENTIABLE MANIFOLD. Given a positive integer n and a non-negative integer m, let V be a C^m-differentiable n-dimensional manifold. Let (U_i, ϕ_i), $i \in I$, an atlas in V. A K-valued function f defined on V is said to be C^m-differentiable if, for every i of I and for every multi-index α with $|\alpha| \leq m$, $D^\alpha(f \circ \phi_i^{-1})$ exists and is continuous in the open set $\phi_i(U_i)$ of R^n. We write $C^m(V)$ to denote the linear space over K of all K-valued functions defined on V which are C^m differentia-

ble. For every compact set L of V contained in U_i, we set

$$|f|_{L,i} = \sum_{|\alpha| \leq m} \sup \{|D^{\alpha}(f \circ \phi_i^{-1})(x)| : x \in \phi_i(U)\}.$$

Then $|.|_{L,i}$ is a family of seminorms on $C^m(V)$ when $i \in I$ and L runs through all compact sets contained in U_i. We suppose $C^m(V)$ endowed with the locally convex topology defined by this family of seminorms. It is not difficult to check that $C^m(V)$ is complete. Moreover, if we use an equivalent atlas to (U_i, ϕ_i) $i \in I$, to define $C^m(V)$, we get the same locally convex space.

Let H be a compact subset of V. We denote by $\mathcal{D}^m(H)$ the subspace of $C^m(V)$ of all those functions with support contained in H. If M is the closure of the interior of H it is obvious that $\mathcal{D}^m(H)$ coincides with $\mathcal{D}^m(M)$. Let $\mathcal{D}^m(V)$ be the linear subspace of $C^m(V)$ of all those functions with compact support. Let $\mathcal{H}$ be the family of all compact subsets of V. We suppose $\mathcal{D}^m(V)$ endowed with the inductive limit topology generated by the family of locally convex spaces $\{\mathcal{D}^m(H) : H \in \mathcal{H}\}$.

Given $i \in I$, we define a cube in U_i as we did in §1, Section 22. Given a compact subset H of V with non-void interior, let M be the closure of H. For every z of M we find i_z in I and a cube Q_{i_z} in U_{i_z} such that z belongs to $\overset{\circ}{Q}_{i_z}$. Since M is compact we find points $z_1, z_2, \ldots, z_r$ in M such that $\bigcup_{j=1}^{r} \overset{\circ}{Q}_{i_{z_j}} \supset M$. We set

$$Q_{i_{z_j}} \cap M = P_j,\ \phi_{i_{z_j}} = \psi_j,\ U_{i_{z_j}} = V_j,\ j = 1,2,\ldots, r.$$

(1) *If j is an integer with* $1 \leq j \leq r$, *then* $\mathcal{D}^m(P_j)$ *is isomorphic to* $C^m(I)$.

Proof. We proceed as we did in the proof of §1, 22.(1) to obtain that $\mathcal{D}^m(P_j)$ is isomorphic to $\mathcal{D}^m(\psi_j(P_j))$. The result follows easily according to 5.(35).

If $(f_1, f_2, \ldots, f_r)$ is any element of $\prod_{j=1}^{r} \mathcal{D}^m(P_j)$ we write

$$Z(f_1, f_2, \ldots, f_r) = f_1 + f_2 + \ldots + f_r.$$

Then $Z : \prod_{j=1}^{r} \mathcal{D}^m(P_j) \longrightarrow \mathcal{D}^m(H)$ is linear and continuous.

If g_j is the function defined in §1, Section 22, it is clear that g_j belongs to $\mathcal{D}^m(Q_{i_{z_j}})$ and $g_j(z) > 0$ when z belongs to $\overset{\circ}{Q}_{i_{z_j}}$, $1 \leq j \leq r$. If g is any element of $\mathcal{D}^m(H)$ we write $T_j\, g$ to denote the element of $\mathcal{D}^m(P_j)$ such that

$$T_j g(z) = \frac{g(z)g_j(z)}{\Sigma g_h(z)}, \; z \in M, \; T_j g(z) = 0, \; z \in V \sim M.$$

$T_j : \mathcal{D}^m(H) \longrightarrow \mathcal{D}^m(P_j)$ is obviously linear. Since

$$\Sigma g_j(z) > 0, \; z \in M,$$

and M is compact it follows that T_j is continuous. If $Tg = (T_1g, T_2g, \ldots, T_rg)$, we have that $T : D^m(H) \longrightarrow \prod_{j=1}^{r} \mathcal{D}^m(P_j)$ is continuous.

(2) *T is an isomorphism from $\mathcal{D}^m(H)$ into $\prod_{j=1}^{r} \mathcal{D}^m(P_j)$.*

Proof. Proceed analogously as in §1,22.(2).

(3) *$\mathcal{D}^m(H)$ is isomorphic to a complemented subspace of $\prod_{j=1}^{r} \mathcal{D}^m(P_j)$.*

Proof. T o Z is a continuous projection from $\prod_{j=1}^{r} \mathcal{D}^m(P_j)$ into itself such that

$$T \circ Z\Big(\prod_{j=1}^{r} \mathcal{D}^m(P_j)\Big) = T(\mathcal{D}^m(H)).$$

According to (2), $T(\mathcal{D}^m(H))$ is isomorphic to $\mathcal{D}^m(H)$ and therefore $\mathcal{D}^m(H)$ is isomorphic to a complemented subspace of $\prod_{j=1}^{r} \mathcal{D}^m(P_j)$.

(4) *$C^m(I) \times C^m(I)$ is isomorphic to $C^m(I)$.*

Proof. Consider two non-void disjoint open set Ω_1 and Ω_2 in R^n. According to 5.(34), we have that

$$C^m(I) \simeq C_0^m(\Omega_1 \cup \Omega_2) \simeq C_0^m(\Omega_1) \times C_0^m(\Omega_2) \simeq C^m(I) \times C^m(I).$$

(5) *$\mathcal{D}^m(H)$ is isomorphic to a complemented subspace of $C^m(I)$.*

Proof. It is an immediate consequence from 5.(35), (3) and (4).

(6) *$C^m(I)$ is isomorphic to a complemented subspace of $\mathcal{D}^m(H)$.*

Proof. Proceed as in §1,22.(4).

Now we have the first fundamental result of this section:

(7) *Let V be an n-dimensional C^m differentiable manifold. Let H be a compact subset of V with non-void interior. Then $\mathcal{D}^m(H)$ is isomorphic to $C^m(I)$*

Proof. Since

$$c_0 \hat{\otimes}_\varepsilon C^m(I) \simeq c_0 \hat{\otimes} \mathcal{D}^m(I) \simeq \mathcal{D}^m(I) \simeq C^m(I)$$

it is enough to apply (5), (6) and 1.(5) to arrive to the conclusion.

(8) *Let* V *and* Ω *be* n-*dimensional* C^m *differentiable compact manifolds. Then* $C^m(V)$ *is isomorphic to* $C^m(\Omega)$.

Proof. It is enough to apply (7) having in mind that $\mathcal{D}^m(V) = C^m(V)$ and $\mathcal{D}^m(\Omega) = C^m(\Omega)$.

Now suppose that V is non compact and countable at the infinity. We proceed as we did in §1, Section 22, to extract from the atlas (U_i,ϕ_i), $i \in I$, a sequence of charts

(9) $$(W_1,\alpha_1),\ (W_2,\alpha_2),\dots,(W_r,\alpha_r),\dots$$

such that there exists a cube Q_r in W_r, $r = 1,2,\dots$ such that

(10) $$\{\mathring{Q}_r : r = 1,2,\dots\}$$

is a locally finite covering of V. We construct the function k_r as in §1, Section 22.

(11) $C^m(V)$ *is isomorphic to a complemented subspace of* $C^m(I)^N$.

Proof. See §1,22.(9).

Since the covering (8) is locally finite we find a subsequence of (9).

$$(W_{r_1},\alpha_{r_1}),(W_{r_2},\alpha_{r_2}),\dots,(W_{r_p},\alpha_{r_p}),\dots$$

such that the set Q_{r_1}, $Q_{r_2},\dots,Q_{r_p},\dots$ are pairwise disjoint. We set

$$V_j = W_{r_j}, \beta_j = \alpha_{r_j},\ H_j = Q_{r_j},\ j = 1,2,\dots$$

Let L_r be a cube in V_r contained in the interior of H_r, $r = 1,2,\dots$

(12) $C^m(I)^N$ *is isomorphic to a complemented subspace of* $C^m(V)$.

Proof. See §1,22.(10).

Now we arrive to another fundamental result of this section:

(13) *Let* V *be an* n-*dimensional* C^m *differentiable manifold. If* V *is non com-*

pact and countable at the infinity, then $C^m(V)$ *is isomorphic to* $C^m(I)^N$.

Proof. It is an immediate consequence from (11), (12) and §1,2.(1).

(14) $\mathcal{D}^m(V)$ *is isomorphic to a complemented subspace of* $C^m(I)^{(N)}$.

Proof. See §1,22.(12).

(15) $C^m(I)^{(N)}$ *is isomorphic to a complemented subspace of* $\mathcal{D}^m(V)$.

Proof. See §1,22.(13).

The last fundamental result of this section is as follows:

(16) *Let* V *be an* n*-dimensional* C^m *differentiable manifold. If* V *is non compact and countable at infinite, then* $\mathcal{D}^m(V)$ *is isomorphic to* $C^m(I)^{(N)}$.

Proof. It is an immediate consequence from (14), (15) and §1,2.(2).

Results (13) and (16) can be found in VALDIVIA [27]. (7) appears here for the first time and (8) is due to MITIAGIN [1].

7. SPACES OF FUNCTIONS DEFINED ON A C^m-DIFFERENTIABLE MANIFOLD WITH BOUNDARY. Given a positive integer n we set R^n_+ to denote

$$\{(x_1,x_2,\ldots,x_n) \in R^n : x_1 \geq 0\}$$

If L is a compact subset of R^n_+ we denote by $\mathcal{D}^m_+(L)$ the linear space over K of all the K-valued functions defined on R^n_+ such that for every multi-index α with $|\alpha| \leq m$, $D^\alpha f$ is defined and continuous on R^n_+ and the support of f is contained in L. We set

$$|f| = \sum_{|\alpha| \leq m} \sup \{|D^\alpha f(x)| : x \in L\}.$$

Then $|.|$ is a norm on $\mathcal{D}^m_+(L)$. We suppose $\mathcal{D}^m_+(L)$ endowed with the topology derived from this norm.

We suppose that L has non-void interior. We find a positive number b such that $\sup\{\|x\| : x \in L\} < b$. A,B and D have the same meaning as in §1, Section 23. Let β be a linear continuous extension operator from $C^m(B)$ into $C^m(A)$. We set

$$\gamma f = g\,\beta\, f,\ f \in \mathcal{D}^m_+(L).$$

Then γ is a linear continuous extension operator from $\mathcal{D}^m_+(L)$ into $\mathcal{D}^m(M)$ with

$M = D \cup L$.

(1) $\mathcal{D}_+^m(L)$ *is isomorphic to a complemented subspace of* $C^m(I)$.

Proof. See §1,23.(1).

(2) $C^m(I)$ *is isomorphic to a complemented subspace of* $\mathcal{D}_+^m(L)$.

Proof. Let P and Q be two n-dimensional compact cubes such that $P \subset \overset{\circ}{Q}, Q \subset \overset{\circ}{L}$ Let λ be a linear continuous extension operator from $C^m(P)$ into $\mathcal{D}^m(Q)$. Then λ is a linear continuous extension operator from $C^m(P)$ into $\mathcal{D}_+^m(L)$. Consequently the subspace of $\mathcal{D}_+^m(L)$ of all those functions which vanish on P is a topological complement of $\lambda(C^m(P))$ in $\mathcal{D}_+^m(L)$. Since

$$\lambda(C^m(P)) \simeq C^m(P) \simeq C^m(I)$$

the conclusion follows.

(3) $\mathcal{D}_+^m(L)$ *is isomorphic to* $C^m(I)$,

Proof. Since

$$c_0 \hat{\otimes}_\varepsilon C^m(I) \simeq C^m(I)$$

we apply (1), (2) and 1.(5) to obtain the conclusion.

We suppose now that V is an n-dimensional C^m differentiable manifold with boundary. Let (U_i, ϕ_i) $i \in I$, be an atlas on V such that $\phi_i(U_i)$ is an open subset in R_+^n, $i \in I$. If z belongs to the boundary of V and $z \in U_i$, then first coordinate of $\phi_i(z)$ vanishes. We define now the space $C^m(V)$ as in Section 6 by substituing R^n by R_+^n. If H is a compact subset of V, $\mathcal{D}^m(H)$ is the subspace of $C^m(V)$ of all those functions with support contained in H. $\mathcal{D}^m(V)$ is defined as in the former section. The definition of a cube in U_i is the same as in §1, Section 22. We use here the same notations of Section 6.

(4) *If* j *is an integer with* $1 \leq j \leq r$, *then* $\mathcal{D}^m(P_j)$ *is isomorphic to* $C^m(I)$.

Proof. It is analogous to §1,22.(1) by substituing $\mathcal{D}(\psi_i(P_j))$ by $\mathcal{D}_+^m(\psi_j(P_j))$ and having in mind that this last space is isomorphic to $C^m(I)$.

T has the same meaning as in the former section. Results (5) and (6) can be proved analogously to 22.(2) and 22.(3) paragraph 1.

(5) T *is an isomorphism from* $\mathcal{D}^m(H)$ *into* $\prod_{j=1}^{r} \mathcal{D}^m(P_j)$.

(6) *$\mathcal{D}^m(H)$ is isomorphic to a complemented subspace of $C^m(I)$.*

(7) *$C^m(I)$ is isomorphic to a complemented subspace of $\mathcal{D}^m(H)$.*

Proof. See §1,23.(7).

We arrive now to the first fundamental results of this section:

(8) *Let V be an n-dimensional C^m-differentiable manifold with boundary. Let H be a compact subset of V with non-void interior. Then $\mathcal{D}^m(H)$ is isomorphic to $C^m(I)$.*

Proof. Since

$$c_0 \hat{\otimes}_\varepsilon C^m(I) \simeq C^m(I)$$

we apply (6), (7) and 1.(5) to obtain the conclusion.

(9) *Let V and Ω be two n-dimensional C^m differentiable compact manifolds with boundary. Then $C^m(V)$ is isomorphic to $C^m(\Omega)$.*

Proof. It is enough to apply (8) having in mind that

$$\mathcal{D}^m(V) = C^m(V) \text{ and } \mathcal{D}^m(\Omega) = C^m(\Omega)$$

We suppose now that V is non compact and countable at infinity. We proceed as we did in §1, Section 22, and extract from the atlas (U_i,ϕ_i), $i \in I$, a sequence of charts

$$(W_1,\alpha_1),\ (W_2,\alpha_2),\ldots,(W_r,\alpha_r),\ldots$$

such that there exists a cube Q_r in W_r such that $\{\overset{\circ}{Q}_r : r = 1,2,\ldots\}$, is a locally finite covering of V. For every positive integer r let h_r be an element of $\mathcal{D}_+(\alpha_r(Q_r))$such that $h_r(x) > 0$ for every x of the interior of $\alpha_r(Q_r)$ in R^n_+. We define k_r as we did in §1, Section 22.

Result (10) can be proved analogously to §1,22.(9).

(10) $C^m(V)$ is isomorphic to a complemented subspace of $C^m(I)^N$.

We define V_j,β_j and H_j as we did in the former section. Then result (11) can be proved analogously to §1, 22(10) by substituing $\mathcal{D}(\beta_r(H_r))$ by $\mathcal{D}^m_+(\beta_r(H_r))$.

(11) *$C^m(I)^N$ is isomorphic to a complemented subspace of $C^m(V)$.*

Now we arrive to another fundamental result of this section.

(12) *Let V be an n-dimensional C^m differentiable manifold with boundary.*

If V *is non compact and countable at infinity then* $C^m(V)$ *is isomorphic to* $C^m(I)^N$.

Proof. It is an immediate consequence from (10), (11) and §1,2.(1).

Results (13) and (14) can be proved analogously to 22(12) and 22.(13) of paragraph 1, respectively.

(13) $\mathcal{D}^m(V)$ *is isomorphic to a complemented subspace of* $C^m(I)^{(N)}$.

(14) $C^m(I)^{(N)}$ *is isomorphic to a complemented subspace of* $\mathcal{D}^m(V)$.

We arrive to the last fundamental result of this section.

(15) *Let* V *be an* n-*dimensional* C^m *differentiable manifold with boundary. If* V *is non compact and countable at infinity, then* $\mathcal{D}^m(V)$ *is isomorphic to* $C^m(I)^{(N)}$.

Proof. It is an immediate consequence from (13), (14) and §1,2.(2).

Result (9) is due to MITIAGIN [1]. All the other results in this section appear here for the first time.

§3. SPACES OF CONTINUOUS FUNCTIONS

1. THE THEOREM OF BORSUK. If X is a Hausdorff topological space we denote by C(X) the linear space over K of all those functions defined on X with values in K which are continuous. We suppose C(X) endowed with the topology of uniform convergence on the compact subset of X. C*(X) denotes the linear subspace of C(X) of all bounded functions. We suppose C*(X) endowed with the topology of the uniform convergence on X. It is obvious that C*(X) is a Banach space. If X is compact, then C(X) coincides with C*(X).

Suppose now X metrizable and its topology defined by the metric d. If B is a non-void subset of X and if x belongs to X we set d(x,B) to denote the distance from x to B.

(1) *If* A *is a non-void closed subset of* X, *then there is a continuous linear extension operator* T *from* C*(A) *into* C*(X).

Proof. For every x of $X \sim A$ we set

$$B_x = \{z \in X \sim A : 4d(x,z) < d(x,A)\}.$$

According to the theorem of A.H. STONE, $X \sim A$ is paracompact (cf. KELLEY

[1], Chapter V, p. 160) and therefore there is a locally finite refinement $\mathcal{U}$ of

(2) $\{B_x : x \in X \sim A\}$.

For every U of $\mathcal{U}$ we determine a point x(U) in $X \sim A$ such that U is contained in $B_{x(U)}$. Take a point a(U) in A with

$$4d(x(U), a(U)) < 5d(x(U), A)$$

Let δ be a positive number. We shall check that if x belongs to U, z belongs to A and $3d(z,x) < \delta$, then $d(z,a(U)) < \delta$. Indeed

$$4d(x(U),x) < d(x(U),A) \leqslant d(x(U),z)$$

$$\leqslant d(x(U),x) + d(x,z) < d(x(U),x) + \frac{\delta}{3}$$

and thererefore $d(x(U),x) < \frac{\delta}{9}$ and thus

$$d(x(U),a(U)) < \frac{5}{4} d(x(U),A) < \frac{5}{4} d(x(U),x) + \frac{5\delta}{12} < \frac{5\delta}{36} + \frac{5\delta}{12} = \frac{5\delta}{9},$$

then

$$d(z,a(U)) \leqslant d(z,x)+d(x,x(U)) + d(x(U), a(U)) <$$

$$\frac{\delta}{3} + \frac{\delta}{9} + \frac{5\delta}{9} = \delta.$$

If U belongs to $\mathcal{U}$ we set f_U to denote the function defined on $X \sim A$ with

$$f_U(x) = d(x, X \sim U), \ x \in X \sim A.$$

Since $\mathcal{U}$ is locally finite we have that

$$\sum_{U \in \mathcal{U}} f_U = h$$

is a continuous function defined on $X \sim A$. If x belongs to $X \sim A$ there is V in $\mathcal{U}$ such that x belongs to V. Consequently $f_V(x) \neq 0$. Thus h does not vanish in $X \sim A$. We set

$$g_U = \frac{f_U}{h}, \ U \in \mathcal{U}.$$

Then g_U is continuous on $X \sim A$. For every x of $X \sim A$, we have that $g_U(x) \geqslant 0$ and $\sum_{U \in \mathcal{U}} g_U(x) = 1$, the support of g_U being a subset of $B_{x(U)}$, i.e., $\{g_U : U \in \mathcal{U}\}$ is a continuous partition of the unity subordinated to the covering (2) of $X \sim A$. If f belongs to $C^*(A)$ we set Tf to denote the function defined in X such that

$$Tf(x) = f(x),\ x \in A,\ Tf(x) = \sum_{U \in \mathcal{U}} f(a(U))g_U(x),\ x \in X \sim A.$$

It is obvious that Tf is continuous on $X \sim A$. Let us see that Tf is continuous on A. Take any point z in A and $\varepsilon > 0$. Since f is continuous on A there is $\delta > 0$ such that

$$|f(x) - f(x)| < \varepsilon,\ x \in A,\ d(x,z) < \delta.$$

Suppose that x is any point of $X \sim A$ with $d(z,x) < \frac{\delta}{3}$. Let U be any element of $\mathcal{U}$. If x belongs to U we have that $d(z,a(U)) < \delta$ and therefore $|f(z) - f(a(U))| < \varepsilon$. If x does not belong to U then $g_U(x) = 0$ and thus

$$|Tf(z) - Tf(x)| = |f(z) - \sum_{U \in \mathcal{U}} f(a(U))g_U(x)|$$

$$= |\sum_{U \in \mathcal{U}} (f(z)-f(a(U)))g_U(x)| \leq \sum_{U \in \mathcal{U}} |f(z)-f(a(U))|\ g_U(x)$$

$$\leq \varepsilon \sum_{U \in \mathcal{U}} g_U(x) = \varepsilon$$

from where the continuity of Tf on A follows.

Obviously $T : C^*(A) \longrightarrow C^*(X)$ is a linear extension operator. If $k = \sup\{|f(y)| : y \in A\}$ we have that

$$|Tf(x)| \leq \sum_{U \in \mathcal{U}} |f(a(U))|\ g_U(x) \leq k \sum_{U \in \mathcal{U}} g_U(x) = k$$

for every x of $X \sim A$. Therefore T is continuous.

Result (1) is due to BORSUK [1]. The former proof can be found in ARENS [1].

Even if X is compact the metrizability condition on X can not be omitted in (1). For instance, let N be the topological space of natural numbers endowed with the discrete topology. We set X to denote the STONE-ČECH compactification of N. Let A be the closed subset $X \sim N$ of X. Every function f of C(X) is determined by the bounded sequence (f(m)). f vanishes in A if and only if $\lim f(m) = 0$. Thus, C(X) is isomorphic to ℓ^∞ and if E denotes the subspace of C(X) of all those functions which vanish in A, then E is isomorphic to c_0. Suppose the existence of a continuous linear extension operator T from C(A) into C(X). The T(C(A)) is a subspace of C(X) with E as topological complement. Then c_0 has a topological complement in ℓ^∞ which is not true (cf. KÖTHE [1], Chapter Six, § 31. Section 2).

2. PRODUCTS OF PROBABILITIES. Let X_m be a metrizable separable topological space. Let $\mathcal{B}_m$ be the σ-algebra of Borel in X_m and let $\alpha_m: \mathcal{B}_m \longrightarrow [0,1]$ a probability, m = 1,2,... α_m is called a Borel probability on X_m. Now we shall define a Borel probability on $X = \prod_{m=1}^{\infty} X_m$. A cylinder in X is a subset Y of the form $\prod_{m=1}^{\infty} Y_m$ with Y_m contained in X_m, m = 1,2,..., and such that there is a positive integer r verifying $Y_m = X_m$, m = r, r+1,... If for every positive integer m, Y_m is a Borel subset of X_m, then Y is a Borel subset of X and we call it a Borel cylinder of X. If Y_m is open, m = 1,2,..., we say that Y is an open cylinder of X.

The family M of all Borel cylinders of X is an algebra of sets, Obviously the σ-algebra generated by **M** is contained in the σ-algebra of Borel $\mathcal{B}$ of X. On the other hand, X is metrizable and separable and therefore every open subset of X is countable union of open cylinders of X and therefore the α-algebra generated by M coincides with $\mathcal{B}$.

There is a probability α defined on $\mathcal{B}$, i.e., α is a Borel probability on X such that

$$\alpha(Y) = \lim_r \prod_{m=1}^{r} \alpha_m(Y_m) = \prod_{m=1}^{\infty} \alpha_m(Y_m)$$

for every Borel cylinder $Y = \prod_{m=1}^{\infty} Y_m$ in X. α is called the product probability and also the tensor product of the probabilities $\alpha_1, \alpha_2, \ldots, \alpha_m, \ldots$ and it is denoted by $\bigotimes_{m=1}^{\infty} \alpha_m$. (cf. HALMOS [1], Chapter VII, §38.)

For every positive integer m, let D_m be the set {0,1} of two elements endowed with the discrete topology. D_m is metrizable and separable. The Borel σ-algebra of D_m is constitued by

$$\emptyset, \{0\}, \{1\}, \{0,1\}.$$

We set

$$\mu_m(\emptyset) = 0, \mu_m(\{0\}) = \mu_m(\{1\}) = \frac{1}{2}, \mu_m(\{0,1\}) = 1.$$

Then μ_m is a Borel probability on D_m. Let μ the tensor product of the probabilities $\mu_1, \mu_2, \ldots, \mu_m, \ldots$ Then μ is a Borel probability on $D = \prod_{m=1}^{\infty} D_m$. D is called the Cantor space.

Given the numbers $t_1, t_2, \ldots, t_m$ where t_j takes the value zero or one, $j = 1,2,\ldots,m$, we denote by $A(t_1,t_2,\ldots,t_m)$ the open cylinder in D

$$\{x = (x_n) \in D: x_j, t_j, j = 1,2,\ldots m\}.$$

Then

$$\mu\,(A(t_1,t_2,\ldots,t_m)) = \frac{1}{2^m}$$

(1) *If* $t = (t_m)$ *is a point of* D, *then* $\mu(\{t\}) = 0$.

Proof. It is obvious that

$$\{t\} = \bigcap_{m=1}^{\infty} A(t_1,t_2,\ldots,t_m)$$

Since μ is a finite measure and since the sequence of sets $(A(t_1,t_2,\ldots t_m))$ is decreasing we have that

$$\mu(\{t\}) = \lim \ \mu(A(t_1,t_2,\ldots,t_m)) = \lim \frac{1}{2^m} = 0$$

(2) *Let* E *and* F *be topological spaces. Let* f *be a continuous mapping from* E *into* F. *If* A *is a Borel subset of* F, *then* $f^{-1}(A)$ *is a Borel subset of* E.

Proof. Let A be the family of all subsets A of F such that $f^{-1}(A)$ is a Borel subset of E. Since f is continuous the open subsets of F belong to A . On the other hand, if B and B_m belongs to A, then

$$f^{-1}(F \sim B) = E \sim f^{-1}(B),\ f^{-1}\left(\bigcup_{m=1}^{\infty} B_m\right) = \bigcup_{m=1}^{\infty} f^{-1}(B_m)$$

and therefore,

$$F \sim B,\ \bigcup_{m=1}^{\infty} B_m \in A$$

from where it follows that A is a σ-algebra. Clearly A contains the Borel σ-algebra of F. The proof is complete.

In what follows I denotes the closed interval $[0,1]$ and λ the Lebesgue measure on R restricted to the family of the Borel sets of I, i.e., λ is the Lebesgue measure on I.

Let $h;\ D \longrightarrow I$ be the mapping defined by

$$h(x) = \sum_{m=1}^{\infty} \frac{x_m}{2^m},\ x = (x_m) \in D$$

h is called the Cantor mapping.

(3) $h : D \longrightarrow I$ *is continuous.*

Proof. If $z = (z_m)$ is a point of D, given $\varepsilon > 0$ we find a positive integer r such that $\frac{1}{2^r} < \varepsilon$. The point z belongs to the open cylinder $A(z_1, z_2, \ldots, z_r)$. On the other hand, if $x = (x_m)$ belongs to that cylinder, we have that $x_j = z_j$, $j = 1, 2, \ldots, r$, and therefore

$$|h(x)-h(z)| \leq \sum_{m=r+1}^{\infty} \frac{|x_m - z_m|}{2^m} \leq \frac{1}{2^r} < \varepsilon$$

from where the continuity of h in z follows.

(4) $h : D \longrightarrow I$ *is onto,*

Proof. If u is any point of I we take a dyadic expansion of u of the form $\Sigma \frac{u_m}{2^m}$ where u_j is zero or one, Then $x = (u_m)$ is a point of D and $h(x) = u$.

Let M be the subset of D of all those elements $x = (x_m)$ such that there exists a positive integer r, depending on x, with $x_r = x_{r+1} = x_{r+2} = \ldots$ Let h_1 be the restriction of h to the topological subspace $D \sim M$ of D.

(5) h_1 *is an homeomorphism from* $D \sim M$ *onto* $h(D \sim M)$.

Proof. According to (3), h_1 is continuous. Let $x = (x_m)$ and $y = (y_m)$ be two different elements of $D \sim M$. Let r be the first positive integer such that $x_r \neq y_r$. The sequence

(5) $$x_{x+1}-y_{r+1}, x_{r+2}-y_{r+2}, \ldots$$

takes only the values 0,1 or -1. On the other hand, it follows from the definition of M that, if not all the terms of (5) are one, then they can not be equal. Consequently

$$\left| \sum_{m=r+1}^{\infty} \frac{x_m - y_m}{2^m} \right| < \sum_{m=r+1}^{\infty} \frac{1}{2^m} = \frac{1}{2^r}.$$

Then

$$|h(x)-h(y)| = \left| \sum_{m=r}^{\infty} \frac{x_m - y_m}{2^m} \right| > \frac{1}{2^r} - \left| \sum_{m=r+1}^{\infty} \frac{x_m - y_m}{2^m} \right|$$

and therefore h_1 is injective,

If $D \sim M$, let A be a neighbourhood of the point $z = (z_m)$. There is a positive integer r such that $(D \sim M) \cap A(z_1,z_2,\dots,z_r)$ is contained in A. If

$$a = \frac{z_1}{2} + \frac{z_2}{2^2} + \dots + \frac{z_r}{2^r}, b=a+ \frac{1}{2^r},$$

we set

$$B = h\,(D \sim M) \cap \,]a,b[$$

We have that

$$a < \sum_{m=1}^{\infty} \frac{z_m}{z^m} = h(z) < b$$

and therefore B is a neighbourhood of $h(z)$ in $h(D \sim M)$. If u is any point of B we find an element $y = (y_m)$ in $D \sim M$ such that $h(y) = u$. Then

$$a = 0_1 z_1 z_2 \dots z_r,\ b = 0_1 z_1 z_2 \dots z_r 11\dots,\ u = 0_1 y_1 y_2 \dots y_m \dots,$$

in their dyadic expansion and consequently $y_j = z_j$, $j = 1,2,\dots,r$; thus y belongs to the cylinder $A(z_1,z_2,\dots,z_r)$. Then $h(A)$ contains B and the conclusion follows.

(6) *Let E be a Hausdorff topological space. Let F be a subspace of E such that $E \sim F$ is countable. If $\mathcal{B}$ is the Borel σ-algebra of E and*

$$\mathcal{A} = \{B \cap F : B \in \mathcal{B}\}$$

then $\mathcal{A}$ is the Borel σ-algebra of F and $\mathcal{A} \subset \mathcal{B}$.

Proof. Let $\mathcal{M}$ be Borel σ-algebra of F. It is immediate that $\mathcal{A}$ is a σ-algebra in F containing the family of open sets of F and therefore $\mathcal{A} \supset \mathcal{M}$. If M belongs to $\mathcal{M}$ there is an element Q in $\mathcal{B}$ such that M coincides with $Q \cap F$. Since $Q \sim M$ is countable it follows that M belongs to $\mathcal{B}$.

Let $\mathcal{P}$ be the family of all subset of $E \sim F$. If we set

$$\mathcal{N} = \{H \cup P\} : H \in \mathcal{M},\ P \in \mathcal{P}\}$$

we have that $\mathcal{N}$ is contained in $\mathcal{B}$ and every open set of E belongs to $\mathcal{N}$. It is obvious that $\mathcal{N}$ is a σ-algebra. Consequently $\mathcal{B}$ coincides with $\mathcal{N}$. If B belongs to $\mathcal{B}$ we find elements H in $\mathcal{M}$ and P in $\mathcal{P}$ such that $H \cup P = B$. Then $B \cap F = (H \cup P) \cap F = H$ and therefore $\mathcal{A}$ is contained in $\mathcal{M}$. Obviously $\mathcal{A} \subset \mathcal{B}$. The proof is complete.

(7) *h maps every Borel set of D in a Borel set of I.*

Proof. It is obvious that the set M is countable. Let B be a Borel set in D. According to (6), $B \cap (D \sim M)$ is a Borel set in $D \sim M$ and, having in mind (2) and (5), it follows that $h(B \cap (D \sim M)$ is a Borel set in $h(D \sim M)$. Since h is onto, $I \sim h(D \sim M)$ is countable and, according to (6), $h(B \cap (D \sim M))$ is a Borel set in I. Since the complement of $h(B \cap (D \sim M))$ in $h(B)$ is a countable set, it follows that $h(B)$ is a Borel set of I.

(8) *If B is a Borel set of D, then $\mu(B)$ coincides with $\lambda(h(B))$.*

Proof. Let A be the Borel σ-algebra of I. Let A be any element of A. Since $h : D \longrightarrow I$ is continuous, $h^{-1}(A)$ is a Borel set of D. We set

$$\nu(A) = \mu(h^{-1}(A)),\ A \in A.$$

It is not difficult to check that ν is a Borel probability on I. Given the cylinder $A(z_1,z_2,\ldots,z_m)$ of D, we have that $h(A(z_1,z_2,\ldots,z_m))$ coincides with the interval $[a,b]$ with

$$a = \frac{z_1}{2} + \frac{z_2}{2^2} + \ldots + \frac{z_m}{2^m},\ b = a + \frac{1}{2^{m+1}} + \frac{1}{2^{m+2}} + \ldots = a + \frac{1}{2^m}$$

and thus

$$\frac{1}{2^m} = \mu(A(z_1,z_2,\ldots,z_m)) = b - a = \lambda([a,b]).$$

Since M is countable and the restriction h_1 of h to $D \sim M$ is injective, it follows that the complement of $A(z_1,z_2,\ldots,z_m)$ in $h^{-1}([a,b])$ is a countable set from where it follows

$$\mu(A(z_1,z_2,\ldots,z_m)) = \mu(h^{-1}([a,b])).$$

Consequently $\nu([a,b]) = b - a = \lambda([a,b])$. Given any point u of I, let $x = (x_1,x_2,\ldots,x_m,\ldots)$ be an element of D such that $h(x) = u$. Then a dyadic expansion of u is

(9) $$0,x_1,x_2,\ldots,x_m\ldots$$

If u is distinct from zero and one and if a dyadic expansion of u is $0,y_1 y_2\ldots y_p 00\ldots$, $y_p = 1$, then u admits also the dyadic expansion $0,y_1,y_2\ldots y_{p-1}\,011\ldots$ and both expansions are the unique dyadic expansion of u. Consequently $h^{-1}(u)$ has two points. If u is zero, one or if u does not belong to the class quoted above, then (9) is the unique dyadic expansion of u and thus, $h^{-1}(u)$ has only one point. Therefore, if B is a countable set of I, $\nu(B)$ is zero.

We take now a point u of I with the dyadic expansion $0,z_1,z_2,\ldots,z_p$ 00 ... If u is zero, then

$$\nu([u,1]) = \mu(D) = 1 = \lambda([u,1]).$$

If $u \neq 0$, let us suppose $z_p = 1$. We set

$$u_1 = 0,z_1,z_2,\ldots,z_p\ 11\ldots$$

Then

$$\nu([u,u_1]) = \mu(A(z_1,z_2,\ldots,z_p)) = \lambda([u,u_1]).$$

If $u_1 \neq 1$ there is a positive integer $r < p$ such that $z_r = 0$, $z_{r+1} = z_{r+2} = \ldots = z_p = 1$. Then

$$u_1 = 0,z_1\ z_2\ldots\ z_{r-1}\ 1\ 0\ 0\ \ldots$$

We set

$$u_2 = 0,z_1\ z_2\ldots z_{r-1}\ 1\ 1\ 1\ \ldots$$

We have that

$$\nu([u_1,u_2]) = \mu(A(z_1,z_2,\ldots,z_{r-1},1)) = \lambda([u_1,u_2]).$$

Following the same path we obtain a positive integer q such that $u_q = 1$ and

$$\nu([u_{q-1},u_q]) = \lambda([u_{q-1},u_q])$$

from where it follows

$$\begin{aligned}\nu([u,1]) &= \nu([u,u_1[) + \nu([u_1,u_2[) + \ldots + \nu([u_{q-1},1])\\ &= \nu([u,u_1]) + \nu([u_1,u_2]) + \ldots + \nu([u_{q-1},u_q])\\ &= \lambda([u,u_1]) + \lambda([u_1,u_2]) + \ldots + \lambda([u_{q-1},1]) = \lambda([u,1]).\end{aligned}$$

If a and b are two points of I, $a < b$, with exact dyadic expansions, then

$$\nu([a,b]) = \nu([a,1]) - \nu([b,1]) = \lambda([a,1]) - \lambda([b,1]) = \lambda([a,b]).$$

Suppose now that a and b, a < b, are two points of I. We take a_n and b_n in I with exact dyadic expansions such that

$$a < a_{n+1} < a_n < b_n < b_{n+1} < b, \; n = 1,2,\ldots$$

$$\lim a_n = a, \quad \lim b_n = b$$

Then the sequence of intervals $([a_n,b_n])$ is increasing and

$$]a,b[\; = \bigcup_{n=1}^{\infty} [a_n,b_n]$$

and thus

$$\nu(]a,b[) = \lim \nu(]a_n,b_n[) = \lim ([a_n,b_n]) = \lambda(]a,b[)$$

from where it follows that λ and ν coincide in all the open, closed and half-open intervals of I. We set

$$Q = \{[a,b[\; : 0 \leq a < b < 1\} \cup \{[a,1] : 0 < a < 1\}.$$

The disjoint finite unions of elements of Q form an algebra S of I in which λ and ν coincide. The Borel σ-algebra A of I is generated by S. Now we apply the theorem of extension of measures (cf. HALMOS [1], Chapter III, §13) and we obtain that ν coincides with λ on A. Finally, if B is a Borel set of D we have that

$$\mu(B) = \mu(h^{-1}(h(B))) = \nu(h(B)) = \lambda(h(B)).$$

The sets M and $I \sim h(D \sim M)$ are countable infinite and thus there is an injective mapping ρ from $I \sim h(D \sim M)$ onto M. We set

$$k(x) = h_1^{-1}, x \in h(D \sim M), \; k(x) = \rho(x), \; x \in I \sim h(D \sim M)$$

(10) *If* $f : D \longrightarrow K$ *is μ-medible, then* f o k *is λ-medible.*

Proof. Let A be an open subset of K. Then $f^{-1}(A)$ is a Borel subset of D and consequently $h(f^{-1}(A))$ is a Borel set of I. Obviously $h(f^{-1}(A))$ and $k^{-1}(f^{-1}(A))$ save is a set of points which is at most countable. Thus $(f \circ k)^{-1}(A)$ is a Borel set of I and the conclusion follows.

(11) *If* $f : D \longrightarrow K$ *is μ-integrable, then* f o k *is λ-integrable and*

$$\int_D f d\mu = \int_I f \circ k \, d\lambda$$

Proof. First we suppose that f is the characteristic function of a Borel set B of D. Then f o k is the characteristic function of the set k^{-1} (B). Since the set

$$(k^{-1}(B) \sim h(B)) \cup (h(B) \sim k^{-1}(B))$$

is countable it follows that

$$\lambda(k^{-1}(B)) = \lambda(h(B)) = \mu(B)$$

(see (8)) and, consequently f o k is λ-integrable and

$$\mu(B) = \int_D f \, d\,\mu = \lambda(k^{-1}(B)) = \int_I f \circ k \, d\lambda.$$

from where it follows that the property is true for simple functions.

If f is a non-negative μ-integrable function there exists an increasing sequence (f_m) of non-negative simple functions that converges to f pointwise. Then

$$\int_D f \, d\mu = \lim \int_D f_m \, d\,\mu.$$

The functions of the sequence $(f_m \circ k)$ are λ-measurable and non-negative. This sequence is increasing and converges to f o k pointwise. Then

$$\int_I f \circ k \, d\lambda = \lim \int_I f_m \circ k \, d\,\lambda.$$

From

$$\int_D f_m \, d\mu = \int_I f_m \circ k \, d\,\lambda$$

it follows that

$$\int_D f \, d\mu = \int_I f \circ k \, d\,\lambda.$$

If f is any μ-integrable real function we write $f = f_1 - f_2$, f_1 and f_2 being non-negative μ-integrable functions. Then

$$\int_D f \, d\,\mu = \int_D f_1 \, d\,\mu - \int_D f_2 \, d\,\mu = \int_I f_1 \circ k \, d\,\lambda - \int_I f_2 \circ k \, d\,\lambda$$
$$= \int_I f \circ k \, d\,\lambda$$

Finally if K is the field of complex numbers and f is any μ-integrable function, we write $f = f_1 + if_2$, f_1 and f_2 being real functions. Then

$$\int_D f \, d\mu = \int_D d_1 \, d\mu + i \int_D f_2 \, d\mu$$

$$= \int_I f_1 \circ k \, d\lambda + i \int_I f_2 \circ k \, d\lambda = \int_I f \circ k \, d\lambda.$$

3. THE INTEGRAL OPERATOR OF MILUTIN. For every real number α with $0 \leq \alpha \leq 1$ we set

$$g_\alpha(x) = \alpha^2 x + \alpha(1-x), x \in I.$$

Let f_α be the characteristic function of the set $[0,\alpha]$. We take a point (a, b) in $I \times I$ with $b \neq 0$. Since f_b is λ-measurable, g_α is continuous and $g_\alpha(I)$ is contained in I it follows that $f_b \circ g_\alpha$ is well defined in I and is λ-measurable. Since this function is bounded in I, it is λ-integrable. We set

$$J(\alpha) = \int_0^a f_b \circ g_\alpha \, d\lambda, \ \alpha \in I.$$

If a = 0, then $J(\alpha) = 0$ for every α of I and thus J is continuous. If b=1, then $f_b \circ g_\alpha(x) = 1$, $x \in I$, and consequently $J(\alpha) = a$ for every α of I and this the function J is continuous. If $0 < b < 1$ and $a > 0$ we set $F(\alpha) = g_\alpha(a)-b$. The equation $F(\alpha) = 0$ has two real solutions whose product is negative. Since

$$F(0) = -b < 0, \ F(1) = 1 - b > 0$$

the positive solution α_1 verifies $0 < \alpha_1 < 1$.

If $0 \leq \alpha \leq b$ we have that

$$\alpha^2 \leq \alpha^2 x + \alpha(1-x) \leq \alpha \leq b$$

and therefore

$$f_b \circ g_\alpha(x) = 1, \text{ for every } x \text{ of } I,$$

and thus $J(\alpha) = a$. If $b \leq \alpha \leq \alpha_1$ we have that

$$g_\alpha(0) = \alpha > b, \ g_\alpha(a) \leq g_{\alpha 1}(a) = b,$$

and therefore there is a point $x(\alpha)$ of I such that

$$0 < x(\alpha) \leq a, \ g_\alpha(x(\alpha)) = b$$

It is immediate that $x(\alpha) = \frac{\alpha-b}{\alpha-\alpha^2}$. The function g_α is strictly decreasing in I from where it follows that

$$g_\alpha(x) > b,\ x < x(\alpha),\ g_\alpha\ (x) \leqslant b,\ x(\alpha) \leqslant x \leqslant a$$

Consequently

$$J(\alpha) = \int_{x(\alpha)}^{a} f_b \circ g_\alpha \, d\lambda = a - \frac{\alpha-b}{\alpha-\alpha^2}$$

If $\alpha_1 < \alpha \leqslant 1$ and $0 \leqslant x \leqslant a$ it follows that

$$g_\alpha(x) > g_{\alpha 1}(x) \geqslant g_{\alpha 1}(a) = b$$

and thus $f_b \circ g_\alpha(x) = 0,\ 0 \leqslant x \leqslant a$, and therefore $J(\alpha) = 0$.

Then J is continuous on the intervals

$$[0,b],\ ]b,\alpha_1],\ ]\alpha_1,1].$$

Moreover

$$J(b) = a = \lim_{\alpha\to b+0} a - \frac{\alpha-b}{\alpha-\alpha^2},$$

$$J(\alpha_1) = a - \frac{\alpha_1-b}{\alpha_1-\alpha_1{}^2} = 0$$

and consequently J is continuos on I. So we can conclude:

(1) *The function* J *is continuous on* I.

Let f be a continuous function defined on D × D with values in K. Let f_j be a μ-measurable mapping from D into D, j = 1,2,. We set (f_1,f_2) to denote the mapping from D into D × D such that

$$(f_1,f_2)(x) = (f_1(x),f_2(x)),\ x \in D$$

(2) $f \circ (f_1,f_2)$ *is* μ-*measurable.*

Proof. Let P be an open subset of K. Then $f^{-1}(P)$ is an open subset of D × D and since, this topological space is metrizable and separable, there are open subsets A_n and B_n of D, n = 1,2,..., such that

$$f^{-1}(P) = \bigcup_{n=1}^{\infty} A_n \times B_n$$

Then

$$(f \circ (f_1,f_2))^{-1}(P) = (f_1,f_2)^{-1}(f^{-1}(P))$$

$$= \bigcup_{n=1}^{\infty} (f_1,f_2)^{-1}(A_n \times B_n) = \bigcup_{n=1}^{\infty} (f_1^{-1}(A_n) \cap f_2^{-1}(B_n))$$

which is a Borel set of D. This completes the proof.

Let H be a compact topological space. $C(H)$ is the linear space over K of all the K-valued functions which are defined and continuous on H. We suppose $C(H)$ endowed with the norm $||.||$ such that

$$||f|| = \sup \{|f(x)| : x \in H\}, \ f \in C(H).$$

We also represent by $||.||$ the norm of the Banach dual of $C(H)$.

For every α of I and every f of $C(D \times D)$, we set $T_\alpha f$ to denote the function defined on D with values in K such that

$$T_\alpha f(x) = f(x, k \circ g_\alpha \circ h(x)), \ x \in D.$$

Then $T_\alpha f$ coincides with $f \circ (f_1,f_2)$, f_1 being the identity mapping from D into itself and f_2 being the mapping $k \circ g_\alpha \circ h$. It is obvious that f_1 is μ-measurable. On the other hand, if A is a Borel set of D we have that $k^{-1}(A)$ is a Borel set of I and, since g_α is continuous, $g_\alpha^{-1}(k^{-1}(A))$ is a Borel set of I, from where it follows that $(k \circ g_\alpha \circ h)^{-1}(A)$ is a Borel set of D. Then f_2 is μ-measurable. We apply (2) to obtain that $T_\alpha f$ is μ-measurable. Since $T_\alpha f$ is bounded on D, it follows that $T_\alpha f$ is μ-integrable. We set

$$Tf(\alpha) = \int_D T_\alpha f \, d\mu, \ \alpha \in I$$

Then T is a linear mapping from $C(D \times D)$ into the linear space over K $A(I)$ of all the K-valued functions which are defined and bounded on I. The operator T is called the integral operator of Milutin. We suppose $A(I)$ endowed with the norm $||.||$ such that

$$||g|| = \sup \{|g(x)| : x \in I\}, \ g \in A(I)$$

Then $C(I)$ is a closed subspace of $A(I)$.

(3) T *is a continuous operator with norm one.*

Proof. If f is any element of $C(D \times D)$ we have that

$$||Tf|| = \sup\{|\int_D T_\alpha f \, d\mu| : \alpha \in I\} \leq \int_D ||f|| \, d\mu = ||f||.$$

On the other hand, if f coincides with the function which takes the value one on each point of $D \times D$, it follows that

$$|Tf(\alpha)| = |\int_D T_\alpha f \, d\mu| = 1, \ \alpha \in I,$$

and the conclusion follows.

We represent by U the set of all the elements $u = (u_m)$ of D such that there is a positive integer r, depending on u, with $1 = u_r = u_{r+1} = \ldots$

For every $u = (u_m)$ of U, we set

$$U_u = \{(x_m) \in D : x_m \leqslant u_m, \ m = 1,2,\ldots\}$$

and we denote by h_u the characteristic function of U_u.

(4) *If* $u = (u_m)$ *belongs to* U, *then* h_u *is a continuous function on* D

Proof. Let $u^r = (u_m^{(r)})$, $r = 1,2,\ldots$, a sequence in U_u converging to $v = (v_m)$ in D. Then

$$u_m^{(r)} \leqslant u_m, \ m,r = 1,2,\ldots$$

and thus $v_m \leqslant u_m$, $m = 1,2,\ldots$, from where it follows that v belongs to U_u. Then U_u is a closed subset of D. Let $w = (w_m)$ any element of U_u. Let r be a positive integer such that $1 = u_r = u_{r+1} = \ldots$ The cylinder $A(w_1, w_2, \ldots, w_r)$ is a neighbourhood of w which is included in U_u. Therefore U_u is open. Now the conclusion follows.

(5) *If* $u = (u_m)$ *belongs to* U, $h_u \circ k$ *coincides almost everywhere with the characteristic function of* $[0, h(u)]$.

Proof. Since h_u is the characteristic function of U_u, $h_u \circ k$ is the characteristic function of $k^{-1}(U_u)$. The set

$$(k^{-1}(U_u) \sim h(U_u)) \cup (h(U_u) \sim k^{-1}(U_u))$$

is countable, from where it follows that $h_u \circ k$ coincides almost everywhere with the characteristic function of $h(U_u)$. If $v = (v_m)$ belongs to U_u have that

$$h(v) = \Sigma \frac{v_m}{2^m} \leqslant \Sigma \frac{u_m}{2^m} = h(u)$$

and thus $h(U_u)$ is contained in $[0,h(u)]$. Let t be any point of $[0,h(u)[$. Let $0,x_1 x_2,\ldots x_m\ldots$ be a dyadic expansion of t. We suppose that $x=(x_m)$ is not in U_u. Let r be the first positive integer such that $x_r > u_r$. Then

$$h(u) = \sum_{j=1}^{r-1} \frac{u_j}{2^j} + \sum_{j=r+1}^{\infty} \frac{u_j}{2^j} \leqslant \sum_{j=1}^{r-1} \frac{u_j}{2^j} + \frac{1}{2^r}$$

$$\leqslant \sum_{j=1}^{r-1} \frac{u_j}{2^j} + \sum_{j=r}^{\infty} \frac{x_j}{2^j} = t$$

which is a contradiction. So we can assure that $h(U_u) = [0,h(u)]$ and the proof is complete.

For every u and v of U we set $h_u \otimes h_v$ to denote the tensor product of mapping h_u and h_v, i.e., if (x,y) belongs to $D \times D$ we have that

$$h_u \otimes h_v(x,y) = h_u(x)h_v(y)$$

Obviously $h_u \otimes h_v$ belongs to $C(D \times D)$.

Now we fix u and v in U and set $a = h(u)$, $b = h(v)$. Then (a,b) is a point of $I \times I$ with $b \neq 0$.

(6) *For every* α *of* I.

(7) $(T_\alpha(h_u \otimes h_v)) \circ k$

coincides almost everywhere with $f_a \times (f_b \circ g_\alpha)$.

Proof. If x is any point of I we have that

$$T_\alpha(h_u \otimes h_v) \circ k(x) = h_u(k(x)) \times h_v(k \circ g_\alpha \circ h \circ k(x)$$

and, since $h \circ k$ coincides with the identity function on I save at most on a countable set, it follows that (7) coincides almost everywhere with

$$(h_u \circ k) \times (h_v \circ k \circ g_\alpha)$$

and, according to (5), this function coincides almost everywhere with $f_a \times (f_b \circ g\ )$. The proof is complete.

(8) *The function* $T(h_u \otimes h_v)$ *belongs to* $C(I)$.

Proof. According to 2.(11), we have that

$$T(h_u \otimes h_v)(\alpha) = \int_D T_\alpha(h_u \otimes h_v)\, d\mu = \int_I T_\alpha(h_u \otimes h_v) \circ k \, d\lambda$$

for every α of I. By (6),

$$T(h_u \otimes h_v)(\alpha) = \int_I f_a \times (f_b \circ g_\alpha)\, d\lambda = \int_0^a f_b \circ g_\alpha \, d\lambda = J(\alpha)$$

Now we apply (1) to reach the conclusion.

(9) T *maps* $C(D \times D)$ *into* $C(I)$.

Proof. Let H be the linear hull of

$$\{h_u \otimes h_v : u, v, \in U\}.$$

If we take $u = v = (1,1,\ldots)$, then for every (x,y) of $D \times D$ we have $h_u \otimes h_v$ $(x,y) = 1$ and therefore H contains all constant functions.

Given u and v in U, we have that $h_u \otimes h_v$ is the complex function conjugate of $h_u \otimes h_v$ since h_u and h_v are the characteristic functions of U_u and U_v respectively. Therefore, if a function belongs to H, its complex conjugate function belongs to H.

Let (t,x) and (y,z) be two different points of $D \times D$. If $t = (t_m)$ is distinct from $y = (y_m)$, let r be the first positive integer with $t_r \neq y_r$. Suppose $t_r < y_r$. By taking

$$u = (t_1, t_2, \ldots, t_r, 1, 1, \ldots)$$

we have that $t \in U_u$, y U_u, $y \notin U_u$. By taking $v = (1,1\ldots)$ it follows that

$$h_u \otimes h_v\,(t,x) = h_u(t)h_v(x) = h_u(t) = 1,$$

$$h_u \otimes h_v\,(y,z) = h_u(y)h_v(z) = h_u(y) = 0$$

and therefore H separates the points (t,x) and (y,z). If $t = y$, then $x \neq z$ and proceedings as we did before, we obtain that H separates the points (t,x) and (y,z).

Now suppose that $u = (u_m)$, $v = (v_m)$, $w = (w_m)$ and $z = (z_m)$ are four points of U. We set

$$x_m = \min(u_m, v_m), y_m = \min(w_m, z_m),\ m = 1,2,\ldots$$

If $x = (x_m)$ and $y = (y_m)$ we have that x and y belong to U and

$$(h_u \otimes h_w) \times (h_v \otimes h_z) = h_x \otimes h_y$$

from where it follows that H is a linear subalgebra of $C(D \times D)$. We apply Stone-Weierstrass' theorem to obtain that H is dense in $C(D\times D)$.

Let f be an element of $C(D \times D)$. Let (f_m) be a sequence in H converging to f in $C(D \times D)$. According to (8), Tf_m belongs to $C(I)$, $m = 1,2,\ldots$ Since T is continuous, the sequence (Tf_m) converges to Tf in $A(I)$ from where it follows that Tf belongs to $C(I)$.

Consequently $T : C(D \times D) \longrightarrow C(I)$ is a continuous linear operator.

The operator T has been introduced by MILUTIN [1].

4. SOME PROPERTIES ON COMPLEMENTATION. In what follows X denotes a metrizable compact topological space. A subset A of X is said to be perfect if it is closed and, if provided with the topology induced by X, has no isolated points. A point x of X is a condensation point of X if every neighbourhood of x is not a countable set.

(1) *If X is not countable, the set B of the condensation points of X is non-void and perfect.*

Proof. Suppose B void. Given any point x of X, let U_x be a countable open neighbourhood of X. The family $\{U_x : x \in X\}$ covers X and, since this space is compact, there are points $x_1,x_2,\ldots,x_p$ in X such that

$$X = \bigcup_{j=1}^{p} U_{x_j}$$

Then X is countable and this is a contradiction. Suppose that z is an isolated point of the topological space B. Then there is a closed neighbourhood V of z in X such that $V \cap B = \{z\}$. Let $\{V_n : n = 1,2,\ldots\}$ be a fundamental system of open neighbourhoods of z in X. We set $W_n = V \sim V_n$, $n = 1,2,\ldots$ Since z is a condensation point of X we have that V is not countable. We have that

$$\bigcup_{n=1}^{\infty} W_n = V \sim \{z\}$$

and therefore there is a positive integer r such that W_r is not countable. The topological subspace W_r of X is compact and metrizable. Consequently there is a point u in W_r which is a condesation point of this space. Obviously u is a condensation point of X and, therefore belongs to B. Then

$$u \neq z, \ u \in V \cap B$$

and that is a contradiction. Finally, if y is a point of the closure of B in X and if Z is an open neighbourhood of y, there is a point x in B $\cap$ Z. Then Z is a neighbourhood of x and consequently it is not countable, which implies that y is a condensation point of X.

Therefore B is perfect.

(2) *If X is not countable, the Cantor set D is homeomorphic to a subspace of X.*

Proof. Suppose X endowed with a metric d compatible with its topology. By (1), there are two disjoint closed balls V_0 and V_1 which are not countable and with radii less than 1. Applying (1) to the non-countable metrizable compact space V_0, we can find two non-countable disjoint closed balls V_{00} and V_{01} in V_0 of radii less than $\frac{1}{2}$. Analogously we determine two non-countable disjoint closed ball V_{10} and V_{11} in V_1 of radii less than $\frac{1}{2}$. Proceeding by recurrence, suppose we have obtained in X the non-countable closed ball $V_{p_1 p_2 \ldots pn}$ with $p_j = 0$ or $p_j = 1$, $j = 1,2,\ldots,n$, of radius less than $\frac{1}{n}$. We apply (1) again to that ball, to obtain two non-countable disjoint closed balls

$$V_{p_1 p_2 \cdots p_n 0} \quad \text{and} \quad V_{p_1 p_2 \cdots p_n 1}$$

in $V_{p_1 p_2 \cdots p_n}$ with radii less than $\frac{1}{n+1}$.

If $x = (x_m)$ is any point of D.

$$V_{x_1} \cap V_{x_1 x_2} \cap \cdots \cap V_{x_1 x_2 \cdots x_n} \cdots$$

is a set formed by an unique point Y(x). Then Y is a mapping from D into X, obviously injective. If V is any neighbourhood of Y(x), there is a positive integer q with $V_{x_1 x_2 \ldots x_q} \subset V$. Then $Y^{-1}(V)$ contains the open cylinder $A(x_1,x_2,\ldots,x_q)$ of D. Since x belongs to this cylinder, it follows that Y is continuous in x. Finally, since D is compact, we obtain that Y is an homeomorphism from D onto the subspace Y(D) of X.

(3) *If* X *is non-countable, then* C(D) *is isomorphic to a complemented subspace of* C(X).

Proof. Let Y be the mapping constructed in the former proof. If f belongs to C(Y(D)) we set Lf = f o Y. Then L is an isomorphism from C(Y(D)) onto C(D). Now we apply Borsuk's theorem to obtain a continuous linear extension operator Q from C(Y(D)) into C(X). Then C(D) is isomorphic to Q(C(Y(D)) and this space has the subspace of those functions which vanish in Y(D) as topological complement in C(X). The proof is complete.

For every α of I, let g_α be the function introduced in the former section. The graph of g_α is a segment of line having as initial and final points $(0,\alpha)$ and $(1,\alpha^2)$ respectively. If we fix a point x in the interval I we have that $g_\alpha(x)$ is a function of α which is strictly increasing and takes all the values of the interval I. Consequently, given the point (x,y) of I $\times$ I, there is an unique α in I such that $g_\alpha(x) = y$. We denote by $\psi(x,y)$ this value of α.

(4) ψ *is a continuous mapping from* I$\times$I *onto* I.

Proof. Given any α of I and taking any point (x,y) in the segment of line considered above, we have that $\psi(x,y) = \alpha$. Thus ψ: I $\times$ I $\longrightarrow$ I is a mapping onto. Now we show that ψ is continuous. Take ε such that $0 < \varepsilon < 1$. Then $\psi^{-1}([0,\varepsilon[)$ coincides with the interior in I $\times$ I of the trapezium whose vertices are the points (0,0), (1,0), $(1,\varepsilon^2)$ and $(0,\varepsilon)$. $\psi^{-1}(]1-\varepsilon,1])$ is the interior in I $\times$ I of the trapezium whose vertices are $(0,1-\varepsilon)$, $(1,(1-\varepsilon)^2)$, (1,1) and (0,1). Finally, if α verifies $0 < \varepsilon < 1$ and if we take ε such that $0 < \alpha-\varepsilon < \alpha+\varepsilon < 1$, then $\psi^{-1}(]\alpha-\varepsilon,\alpha+\varepsilon[)$ is the interior in I $\times$ I of the trapezium whose vertices are $(0,\alpha-\varepsilon)$, $(1,(\alpha-\varepsilon)^2)$, $(1,(\alpha+\varepsilon)^2)$ and $(0,\alpha+\varepsilon)$. Consequently, if A is an open set of I, $\psi^{-1}(A)$ is an open set of I$\times$ I. This completes the proof.

If f is any element of C(I), we set Sf to denote the element of C(D$\times$D) such that

$$Sf(u,v) = f(\psi(h(u),h(v))), (u,v) \in D \times D$$

It is obvious that S is an isomorphism from C(I) into C(D $\times$ D) and therefore S(C(I)) is isomorphic to C (I)

If f is any element of C(I) and $0 < \alpha < 1$ we have that

$$T_\alpha(Sf)(x) = f(\psi(h(x),h \circ k(\alpha^2 h(x)+\alpha(1-h(x)))$$

$$=f(\psi(h(x),h \circ k \circ g_\alpha (h(x))$$

for every x of D. Since h o k and the identity mapping on I take the same values save in a countable set and, since g_α is strictly decreasing on I, it follows that

$$f(\psi(h(x),h \circ k\ g_\alpha(h(x)) = f(\psi(h(x),g_\alpha(h(x))) = f(\alpha)$$

save in a countable set of points x. Consequently

$$T(Sf)(\alpha) = \int_D T_\alpha(Sf)\, d\mu = \int_D f(\alpha)\, d\mu = f(\alpha)$$

from where it follows that S o T is a continuous projection from C(D × D) into itself.

(5) C(I) *is isomorphic to a complemented subspace of* C(D).

Proof. C(I) is isomorphic to S(C(I)) which coincides with S o T (C (D × D)). Since S o T is a continuous projection from C(D × D) into itself, it follows that C(I) is isomorphic to a complemented subspace of C(D × D). On the other hand, the mapping ϕ from D × D onto D defined by

$$\phi((x_m),(y_m)) = (x_1,y_1,x_2,y_2,\ldots,x_m,y_m,\ldots),(x_m),(y_m) \in D,$$

is obviously an homeomorphism. If we set

$$\rho f = f \circ \phi,\ f \in C\ (D),$$

is an isomorphism from C(D) onto C(D × D). The proof is complete.

(6) C(I) *is isomorphic to* C(D).

Proof. Since the topological space I is compact, metrizable and non-countable, we apply (3) to obtain that C(D) is isomorphic to a complemented subspace of C(I). The results obtained in §2, Section 5, imply that C(I) is isomorphic to $c_0 \hat{\otimes}_\varepsilon$ C(I). Having in mind (5) and §1,2.(9), the conclusion follows.

(7) *If* X *is non-countable, then* C(I) *is isomorphic to a complemented subspace of* C(X).

Proof. It is an immediate consequence from (3) and (6).

5. THE THEOREM OF MILUTIN. Let X be a compact topological space. We set $M(X)$ to denote the conjugate space of the Banach space $C(X)$. A linear form u on $C(X)$ is said to be positive if for every g of $C(X)$, $g \geq 0$, one has that $u(g) \geq 0$. If s is a point of X, we set δ_x to denote the evaluation of x, i.e., δ_x is the linear form on $C(X)$ such that

$$\langle f,\delta_x\rangle = f(x),\ f \in C(X).$$

Obviously δ_x belongs to $M(X)$.

ϕ is the element of $C(D \times D)$ such that $\phi(x) = 1$, $x \in D\times D$. The integral operator of Milutin T maps ϕ in the element of $C(I)$ which takes the value one in every point of I. Moreover it is obvious that, if f belongs to $C(D\times D)$ and $f \geq 0$, then $Tf \geq 0$. $Z : M(I) \longrightarrow M(D \times D)$ is the transposed mapping of T.

(1) *Given any point* z *of* I *there is a Borel probability* μ_z *in* $D \times D$ *with*

$$Tf(z) = \int_{D\times D} f\, d\mu_z,\ f \in C(D \times D).$$

Proof. Let g be an element of $C(D \times D)$. $Z\delta_z$ is an element of $M(D\times D)$ such that

$$\langle Z\delta_z, g\rangle = \langle\delta_z, Tg\rangle = Tf(z) \geq 0$$

and therefore $Z\delta_z$ is a positive linear form on $C(D \times D)$. We apply Riesz's representation theorem (cf. RUDIN [1], Chapter 6, p. 130-133) to obtain a Borel measure μ_z on $D \times D$ such that

$$\langle f, Z\delta_z\rangle = \int_{D\times D} f\, d\mu_z,\ f \in C(D \times D).$$

Then

$$Tf(z) = \langle Tf,\delta_z\rangle = \langle f, Z\delta_z\rangle = \int_{D\times D} f\, d\mu_z,\ f \in C(D \times D).$$

Finally

$$\mu_z(D \times D) = \int_{D\times D} \phi\, d\mu_z = \langle T\phi,\delta_z\rangle = T\phi(z) = 1$$

from where it follows that μ_z is a probability. The proof is complete

Let A_m and B_m be copies of D and I respectively, m = 1,2,... We set

$$A = \prod_{m=1}^{\infty} A_m \times A_m, \quad B = \prod_{m=1}^{\infty} B_m,$$

$$C_m = \prod_{j=1}^{m} A_j \times A_j, \quad E_m = \prod_{j=m+1}^{\infty} A_j \times A_j$$

For every positive integer m we set $T_m : C(A_m \times A_m) \longrightarrow C(B_m)$ to denote the integral operator of Milutin. $S_m : C(B_m) \longrightarrow C(A_m \times A_m)$ is the mapping S defined in the former section. If z_m belongs to B_m we apply (1) to obtain a Borel probability μz_m on $A_m \times A_m$ such that

$$T_m f(z_m) = \int_{A_m \times A_m} f \, d\mu z_m, \quad f \in C(A_m \times A_m)$$

If $z = (z_m)$, we set μ_z to denote the probability $\bigotimes_{m=1}^{\infty} \mu z_m$ on A. For every positive integer r we write

$$\mu_z^{\,r} = \bigotimes_{m=r+1}^{\infty} \mu z_m$$

Then

$$\mu_z = \mu z_1 \otimes \mu z_2 \otimes \ldots \otimes \mu z_r \otimes \mu_z^{\,r}$$

If $\psi : I \times I \longrightarrow I$ is the mapping defined in the former section, we set ψ_m instead of ψ when $I = B_m$, $h_m : A_m \longrightarrow B_m$ is the Cantor mapping. Denote by ϕ_m the function defined on $A_m \times A_m$ which takes the value one in every point.

Let X_m be a compact Hausdorff topological space, m = 1,2,... We set $X = \prod_{m=1}^{\infty} X_m$. If

$$f_j \in C(X_j), \; j = 1,2,\ldots, r,$$

we write

$$(2) \qquad f_1 \otimes f_2 \otimes \ldots \otimes f_r$$

to denote the element of $C(X)$ such that, if $x = (x_m)$ is any point of this space,

$$f_1 \otimes f_2 \otimes \ldots \otimes f_r(x) = f_1(x_1)f_2(x_2)\ldots f_r(x_r)$$

Let E be the linear subspace of $C(X)$ generated by all the functions of the form (2).

(3) E *is a subalgebra of* $C(X)$ *containing the constant functions*

Proof. Let r and s be positive integers with $r \leqslant s$. If

$$f_j \in C(X_j),\ j = 1,2,\ldots,\ r,\ g_j \in C(X_j),\ j = 1,2,\ldots,s,$$

we have that

$$(f_1 \otimes f_2 \otimes \ldots \otimes f_r) \times (g_1 \otimes g_2 \otimes \ldots . \otimes g_s)$$

$$= (f_1 g_1) \otimes (f_2\, g_2) \otimes \ldots \otimes (f_r g_r) \otimes g_{r+1} \otimes \ldots \otimes g_s$$

from where it follows easily that E is a subalgebra of $C(X)$. On the other hand, if b belongs to K, the function defined on X taking the value b in every point belongs obviously to E. The proof is complete.

(4) E *separates the point of* X.

Proof. Let $x = (x_m)$ and $y = (y_m)$ be different points of X. We find a positive integer w with $x_r \neq y_r$. Let f_r be an element of $C(X_r)$ such that $f_r(x_r) = 0$, $f_r(y_r) = 1$. We set β_j to denote the function defined on X_j which takes the value one in every point, $j = 1,2,\ldots$ If

$$f = \beta_1 \otimes \beta_2 \otimes \ldots \otimes \beta_{r-1} \otimes f_r$$

we have that

$$f(x) = \beta_1(x_1)\beta_2(x_2)\ldots\beta_{r-1}(x_{r-1})f_r(x_r) = 0,$$

$$f(y) = \beta_1(y_1)\beta_2(y_2)\ldots\beta_{r-1}(y_{r-1})f_r(y_r) = 1$$

and the conclusion follows.

(5) *If* f *belongs to* E, *the complex conjugate functions of* E *belongs also to* E.

Proof. If $f_j \in C(X_j)$, $j = 1,2,\ldots,r$, and if g_j is the complex conjugate function of f_j, it is obvious that $g_1 \otimes g_2 \otimes \ldots \otimes g_r$ is the complex conjugate function of $f_1 \otimes f_2 \otimes \ldots \otimes f_r$ and the conclusion follows.

(6) E *is dense in* $C(X)$.

Proof. Since (3), (4) and (5) are verified, we apply Stone-Weierstrass's theorem to obtain the conclusion.

If f belongs to $C(A)$, let Vf be the function defined on B such that

$$Vf(z) = \int_A f \, d\mu_z, \ z \in B.$$

We have that

$$|Vf(z)| \leq \int_A ||f|| \, d\mu_z = ||f||$$

and consequently $V: C(A) \longrightarrow A(B)$ is a continuous linear operator.

(7) *If*

$$f_j \in C(A_j \times A_j), \ j = 1,2,\ldots,r,$$

Then

$$V(f_1 \otimes f_2 \otimes \ldots \otimes f_r) = T_1 f_1 \otimes T_2 f_2 \otimes \ldots \otimes T_r f_r$$

Proof. If $z = (z_m)$ belongs to B, we apply Fubini's theorem to obtain that

$$V(f_1 \otimes f_2 \otimes \ldots \otimes f_r)(z) = \int_A f_1 \otimes f_2 \otimes \ldots \otimes f_r \, d\mu_z$$

$$= \int_A f_1 \otimes f_2 \otimes \ldots \otimes f_r \otimes \phi_{r+1} \, d\mu_z = \int_{C_r} f_1 \otimes f_2 \otimes$$

$$\ldots \otimes f_r \, d\mu z_1 \otimes \mu z_2 \otimes \ldots \otimes \mu z_r \int_{E_r} \phi_{r+1} \, d\mu_z{}^r$$

$$= \int_{C_r} f_1 \otimes f_2 \otimes \ldots \otimes f_r \, d\mu_{z_1} \otimes \mu_{z_2} \otimes \ldots \otimes \mu_{z_r} =$$

$$= \int_{A_1 \times A_1} f_1 \, d\mu_{z_1} \ldots \int_{A_r \times A_r} f_r \, d\mu_{z_r}$$

$$= T_1 f_1(z_1) T_2 f_2(z_2) \ldots T_r f_r(z_r) = T_1 f_1 \otimes T_2 f_2 \otimes \ldots \otimes T_r f_r(z)$$

(8) $V(C(A))$ *is contained in* $C(B)$.

Proof. In (6), we take $X_m = A_m \times A_m$, $m = 1,2,\ldots$. Let f be any element of $C(A)$. We find a sequence (f_m) in E converging to f in $C(A)$. According to (7), Vf_m belongs to $C(B)$, $m = 1,2,\ldots$ Since V is continuous, we have that (Vf_m) converges to Vf in $A(B)$. Then Vf belongs to $C(B)$.

For every f of $C(B)$, let Wf be the element of $C(A)$ such that if $z = ((x_m,y_m))$, $x_m,y_m \in A_m$, $m = 1,2,\ldots$, then

$$Wf(z) = f((\psi_m(h_m(x_m),h_m(y_m)))).$$

It is obvious that W is an isomorphism from $C(B)$ into $C(A)$ and therefore $C(B)$ is isomorphic to $W(C(B))$.

Now take $X_m = B_m$, $m = 1,2,\ldots$,in (6). If

$$f_j \in C(B_j),\ j = 1,2,\ldots,r,\ z = ((x_m,y_m)) \in A,$$

then

$$W(f_1 \otimes f_2 \otimes \ldots \otimes f_r)(z) = f_1 \otimes f_2 \otimes \ldots \otimes f_r((\psi_m(h_m(x_m),$$

$$h_m(y_m)))) = f_1(\psi_1(h_1(x_1),h_1(y_1))) \times \ldots \times f_r(\psi_r(h_x(x_r),h_r(y_r)))$$

$$= S_1f_1(x_1,y_1)\times\ldots\times S_rf_r(x_r,y_r) = S_1f_1 \otimes S_2f_2 \otimes \ldots \otimes S_rf_r(z)$$

and therefore

$$W(f_1 \otimes f_2 \otimes \ldots \otimes f_r) = S_1f_1 \otimes S_2f_2 \otimes \ldots . \otimes S_rf_r$$

from where it follows

$$V \circ W(f_1 \otimes f_2 \otimes \ldots \otimes f_r)$$

$$= V(S_1f_1 \otimes S_2f_2 \otimes \ldots \otimes S_rf_r) = f_1 \otimes f_2 \otimes \ldots \otimes f_r,$$

according to (7). Consequently $V \circ W$ is the identity mapping on E. If g is any element of $C(B)$ we find a sequence (g_m) in E converging to g in $C(B)$. Then

$$V \circ W(g) = \lim V \circ W(g_m) = \lim g_m = g.$$

from where it follows that $W \circ V$ is a continuos projection from $C(A)$ into itself.

(9) $C(B)$ *is isomorphic to a complemented subspace of* $C(D)$.

Proof. $C(B)$ is isomorphic to $W(C(B))$ which coincides $W \circ V(C(A))$.

Since W o V is a continuos projection from $C(A)$ into itself, it follows that $C(B)$ is isomorphic to a complemented subspace of $C(A)$. On the other hand, since A_m is homeomorphic to $\{0,1\}^N$, it follows that A is homeomorphic to $(\{0,1\}^N \times \{0,1\}^N)^N$ which in turn is homeomorphic to $\{0,1\}^N = D$. If Y is an homeomorphism from A onto D and if we set

$$\rho f = f \circ Y, \ f \in C(A),$$

then ρ is an isomorphism from $C(D)$ onto $C(A)$. Consequently $C(B)$ is isomorphic to a complemented subspace of $C(D)$.

(10) $C(B)$ *is isomorphic to a complemented subspace of* $C(I)$.

Proof. It is an immediate consequence from 4.(6) and (9).

Now we arrive to the fundamental result of this section which is Milutin's theorem.

(11) *Let* X *be a metrizable compact topological space. If* X *is non-countable, then* $C(X)$ *is isomorphic to* $C(I)$.

Proof. X is homeomorphic to a closed subspace F of B (cf. KELLEY [1], Chapter 4, p. 125). Applying Borsuk's theorem, we obtain a continuous linear extension operator $\beta: C(F) \longrightarrow C(B)$. Then $C(X)$ is isomorphic to the subspace $\beta(C(F))$ of $C(B)$. The subspace of $C(B)$ of all those functions vanishing in F is a topological complement of $\beta(C(F))$. Therefore $C(X)$ is isomorphic to a complemented subspace of $C(B)$. According to (10), $C(X)$ is isomorphic to a complemented subspace of $C(I)$. By 4.(7), $C(I)$ is isomorphic to a complemented subspace of $C(X)$. Since $c_0 \hat{\otimes}_\varepsilon C(I)$ is isomorphic to $C(I)$, we apply §1.2(9) to reach the conclusion.

Result (11) can be found in MILUTIN [1]. For other results related with Milutin's theorem we refer to PELZCINSKY [1].

6. SPACES OF CONTINUOUS FUNCTIONS DEFINED ON CERTAIN K_R-SPACES. Let X be a topological space. X is said to be submetrizable if there is a topology T on X, which is coarser than the initial topology, such that $X[T]$ is metrizable. X is a k_R-space if every function $f : X \longrightarrow K$, whose restriction to every compact of X is continuous, is continuous. We recall the definition of

$\mathcal{C}(X)$: linear space over K of all the K-valued continuous functions defined on X, endowed with the topology of the uniform convergence on every compact subset of X.

$\mathcal{C}^*(X)$ is the linear subspace of $\mathcal{C}(X)$ of all the bounded functions endowed with the topology of uniform convergence.

(1) *Let X be a submetrizable topological space with an increasing fundamental sequence of compact subsets* (H_m) *such that* $H_{m+1} \sim H_m$ *is non-countable,* $m = 1,2,\dots$ *If X is a* k_R*-space, then* $C(X)$ *is isomorphic to* $C(I)^N$.

Proof. We can suppose H_1 non-countable. Let $\mathcal{T}$ be a topology on X, coarser than initial one, such that $X[\mathcal{T}]$ is metrizable. In every compact subset of X, $\mathcal{T}$ coincides with the topology of X. For every positive integer m we apply Borsuk's theoren to obtain a continuoś linear extension operator $Z_m : \mathcal{C}(H_m) \longrightarrow \mathcal{C}^*(X[\mathcal{T}])$. Let f be any element of $\mathcal{C}(X)$. We set f_1 to denote the restriction of f to H_1. Proceeding by recurrence, suppose that we have obtained $f_1, f_2,\dots,f_m$ for a positive integer m. Then f_{m+1} is the restriction to H_{m+1} of

$$f-Z_1f_1-Z_2f_2-\dots-Z_mf_m$$

We set $E_1 = \mathcal{C}(H_1)$ and, for every positive integer m, let E_{m+1} be the subspace of $\mathcal{C}(H_{m+1})$ of all those functions vanishing in H_m. We write E instead of $\prod_{m=1}^{\infty} E_m$.

Let $Z : \mathcal{C}(X) \longrightarrow E$ be the mapping defined by

$$Zf = (f_1,f_2,\dots,f_m,\dots),\ f \in \mathcal{C}(X)$$

It is obvious that Z is linear, injective and continuos. On the other hand, given the element $(g_1,g_2,\dots,g_m,\dots)$ of E, if $g = \Sigma Z_m g_m$, then g belongs to $\mathcal{C}(X)$ and $Zg = (g_1,g_2,\dots,g_m,\dots)$. Therefore Z is onto. We apply the open mapping theorem to obtain that Z is an isomorphism.

According to Milutin's theorem E_1 is isomorphic to $\mathcal{C}(I)$. Let m be an integer larger than one. If f belongs to $\mathcal{C}(H_m)$, let $T_m f$ be the restriction of $Z_m f$ to H_{m+1}. Then $T_m : \mathcal{C}(H_m) \longrightarrow \mathcal{C}(H_{m+1})$ is a continuous linear extension operator. Consequently E_{m+1} has $T_m(\mathcal{C}(H_m))$ as topological complement in $\mathcal{C}(H_{m+1})$ and, according to Milutin's theorem again, it follows that E_{m+1}

is isomorphic to a complemented subspace of $C(I)$. In the topological space H_{m+1} we find a closed neighbourhood B_x of x which does not meet H_m for every x of $H_{m+1} \sim H_m$. Since $H_{m+1} \sim H_m$ is metrizable and separable, there is a countable subset A in this space such that

$$\cup \{B_x : x \in A\} = H_{m+1} \sim H_m.$$

Consequently there is a point z in A such that B_z is non-countable. Let k be a function defined on H_{m+1} which is real and continuous and takes the value zero in H_m and the value one in B_z. We apply Borsuk's theorem to obtain a continuous linear extension operator $Y_m : C(B_z) \longrightarrow C(H_{m+1})$. For every f of $C(B_z)$ we set

$$L_m f = k\, Y_m f.$$

Then $L_m(C(B_z))$ is a subspace of E_{m+1} isomorphic to $C(B_z)$ and, by virtue of Milutin's theorem, isomorphic to $C(I)$. On the other hand, $L_m(C(B_z))$ has in E_{m+1} the subspace of all those functions vanishing in B_z as topological complement. Consequently $C(I)$ is isomorphic to a complemented subspace of E_{m+1}. Since $C(I)$ is isomorphic to $c_0 \hat{\otimes}_\varepsilon C(I)$, we apply § 2,1.(5) to obtain that E_{m+1} is isomorphic to $C(I)$. Now the conclusion follows.

(2) *Let $F \neq \{0\}$ be a separable Fréchet space. If G is the topological dual of F, endowed with the topology of the uniform convergence on every compact subset of F, then $C(G)$ is isomorphic to $C(I)^N$.*

Prof. Since F is separable, there is a dense subspace P in F with countable algebraic basis. Then $\sigma(G,P)$ is a metrizable topology on G, coarser than the initial one. Let (U_m) be a fundamental system of neighbourhoods of the origin in F, which we suppose absolutely convex and closed, such that

$$U_{m+1} \subset \overset{\circ}{U}_m, \quad m = 1,2,\ldots$$

Let V_m be the polar set of U_m in G. Then V_m is $\sigma(G,F)$-compact in G (cf. KÖTHE [1], Chapter Four, §21, Section 7). Since F is a Fréchet space, F is barrelled and consequently (V_m) is an increasing fundamental sequence of compact subsets of G. According to Banach-Dieudonné's theorem, G is a k_R-space. (cf. KÖTHE [1], Chapter Four, §21, Section 10). Given a positive integer m we find a vector x_m in $\overset{\circ}{U}_m \sim U_{m+1}$. Then there is an element u_m in V_{m+1} such that $|u_m(x_m)|>1$. Consequently there is a real number δ, with $0<\delta<1$, such that

$$\lambda\,|u_m(x_m)| > 1,\ \lambda \in\,]\,1-\delta,\ 1].$$

Therefore

$$\lambda u_m \in V_{m+1} \sim V_m,\ \lambda \in\,]\,1-\delta,\ 1],$$

and thus the set $V_{m+1} \sim V_m$ is not countable. We apply (1) to reach the conclusion.

Results (1) and (2) can be found in VALDIVIA [28].

7. SPACES OF CONTINUOUS FUNCTIONS WITH COMPACT SUPPORT DEFINED ON CERTAIN LOCALLY COMPACT SPACES. Let H be a compact subset of a topological space X. We denote by $C_o(H)$ the subspace of $C(X)$ of all those functions with support contained in H. Let L be the family of all compact subset of X. Denote by $H(X)$ the linear subspace of $C(X)$ of all those functions with compact sup-pport. We suppose this space endowed with the locally convex topology such that $H(X)$ coincides with the inductive limit of the family of Banach spaces

$$\{C_o(H) : H \in L\}$$

(1) *Let* H *be a compact and metrizable subspace of* X. *If* $\overset{\circ}{H}$ *is non-countable, then* C(I) *is isomorphic to a complemented subspace of* $C_o(H)$.

Proof. Since $\overset{\circ}{H}$ is metrizable and separable, its topology has a countable basis. Therefore there is a non-countable compact subset A in $\overset{\circ}{H}$. Since A and $H \sim \overset{\circ}{H}$ are disjoint closed subset in H there is a continuous real function ϕ on H taking the value one on A and zero on $H \sim \overset{\circ}{H}$. Let $Z : C(A) \longrightarrow C(H)$ be a continuous linear extension operator. For every f of $C(A)$, we set

$$Yf(x) = \phi(x)Zf(x),\ x \in H,$$

$$Yf(x) = 0,\ x \in X \sim H.$$

Then $Y : C(A) \longrightarrow C_o(H)$ is a continuos linear extension operator. According to Milutin's theorem we have that

$$C(I) \simeq C(A) \simeq Y(C(A))$$

Finally $Y(C(A))$ has the subspace of all those functions of $C_0(H)$ vanishing in A as topological complement in $C_0(H)$. The proof is complete.

(2) *Let* H *be a metrizable and compact subset of* X. *If* H *is non-countable, then* $C_0(H)$ *is isomorphic to a complemented subspace of* $C(I)$.

Proof. For every element f of $C_0(H)$, we set Yf to denote the restriction of f to H. Then Y is an isomorphism from $C_0(H)$ into $C(H)$. We set E to denote the subspace $Y(C_0(H))$ of $C(H)$.

If $H \sim \overset{\circ}{H}$ is a void set, then

$$C_0(H) \simeq E = C(H) \simeq C(I)$$

If $H \sim \overset{\circ}{H}$ is non-void, let $Z : C(H \sim \overset{\circ}{H}) \longrightarrow C(H)$ be a continuous linear extension operator. Then E is a subspace of $C(H)$, having $Z(C(H \sim \overset{\circ}{H}))$ as topological complement. The conclusion follows easily.

(3) *Let* H *be a metrizable and compact subset of* X. *If* $\overset{\circ}{H}$ *is non-countable, then* $C(I)$ *is isomorphic to* $C_0(H)$.

Prof. Since $c_0 \hat{\otimes}_\varepsilon C(I)$ is isomorphic to $C(I)$, it is enough to apply (1), (2) and §2.1,(5) to obtain the conclusion.

In what is left in this chapter we suppose X Hausdorff and having a sequence (H_m) of metrizable compact subsets such that the following conditions are satisfied:

a) The set $H_{m+1} \sim H_m$ is non-countable, $m = 1,2,\dots$;

b) $H_m \subset \overset{\circ}{H}_{m+1}$, $m = 1,2,\dots$;

c) $\bigcup_{m=1}^{\infty} H_m = X$

Set $H_0 = \emptyset$ and suppose, without loss of generality, that $\overset{\circ}{H}_m \sim H_{m-1}$ is non-countable, $m = 1,2,\dots$ We write E_m and E instead of $C_0(H_{m+1} \sim \overset{\circ}{H}_{m-1})$, $m=1,2,\dots$, and $\bigoplus_{m=1}^{\infty} E_m$ respectively. Let $T : E \longrightarrow \mathcal{H}(X)$ be the mapping defined by

$$T(f_1,f_2,\dots,f_m,\dots) = \Sigma f_m, (f_1,f_2,\dots,f_m,\dots) \in E.$$

It is immediate to check that T is linear and continuous. For every positive integer m, let d_m be a metric on H_{m+2} compatible with its topology. We set

$$\psi'_m(x) = d_m(x, H_{m-1} \cup (H_{m+2} \sim \overset{\circ}{H}_{m+1})), \; x \in H_{m+2},$$

$$\psi_m(x) = 0, \; x \in X \sim H_{m+2}.$$

Since $\overset{\circ}{H}_{m+2} \supset H_{m+1}$, ψ_m is a continuous function on X. If x is any point of X, let m be the first positive integer with $x \in H_m$. Then x belongs to $\overset{\circ}{H}_{m+1}$ and therefore $\psi_m(X) > 0$. On the other hand, $\psi_m(z) = 0$ for every z of H_{m-1}. Consequently $\Sigma\, \psi_m$ is a continuous function on X such that $\Sigma\, \psi_m(y) > 0$ for every y of X. We set

$$\phi_r = \frac{\psi_r}{\Sigma\, \psi_m}, \; r = 1,2,\ldots$$

For every f of $H(X)$, we write

$$Zf = (f\phi_1, f\phi_2, \ldots, f\phi_m, \ldots).$$

Then $Z : H(X) \longrightarrow E$ is a continuous linear operator. Since

$$T \circ Zf = \Sigma f\phi_m = f\, \Sigma\, \phi_m = f$$

it follows that Z is an isomorphism into. Consequently $Z(H(X))$ is isomorphic to $H(X)$.

(4) $H(X)$ *is isomorphic to a complemented subspace of* $C(I)^{(N)}$.

Proof. The mapping $Z \circ T$ is a continuous projection from E into itself such that $Z \circ T(E)$ coincides with $Z(H(X))$. Therefore $H(X)$ is isomorphic to the complemented subspace $Z(H(X))$ of E. According to (3), E_m is isomorphic to $C(I)$, m = 1,2,..., from where the conclusion follows.

For every positive integer m, we find a non-countable compact subset A_m in $\overset{\circ}{H}_m \sim H_{m-1}$. Proceeding as in the proof of (1), we obtain a continuous linear extension operator Y_m from $C(A_m)$ into $C_0(H_m \sim \overset{\circ}{H}_{m-1})$. We write F_m instead of $C(A_{2m+1})$. We set $F = \bigoplus_{m=1}^{\infty} F_m$. Let S be the mapping from F into $H(X)$

defined by

$$S(f_1,f_2,\ldots,f_m,\ldots) = \Sigma\, Y_{2m+1} f_m,\ (f_1,f_2,\ldots,f_m,\ldots) \in F.$$

It is immediate that S is an isomorphism from F into $H(X)$.

(5) $C(I)^{(N)}$ *is isomorphic to a complemented subspace of* $H(X)$.

Proof. According to Milutin's theorem, F_m is isomorphic to $C(I)$, m=1,2,..., and therefore S(F) is isomorphic to $C(I)^{(N)}$. Since the functions of $H(X)$ which vanish in $\overset{\infty}{\cup} A_{2m+1}$ constitute a complemented subspace of S(E), the conclusion follows.

Now we arrive to the fundamental result of this section.

(6) *Let* X *be a Hausdorff topological space which has a sequence* (H_m) *of metrizable compact subset verifying the following conditions:*

a) $H_{m+1} \sim H_m$ *is non-countable*, m = 1,2,...;

b) $H_m \subset \overset{\circ}{H}_{m+1}$, m = 1,2,...;

c) $\bigcup_{m=1}^{\infty} H_m = X$.

Then $H(X)$ *is isomorphic to* $C(I)^{(N)}$.

Proof. It is an immediate consequence from (4), (5), and §1,2.(2).

Result (6) can be found in VALDIVIA [29].

REFERENCES

ADASH,N.: [1] Tonelierte Räume und zwei Sätze von Banach. Math. Ann. 186, 209 - 214 (1970).

- [2] Vollstandigkeit und der Graphensatz. J. reine angew. Math. 249, 217 - 220 (1971).

AMEMIYA, I., u. KOMURA,Y.: [1] Uber nicht - vollstandige Montelraume. Math. Ann. 177, 273 - 220 (1968).

ARENS, R.: [1] Extensions of functions on fully normal spaces. Pacific J. Math. 2, 11 - 12 (1952).

ARIAS DE REYNA, J.: [1] Dense hyperplanes of first cathegory. Math. Ann. 249, 111 - 114 (1980).

BIERSTEDT, K., u. MEISE, R.: [1] Bemerkungen über die Approximationseigens_ chaft lokalkonvexer Funktionenräume. Math. Ann. 209, 99 - 107 (1974).

BORSUK,K.:[1]Über isomorphic der Funktionalraume. Bull. Int. Acad. Pol. Sci. 1 - 10 (1933).

BOURBAKI, N.: [1] Éléments de mathématique, Livre V, Espaces vectoriels topologiques, 2 Vols. Act. Sci. et Ind. Vols. 1189, 1229 (1953, 1955).

- [2] Éléments de mathématique, Livre III, Topologie Génerale, Cha_ pitre 9. Act. Sci. et Ind. 1045 (1958).

CHOQUET, G.: [1] Theory of capacities. Ann. Inst. Fourier 5, 131 - 295, (1953).

CORSON, H. H.: [1] The weak topology of a Banach space. Trans. Amer. Math. Soc. 101, 1 - 15 (1961)

CROFTS, G.: [1] Concerning perfect Fréchet spaces and diagonal transformations. Math. Ann. 182, 67 - 76 (1969).

DAVIS, W. J., FIGIEL, T., JOHNSON, W. B., and PELCZYNSKI : [1] Factoring weakly compact operators. Journal of Functional Analysis 17, 311 - 327 (1974).

DE WILDE, M.: [1] Réseaux dans les espaces linéaires à semi - normes. Mém. Soc. R. Sc. Liège 18, 2 (1969).

- [2] Closed graph theorems and webbed spaces. Pitman, London, San Francisco, Melbourne, 1978.

- [3] Critères de densité et séparation dans les limites projectives et inductives dénombrables. Bull. Soc. R. Sc. Liège, 41, 3 - 4,

155 - 162 (1972).

DE WILDE, M., et SUNYACH, C.: [1] Un théorème de selection pour des applications à graphe borelién. C.R. Acad. Sc. Paris 269, 273 - 274 (1969).

DE WILDE, M., and HOUET, C.: [1] On increasing sequences of absolutely convex sets in locally convex spaces. Math. Ann. 192, 257 - 261 (1971)

DIEUDONNÉ, J.: [1] Sur les propriétés de permanence de certains espaces vectoriels topologiques. Ann. Soc. Polon. Math. 25, 50 - 55 (1952).

\- [2] Sur les espaces de Montel metrizables. C. R. Acad. Sc. Paris 238, 194 - 195 (1954).

DIEUDONNÉ, J., et GOMES, A. P.: [1] Sur certains espaces vectoriels topologiques. C. R. Acad. Sc. Paris 230, 1129 - 1130 (1950).

DIEROLF, P.: [1] L'espace $\mathring{\mathcal{B}}$ (Ω) et les distributions sommables. C. R. Acad. Sc. Paris 288, 197 - 199 (1979).

DIEROLF, P., and VOIGT, J.: [1] Calculation of the bidual for some function spaces. Integrable distributions. Math. Ann. 253, 63 - 87 (1980)

DIEROLF, P., DIEROLF, S., and DREWNOWSKI, L.: [1] Remarks and examples concerning unordered Baire - like and ultrabarrelled spaces. Coll. Math. 39, 109 - 116 (1978).

DIEROLF, S., et LURJE, P.: [1] Deux exemples concernant des espaces (ultra)bornologiques. C. R. Acad. Sc. Paris 282, 1347 - 1350 (1976).

DUBINSKI, E.: [1] Echelon spaces of order ∞. Proc. Amer. Math. Soc. 16, 1178 - 1183 (1965).

\- [2] Perfect Fréchet spaces. Math. Ann. 174, 186 - 194 (1967).

EBERHARDT, V.: [1] Durch Graphensätze definierte lokalkonvexe Räume. Diss. München 1972

\- [2] Der Graphensatz von A. P. und W. Robertson für s - Räume. Manusc. Math. 4, 255 - 262 (1970).

FENSKE, CH., u SCHOCK, E.: [1] Uber die diametrale Dimension von lokalconvexen Räume. Gesellsch. Math. Datenverarbeitung Bonn 1969.

FLEISSNER, W.G., and KUNEN, K.: [1] Barely Baire spaces. Fund. Math., 51, 229 - 240 (1978).

FLORET, K.: [1] Folgenretraktive Sequenze mit kompakten abbildungen. J. reine angew. Math. 259, 65 - 85 (1973).

FROLIK, Z.: [1] On the descriptive theory of sets. Czech. Math. J. 13(88) 335 - 359 (1963).

GARLING, D. J. H.: [1] A generalized form of inductive limit topology for vector spaces. Proc. London Math. Soc. 14, 1 - 28 (1964).

GILLMANN, L., and JERISON, M.: [1] Rings of Continuous functions. Van Nostrand, Princeton 1960.

GROTHENDIECK, A.: [1] Produits tensoriels topologiques et espaces nucléaires. Mem. Amer. Math. Soc. No. 16 (1955).

- [2] Sur les espaces (F) et (DF). Summa Brasil. Math. 3, 57 - 123 (1954)

HALMOS, P.R.: [1] Measure theory. Van Nostrand, Princeton 1950.

HESTENESS, M. R.: [1] Extension of the range of a differentiable function. Duke Math. J. 8, 183 - 192 (1941).

HORVÁTH, J.: [1] Topological vector spaces and distributions I. Addison-- Wesely, Readyng, Massachusetts 1966.

- [2] Locally convex spaces. Lecture Notes in Mathematics, 331. Summer school on topological vector spaces, 41 - 83, Berlin. Heidelberg. New York 1973.

IYAHEN, S.O.: [1] On the closed graph theorem. Israel J. Math. 10, 96 - 105 (1971).

JAMES, R.C.: [1] Bases and reflexivity of Banach spaces. Ann. of Math. 52, 518 - 527 (1950).

- [2] Separable conjugate spaces. Pacific J. Math. 10, 563 - 571 (1960).

JAMESON, G. J. O.: [1] Topology and normed spaces. Chapmann and Hall, London 1974.

JARCHOW, H.: [1] Locally convex spaces. B.G. Teubner Stuttgart 1981.

- [2] Die Universalitat des Raumes c_o für die Klasse der Schwartz - Raüme. Math. Ann. 203, 211 - 214 (1973).

KALTON, N. J.: |1| Some forms on the closed graph theorem. Proc Cambridge Phil. Soc. 70, 401 - 408 (1971).

KELLEY, J.L.: [1] General Topology. Springer - Verlag, New York, Heidel - berg, Berlin 1955.

KÖTHE, G.: [1] Topological Vector Spaces I. Springer-Verlag, New York Heidelberg Berlin 1969.

- [2] Topological Vector Spaces II. Springer - Verlag, New York Heidelberg Berlin 1979.

- [3] Die Stufenräume, eine einfache Klasse linearer vollkommener Raüme. Math. Z. 51, 317 - 345 (1948).

KŌMURA, T.: [1] On linear topological spaces. Kumamoto J. Science, Series A, 5, Nr. 3, 148 - 157 (1962).

- [2] Some examples on linear topological spaces. Math. Ann. 153, 150 - 162 (1964).

KŌMURA, T., et KÒMURA, Y.: [1] Sur les espaces parfaits de suites et leurs généralisations. J. Math. Soc. Japan 15, 319 - 338 (1963).

LEVIN, M. and SAXON, S.: [1] A note on the inheritance of properties of locally convex spaces by subspaces of countable codimension. Proc. Amer. Math. Soc. 29, 97 - 102 (1971).

LICHTESTEIN, L.: [1] Eine elementare Bemerkung zur reellen Analysis. Math. Z. 30, 794 - 795 (1929).

LINDENSTRAUSS, J.: [1] On Jame's paper "Separable conjugate spaces". Israel J. Math. 9, 279 - 284 (1971).

MAcINTOSH, A.: [1] On the closed graph theorem. Proc. Amer. Math. Soc. 20 397 - 404 (1969).

MAHOWALD, M.: [1] Barrelled spaces and the closed graph theorem J. London Math. Soc. 36, 108 - 110 (1969).

MARQUINA, A.: [1] A note on the closed graph theorem. Archiv. Math. 28, 82 - 85 (1977).

MARTINEAU, A.: [1] Sur des théorèmes de S. Banach et L. Schwartz concernant le graphé fermé. Studia Math. 30, 43 - 51 (1968).

- [2] Sur le theoreme de graphé fermé. C.R. Acad. Sci. Paris 263, 870 - 871 (1966).

MILUTIN, A. A.: [1] Isomorphisms of spaces of continuous functions on compacta of power continuum. Tieoria Funct. (Kharkov), 2, 150 - 156 (1966) (Russian).

MITIAGIN, B.S.: [1] Approximative dimension and bases in nuclear spaces. Uspehi Math. Nauk 164, 63 - 132 (1961) (Russian).

NEUS, H.: [1] Uber die Regularitätsbegriffe induktiver lokalkonvexer Sequenzen. Manuscr. Math. 25, 135 - 145 (1978).

OGRODZKA, Z.: [1] On simultaneous extension of infinitely differentiable functions. Studia Math. 28, 191 - 207 (1967).

OXTOBY, J.C.: [1] Cartesian products of Baire spaces. Fund. Math. 49, 157 - 166 (1961).

PFISTER, H.: [1] Bemerkungen zum Satz über die Separabilität der Fréchet - Montel - Räume. Archiv Math. 28, 86 - 92 (1976).

PELCZYNSKI, A.: [1] Linear extensions, linear averaging and their application to linear topological calssification of spaces of continuous functions. Rozprawy Mathematycne, 58 (1968).

PÉREZ CARRERAS, P. and BONET, J.: [1] Espacios tonelados. Univ. Sevilla 1980.

PERSSON, A: [1] A remark on the closed graph theorem in locally convex spa spaces. Math. Scand. 19, 54 - 58 (1966).

PIETSCH, A.: [1] Nuclear locally convex spaces. Springer - Verlag, Berlin Heidelberg New York, 1972.

- [2] Zur Theorie der topologischen Tensorprodukte. Math. Nach. 25, 19 - 31 (1963).

[3] Verallgemeinerte vollkommene Folgenräume. Berlin 1962.

POWELL, M.: [1] On Komura's closed graph theorem. Trans. Amer. Math. Soc. 211, 391 - 426 (1975).

PTÁK, V.: [1] Completeness and the open mapping theorem. Bull. Soc. Math. France 86, 41 - 74 (1958).

RAÍKOV, D. A.: [1] Double closed graph theorem for topological linear spaces. Siberian Math. J. (Trans. for Russian) 7, 2, 287 - 300 (1966).

- [2] Completeness in locally convex spaces. Uspeki Math. Nauk. 14. 1, 223 - 229 (1958) (Russian).

RETAKH, V.S.: [1] Subspaces of a countable inductive limit. Soviet Math. Dokl. 11, 1384 - 1386 (1970).

ROBERTSON, W.: [1] Completions of topological vector spaces. Proc. London Math. Soc. 8, 241 - 257 (1958).

ROBERTSON, A. P., and ROBERTSON, W. J.: [1] Topological vector spaces. Cambridge Tracts 53 (1964).

\- [2] On the closed graph theorem. Proc. Glasgow Math. Ann. Ass. 3, 9 - 12 (1956).

ROELCKE, W.: [1] On the finest locally convex topology agreeing with a given topology on a sequence of absolutely convex sets. Math. Ann. 198, 57 - 80 (1972).

ROGERS, C. A.: [1] Analytic Sets in Hausdorff spaces. Mathematika 11, 1 - 8 (1968).

RUCKLE, W. H., and SWART, J.: [1] Schwartz topologies on sequence spaces. Math. Ann. 230, 91 - 95 (1977).

RUDIN, W.: [1] Homogeneity problems in the theory of Čech compactifications. Duke Math. J. 23, 409 - 419 (1956).

\- [2] Real and amplex Analysis. MacGraw - Hill, New York 1970.

RUESS, W.: [1] Generalized inductive limit topologies and barrellednes properties. Pacific J. Math. 63, 449 - 516 (1976).

SAXON, S. A.: [1] Nuclear and product spaces, Baire - like spaces and the strongest locally convex topology. Math. Ann. 197, 87 - 106 (1972).

SCHAEFER, H. H.: [1] Topological Vector Spaces. Macmillan, New York 1966.

SCHWARTZ, L.: [1] Théorie des distributions. Hermann, Paris 1966.

[2] Sur le théorème du graphe fermé. C. R. Acad. Sc. Paris 263, 602 - 605 (1966).

SEELEY, R. T.: [1] Extension of C^∞ functions defined in a half - space. Proc. Amer. Mat. Soc. 15, 625 - 626 (1964).

TODD, A., and SAXON, S.: [1] A property of locally convex Baire spaces. Math. Ann. 206, 23 - 35 (1973).

TSIRULNIKOV, B.: [1] Sur les topologies tonelée et bornologique associées à une topologie d'un espace localmente convexe. C. R. Acad. Sci. Paris 288, 821 - 822 (1979).

VALDIVIA, M.: [1] Sucesiones de conjuntos convexos en los espacios vectoriales topologicos. Rev. Mat. Hispano - Americana 26, 92 - 99 (1976).

\- [2] Absolutely convex sets in barrelled spaces. Ann. Inst. Fourier 21, 3 - 13 (1971).

\- [3] A hereditary property in locally convex spaces. Ann. Inst. Fourier 21, 1 - 2 (1971).

\- [4] On final topologies. J. reine angew. Math. 251, 193 - 199 (1971).

\- [5] On subspaces of countable codimension of a locally convex space. J. reine angew. Math. 256, 185 - 189 (1972).

- [6] Some examples on quasi - barrelled spaces. Ann. Inst. Fourier 22, 21 - 26 (1971).

- [7] Sur certains hyperplans qui ne sont pas ultra - bornologiques dans les espaces ultra - bornologiques. C. R. Acad. Sci. Paris 284, 935 - 937 (1977).

- [8] On the closed graph theorem in topological spaces. Manuscr. Math. 23, 173 - 184 (1978).

- [9] Sobre una cierta clase de espacios topológicos. Collectanea Math. 28, 9 - 20 (1977).

- [10] On suprabarrelled spaces. Lect. Notes in Math. 843. Func. Anal. Holomorphy and Approximation Theory, Rio de Janeiro 1978. 572 - 580 Springer, Berlin Heidelberg New York 1981.

- [11] Algunos nuevos resultados sobre el teorema de la gráfica cerrada. Rev. Mat. Hispano-Amer. 39, 27 - 47 (1979).

- [12] Sobre el teorema de la gráfica cerrada. Collect. Math., 22, 51 - 72 (1971).

- [13] On quasi - normable echelon spaces. Proc. Edimbourgh Math. Soc. 73 - 80 (1981).

- [14] Algunas propiedades de los espacios escalonados. Rev. Real Acad. Cienc. Exáctas Físicas y Naturales, Madrid, 73, 389 - 400 (1979).

- [15] A characterization of echelon Köthe - Schwartz spaces. North - Holland Math. Studies 35, Notas de Matemática (66), 409 - 419 (1979).

- [16] Representaciones de los espacios $\mathcal{D}(\Omega)$ y $\mathcal{D}'(\Omega)$. Rev. Real Acad. Cienc. Exáctas, Físicas y Naturales, de Madrid 72, 385 - 414 (1978).

- [17] Una representación del espacio $\mathcal{D}_+(\Omega)$. Rev. Real Acad. Cienc. Exáctas, Físicas y Naturales, de Madrid 72 560 - 571 (1978).

- [18] A representation of the space $\mathcal{D}(K)$. J. reine angew. Math. 320, 97 - 98 (1980).

- [19] Sobre el espacio $\mathcal{B}_o(\Omega)$. Rev. Real Acad. Cienc. Exáctas, Físicas y Naturales, de Madrid 74, 838 - 863 (1980).

- [20] On certain infinitely differentiable function spaces. Lect. Notes in Math. 822, Sém. Pierre Lelong - Henri Skoda (Analyse) Ann. 1978/79, 310 - 316, Berlin Heidelberg New YorK 1980.

- [21] A representation of the space $\mathcal{O}_M$. Math. Z. 177, 463 - 478 (1981).

- [22] On the space $\mathcal{D}_{Lp}$. Math. Anal. and applic. Part. B, Advances in Mathematics suplementary Studies, V. 7 B, 759 - 767 Acad. Press 1981.

- [23] El teorema general de la gráfica cerrada en los espacios vectoriales topológicos. Rev. Real Acad. Cienc. Exáctas, Físicas y Naturales de Madrid 62, 545 - 551 (1968).

- [24] Mackey convergence and the closed graph theorem. Archiv Math. 25, 649 - 656 (1974).

- [25] On certain barrelled normed spaces. Ann. Inst. Fourier 29, 39 - 56 (1979).

- [26] Cocientes de espacios escalonados. Rev. Real Acad. Cien. Exác tas, Físicas y Naturales, de Madrid 73, 169 - 183 (1979).

- [27] Representaciones de los espacios $\mathcal{C}^m(V)$ y $\mathcal{D}^m(V)$. Rev. Real Acad. Cienc. Exáctas, Físicas y Naturales, de Madrid 75, 589 - 596 (1981).

- [28] Sobre ciertos espacios de funciones continuas. Rev. Real Acad. Cienc. Exáctas, Físicas y Naturales, de Madrid 73, 485 - 490 (1979)

- [29] Espacios de medidas de Radon. Rev. Real Acad. Cienc. Exáctas Físicas y Naturales, de Madrid 74, 91 - 98 (1980).

VOGT, D.: [1] Sequence space representations of spaces of functions and distributions. Preprint.

WHITNEY, H.: [1] Analytic extension of differentiable functions defined in closed sets. Trans. Amer. Math. Soc. 36, 63 - 89 (1934).

AUTHOR INDEX